# Student Study Guide

for use with

# Biology

## Eighth Edition

**Sylvia S. Mader**

Boston Burr Ridge, IL Dubuque, IA Madison, WI New York San Francisco St. Louis
Bangkok Bogotá Caracas Kuala Lumpur Lisbon London Madrid Mexico City
Milan Montreal New Delhi Santiago Seoul Singapore Sydney Taipei Toronto

The McGraw·Hill Companies

Student Study Guide for use with
BIOLOGY, EIGHTH EDITION
SYLVIA S. MADER

Published by McGraw-Hill Higher Education, an imprint of The McGraw-Hill Companies, Inc., 1221 Avenue of the Americas, New York, NY 10020. 

This book is printed on recycled, acid-free paper containing 10% postconsumer waste.

1 2 3 4 5 6 7 8 9 0 QPD/QPD 0 9 8 7 6 5 4 3

ISBN 0-07-241883-4

www.mhhe.com

# Contents

# To the Student

## The Study Guide

The *Study Guide* is designed to accompany your text, *Biology,* eighth edition, by Sylvia S. Mader. A number of different approaches are used to help you achieve mastery of the chapter concepts.

**Chapter Review:** Each chapter begins with a brief summary based on chapter concepts. If you read this summary before studying the textbook chapter, you will gain a quick overview of the content of the chapter before studying it in-depth.

**Study Exercises:** In this section, the exercises are tied to the chapter concepts, and one or two exercises are provided to help you understand each concept. The varied nature of the exercises—fill-ins, true-false, selecting the correct correlations, completion of tables, matching, and so forth—will sustain your interest. Each chapter also contains one or more labeling exercises.

**KeyWord CrossWord:** These crossword puzzles help you learn the boldface terms that appear in the chapter.

**Chapter Test:** This section helps you assess your mastery of chapter concepts. The test contains about twenty multiple-choice questions and two critical thinking questions, which require a written answer. The chapter test ends with directions for computing your score; if you are not satisfied with your score, more study is needed.

**Games:** Occasionally, a game is provided to help you learn the background information needed to understand a concept. Each type of game is different and provides a way for you to assess whether you are a winner.

## The Textbook

*Biology* also has many student aids that you should be sure to use. These will help you accentuate the benefit you receive from the *Study Guide.*

**Part Introduction:** An introduction for each part highlights the central ideas of that part and specifically tells you how the topics within each part contribute to biological knowledge.

**Chapter Concepts:** Each chapter begins with an integrated outline that identifies and numbers the major topics of the chapter and lists the concepts for each topic.

**Chapter Introductions:** Each chapter has an introduction on the chapter opening page that sparks interest in the themes for the chapter.

**Internal Summary Statements:** Internal summaries stress the chapter's key concepts. These appear at the ends of major sections and help you focus your study efforts on the basics.

**Illustrations and Tables:** The illustrations and tables in *Biology* are consistent with multicultural educational goals. Often, it is easier to understand a given process by studying a drawing, especially when it is carefully coordinated with the text. Every illustration appears on the same or facing page to its reference.

**Connecting the Concepts:** These appear at the close of the text portion of the chapter, and they stimulate critical thinking by showing how chapter concepts are related to other concepts in the text.

**End-of-Chapter Pedagogy:** The numbered major topics are repeated in the *Summary,* which reviews the concepts for each topic. *Reviewing the Chapter* is a series of study questions whose sequence follows that of the chapter. *Testing Yourself* consists of objective questions that allow you to test your ability to answer recall-based questions. Answers to *Testing Yourself* questions are given in Appendix A. *Thinking Scientifically* is a set of critical thinking questions based on biological concepts. *Bioethical Issue* presents current bioethical topics and asks questions to help students center their thoughts and arrive at an opinion. *Understanding the Terms* provides a page reference for boldfaced terms in the chapter. A matching exercise allows you to test your knowledge of the terms. *Online Learning Center* gives the Mader Home Page address for the chapter.

**Appendices and Glossary:** The appendices contain optional information. *Appendix A* is the answer key to the objective *Testing Yourself* and *Understanding the Terms* questions at the end of each chapter; *Appendix B* is a classification of organisms; *Appendix C* explains the metric system; and *Appendix D* is an expanded periodic table of chemical elements. The glossary defines the boldface terms in the text. These terms are the ones most necessary for the successful study of biology.

**Online Learning Center: http://www.mhhe.com/maderbiology8/**

The Mader *Biology* Online Learning Center offers access to a wide variety of tools to help students learn biological concepts and to reinforce their knowledge. Online study aids such as practice quizzes, interactive activities, animations, labeling exercises, flashcards, and much more are organized according to the major sections of each chapter. There is even an online tutorial service!

# HELPFUL STUDY HINTS

Learning how to study efficiently is a prerequisite to being a successful student. The following suggestions will help you study more productively. Please realize that this list is not all-inclusive and that it may not apply to all individuals who come to this course with different backgrounds, interests, and abilities.

1. **Be motivated.** You cannot acquire this attitude from an instructor or a textbook; it must come from within yourself. Motivation requires commitment, discipline, and perseverance, even when you do not feel like studying. Maintain high expectations. Think positively.
2. **Set aside several hours** at a particular time of the day to go over old and new material. Preview and review your assigned material; repetition of the material and consistency in your study habits combine for successful learning. Review every chance you have. Try to stay one lecture ahead of the instructor. Study in a quiet, well-lit room away from distractions.
3. **Try to grasp the big picture** as well as the details in every chapter. Try to understand how this chapter is related to the rest of the textbook. Whenever possible, try to attach concepts to "sensible" images and build upon them for better recall. When necessary, make up rhymes by using the first letter of each word. Look for distinguishing characteristics when analyzing abstract concepts.
4. **Continually ask yourself questions.** Although it may be boring to remember some things, asking questions will keep you mentally alert. After each paragraph or main idea, you should ask: "What does this mean?" "How does this relate to what I've already learned?" "How can I put that idea into my words?" or "How can I apply that concept?" Try to relate the information to your past experiences. Make sure you know the definitions of boldfaced or italicized words in the text. Change the statements in each paragraph into questions and answers. Draw a line on a paper, and write the question on one side and the answer on the other side.
5. **Have a positive attitude.** Even though all of us have personal problems that may interfere with learning, look at learning as a growing experience. A positive attitude means that you do not give up. Sometimes, you will have to dig the information out of the textbook by yourself. After all, that is what learning is ultimately all about: it is a lifelong process that we have to achieve basically by ourselves.
6. **Have someone ask you questions.** Asking and responding to questions from another person will improve your thinking and understanding of the material. Cover up definitions of terms and see if you can define the terms correctly. Ask the instructor if you can use a tape recorder in class and then listen to the lecture again at another time. Don't be embarrassed to ask the instructor during or after class about concepts you do not understand. Most instructors will be glad to help you. Use any other available study aids. Perhaps rewriting your notes into an outline format will be helpful as well.
7. **Learn as much as possible while in class.** You may want to avoid taking notes in class. Instead, listen very attentively, follow along with the instructor from the material in the textbook, and highlight key words that the instructor mentions. If you have previewed the lecture material, you will know where to find the information being discussed in class. If you do take lecture notes, be selective; do not try to take notes on everything mentioned. Abbreviate words and fill them in immediately after class. Review the same material from the textbook as soon as possible. Try to grasp the big picture given in class.
8. **When taking the test, try to relax** by taking several deep breaths or by tightening your muscles and then slowly relaxing them. When reading the questions, make sure you understand exactly what the teacher wants to know AND read carefully to understand any limitations placed upon the question. On essay questions that are very broad, limit your answers to what the teacher felt was relevant in the class. Make a quick outline of the salient, major points and then fill in the details as you write your essay. When exams are returned, review them so that you can learn from your mistakes. That information may be on the final comprehensive test.

# 1

# A View of Life

## Chapter Review

Although living things are diverse, they share certain characteristics. The organization of living things is exhibited by the smallest unit of life, the cell. In multicellular organisms, similar cells make up a tissue, tissues form organs, organs are part of organ systems, and organ systems work together within an organism. Each level of organization has emergent properties that cannot be accounted for by simply adding up the properties of the previous levels.

Living things acquire materials and energy from the environment that are used during **metabolism,** a process that maintains **homeostasis.** Living things respond to the environment. When they **reproduce** and develop, genetic changes are passed on that result in **adaptation** to the environment.

**Evolution** explains both the unity of life (similar characteristics) and the diversity of life (adaptation to different environments). Adaptations allow organisms to play diverse roles in **ecosystems** where they interact with each other and the physical environment.

Biologists classify organisms into a particular **domain** (most distantly related), **kingdom, phylum, class, order, family, genus,** and **species.** Each organism is given a binomial name consisting of the genus and species. Domain Archaea and domain Bacteria contain unicellular prokaryotes that lack a membrane-bounded nucleus; domain Eukarya contains kingdoms Protista (unicellular eukaryotes, various modes of nutrition—e.g., protozoans and algae), Fungi (usually multicellular, absorb food—e.g., mushrooms), Plantae (multicellular **photosynthesizers**), and Animalia (multicellular, ingest food).

The **scientific process** involves using the scientific method, from which scientists gain information about the natural world. This method uses accumulated scientific data to arrive at a **hypothesis** that is tested by observation and experimentation. **Inductive reasoning** is used to arrive at a hypothesis, and **deductive reasoning** is used to decide what types of observations and experiments are appropriate. Scientists analyze **data** to reach a **conclusion.** Data should be observable and objective. The conclusion either supports or does not support the hypothesis.

Science often employs controlled experiments that involve two elements of interest: the **experimental variable** and the **dependent variable.** The experimenter deliberately alters the experimental variable. The group being tested in the experiment is compared with a control group, or **control.** Significant differences in the results between the experimental group and the control group contribute to the investigation's conclusions. Descriptive research is based on observational data. Although a control group is not used, the steps of the scientific method are still employed. Experiments and observations must be repeatable by other scientists.

The ultimate aim of all scientific investigations is to construct **scientific theories.** Scientific theories are conceptual ways of understanding the natural world. Scientists generally accept theories because they are so well supported by observations, experiments, and data.

## Study Exercises

Study the text section by section as you answer the questions that follow.

### 1.1 How to Define Life (pp. 2–5)

- There are various levels of biological organization. At each higher level properties emerge that cannot be explained by the sum of the parts.
- Although life is diverse, it can be defined by certain common characteristics.
- All living things have characteristics in common because they are descended from a common ancestor; but they are diverse because they are adapted to different environments.

1. List the following levels of biological organization in order, from smallest to largest:
cell, community, ecosystem, molecule, organ, organism, organ system, population, tissue

a. ____________________ (smallest)

____________________

____________________

____________________

____________________

____________________

____________________

____________________

____________________ (largest)

Living things have b.__________________ that cannot be accounted for by simply summing the parts. Among the levels of biological organization, the properties of life first emerge at the level of the c.__________________.
Different groups of organisms interact with the environment at the level of the d.__________________.

2. Match these characteristics of life with the examples that follow.
Living things are organized.
Living things take materials and energy from the environment.
Living things respond to stimuli.
Living things reproduce.
Living things adapt to the environment.
Living things are homeostatic.
Living things grow and develop.

Frogs have a life cycle that includes an egg, a larva (tadpole) that undergoes the process of metamorphosis, and an adult. a.__________________

Humans immediately remove their hands from a hot object. b.__________________

All living things are composed of cells. c.__________________

A flounder is a flattened fish that lives on the bottom of bodies of water, while a tuna is a streamlined fish that swims in the open sea. d.__________________

Most cells use the sugar glucose as an energy source. e.__________________

3. Homeostasis refers to keeping a.__________________ relatively stable, such as body b.__________________.

4. Label each of the following statements as an example of either the unity of life (U) or the diversity of life (D):
a. _____ Fungi absorb food; plants carry out photosynthesis.
b. _____ Homeostasis, metabolism, and evolution are characteristics of living things.
c. _____ Life began with single cells.
d. _____ Living things consist of cells.
e. _____ Maple trees have broad, flat leaves, and pine trees have needlelike leaves.

## 1.2 How the Biosphere Is Organized (pp. 6–7)

- The biosphere is made up of ecosystems, where living things interact with each other and with the physical environment.
- Coral reefs and tropical rain forests, noted for their diversity, are endangered due to human activities.

5. Study the diagram that follows and then answer the questions.
a. Which type of population in a food chain does every other population rely upon? __________
b. What is the ultimate source of energy for a food chain? __________
c. What type of population in a food chain processes organic remains and returns inorganic nutrients to the ecosystem? __________
d. How do producers (i.e., plants) interact with the physical environment? __________
e. How do producers (i.e., plants) interact with the biological environment? __________

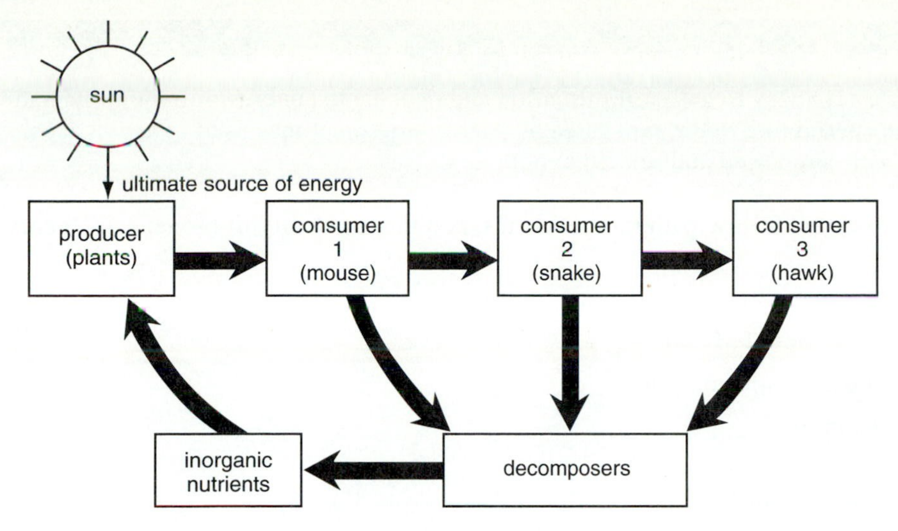


6. Indicate whether these statements are true (T) or false (F).
   a. _____ As more ecosystems are converted to towns and cities, fewer of the natural cycles are able to function adequately.
   b. _____ Coral reefs play no protective role in the biosphere, and their destruction is of little concern to humans.
   c. _____ The present rate of extinction is normal and about the same as at any other time in the history of the Earth.

## 1.3 How Living Things Are Classified (pp. 8–9)

- Living things are classified into taxonomic categories according to their evolutionary relationships.

7. List the following levels of classification in order, from smallest (least inclusive) to largest (most inclusive): class, family, domain, genus, kingdom, order, phylum, species

   ____________________ (smallest)

   ____________________

   ____________________

   ____________________

   ____________________

   ____________________

   ____________________

   ____________________ (largest)

8. Name the kingdoms described in each of the following statements:
   a. ____________________ multicellular; ingest food
   b. ____________________ absorb food; includes molds and mushrooms
   c. ____________________ photosynthesize food; includes ferns
   d. ____________________ absorb, photosynthesize, or ingest food; includes protozoans and algae
9. Name the domain occupied by each of the following:
   a. ____________________ plants
   b. ____________________ bacteria
   c. ____________________ animals
   d. ____________________ fungi
   e. ____________________ unicellular prokaryotes

## 1.4 The Process of Science (pp. 10–14)

- The scientific process is a way to gather information and to come to conclusions about the natural world.
- Various conclusions pertaining to the same area of interest are sometimes used to arrive at scientific theories, concepts that join well-supported and related hypotheses.

10. Match the descriptions that follow with these theories: cell theory, theory of biogenesis, theory of evolution, gene theory.

    a. Common descent with modification of form. ___________

    b. All organisms are composed of cells. ___________

    c. Organisms inherit coded information. ___________

    d. Life comes only from life. ___________

11. Place the terms *conclusion, experiment/observations, hypothesis, observation,* and *theory* in the correct boxes of the adjacent diagram:

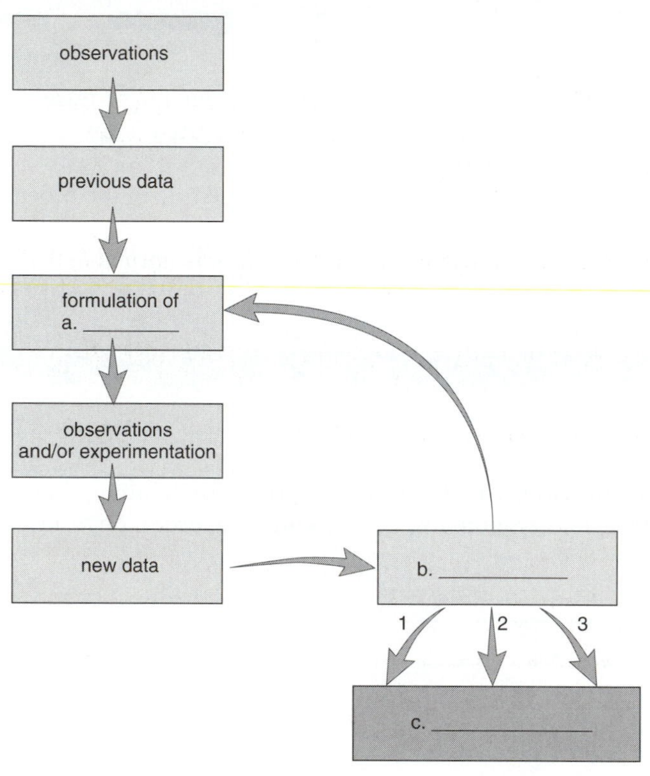


12. Each of these statements demonstrates a step of the scientific method used in a study. Match the steps from question 11 to these examples.

    To spawn, striped bass migrate up a major river. In 1991, a dam was built on this river. Fishers reported catching fewer striped bass in the river in 1993. A study of the fish population in the river revealed decreased numbers of striped bass. a. ______________________ Most likely, the dam is blocking the migration and spawning of striped bass. b. ______________________ A fish ladder is built along the dam. Scientists observe that the ladder helps striped bass clear the dam and spawn upriver. c. ______________________ Survival of striped bass requires free migration of fish upriver to spawn. d. ______________________ Construction of the fish ladder improves fishing. More fish are caught the following year, and studies indicate that the number of striped bass is increasing. e. ______________________

13. Biologists are doing a study on athletes' intake of steroids. Which of these would be the experimental variable (E) and which would be the dependent variable (D)?

    a. _____ average gain in body weight of a group of athletes

    b. _____ intake of anabolic steroids by an athlete

14. In the study in question 13, how would a control group of athletes differ from the experimental group? _______

______________________________________________

15. Over a six-month period, the weight gain of the athletes in the experimental group in question 13 averaged 11%. Members of the control group experienced a gain of 4%.

a. What is the benefit of mathematical data like this? ______________________________

______________________________________________

b. State the conclusion of this experiment. ______________________________

______________________________________________

16. Place the appropriate letter(s) next to each statement. (Some are used more than once.)

O—observational research only
E—experimental research only
E,O—both experimental and observational research

a. _____ Steps of the scientific method are used.
b. _____ A control group is employed.
c. _____ Much of the data required are purely observational.
d. _____ A hypothesis is disproven or supported.
e. _____ Observations are made.

# Chapter Test

## Objective Questions

Do not refer to the text when taking this test.

____ 1. The binomial name *Notorcytes typhlops* refers to taxonomic levels of
a. class and order.
b. genus and species.
c. kingdom and phylum.
d. order and phylum.

____ 2. Select the smallest, least inclusive taxonomic level among the following choices:
a. class
b. genus
c. order
d. specific epithet

____ 3. Select the largest, most inclusive taxonomic level among the following choices:
a. class
b. order
c. domain
d. phylum

____ 4. Each is a general characteristic of life EXCEPT
a. the ability to respond.
b. reproduction and development.
c. organization.
d. classification.

____ 5. The lowest level of organization to have the characteristics of life is the ______ level.
a. atomic
b. cellular
c. molecular
d. population

____ 6. The term *metabolism* refers best to
a. chemical and energy transformations.
b. maintenance of internal conditions.
c. the ability to respond to stimuli.
d. the lack of reproduction.

____ 7. Plants are unique among living things in that they are
a. multicellular and absorb food.
b. unicellular and ingest food.
c. multicellular and photosynthesize.
d. All of these are correct.

____ 8. Bacteria belong to the domain
a. Animalia.
b. Fungi.
c. Bacteria.
d. Protista.

____ 9. Changes in ______ account for the ability of a species to evolve.
a. physical factors
b. ecosystems
c. genes
d. sunlight

____ 10. Through evolution, populations can
a. adapt to the environment.
b. change their level of organization.
c. fail to reproduce.
d. eliminate cell structure.

____ 11. Which of these associations are incorrect?
a. data—factual information
b. deductive reasoning—general hypothesis to prediction
c. hypothesis—final conclusion
d. inductive reasoning—specific data to general hypothesis

____ 12. The ________ variable deals with the observable effects of an experiment.
a. control
b. dependent
c. experimental
d. independent

____ 13. Valid scientific results should be repeatable by other scientific investigators.
a. true
b. false

____ 14. The statement with the greatest acceptance and predictive value from scientists is the
a. hypothesis.
b. induction.
c. observation.
d. scientific theory.

## Critical Thinking Questions

Answer in complete sentences.

15. Why is any level of organization not a mere sum of its parts?

16. Testing a hypothesis is the central core of the scientific process. Explain.

**Test Results:** ______ number correct ÷ 16 = ______ × 100 = ______ %

## Exploring the Internet

The Online Learning Center at *www.mhhe.com/maderbiology8* has additional study material and practice quizzes that can help you master the content of this chapter. You can also find links to websites exploring additional topics in biology. Access to the Online Learning Center is free for those who have purchased a new textbook.

# ANSWER KEY

## STUDY EXERCISES

**1. a.** molecule, cell, tissue, organ, organ system, organism, population, community, ecosystem **b.** emergent properties **c.** cell **d.** ecosystem **2. a.** Living things grow and develop **b.** Living things respond to stimuli **c.** Living things are organized **d.** Living things adapt to the environment **e.** Living things take materials and energy from the environment **3. a.** internal conditions **b.** temperature **4. a.** D **b.** U **c.** U **d.** U **e.** D **5. a.** producer **b.** sun **c.** decomposers **d.** They use solar energy and inorganic nutrients. **e.** They are eaten and supply energy to consumers. **6. a.** T **b.** F **c.** F **7.** species, genus, family, order, class, phylum, kingdom, domain **8. a.** Animalia **b.** Fungi **c.** Plantae **d.** Protista **9. a.** Eukarya **b.** Bacteria **c.** Eukarya **d.** Eukarya **e.** Bacteria and Archaea **10. a.** evolution **b.** cell **c.** gene **d.** biogenesis **11. a.** observation **b.** hypothesis **c.** experiment/observation **d.** conclusion **e.** theory **12. a.** observation **b.** hypothesis **c.** experiment/observations **d.** conclusion **e.** observation **13. a.** D **b.** E **14.** Control group would not take steroids. **15. a.** It is objective. **b.** Steroids cause weight gain. **16. a.** E, O **b.** E **c.** O **d.** E, O **e.** E, O

## CHAPTER TEST

**1.** b **2.** b **3.** c **4.** d **5.** b **6.** a **7.** c **8.** c **9.** c **10.** a **11.** c **12.** b **13.** a **14.** d **15.** At each level, there are emergent properties not seen at lower levels of organization. For example, the cell, the basic living unit, is more than just a combination of chemicals. **16.** No matter the sequence of steps followed, scientists always test a hypothesis. The testing can take the form of making more observations rather than doing an experiment.

PART I THE CELL

# 2

# BASIC CHEMISTRY

## CHAPTER REVIEW

Life has a chemical basis. All **matter,** living or nonliving, consists of **elements** composed of discrete units called **atoms.** The atom of each kind of element has its own arrangement of three subatomic particles: protons, neutrons, and electrons. The **protons** (positively charged) and **neutrons** (neutral) are located in the nucleus of the atom. The atoms of an element can differ by their number of neutrons. The **electrons** (negatively charged) are located in energy levels at varying distances from the nucleus. In a neutral atom, the number of protons equals the number of electrons; this is the atomic number of the atom (element). Atoms with the same atomic number but a different number of neutrons are called **isotopes.**

The arrangement and behavior of electrons—specifically the outermost electrons—determine the chemical properties of that element. The electrons are located in **electron shells** around the nucleus. The first shell is closest to the nucleus, the second shell is the next closest, and so forth. The potential energy of an electron increases with the increasing distance of these energy levels from the nucleus. **Orbitals** are the volumes of space where electrons are found within these electron shells. The first shell holds a maximum of two electrons, the second shell holds a maximum of eight electrons, and the third shell holds eight electrons when it is the outer shell. This description applies to atoms of an atomic number less than 20.

The number of electrons in the outer shell determines the chemical reactivity of an atom. Atoms with two or more shells are most stable when the outer shell has eight electrons. For atoms with only one shell (i.e., H, atomic number of 1), the stable electron configuration is two. Atoms fulfill their stable, outer-shell electron configuration by either sharing or transferring electrons with other atoms; that is, they form bonds. Two major types of bonds can form: ionic and covalent. **Ionic bonds** form by the transfer of electrons between an electron donor and an electron acceptor. Following an ionic reaction, the donor atoms are positively charged **ions,** and the acceptor atoms are negatively charged ions. An ionic bond is the attraction between oppositely charged ions.

By contrast, **covalent bonds** develop from the sharing of electrons between atoms. A covalent bond is often represented by overlapping circular representations of the outer shells or by drawing a straight line between the atoms. The three-dimensional shape of a covalent molecule often determines the role it plays in cells and organisms.

Water is an important biological **compound,** making life possible on earth. Water **molecules** exhibit **polar covalent bonding.** The electrons between atoms (H and O) are shared, but not equally. The shared electrons spend more time near the oxygen atom. This establishes a positive pole (hydrogen atoms) and a negative pole (oxygen atoms) and promotes **hydrogen bonding** between the water molecules. This weak bond between the molecules is the source of many remarkable properties of water, such as its role as a universal solvent, its ability to resist temperature changes and remain a liquid, and its formation of a maximum density at 4°C.

**Acids** (e.g., HCl) are compounds that release hydrogen ions in solution; **bases** (e.g., NaOH) are compounds that take up hydrogen ions or release hydroxide ions in solution. The **pH scale** indicates the relative concentration of hydrogen and hydroxide ions in a solution. The scale ranges from 0 to 14, with 7 a neutral pH. Acids lower the pH of a solution from 7 by releasing hydrogen ions, and bases increase the pH above 7 by an opposite effect. **Buffers** are compounds in solution that resist these changes through chemical reaction, thus stabilizing the pH of a given solution.

# STUDY EXERCISES

Study the text section by section as you answer the questions that follow.

## 2.1 CHEMICAL ELEMENTS (PP. 20–23)

- Matter is composed of 92 naturally occurring elements, each having one type of atom.
- Atoms have subatomic particles: electrons, protons, and neutrons.
- Atoms of the same type that differ by the number of neutrons are called isotopes.
- Atoms are characterized by the number of protons and neutrons in a nucleus and the number of electrons in shells about the nucleus.

1. Name the six elements commonly found in living things. ______________________________________________

______________________________________________________________________________

2. The three most stable subatomic particles in an atom are ________________, __________________, and ________________.

3. An atom has nine protons and nine neutrons in each of its atoms. Its atomic number is a.______________, its atomic mass is b.________________, and the number of electrons in this atom is c.______________.

4. In the periodic table of chemical elements, the horizontal rows are called a.______________ and the vertical columns are called b.______________. The atoms in group 8 are called the c.______________ because they d.______________ react with other atoms.

5. Isotopes have the same atomic a.______________, but they differ in the number of b.______________.

6. At least two forms of the oxygen atom exist in the environment: $^{16}_{8}O$ and $^{18}_{8}O$.

   a. Each atom has eight electrons. $^{16}_{8}O$ and $^{18}_{8}O$ represent ________________ of the element oxygen.

   b. The numbers 16 and 18 represent the ________________.

   c. The atomic number of $^{16}_{8}O$ is ________________.

   d. The atomic number of $^{18}_{8}O$ is ________________.

   e. The number of protons in the atom of $^{16}_{8}O$ is ________________.

   f. The number of neutrons in the atom of $^{18}_{8}O$ is ________________.

   g. How do $^{16}_{8}O$ and $^{18}_{8}O$ differ in number of subatomic particles? ________________.

7. Complete the following table with the correct numbers:

| Isotope | Protons | Neutrons | Atomic Number | Atomic Mass |
|---|---|---|---|---|
| $^{12}_{6}C$ | | | | |
| $^{14}_{6}C$ | | | | |
| $^{31}_{15}P$ | | | | |
| $^{33}_{15}P$ | | | | |

8. Indicate whether these statements are true (T) or false (F):
   a. _____ A neutral atom has the same number of protons and electrons.
   b. _____ The arrangement of protons and neutrons in an atom determines the atom's chemical properties.
   c. _____ The electrons of an atom are located in energy levels (electron shells) at varying distances from the nucleus.
   d. _____ Electrons in the first shell possess more energy than do electrons in the second shell.
   e. _____ When a chlorophyll molecule absorbs solar energy, electrons move to higher energy levels.
9. Electrons are most often found in volumes of space called [a]______________. The orbital at the first energy level is [b]______________ shaped. Of the four orbitals at the second energy level, one is [c]______________ shaped and three are [d]______________ shaped.
10. An atom of an element with an atomic number of 9 has [a]______________ electrons in its first shell and [b]______________ electrons in its second shell.
11. An atom of the element neon (Ne) has eight electrons in its outer shell. An atom of the element sulfur (S) has six electrons in its outer shell.
    a. Which atom is reactive? ______________
    b. Which atom is inert? ______________
    c. Why is there a difference in the reactivity of these two elements? ______________________________________
    ______________________________________________________________________
12. The magnesium atom has an atomic number of 12 and an atomic weight of 24. Draw its simplified atomic structure. Draw small circles to indicate the general distribution of electrons in concentric levels around the nucleus. In the nucleus, indicate the number of protons and neutrons.

## 2.2 Elements and Compounds (pp. 24–26)

- Atoms react with one another by giving up, gaining, or sharing electrons.
- Bonding between atoms results in molecules with distinctive chemical properties and shapes.
- The biological role of a molecule is determined in part by its shape.

13. Which of the following are examples of compounds? ______________
    a. $^{14}C$
    b. H atom
    c. $O_2$, oxygen molecule
    d. NaCl, table salt
    e. $H_2O$, water molecule

14. Calcium (Ca) has an atomic number of 20; chlorine (Cl) has an atomic number of 17.

a. The number of electrons in the outer shell of calcium is ______________.

b. The number of electrons in the outer shell of chlorine is ______________.

c. In a chemical reaction between these two atoms, ______________ calcium atom(s) will react with ______________ chlorine atom(s).

d. Which element will gain electrons in this reaction? ______________

e. What will its charge be after the reaction? ______________

f. Which element will lose electrons in this reaction? ______________

g. What will its charge be after the reaction? ______________

h. What type of bond forms between these two atoms? ______________

i. Write the formula for the compound produced through this chemical reaction. ______________

15. Water is a polar molecule.

a. Indicate on the following diagram which atoms are electronegative (i.e., $\delta^-$) and which are electropositive (i.e., $\delta^+$) in relation to the others. Put a $\delta^-$ charge and $\delta^+$ charges where appropriate.

b. Label a hydrogen bond.

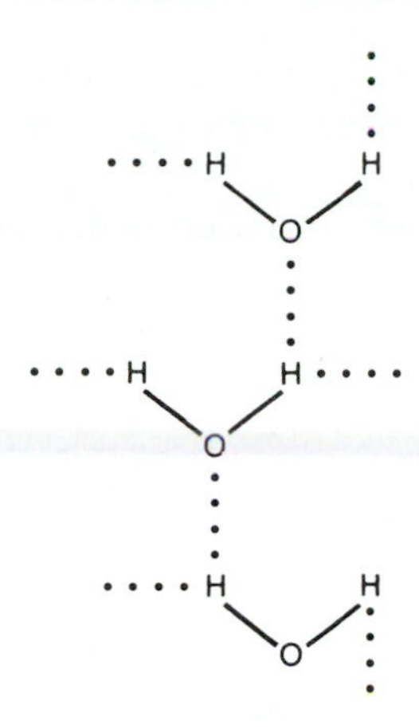


16. Place the appropriate letter next to each statement.

C—covalent bonding I—ionic bonding H—hydrogen bonding

a. _____ Electrons transfer between atoms.
b. _____ Atoms share electrons.
c. _____ This bond is present in sodium chloride.
d. _____ This bond is present in the oxygen molecule.
e. _____ A triple bond of this type is present in nitrogen gas.
f. _____ This bond forms between water molecules.
g. _____ The hydrogen atom in one molecule is attracted to the oxygen atom of another molecule.
h. _____ This bond forms within a water molecule.

## 2.3 Chemistry of Water (pp. 27–30)

- The existence of living things is dependent on the chemical and physical characteristics of water.
- Living things are sensitive to the hydrogen ion concentration [$H^+$] of solutions, which can be indicated by using the pH scale.

17. In each of the pairs of statements that follow, place a check next to the one that correctly describes how hydrogen bonding affects the properties of water. Hydrogen bonding causes water

*Pair 1*

a. _____ to boil at a lower temperature than expected.
b. _____ to boil at a higher temperature than expected.

*Pair 2*

c. _____ to be more dense as ice than as liquid water.
d. _____ to be less dense as ice than as liquid water.

*Pair 3*

e. _____ to absorb heat with a minimal change in temperature.
f. _____ to absorb heat with a maximum change in temperature.

*Pair 4*

g. _____ to be cohesive—the water molecules cling to each other.
h. _____ molecules to shun one another.

18. Refer to the chemical properties of water when answering the following questions:

a. What makes water a good solvent? ______________________________

______________________________

b. How does water moderate temperatures? ______________________________

______________________________

c. What allows ice to float on liquid water? ______________________________

______________________________

19. Place the appropriate letter next to each statement.

A—acid B—base

a. _____ They take up hydrogen ions in solution.
b. _____ HCl is an example.
c. _____ NaOH is an example.
d. _____ They release hydrogen ions in solution.
e. _____ They lower the pH.
f. _____ They raise the pH.

20. Complete the table for the following hydrogen ion concentrations [$H^+$]:

| [$H^+$] | pH | Acid/Base/Neutral |
|---|---|---|
| $1 \times 10^{-7}$ | | |
| $1 \times 10^{-3}$ | | |
| $1 \times 10^{-8}$ | | |

21. Indicate whether these statements are true (T) or false (F):

a. ____ If the pH of blood changes from 7.4 to 7.6, it becomes more acidic.

b. ____ When an acid is added to a solution, the pH decreases.

c. ____ A basic pH indicates that $OH^-$ ions outnumber $H^+$ ions.

d. ____ An acidic pH indicates that $H^+$ ions outnumber $OH^-$ ions.

22. The following questions relate to buffers:

How do living things prevent drastic changes in pH?

a. ________________________________________

________________________________________

b. Complete the following reaction, showing how the carbonic acid buffer system deals with increasing hydrogen ions in the blood:

$$H^+ + HCO_3^- \rightarrow \_\_\_\_\_\_\_\_$$

c. Complete the following reaction, showing how the carbonic acid buffer system deals with decreasing hydrogen ions in the blood:

$$H_2CO_3 \rightarrow \_\_\_\_\_\_\_\_$$

# Chapter Test

## Objective Questions

Do not refer to the text when taking this test.

____ 1. An element has an atomic number of 11 and an atomic weight of 23. The number of neutrons in each atom is

a. 11.
b. 12.
c. 23.
d. 24.

____ 2. The atom of an element has one proton and two neutrons. Its atomic number is

a. 1.
b. 2.
c. 3.
d. 6.

____ 3. The atom of an element has six protons and eight neutrons. The number of electrons in this atom if neutral is

a. 6.
b. 8.
c. 12.
d. 14.

____ 4. The relationship between ${}^{12}_{6}C$ and ${}^{14}_{6}C$ is that they are

a. molecules.
b. isomers.
c. isotopes.
d. polymers.

____ 5. An atom has 11 electrons and 12 neutrons. Its atomic mass is

a. 1.
b. 11.
c. 12.
d. 23.

____ 6. The energy possessed by electrons in the first shell is ____________ than the energy possessed by electrons in the second shell.

a. greater
b. less

____ 7. An element has an atomic number of 14. Its electron distribution over several energy shells is

a. 1–4–8.
b. 1–8–5.
c. 2–8–2.
d. 2–8–4.

____ 8. An element has an atomic number of 13. The number of electrons in each atom's second shell is

a. 1.
b. 2.
c. 4.
d. 8.

____ 9. Select the reactive element by its atomic number.

a. 2
b. 10
c. 12
d. 18

____ 10. Select the most stable element by its atomic number.
a. 1
b. 8
c. 10
d. 16

____ 11. Select the compound.
a. Ca
b. H
c. NaCl
d. $O_2$

____ 12. The atoms of which element tend to lose electrons in a chemical reaction?
a. Cl
b. Mg
c. O
d. S

____ 13. When sodium interacts with chlorine, sodium loses an electron while chlorine gains one. This interaction forms
a. an ionic bond.
b. a condensation synthesis.
c. a condensation.
d. a covalent bond.

____ 14. Select the incorrect association.
a. covalent bond—electrons transferred
b. hydrogen bond—between water molecules
c. ionic bond—charged particles formed
d. polar covalent bond—present in the water molecule

____ 15. Bonds between carbon and hydrogen or oxygen and hydrogen are generally
a. hydrogen bonds.
b. ionic bonds.
c. covalent bonds.
d. weak and highly transient.

____ 16. Each is a property of water EXCEPT
a. easily changed from liquid to gas.
b. good solvent.
c. maximum density at 4°C.
d. molecules are cohesive.

____ 17. Select the correct statement about acids.
a. They cannot be buffered in a solution.
b. They donate hydroxide ions in solution.
c. HCl is an example.
d. They tend to raise the pH.

____ 18. Select the most basic pH of the given hydrogen ion concentrations.
a. $1 \times 10^{-3}$
b. $1 \times 10^{-4}$
c. $1 \times 10^{-9}$
d. $1 \times 10^{-12}$

____ 19. Select the incorrect statement about bases.
a. They can be buffered in solution.
b. They release hydroxide ions in solution.
c. NaOH is an example.
d. They tend to lower the pH.

____ 20. Which of the following is an example of a buffer?
a. carbonic acid
b. hydrogen ion
c. hydroxide ion
d. NaCl

## Critical Thinking Questions

Answer in complete sentences.

21. Element X has an atomic number of 4, whereas element Y has an atomic number of 18. Which element is more reactive, and why?

22. Chemical reactions in the human body produce many acid end products. Yet, the pH of the blood remains remarkably constant. Why?

**Test Results:** ______ number correct ÷ 22 = ______ × 100 = ______ %

## Exploring the Internet

The Online Learning Center at *www.mhhe.com/maderbiology8* has additional study material and practice quizzes that can help you master the content of this chapter. You can also find links to websites exploring additional topics in biology. Access to the Online Learning Center is free for those who have purchased a new textbook.

## Answer Key

### Study Exercises

**1.** carbon, hydrogen, nitrogen, oxygen, phosphorus, sulfur **2.** protons, neutrons, electrons **3. a.** 9 **b.** 18 **c.** 9 **4. a.** periods **b.** groups **c.** noble gases **d.** rarely **5. a.** number **b.** neutrons **6. a.** isotopes **b.** atomic mass **c.** 8 **d.** 8 **e.** 8 **f.** 10 **g.** $^{16}_{8}O$ has eight neutrons in its nucleus; $^{18}_{8}O$ has ten neutrons.

**7.**

| Protons | Neutrons | Atomic Number | Atomic Mass |
|---|---|---|---|
| 6 | 6 | 6 | 12 |
| 6 | 8 | 6 | 14 |
| 15 | 16 | 15 | 31 |
| 15 | 18 | 15 | 33 |

**8. a.** T **b.** F **c.** T **d.** F **e.** T **9. a.** orbitals **b.** spherical **c.** spherical **d.** dumbbell **10. a.** two **b.** seven **11. a.** S **b.** Ne **c.** S has six electrons in its outer shell, and if it reacts to gain two more it will have a stable outer configuration of eight; Ne already has a stable outer shell.

**12.**

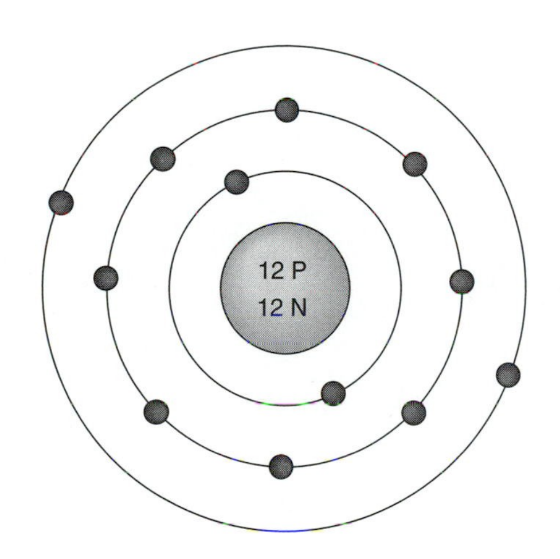


**13.** d, e **14. a.** two **b.** seven **c.** one, two **d.** Cl **e.** –1 **f.** Ca **g.** +2 **h.** ionic **i.** $CaCl_2$

**15.**

hydrogen bonds

**16. a.** I **b.** C **c.** I **d.** C **e.** C **f.** II **g.** II **h.** C **17.** b, d, e, g **18. a.** The partial charges of the water molecule attract and disperse charged particles in solution. **b.** It takes up and releases large amounts of heat without much change in temperature. **c.** Water is most dense at 4°C. It expands as the temperature drops from this point; it is less dense at 0°C, the temperature of ice. **19. a.** B **b.** A **c.** B **d.** A **e.** A **f.** B

**20.**

| pH | Acid/Base/Neutral |
|---|---|
| 7 | neutral |
| 3 | acid |
| 8 | base |

**21. a.** F **b.** T **c.** T **d.** T **22. a.** The pH is stabilized through the action of buffers, chemical systems that absorb either $H^+$ or $OH^-$ to keep the pH steady. **b.** $H_2CO_3$ **c.** $HCO_3^- + H^+$

## Chapter Test

**1.** b **2.** a **3.** a **4.** c **5.** d **6.** b **7.** d **8.** d **9.** c **10.** c **11.** c **12.** b **13.** a **14.** a **15.** c **16.** a **17.** c **18.** d **19.** d **20.** a **21.** Element X is more reactive because its electron arrangement is 2–2. It has only two electrons in its outer shell. If it gives up these two electrons, its outer shell will be stable. The electron arrangement in Y is 2–8–8. It has eight electrons in its outer shell—a stable outer shell. **22.** Buffers in the body take up excess hydrogen ions and thereby act to keep the pH within normal limits.

# 3

# The Chemistry of Organic Molecules

## Chapter Review

Carbon's unique properties permit the formation of many kinds of **organic molecules.** At the molecular level, this variety accounts for the diversity of living things. Many organic molecules have a carbon backbone plus functional groups. Some common functional groups are the hydroxyl, carboxyl, aldehyde, ketone, and amine groups.

Several types of small, organic molecules—sugars, fatty acids, amino acids, and nucleotides—serve as the **monomers** (building blocks) of **polymers** (larger organic molecules). These polymers (e.g., polysaccharides, lipids, proteins, nucleic acids) have important biological functions. When a **dehydration** reaction occurs, two monomers bond chemically as a water molecule is lost. Repetition of this process produces even larger molecules—the polymers—in a cell. The reverse reaction, **hydrolysis,** breaks down polymers into their chemical subunits.

Several classes of organic molecules have biological importance. One of these, the **carbohydrates,** consists of several subclasses: the monosaccharides (e.g., glucose), the disaccharides (e.g., sucrose), and the polysaccharides (e.g., starch). The monosaccharides and disaccharides—the sugars—provide an immediate energy source for organisms. Some polysaccharides store energy (i.e., starch), whereas others contribute structurally (i.e., cellulose).

**Fatty acids** and **glycerol** are the building blocks of fats and oils. Fatty acids may be either saturated or unsaturated. Fats and oils store energy efficiently. **Waxes** and **phospholipids** differ in some of their components compared to fats. These structural differences endow these molecules with different biological abilities. Phospholipids, for example, are a major component of plasma membrane structure and help determine a membrane's properties. **Steroids** are derived from cholesterol; their structure consists of four fused carbon rings.

**Proteins** have a variety of biological functions, such as support, enzymatic, transport, and hormonal regulation. The monomers of these polymers are **amino acids. Peptide bonds** join amino acids within the **polypeptides** of protein molecules. Proteins exhibit several levels of structure. The primary structure of a protein is the order of the amino acids bonded together. Several other structural levels (secondary, tertiary, quaternary) account for the molecule's three-dimensional shape and for the protein's biological properties.

**DNA** and **RNA** are **nucleic acids. Nucleotides,** the monomers of nucleic acids, contain a pentose sugar, phosphate, and nitrogen-containing base. DNA makes up the genes in cells. The DNA molecule is a double helix—it has the appearance of a twisted ladder. Sugar and phosphate molecules make up the sides of the ladder and hydrogen-bonded bases named adenine, guanine, cytosine, and thymine make up the rungs of the ladder. The sequence of bases in DNA stores information regarding the order in which amino acids are to be joined within a protein. RNA conveys this information from the nucleus to the cytoplasm, and therefore is an intermediary in the synthesis of proteins.

The nucleotide ATP is composed of adenosine and three phosphate groups. ATP is a high-energy molecule. Whenever cells need energy, ATP is broken down to ADP + P, and energy is released.

## Study Exercises

Study the text section by section as you answer the questions that follow.

### 3.1 Organic Molecules (pp. 36–38)

- The characteristics of organic compounds depend on the chemistry of carbon.
- Variations in carbon backbones and functional groups account for the great diversity of organic molecules.
- The four classes of organic molecules in cells are carbohydrates, lipids, proteins, and nucleic acids.
- Large organic molecules called polymers form when their specific monomers join together.

1. Indicate whether these statements about a carbon atom are true (T) or false (F):
   a. _____ There are two electrons in its outer shell.
   b. _____ It can bond to other carbon atoms.
   c. _____ It can share two pairs of electrons with another atom.
   d. _____ Chains of 50 atoms are unusual in living systems.
2. Label this diagram using the following functional group names.
   amino
   carboxyl
   hydroxyl
   carbonyl (ketone)

a. ____________________ $R-C(=O)-R$

b. ____________________ $R-C(=O)-OH$

c. ____________________ $R-OH$

d. ____________________ $R-N(H)-H$

3. Place a check next to the functional group(s) that can ionize (take on or give up a hydrogen ion).
   a. _____ amino
   b. _____ carboxyl
   c. _____ hydroxyl
   d. _____ carbonyl (ketone)
4. For each term on the left, write in the corresponding term; the first one is completed for you.

| | |
|---|---|
| polymer | monomer |
| polysaccharide | a. ____________________ |
| fat | b. ____________________ and c. ____________________ |
| protein | d. ____________________ |
| nucleic acid | e. ____________________ |

5. Label this diagram using the following alphabetized list of terms.
   dehydration reaction
   hydrolysis reaction
   monomers
   polymer

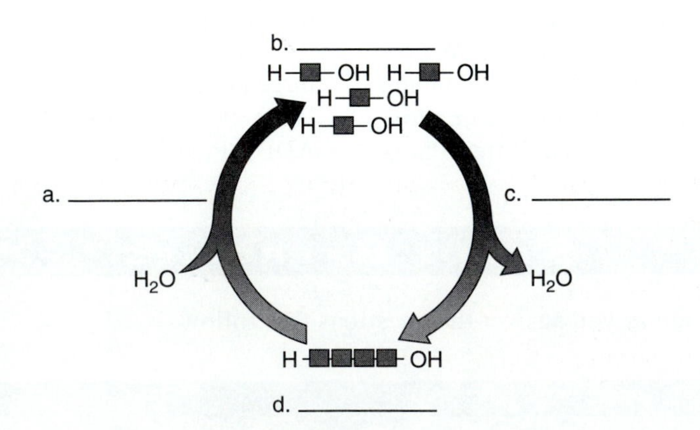


   e. During a hydrolysis reaction, is water added to or taken away from the reactants? __________
   f. During a dehydration reaction, is water added to or taken away from the reactants? __________

## 3.2 Carbohydrates (pp. 39–41)

- Glucose is an immediate energy source for many organisms.
- Some carbohydrates (starch and glycogen) function as stored energy sources.
- Other carbohydrates (cellulose and chitin) function as structural compounds.

6. Write the molecular formula beneath each of these structural formulas by indicating the number of carbons, hydrogens, and oxygens in each.

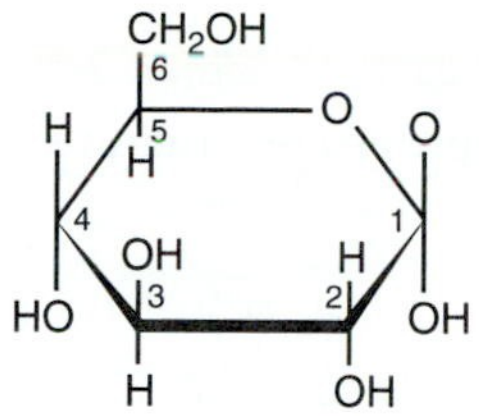

a. 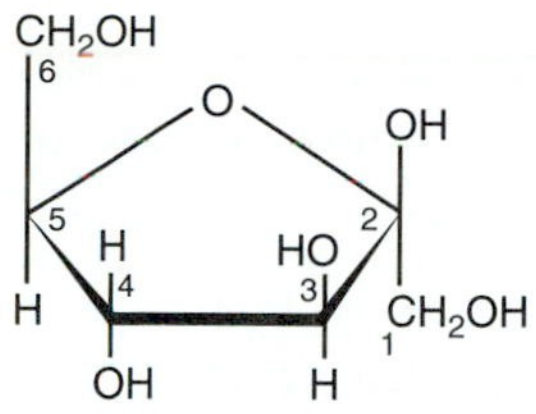 ________________ b. ________________

The term that refers to two structurally dissimilar molecules with the same molecular formula is c. ______________.

7. Complete the following table:

| Carbohydrate | Monosaccharide Composition | Biological Function |
|---|---|---|
| sucrose | | |
| lactose | | |
| maltose | | |
| starch | | |
| glycogen | | |
| cellulose | | |
| chitin | | |

8. a. Which molecules in the first column of the table in question 7 are disaccharides? ______________________

   b. Which are polysaccharides? ______________________________________________

## 3.3 Lipids (pp. 42–45)

- Lipids vary in structure and function.
- Fats function as long-term stored energy sources.
- Cellular membranes are a bilayer of phospholipid molecules.
- Steroids are derived from cholesterol, a complex ring compound.

9. Complete the following table:

| Lipid | Monomers | Biological Functions |
|---|---|---|
| fats and oils | | |
| waxes | | |
| phospholipids | | |

10. Write the word *saturated* or *unsaturated* beneath the appropriate structure.

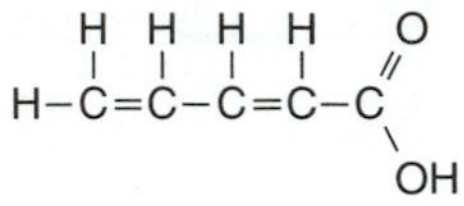

```
   H H H H   O           H H H H   O
   | | | |  //           | | | |  //
 H-C=C-C=C-C           H-C-C-C-C-C
            \            | | | |  \
             OH          H H H H   OH
```

a. 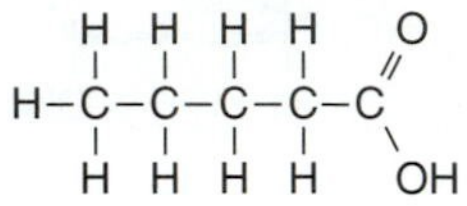 ________________    b. ________________

11. In this representation of a fat, draw a circle around the portion derived from glycerol. Draw lines under the portions derived from fatty acids.

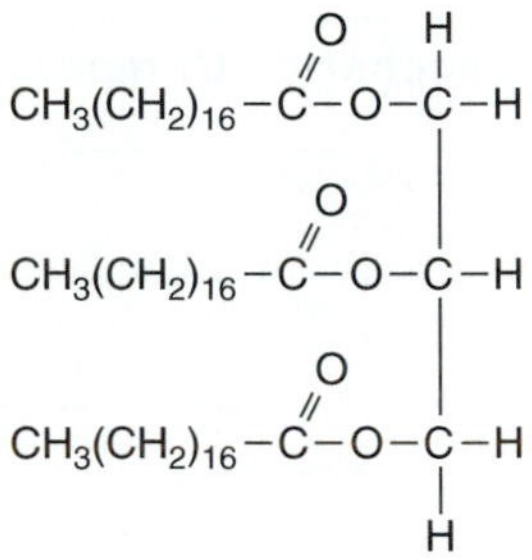

```
                O  H
               //  |
CH3(CH2)16-C-O-C-H
                   |
                O  |
               //  |
CH3(CH2)16-C-O-C-H
                   |
                O  |
               //  |
CH3(CH2)16-C-O-C-H
                   |
                   H
```

12. When phospholipids are placed in water, the a. ______________ face outward and the b. ______________ face each other. This property makes phospholipids suitable molecules to form the c. ______________ of cells.

13. Examples of steroids are a. ______________________, b. ______________________, and c. ____________________.

14. Each steroid differs from other steroids by the ______________ attached to the ring.

## 3.4 Proteins (pp. 46–49)

- Proteins serve many and varied functions such as support, enzymatic, transport, defensive, hormonal regulation, and motion.
- Each protein has levels of structure resulting in a particular shape. Hydrogen, ionic, and covalent bonding, and hydrophobic interactions help maintain a protein's normal shape.
- Environmental conditions can cause a protein to change its shape and no longer function as it did.

15. Complete the following table:

| Protein | Biological Function |
|---|---|
| enzymes | |
| actin, myosin | |
| insulin | |
| hemoglobin | |

## Peptides (p. 46)

16. Label this diagram using the following alphabetized list of terms. One term is used twice.

amino acid
amino group
carboxyl (acid) group
peptide bond

a. ____________ b. ____________

e. ____________ c. ____________

dehydration reaction

hydrolysis reaction

$R$ = rest of the molecule

d. ____________

## Shape of Proteins (p. 49)

17. Study this representation of a polypeptide.

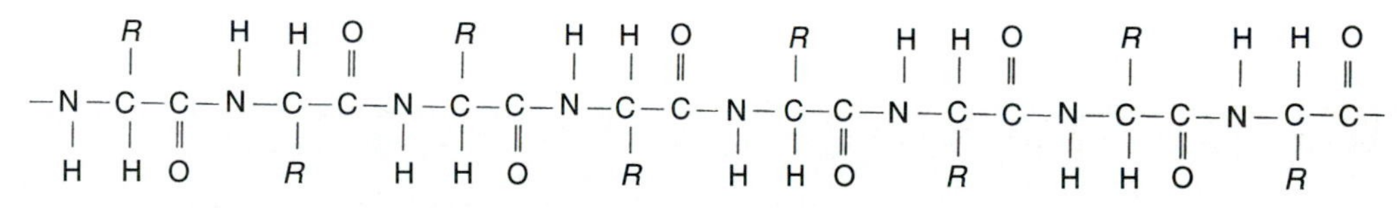

a. This is the ____________ structure of a protein.

b. What are $R$ groups? ____________

c. What shapes do the secondary structure of a protein normally assume? ____________

d. What type of bond between amino acids is necessary to maintain secondary shape? ____________

e. How does the tertiary shape of a globular protein come about? ____________

f. What would cause a protein to have a quaternary shape? ________________________________________

________________________________________________________________________________

________________________________________________________________________________

## 3.5 Nucleic Acids (pp. 50–52)

- Genes are composed of DNA (deoxyribonucleic acid). DNA specifies the correct ordering of amino acids in proteins, with RNA as an intermediary.
- The nucleotide ATP serves as a carrier of chemical energy in cells.

18. Both DNA and RNA are polymers of ____________.
19. On this diagram, label the following components of a nucleotide.
    nitrogen-containing base
    phosphate
    sugar

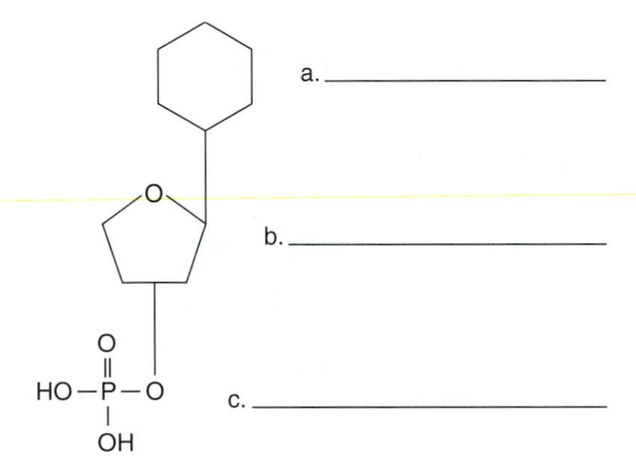


20. Refer to the following diagram of a strand of nucleotides to answer questions *a–d*.
    a. What molecule is represented by S? ____________
    b. What molecule is represented by B? ____________
    c. How many different types of B are in DNA? ____________
    d. What type of bond is represented by the lines? ____________

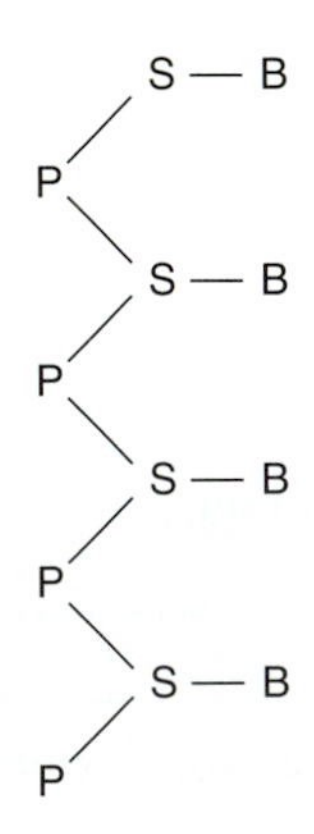

21. a. Complete the following table to distinguish DNA from RNA:

| | DNA | RNA |
|---|---|---|
| Sugar | | |
| Bases | | |
| Strands (how many?) | | |
| Helix (yes or no) | | |

b. What are the functions of DNA and RNA? ______________________________________________

_______________________________________________________________

## ATP (Adenosine Triphosphate) (p. 52)

22. ATP is a(n) [a] ____________; its structure consists of three [b] ____________ groups attached to the five-carbon [c] ____________ of the molecule.
23. Complete this reaction: ATP → ADP + Ⓟ + [a] ____________. When cells need [b] ____________, they break down the molecule [c] ____________.

## KeyWord CrossWord

Review key terms by completing this crossword puzzle, using the following alphabetized list of terms:

*amino acid*
*carbohydrate*
*DNA*
*enzyme*
*hydrolysis*
*hydrophilic*
*hydrophobic*
*isomer*
*lipid*
*nucleic acid*
*nucleotide*
*organic*
*peptide*
*phospholipid*
*polymer*
*protein*
*RNA*
*steroid*

*Across*

1 organic molecule that has an amino group and an acid group, and that covalently bonds to produce protein molecules (two words)

6 monomer of DNA and RNA consisting of a five-carbon sugar bonded to a nitrogen-containing base and a phosphate group

7 nucleic acid polymer produced from covalent bonding of nucleotide monomers that contain the sugar ribose; carries information for protein synthesis from DNA

12 splitting of a compound by the addition of water, with the $H^+$ being incorporated in one fragment and the $OH^-$ in the other

13 type of lipid molecule having four interlocking rings; examples are cholesterol, estrogen, and testosterone

14 type of molecule that does not interact with water because it is nonpolar

15 molecule having the same structure as a fat except that a group that contains phosphate replaces one bonded fatty acid; an important component of plasma membranes

18 class of organic compounds that tend to be soluble in nonpolar solvents such as alcohol; includes fats and oils

*Down*

2 molecules with the same molecular formula but different structure, and therefore shape

3 type of molecule that contains carbon and hydrogen; it usually also contains oxygen

4 class of organic compounds consisting of carbon, hydrogen, and oxygen atoms; includes monosaccharides, disaccharides, and polysaccharides

5 nucleic acid polymer produced from covalent bonding of nucleotide monomers that contain the sugar deoxyribose; the genetic material of nearly all organisms

8 polymer of nucleotides; includes both DNA and RNA (two words)

10 organic catalyst, usually a protein molecule, that speeds chemical reactions in living systems

11 type of molecule that interacts with water by dissolving in water or by forming hydrogen bonds with water molecules

15 macromolecule consisting of covalently bonded monomers

16 a polymer having, as its primary structure, a sequence of amino acids united through covalent bonding

17 a series of amino acids joined by covalent bonding

# Chapter Test

## Objective Questions

Do not refer to the text when taking this test. For questions 1–8 match the descriptions to the following classes:

a. carbohydrates
b. fats and oils
c. proteins
d. nucleic acids

____ 1. sucrose is a member
____ 2. glycerol is a building block
____ 3. specify the sequence of amino acids in a protein
____ 4. contains the bases uracil and adenine
____ 5. insulin is a member
____ 6. triglycerides are members
____ 7. exhibit a primary, secondary, and tertiary structure
____ 8. some have enzymatic roles

____ 9. Select the functional group that can ionize.
a. amino
b. carboxyl
c. hydrogen
d. hydroxyl

____ 10. What is the relationship between glucose and fructose?
a. disaccharides
b. isomers
c. isotopes
d. polysaccharides

____ 11. The products from the hydrolysis of sucrose are
a. fructose and galactose.
b. fructose and glucose.
c. galactose and glucose.
d. galactose and lactose.

____ 12. Select the molecule with mainly a structural role.
a. cellulose
b. glycogen
c. starch
d. sucrose

____ 13. Select the false statement.
a. Fats provide short-term energy to organisms.
b. Hydroxyl groups are polar.
c. Saturated fatty acids do not have double bonds.
d. Cellulose is a chain of glucose molecules.

____ 14. Select the true statement about waxes.
a. They are hydrophobic.
b. They are liquids at room temperature.
c. They are similar in structure to steroids.
d. They consist of short-term fatty acids.

____ 15. The alpha helix refers to a protein's ____________ structure.
a. primary
b. secondary
c. tertiary
d. quaternary

____ 16. Select the smallest structure.
a. amino acid
b. dipeptide
c. polypeptide
d. protein

____ 17. Which of the following is NOT a common function of some proteins?
a. energy storage
b. hormonal regulation
c. structural component
d. transport

____ 18. The opposing reaction to a dehydration reaction is
a. condensation.
b. hydrolysis.
c. monomers.
d. polymerization.

____ 19. Select the base NOT present in DNA.
a. C
b. G
c. T
d. U

____ 20. The monomers of proteins are
a. amino acids.
b. fatty acids.
c. monosaccharides.
d. nucleotides.

## Critical Thinking Questions

Answer in complete sentences.

21. What are the similarities and differences between glycogen and starch?

22. How does the primary structure of a polypeptide determine its secondary structure?

**Test Results:** ______ number correct ÷ 22 = ______ × 100 = ______ %

## Exploring the Internet

The Online Learning Center at *www.mhhe.com/maderbiology8* has additional study material and practice quizzes that can help you master the content of this chapter. You can also find links to websites exploring additional topics in biology. Access to the Online Learning Center is free for those who have purchased a new textbook.

## Answer Key

### Study Exercises

**1. a.** F **b.** T **c.** T **d.** F **2. a.** carbonyl (ketone) **b.** carboxyl **c.** hydroxyl **d.** amino **3. a.** amino **b.** carboxyl **4. a.** monosaccharide **b.** fatty acid **c.** glycerol **d.** amino acid **e.** nucleotide **5. a.** hydrolysis **b.** monomers **c.** dehydration **d.** polymer **e.** added to **f.** taken away from **6. a.** $C_6H_{12}O_6$ **b.** $C_6H_{12}O_6$ **c.** isomer

**7.**

| Monosaccharide Composition | Biological Function |
|---|---|
| glucose, fructose | transport sugar in plants |
| glucose, galactose | in milk, energy source |
| glucose, glucose | digestive breakdown product of starch |
| glucose | energy storage in plants |
| glucose | energy storage in animals |
| glucose | plant structure |
| glucose | exoskeleton in crabs, lobsters, insects |

**8. a.** sucrose, lactose, maltose **b.** starch, glycogen, cellulose, chitin

**9.**

| Monomers | Biological Functions |
|---|---|
| three fatty acids, glycerol | long-term energy storage |
| long-chain fatty acid and long-chain alcohol | protective cuticle to prevent water loss in plants |
| glycerol, two fatty acids, phosphate group | plasma membrane structure and properties |

**10. a.** unsaturated **b.** saturated
**11.**

$$\begin{array}{l} \qquad\qquad\qquad\qquad\quad\;\; H \\ \qquad\qquad\qquad\quad O \quad\;\, | \\ CH_3(CH_2)_{16}-\overset{\|}{C}-O-C-H \\ \qquad\qquad\qquad\quad O \quad\;\, | \\ CH_3(CH_2)_{16}-\overset{\|}{C}-O-C-H \\ \qquad\qquad\qquad\quad O \quad\;\, | \\ CH_3(CH_2)_{16}-\overset{\|}{C}-O-C-H \\ \qquad\qquad\qquad\qquad\quad\;\; | \\ \qquad\qquad\qquad\qquad\quad\;\; H \end{array}$$

**12. a.** polar heads **b.** nonpolar tails **c.** plasma membrane **13. a.** cholesterol **b.** estrogen **c.** testosterone **14.** functional groups
**15.**

| Biological Function |
|---|
| catalysts that speed chemical reactions |
| contractile proteins in muscle |
| hormone involved in blood sugar regulation |
| oxygen pigment that transports in blood |

**16. a.** amino acid **b.** amino acid **c.** carboxyl (acid) group **d.** peptide bond **e.** amino group **17. a.** primary **b.** they represent the variable parts of the amino acids (i.e., H, $CH_3$, C chain, C ring) **c.** $\alpha$ (alpha) helix and $\beta$ (beta) sheet **d.** hydrogen **e.** folding and twisting of polypeptide **f.** if it contained more than one polypeptide **18.** nucleotides **19. a.** nitrogen-containing base **b.** sugar **c.** phosphate **20. a.** sugar **b.** nitrogen-containing base **c.** 4 **d.** covalent
**21. a.**

| DNA | RNA |
|---|---|
| deoxyribose | ribose |
| A, T, C, G | A, U, C, G |
| double stranded | single stranded |
| yes | no |

**21. b.** DNA stores information regarding the order of amino acids in a polypeptide (protein); RNA carries this information as an intermediary for the process of protein synthesis. **22. a.** nucleotide **b.** phosphate **c.** sugar **23. a.** energy **b.** energy **c.** ATP

## KeyWord CrossWord

| 1 A | M | 2 I | N | 3 O | | A | 4 C | I | 5 D | | | | | | | | | | | |
|---|---|---|---|---|---|---|---|---|---|---|---|---|---|---|---|---|---|---|---|---|
| | | S | | R | | | A | | 6 N | U | C | L | E | O | T | I | D | E | | |
| | | O | | G | | | 7 R | N | A | | | | | | | | | | | 8 N |
| | | M | | A | | | B | | | | | | | | | | | | | U |
| | | E | | N | | | O | | | 10 E | | | | | | | | | | C |
| | | R | | I | | | H | | | N | | | | | | | | | | L |
| | | | | C | | | Y | | | Z | | | | | 11 H | | | | | E |
| | | | | | | | D | | 12 H | Y | D | R | O | L | Y | S | I | S | | I |
| | | | | | | | R | | | M | | | | | D | | | | | C |
| | | | | | | | A | | | E | | 13 S | T | E | R | O | I | D | | |
| | | | | | | | T | | | | | | | | O | | | | | A |
| | | | | | | | E | | | 14 H | Y | D | R | O | P | H | O | B | I | C |
| | | | | | | | | | | | | | | | H | | | | | I |
| | | | | | | | | | | | | | | | I | | | | | D |
| | | | | | | | | 15 P | H | O | S | 16 P | H | O | L | I | 17 P | I | D | |
| | | | | | | | | O | | | | R | | | I | | E | | | |
| | | | | | | | | L | | | | O | | | C | | P | | | |
| | | | | | | | | Y | | | | T | | | | | T | | | |
| | | | | | | | | M | | | | E | | | | 18 L | I | P | I | D |
| | | | | | | | | E | | | | I | | | | | D | | | |
| | | | | | | | | R | | | | N | | | | | E | | | |

## Chapter Test

**1.** a **2.** b **3.** d **4.** d **5.** c **6.** b **7.** c **8.** c **9.** b **10.** b **11.** b **12.** a **13.** a **14.** a **15.** b **16.** a **17.** a **18.** b **19.** d **20.** a **21.** Both glycogen and starch are polysaccharides with glucose as the monomer. Both are energy-storing molecules, but starch fulfills this role in plants and glycogen does it in some animals. Glycogen exhibits more branching than starch. **22.** The secondary structure of a polypeptide depends on hydrogen bonding between the *R* groups of the amino acids making up the polypeptide. Each particular polypeptide has its own sequence of amino acids and therefore *R* groups.

# 4

# Cell Structure and Function

## Chapter Review

All organisms are composed of cells. **Cells** are the smallest units displaying the properties of life. Cells normally are measured in micrometers because they are so small. Their small size ensures a sufficient amount of plasma membrane to serve the **cytoplasm.**

The two major kinds of cells are **prokaryotic** and **eukaryotic.** They differ by the organization of chromosomal DNA and the presence of **organelles** in the cytoplasm. Prokaryotic cells are divided into two domains, **Archaea** and **Bacteria.** These cells lack a membrane-bounded nucleus and other membranous organelles.

Organisms with eukaryotic cells are members of the domain **Eukarya.** The **nucleus** of eukaryotic cells (plant and animal) is defined by a nuclear envelope, which separates the nucleoplasm from the cytoplasm. The chromosomal material exists as **chromatin** until the cell divides. The **nucleolus** in the nucleus contains ribosomal RNA and the proteins of ribosomal subunits.

The eukaryotic cell contains a variety of structures in the cytoplasm. **Ribosomes** are the site of protein synthesis. They may exist freely or be attached to the **endoplasmic reticulum.** The **endomembrane system** consists of the nuclear envelope, the endoplasmic reticulum, the Golgi apparatus, and vesicles. The **endoplasmic reticulum** provides channels that transport substances through the cell. Substances are processed and packaged by the **Golgi apparatus. Lysosomes** contain enzymes that promote the breakdown of cell substances. Vesicles transport molecules from one part of the system to another.

Some organelles are specialized to handle energy in the cell. **Chloroplasts** are the site of photosynthesis, whereas the **mitochondria** are regions involved in cellular respiration. These organelles may be remnants of prokaryotes that inhabited eukaryotic cells over evolutionary time.

The **cytoskeleton** contains microtubules, intermediate filaments, and actin filaments. They maintain cell shape and assist movement of cell parts.

## Study Exercises

Study the text section by section as you answer the questions that follow.

### 4.1 Cellular Level of Organization (pp. 58–59)

- All organisms are composed of cells, which arise from preexisting cells.
- A microscope is usually needed to see a cell because most cells are small.

1. Check the two statements that are tenets of the cell theory.
   a. _____ All organisms are made up of cells.
   b. _____ Cork cells are living.
   c. _____ Multicellular organisms are living.
   d. _____ Cells come only from preexisting cells.

2. Label each of the following statements with B or T as describing the bright-field light microscope (B) or the transmission electron microscope (T):
   a. _____ focusing by glass lenses
   b. _____ focusing by magnetic lenses
   c. _____ image viewed from photographic film or fluorescent screen
   d. _____ image viewed through the microscope

3. Name several other microscopes or microscopy techniques available to study cells today. ________________
________________________________________________

## Cell Size (p. 59)

- Surface-area-to-volume relationships explain why cells are so small.

4. As the volume of a cell [a]______________, the proportionate amount of cell surface area [b]______________.
5. A large cell requires more [a]______________ and produces more [b]______________ than a small cell. Materials are exchanged at the cell's [c]______________. Because the surface area of a large cell actually [d]______________, cell size stays [e]______________.

## 4.2 Prokaryotic Cells (pp. 62–63)

- Prokaryotic cells do not have a membrane-bounded nucleus or other organelles of eukaryotic cells.

6. Label this diagram of a prokaryotic cell, using the following alphabetized list of terms.

| | | |
|---|---|---|
| cell wall | inclusion bodies | pilus |
| fimbriae | mesosome | plasma membrane |
| flagellum | nucleoid | ribosomes |
| glycocalyx | | |

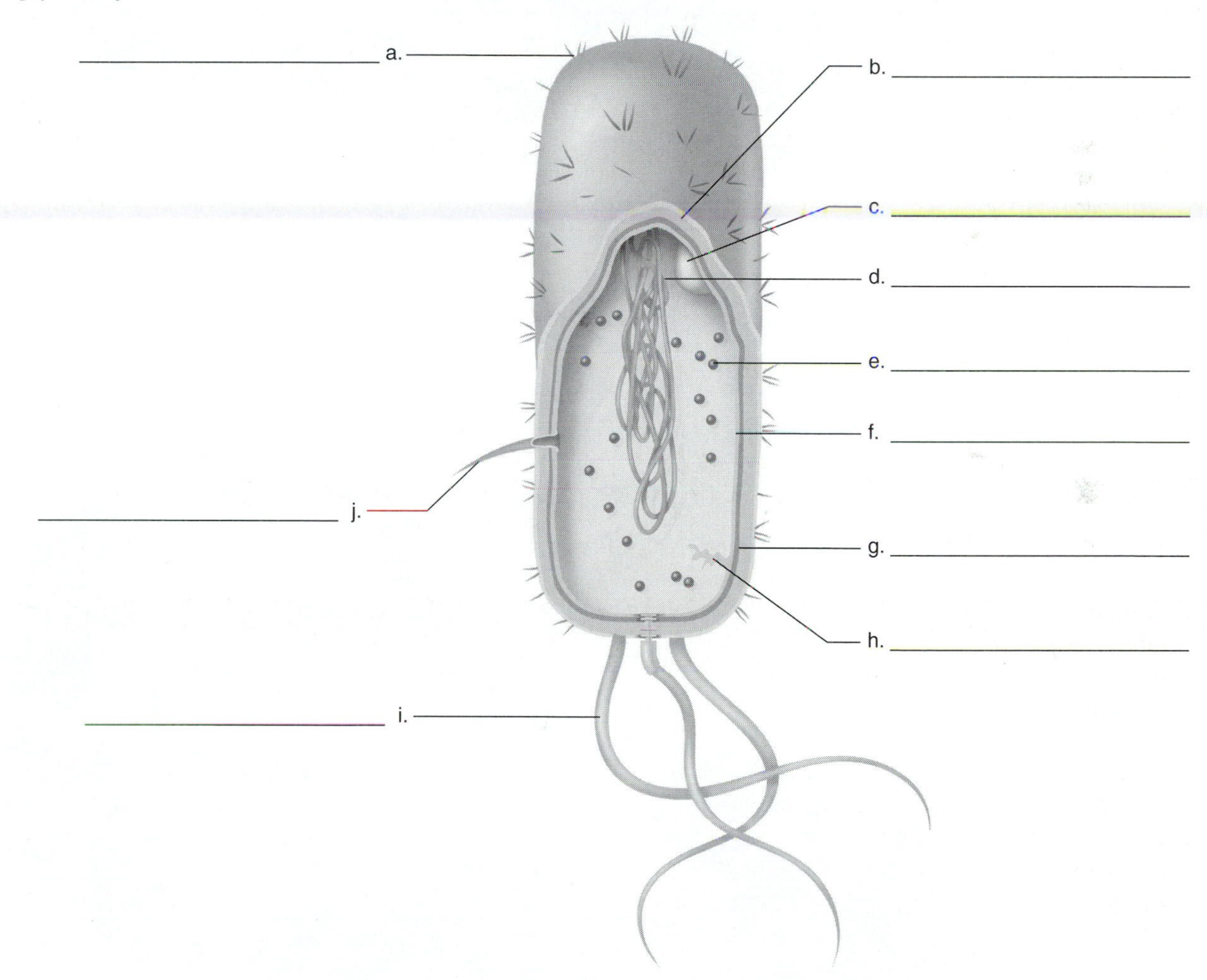


For questions 7–12, match each of the following prokaryotic cell parts to its description:

_____ 7. cell wall
_____ 8. flagellum
_____ 9. glycocalyx
_____10. nucleoid
_____11. pilus
_____12. ribosomes

a. sites of protein synthesis
b. gel-like coating outside the cell wall
c. location of bacterial chromosomes
d. structure that provides support; shapes cell
e. rotating filament that pushes the cell forward
f. hollow appendage that transfers DNA to other cells

# 4.3 Eukaryotic Cells (pp. 64–79)

- Organelles are membrane-bounded compartments specialized to carry out specific functions.

13. Label this diagram of an animal cell using the following alphabetized list of terms.

centriole
Golgi apparatus
lysosome
microtubule
mitochondrion
nucleolus
nucleus
ribosome
rough ER
smooth ER
vacuole

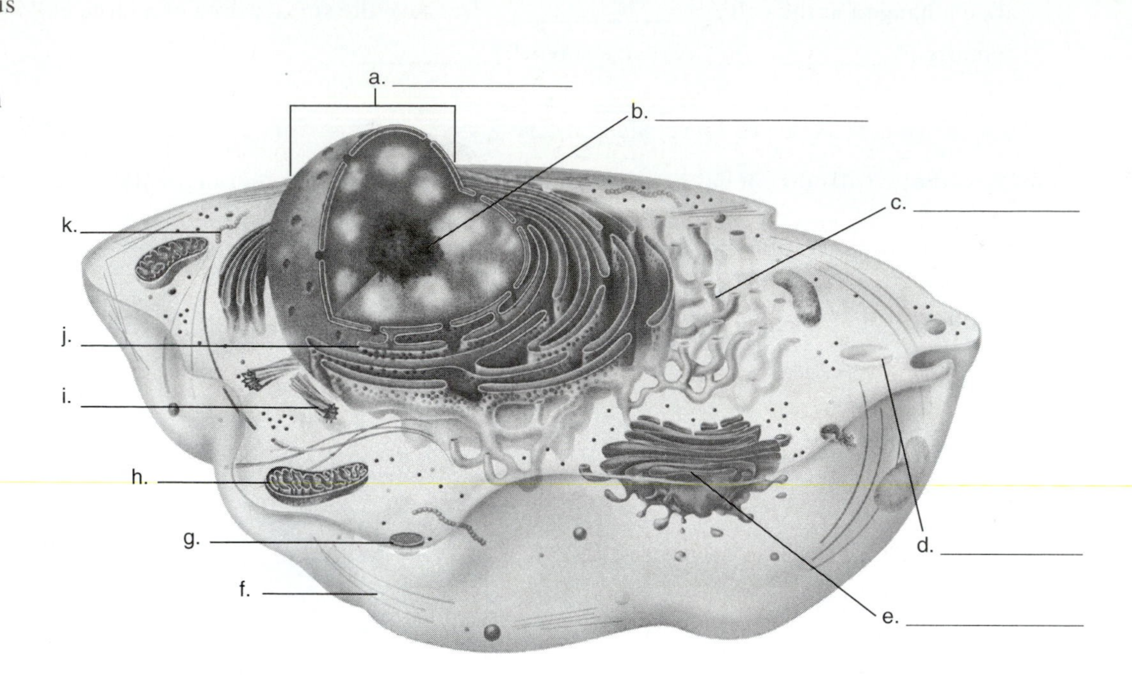


14. Label this diagram of a plant cell using the following alphabetized list of terms.

actin filament
cell wall
central vacuole
chloroplast
chromatin
Golgi apparatus
intracellular space
microtubule
middle lamella
mitochondrion
nuclear envelope
nuclear pore
nucleolus
plasma membrane
ribosome
rough ER
smooth ER

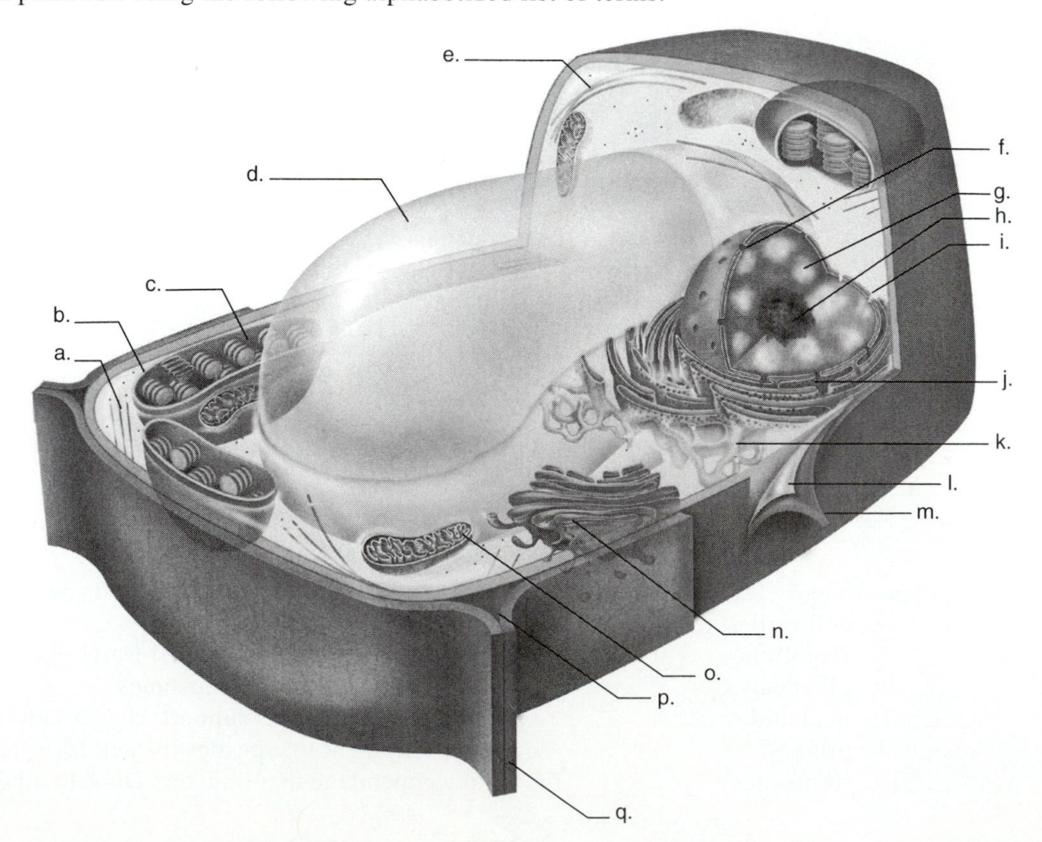

15. Complete this table by writing *yes* (the structure is present) or *no* (it is not present) on the lines provided.

| Cell Part | Prokaryotic | Eukaryotic (animal) |
|---|---|---|
| plasma membrane | | |
| cell wall | | |
| nuclear envelope | | |
| mitochondria | | |
| endoplasmic reticulum | | |
| ribosomes | | |
| centrioles | | |

16. Place the following terms in the appropriate column to compare plant and animal cell structures (some terms are used in both columns).

cell wall  centrioles  chloroplasts  large central vacuole  mitochondria
plasma membrane  small vacuoles

| Animal | Plant |
|---|---|
| | |
| | |
| | |
| | |
| | |
| | |
| | |

17. Which eukaryotic organelles could have evolved from independent prokaryotes that took up residence in early eukaryotic cells?
   a. _____ ribosomes
   b. _____ mitochondria
   c. _____ centrioles
   d. _____ chloroplasts

## The Nucleus and Ribosomes (pp. 68–69)

- Eukaryotic cells have a membrane-bounded nucleus that contains DNA within chromosomes.

18. The nucleus is enclosed by the a.______________, which contains b.______________ that open into the cytoplasm. At the time of cell division, chromatin c.______________ to form chromosomes. Chromatin has a region called the d.______________, where e.______________ is produced.

## The Endomembrane System (pp. 70–72)

19. Explain how these organelles work together.
    a. ribosomes and endoplasmic reticulum ______________________________________________
    ______________________________________________
    b. endoplasmic reticulum and Golgi apparatus ______________________________________________
    ______________________________________________
    c. lysosomes and vesicles ______________________________________________
    ______________________________________________

## Energy-Related Organelles (pp. 74–75)

- Chloroplasts use solar energy to produce organic molecules that are broken down, releasing energy in mitochondria.

For questions 20–25, match each of the following endomembrane system organelles to its description:

| | |
|---|---|
| ____ 20. rough ER | a. contain digestive enzymes |
| ____ 21. smooth ER | b. sorts lipids and proteins and packages them in vesicles |
| ____ 22. lysosomes | c. fuse with plasma membrane to move substances outside cell |
| ____ 23. transport vesicles | d. synthesizes proteins and packages them |
| ____ 24. Golgi apparatus | e. synthesizes lipids |
| ____ 25. secretory vesicles | f. shuttle proteins and lipids to various locations |

26. Chloroplasts use a.______________ energy to synthesize b.______________, which are broken down by c.______________ to produce d.______________ molecules. Photosynthesis occurs in e.______________.

27. Label this diagram of a chloroplast using the following alphabetized list of terms.
granum
inner membrane
outer membrane
stroma
thylakoid space

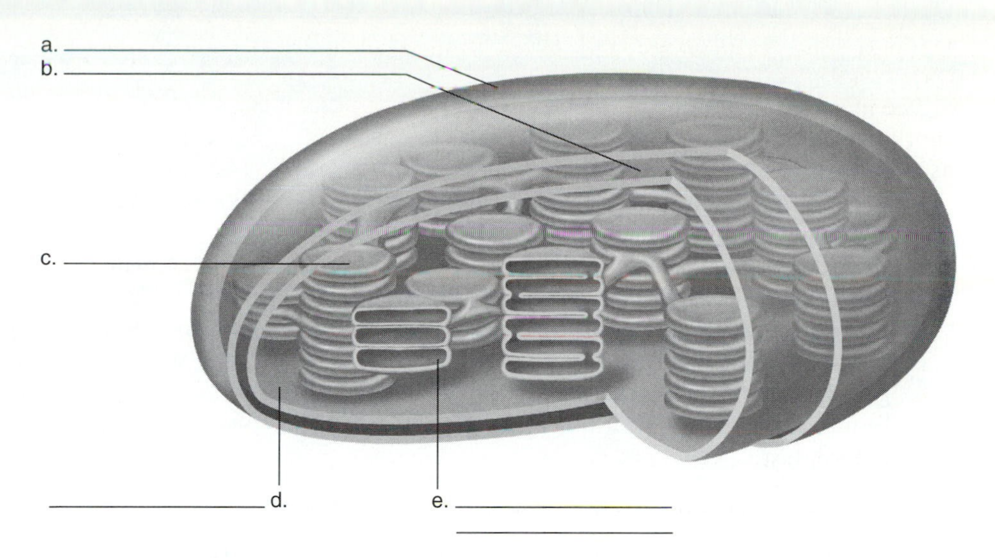


28. Using words, what is the overall equation for photosynthesis? ______________________

29. Label this diagram of a mitochondrion using the following alphabetized list of terms.
cristae
inner membrane
matrix
outer membrane

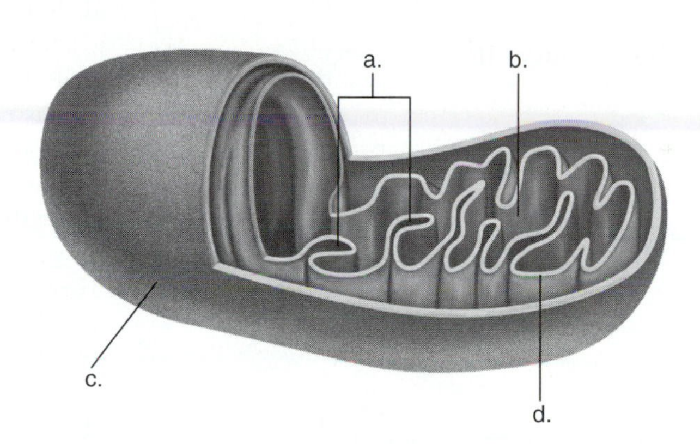


30. Using words, what is the overall equation for cellular respiration? ______________________

## The Cytoskeleton (pp. 76–79)

- The cytoskeleton, a complex network of interconnected filaments and tubules, gives the cell its shape and accounts for the movement of the cell and its organelles.

31. Match the definitions to these terms:
actin filament    intermediate filament    microtubule
a. ______________ small cylinder made of the protein tubulin
b. ______________ long, extremely thin fiber that often interacts with myosin
c. ______________ fibrous polypeptide that varies according to the tissue

32. Microtubules, like actin filaments and intermediate filaments, are able to assemble and a.______________. Microtubules radiate out from the centrosome, the main b.______________ center in a cell. In animal cells, this center contains two c.______________, which have a 9 + 0 pattern of microtubules. Centrosomes have long been associated with the formation of the d.______________ during cell division. Centrioles are believed to give rise to e.______________, which organize cilia and flagella. Cilia and flagella have a(n) f.______________ pattern of microtubules.

## Chapter Test

### Objective Questions

Do not refer to the text when taking this test. In questions 1–8, match each cell part with these descriptions:

a. regulates passage of substances into the cell
b. processing and transport channel
c. contains enzymes for digestion
d. site of protein synthesis
e. location of the nucleolus
f. site of cellular respiration
g. found in plants, not animals
h. maintains cell shape

____ 1. chloroplast
____ 2. cytoskeleton
____ 3. endoplasmic reticulum
____ 4. lysosome
____ 5. mitochondrion
____ 6. nucleus
____ 7. plasma membrane
____ 8. ribosome

____ 9. Cells are normally measured in
a. centimeters.
b. meters.
c. micrometers.
d. millimeters.

____ 10. The minimum distance between two objects before they are seen as one object is known as
a. illumination.
b. magnification.
c. resolution.
d. transmission.

____ 11. Select the structure found in eukaryotic cells but not in prokaryotic cells.
a. plasma membrane
b. cell wall
c. mitochondrion
d. ribosome

____ 12. Select the incorrect association.
a. glycocalyx—coating
b. cell wall—provides support
c. flagellum—movement
d. mesosome—movement

____ 13. The structure that surrounds the cytoplasm in a bacterial cell is the
a. cell wall.
b. nucleoid.
c. plasma membrane.
d. ribosome.

____ 14. How are mitochondria and chloroplasts similar to bacteria?
a. They are bounded by a single membrane.
b. They have a limited amount of genetic material.
c. They lack ribosomes.
d. They are larger than normal cells.

____ 15. Which of the following structures is part of the cell's endomembrane system?
a. chloroplast
b. endoplasmic reticulum
c. mitochondrion
d. nucleolus

____ 16. Plant cells
a. have a cell wall but not a plasma membrane.
b. have chloroplasts but no mitochondria.
c. do not have any centrioles and yet divide.
d. have a large central vacuole but do not have endoplasmic reticulum.

____ 17. Which of these does NOT contain nucleic acid?
a. chromosomes
b. ribosomes
c. chromatin
d. centrioles
e. genes

____ 18. How are mitochondria like chloroplasts?
a. They have the same structure.
b. They both absorb the energy of the sun.
c. They both are concerned with energy.
d. They are both in animal cells.

____ 19. Which of the following cell structures within the cytoplasm is connected to the nuclear envelope?
a. nucleolus
b. chromatin
c. endoplasmic reticulum
d. vacuoles
e. lysosomes

____ 20. Which organelle is used to produce steroid hormones and to detoxify drugs?
a. lysosomes
b. Golgi apparatus
c. mitochondria
d. rough endoplasmic reticulum
e. smooth endoplasmic reticulum

## Critical Thinking Questions

Answer in complete sentences.

21. What would be the effect on a cell if it were suddenly to lose its mitochondria?

22. How would the destruction of the Golgi apparatus affect a cell?

**Test Results:** ______ number correct ÷ 22 = ______ × 100 = ______ %

## Exploring the Internet

The Online Learning Center at *www.mhhe.com/maderbiology8* has additional study material and practice quizzes that can help you master the content of this chapter. You can also find links to websites exploring additional topics in biology. Access to the Online Learning Center is free for those who have purchased a new textbook.

## Answer Key

### Study Exercises

**1.** a, d **2. a.** B **b.** T **c.** T **d.** B **3.** scanning electron microscopy, phase contrast microscopy, video-enhanced contrast microscopy, confocal microscopy **4. a.** increases **b.** decreases **5. a.** nutrients **b.** wastes **c.** surface **d.** decreases proportionately **e.** small **6. a.** fimbriae **b.** glycocalyx **c.** inclusion bodies **d.** nucleoid **e.** ribosomes **f.** plasma membrane **g.** cell wall **h.** mesosome **i.** flagellum **j.** pilus **7.** d **8.** e **9.** b **10.** c **11.** f **12.** a **13. a.** nucleus **b.** nucleolus **c.** smooth ER **d.** vacuole **e.** Golgi apparatus **f.** microtubule **g.** lysosome **h.** mitochondrion **i.** centriole **j.** rough ER **k.** polyribosome **14. a.** actin filament **b.** ribosome **c.** chloroplast **d.** central vacuole **e.** microtubule **f.** nuclear pore **g.** chromatin **h.** nucleolus **i.** nuclear envelope **j.** rough ER **k.** smooth ER **l.** plasma membrane **m.** cell wall **n.** Golgi apparatus **o.** mitochondrion **p.** intracellular space **q.** middle lamella

**15.**

| Prokaryotic | Eukaryotic (animal) |
|---|---|
| yes | yes |
| yes | no |
| no | yes |
| no | yes |
| no | yes |
| yes | yes |
| no | yes |

**16.**

| Animal | Plant |
|---|---|
| centrioles | cell wall |
| mitochondria | mitochondria |
| small vacuoles only | large central vacuole |
| plasma membrane | plasma membrane |
| | chloroplasts |

**17.** b, d **18. a.** nuclear envelope **b.** nuclear pores **c.** condenses **d.** nucleolus **e.** rRNA **19. a.** Proteins are made at the ribosomes located on the endoplasmic reticulum. **b.** Products made at the endoplasmic reticulum are sent to the Golgi apparatus for final processing, packaging. **c.** Vesicles may contain a substance that can be digested after fusion with lysosomes. **20.** d **21.** e **22.** a **23.** f **24.** b **25.** c **26. a.** solar **b.** carbohydrates **c.** mitochondria **d.** ATP/energy **e.** chloroplasts **27. a.** outer membrane **b.** inner membrane **c.** granum **d.** stroma **e.** thylakoid space **28.** solar energy + carbon dioxide + water → carbohydrate + oxygen **29. a.** cristae **b.** matrix **c.** outer membrane **d.** inner membrane **30.** carbohydrate + oxygen → carbon dioxide + water + energy **31. a.** microtubule **b.** actin filament **c.** intermediate filament **32. a.** disassemble **b.** microtubule organizing body **c.** centrioles **d.** spindle **e.** basal bodies **f.** 9 + 2

## Chapter Test

**1.** g **2.** h **3.** b **4.** c **5.** f **6.** e **7.** a **8.** d **9.** c **10.** c **11.** c **12.** d **13.** c **14.** b **15.** b **16.** c **17.** d **18.** c **19.** c **20.** e **21.** The cell would be unable to extract energy from carbohydrates. The ATP harvested by this process would be unavailable for cell functions. Therefore, the cell would die. **22.** The smooth ER packages substances in vesicles. A large portion of these go to the Golgi apparatus for further processing. These vesicles would most likely accumulate in the cell to the point that the cell would be unable to function properly.

# 5

# MEMBRANE STRUCTURE AND FUNCTION

## CHAPTER REVIEW

The sandwich model and its associated unit membrane model preceded the currently accepted **fluid-mosaic model** of membrane structure. According to the fluid-mosaic model, the plasma membrane is a fluid **phospholipid bilayer** with embedded proteins. The hydrophilic polar heads of the phospholipids face the outside and the inside of the cell. The hydrophobic nonpolar tails face each other. The proteins embedded in the plasma membrane form channels and function as receptors, enzymes, and carriers.

Molecules move across a membrane in several ways. By **diffusion,** molecules move down their concentration gradiant. **Osmosis** is the diffusion of water through a differentially permeable membrane. When cells are in an **isotonic solution,** they neither gain nor lose water; when they are in a **hypotonic solution,** they gain water; when they are in a **hypertonic solution,** they lose water. Both diffusion and osmosis are passive transport processes that do not require energy. **Facilitated transport** is also passive and involves carrier proteins moving substances from higher to lower concentrations. **Active transport** moves substances in the opposite direction, with the function of a carrier protein and energy. Larger substances pass through cells by **exocytosis** and **endocytosis.**

An extracellular matrix is a meshwork of polysaccharides and proteins that is now known to influence animal cell shape, movement, and function. Plant cells are bounded by a **cell wall** that is external to the plasma membrane. This boundary is freely permeable. Several types of junctions exist between animal cells: **adhesion junctions, tight junctions,** and **gap junctions**. Between plant cells, **plasmodesmata** perform this function.

## STUDY EXERCISES

Study the text section by section as you answer the questions that follow.

### 5.1 MEMBRANE MODELS (P. 84)

- The present-day fluid-mosaic model proposes that membrane is less rigid and more dynamic than suggested by the previous sandwich model.

1. Indicate whether these statements are true (T) or false (F).
   a. _____ According to the fluid-mosaic model, globular proteins are embedded in the membrane.
   b. _____ According to the unit membrane model, the cell membrane lacks phospholipids.
   c. _____ According to the unit membrane model, membranes differ according to their function.
   d. _____ According to the sandwich model, proteins form an outer and inner layer.
   e. _____ According to the fluid-mosaic model, proteins function as receptors, enzymes, and carriers.

## 5.2 Plasma Membrane Structure and Function (pp. 85–87)

- The plasma membrane contains lipids and proteins, each with specific functions.

2. Label this diagram of the plasma membrane using the following alphabetized list of terms.
   carbohydrate chain
   cholesterol
   cytoskeleton filaments
   glycolipid
   glycoprotein
   integral protein
   peripheral protein
   phospholipid bilayer

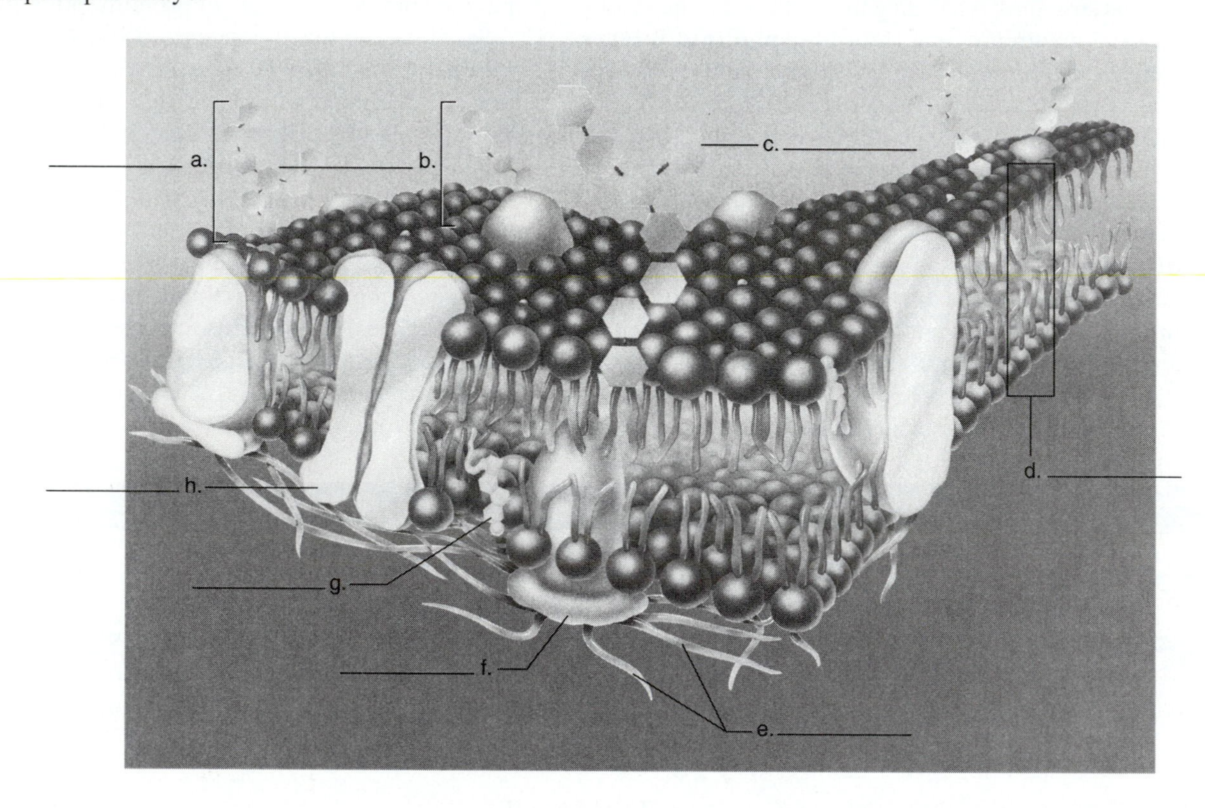


3. The two components of the fluid-mosaic model of membrane structure are [a]_____________ and [b]_____________.
4. [a]_____________ form a bilayer, in which the [b]_____________ heads are at the surfaces of the membranes, and the [c]_____________ tails face each other, making up the interior of the membrane. The lipid [d]_____________, also in the membrane, [e]_____________ the membrane's fluidity.
5. Complete the sentences, using the terms *hydrophilic* and/or *hydrophobic:* Integral proteins are found within the plasma membrane. [a]_____________ regions are embedded within the membrane, and [b]_____________ regions project from both surfaces of the bilayer.
6. Both glycolipids and glycoproteins have a(n) [a]_____________ chain and are active in cell-to-[b]_____________ recognition.

7. Label the diagrams of proteins found in the membrane and state a function on the lines provided:

carrier protein ______________________________

cell recognition protein ______________________________

channel protein ______________________________

enzymatic protein ______________________________

receptor protein ______________________________

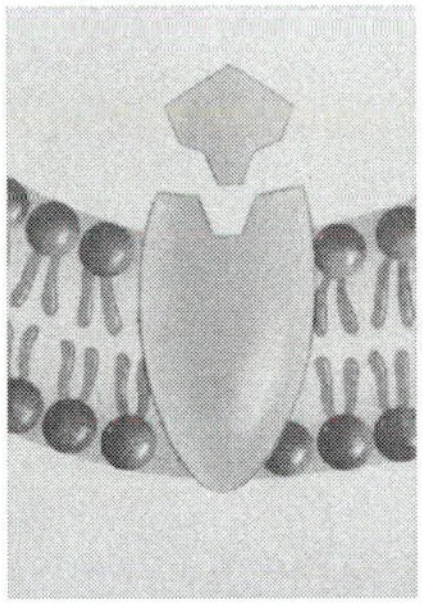

a. ______________
______________

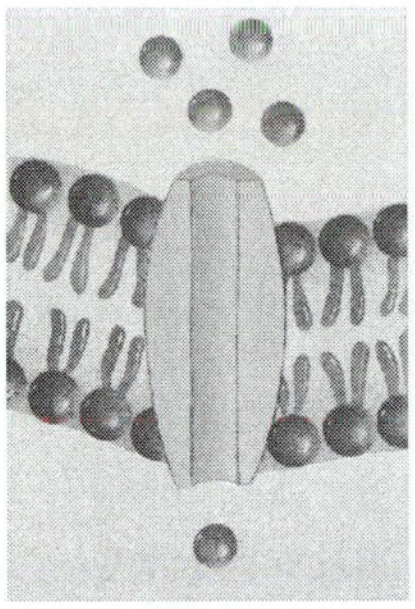

b. ______________
______________

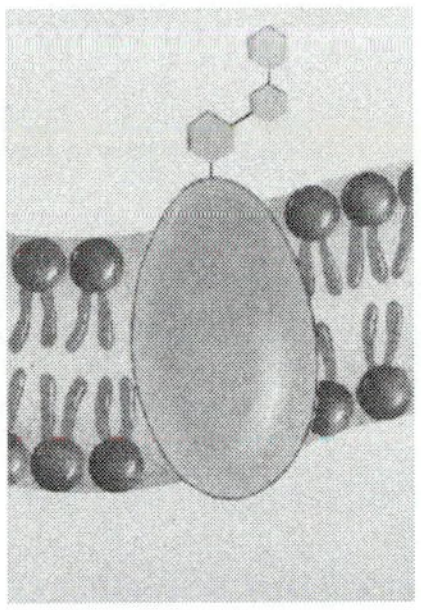

c. ______________
______________

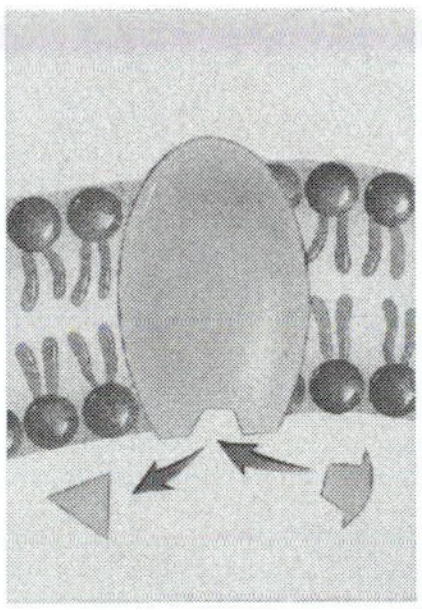

d. ______________
______________

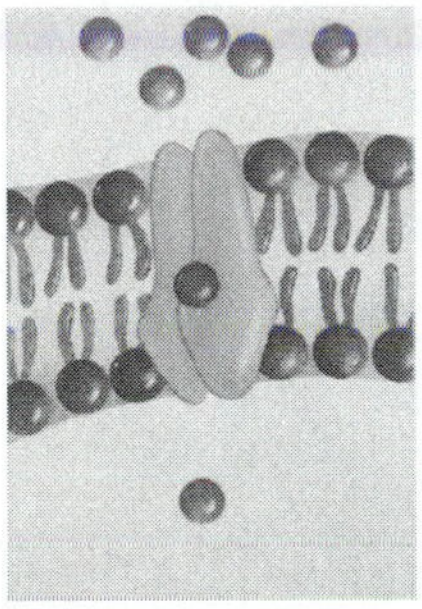

e. ______________
______________

## 5.3 Permeability of the Plasma Membrane (pp. 88–95)

- Small, noncharged molecules tend to pass freely across the plasma membrane.
- Some molecules diffuse (move from an area of higher concentration to an area of lower concentration) across a plasma membrane.
- Water diffuses across the plasma membrane, and this can affect cell size and shape.

8. Place the appropriate letter next to each statement.

D—diffusion O—osmosis

a. _____ Algae in a pond become dehydrated.
b. _____ A hypertonic solution draws water out of cell.
c. _____ A red blood cell bursts in a test tube.
d. _____ Dye crystals spread out in a beaker of water.
e. _____ Gases move across the plasma membrane.
f. _____ Perfume is sensed from the other side of a room.

9. In the following diagram, assume that glucose and water can cross the membrane and that protein cannot.

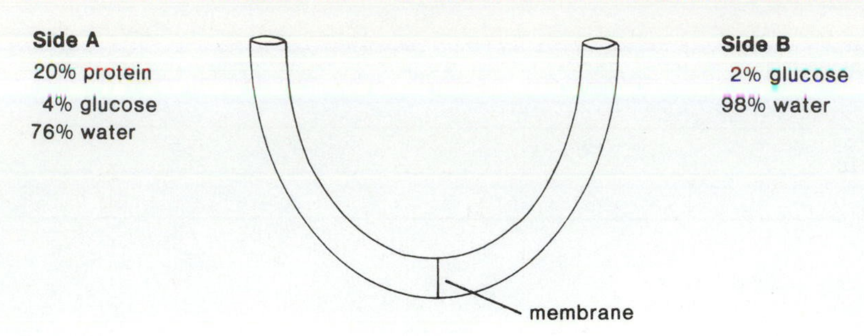


a. Will the amount of water on side A stay the same or increase or decrease with time? ______________

b. Will the amount of protein on side A stay the same or increase or decrease with time? ______________

c. Will glucose cross the membrane toward side A or side B? ______________

d. What will happen to the level of solution on each side of the membrane? ______________________________

______________________________________________________________________________________________

10. Complete the diagram on the left to describe the effect of tonicity on red blood cells.

11. Complete the diagram on the right to describe the effect of tonicity on plant cells.

| Tonicity | Before | After |
|---|---|---|
| Isotonic Solution | | a. |
| b. | | |
| Hypotonic Solution | | c. |

| Tonicity | Before | After |
|---|---|---|
| a. | cell wall<br>water vacuole<br>plasma membrane | |
| b. | | vacuole |
| Hypertonic Solution | | c. |

12. If a solution is 8% solute, it is a.____________% solvent.

If a solution is 99.5% solvent, it is b.__________% solute.

If solution A is 2% solute and solution B is 3% solute, then solution A is c.______________ to solution B, which is d.____________ to solution A.

A solution with 2% solute is e.______________ to solution A.

## Transport by Carrier Proteins (p. 92)

- Carrier proteins assist the transport of some ions and molecules across the plasma membrane.

13. Place the appropriate letter(s) next to each statement.
    F—facilitated transport A—active transport F,A—both
    a. _____ Uses a carrier protein.
    b. _____ Substances travel down a concentration gradient.
    c. _____ Substances travel against a concentration gradient.
    d. _____ Sodium-potassium pump.
    e. _____ Energy is not required.

## Membrane-Assisted Transport (p. 94)

- Vesicle formation takes other substances into the cell, and vesicle fusion with the plasma membrane discharges substances from the cell.

14. Place the appropriate letters next to each statement.
    Ex—exocytosis En—endocytosis
    a. _____ Vesicles formed by Golgi apparatus fuse with the plasma membrane.
    b. _____ Materials leave the cell.
    c. _____ Phagocytosis is an example.
    d. _____ Pinocytosis is an example.
    e. _____ Occurs after receptors bind to receptor proteins.

# 5.4 Modification of Cell Surfaces (p. 96)

- The cell wall of a plant cell supports the cell. The extracellular matrix of animal cells influences their shape, movement, and function.
- The activities of cells within a tissue are coordinated, in part, because cells are linked and directly communicate with one another.

In questions 15–18, match the following descriptions with each type of cell junction:
a. attached to cytoskeleton
b. found only in plants
c. formed from two identical plasma membrane channels
d. where plasma membrane proteins attach to each other

____ 15. adhesion junction
____ 16. gap junction
____ 17. plasmodesmata
____ 18. tight junction

19. Animal cells have a(n) a. _____________ matrix composed of protein and carbohydrate molecules. Collagen and elastin, found in the matrix, are examples of b. _____________ molecules. The matrix has substance and helps animal cells hold their c. _____________. When the gel formed by carbohydrate molecules permits rapid diffusion of chemical signals, it helps animal cells d. _____________.

20. Indicate whether these statements about the cell wall are true (T) or false (F).
    a. _____ It is found in plants but not in animals.
    b. _____ It interferes with plasma membrane function when present.
    c. _____ It is internal to the plasma membrane.
    d. _____ Some woody plants have a primary and secondary cell wall.

# Chapter Test

## Objective Questions

Do not refer to the text when taking this test. In questions 1–7, match the descriptions to each transport process:

a. active transport
b. diffusion
c. exocytosis
d. facilitated transport
e. osmosis
f. phagocytosis
g. pinocytosis

____ 1. small particle or liquid intake into a cell
____ 2. requires vacuole formation
____ 3. carrier molecule, no energy
____ 4. carrier molecule, energy required
____ 5. water enters a hypertonic solution from a cell
____ 6. secretion from the cell
____ 7. dye molecules spread through water

____ 8. Proteins form the nonactive matrix of the plasma membrane.
a. true
b. false

____ 9. Phospholipids are present in the plasma membrane.
a. true
b. false

____10. Hydrophilic regions of proteins protrude from both surfaces of the bilayer.
a. true
b. false

____11. Lipid-soluble molecules pass through the plasma membrane by
a. active transport.
b. diffusing through it.
c. facilitated transport.
d. use of the sodium-potassium pump.

____12. A 2% salt solution is ______________ to a 4% salt solution.
a. hypertonic
b. hypotonic
c. isometric
d. isotonic

____13. Which molecule is directly required for operation of the sodium-potassium pump?
a. ATP
b. $NAD^+$
c. DNA
d. water

____14. In cells, which process moves materials opposite to the direction of the other three?
a. endocytosis
b. exocytosis
c phagocytosis
d. pinocytosis

____15. Which of the following junctions is found only between plant cells?
a. adhesion junctions
b. gap junctions
c. plasmodesmata
d. tight junctions

____16. A small lipid-soluble molecule passes easily through the plasma membrane. Which of these statements is the most likely explanation?
a. A carrier protein must be at work.
b. The plasma membrane is partially composed of lipid molecules.
c. The cell is expending energy to do this.
d. Phagocytosis has enclosed this molecule in a vacuole.

____17. Which of these does NOT require an expenditure of energy?
a. diffusion
b. osmosis
c. facilitated transport
d. None of these require energy.

____18. Which term refers to the bursting of an animal cell?
a. plasmolysis
b. crenation
c. lysis
d. turgor pressure

____19. An animal cell always takes in water when placed in a(n) ______________ solution.
a. hypertonic
b. osmotic
c. isotonic
d. hypotonic

____20. Which of the following is actively transported across plasma membranes?
a. carbon dioxide
b. oxygen
c. water
d. sodium ions

## Critical Thinking Questions

Answer in complete sentences.

21. Why is the plasma membrane considered so important to the cell?

22. What osmotic problem would plant cells experience if they lost their cell walls?

**Test Results:** ______ number correct ÷ 22 = ______ × 100 = ______ %

## Exploring the Internet

The Online Learning Center at *www.mhhe.com/maderbiology8* has additional study material and practice quizzes that can help you master the content of this chapter. You can also find links to websites exploring additional topics in biology. Access to the Online Learning Center is free for those who have purchased a new textbook.

## Answer Key

### Study Exercises

**1. a.** T **b.** F **c.** F **d.** T **e.** T **2. a.** glycolipid **b.** glycoprotein **c.** carbohydrate chain **d.** phospholipid bilayer **e.** cytoskeleton filaments **f.** peripheral protein **g.** cholesterol **h.** integral protein **3. a.** lipids **b.** proteins **4. a.** Phospholipids **b.** hydrophilic (polar) **c.** hydrophobic **d.** cholesterol **e.** regulates **5. a.** Hydrophobic **b.** hydrophilic **6. a.** carbohydrate **b.** cell **7. a.** receptor protein, shaped in such a way that a specific molecule can bind to it. **b.** channel protein, allows molecules to pass across the plasma membrane. **c.** cell recognition protein, functions in cell to cell recognition. **d.** enzymatic protein, catalyzes a specific reaction. **e.** carrier protein, allows selective passage of molecules across the plasma membrane. **8. a.** O **b.** O **c.** O **d.** D **e.** D **f.** D **9. a** increase **b.** stay the same **c.** toward side B **d.** Side A will rise, side B will fall. **10. a.** cell is same size and shape **b.** Hypertonic Solution **c.** cell is bursting **11. a.** Isotonic Solution **b.** Hypotonic Solution **c.** vacuole is much smaller **12. a.** 92 **b.** 0.5 **c.** hypotonic **d.** hypertonic **e.** isotonic **13. a.** F, A **b.** F **c.** A **d.** A **e.** F **14. a.** Ex **b.** Ex **c.** En **d.** En **e.** En **15.** a **16.** c **17.** b **18.** d **19. a.** extracellular **b.** protein **c.** shape **d.** communicate **20. a.** T **b.** F **c.** F **d.** T

### Chapter Test

**1.** g **2.** f **3.** d **4.** a **5.** e **6.** c **7.** b **8.** b **9.** a **10.** a **11.** b **12.** b **13.** a **14.** b **15.** c **16.** b **17.** d **18.** c **19.** d **20.** d **21.** The plasma membrane is very important because it is the outer living boundary of the cell. It provides and regulates the passage of molecules into and out of the cell. **22.** In a hypotonic environment, plant cells would continue to gain water until they burst.

# 6

# Metabolism: Energy and Enzymes

## Chapter Review

Living things can't exhibit any of the characteristics of life without a supply of energy. There are two energy laws basic to understanding energy-use patterns in organisms at the cellular level. The first law says that energy cannot be created or destroyed, but can only be transferred or transformed. The second law states that a usable form of energy cannot be converted completely into another usable form. As a result of these laws, we know that the **entropy** (disorder) of the universe is increasing and that only a constant input of energy maintains the organization of living things.

**Metabolism** is all the reactions that occur in a cell. Only those reactions that result in a negative free energy difference—that is, the products have less usable energy than the reactants—occur spontaneously. Such reactions, called **exergonic reactions,** release energy. **Endergonic reactions,** which require an input of energy, occur because it is possible to **couple** an exergonic process with an endergonic process. For example, glucose breakdown is an exergonic metabolic pathway that drives the buildup of many **ATP** molecules. These ATP molecules then supply energy for cellular work. Thus, ATP goes through a cycle in which it is constantly being built up from, and then broken down to, **ADP** +Ⓟ.

A **metabolic pathway** is a series of reactions that proceed in an orderly, step-by-step manner. Each reaction has a specific **enzyme** that speeds the reaction by forming a complex with its **substrates.** Formation of the enzyme-substrate complex lowers the energy of activation, the amount of energy required to activate the reactants. Any environmental factor that affects the shape of a protein also affects the ability of an enzyme to do its job. (Many enzymes have **cofactors** or **coenzymes** that help them carry out a reaction.)

**Photosynthesis,** which transforms solar energy to chemical energy within carbohydrates, is a metabolic pathway that occurs in chloroplasts. **Reduction** is the gain of hydrogen atoms ($H^+ + e^-$). During photosynthesis, carbon dioxide is reduced to glucose, a carbohydrate. **Cellular respiration,** completed in mitochondria, is a metabolic pathway that transforms the energy of glucose (usually) into that of ATP molecules. **Oxidation** is the loss of hydrogen atoms. During cellular respiration, carbohydrate is oxidized to carbon dioxide and water.

Both photosynthesis and cellular respiration make use of an **electron transport system,** in which electrons are transferred from one carrier to the next with the release of energy that is ultimately used to produce ATP molecules. Chemiosmosis explains how the electron transport system produces ATP. The carriers of this system deposit hydrogen ions (H+) on one side of a membrane. When the ions flow down an electrochemical gradient through an ATPase complex, ATP is formed from ADP and Ⓟ.

## Study Exercises

Study the text section by section as you answer the questions that follow.

### 6.1 Cells and the Flow of Energy (pp. 102–3)

- Energy cannot be created nor destroyed; energy can be changed from one form to another but there is always a loss of usable energy.

1. Indicate whether these statements, related to the energy laws, are true (T) or false (F), and if the statements are false, change them to true statements:
   a. _____ The chemical energy of ATP cannot be transformed into any other type of energy such as kinetic energy. Rewrite: ______________________________
   ______________________________
   b. _____ A cell produces ATP, and therefore cells do not obey the first law of thermodynamics. Rewrite:
   ______________________________

c. _____ Because energy transformations always result in a loss of usable energy, the entropy of the universe is increasing. Rewrite: ______________________________________________

______________________________________________

d. _____ Because our society uses coal as an energy source, it is helping to decrease the entropy of the universe. Rewrite: ______________________________________________

______________________________________________

## 6.2 Metabolic Reactions and Energy Transformations (pp. 104–5)

- In cells, the breakdown of ATP, which releases energy, can be coupled to reactions that require an input of energy.
- ATP goes through a cycle: energy from glucose breakdown drives ATP buildup and then ATP breakdown provides energy for cellular work.

2. Place the appropriate letters next to each statement.

   En—endergonic reaction  Ex—exergonic reaction

   a. _____ Energy is released as the reaction occurs.
   b. _____ Energy is required to make the reaction go.
   c. _____ Reaction used by the body for muscle contraction and nerve conduction.
   d. _____ ATP → ADP + Ⓟ.
   e. _____ ADP + Ⓟ→ ATP.

3. Label this diagram, using these terms:

   ATP
   ADP
   –P
   +P

b. ____________

a. ____________

c. ____________

d. ____________

4. Label each of the following as pertaining to the left (L) or right (R) side of the diagram in question 3. Explain your choice.

   a. _____ cellular respiration. Explain: ______________________________________________

   ______________________________________________

   b. _____ muscle contraction. Explain: ______________________________________________

   ______________________________________________

   c. _____ active transport. Explain: ______________________________________________

   ______________________________________________

5. ATP is the common [a.] ______________________ of cells; when cells require energy, they "spend" ATP. ATP breakdown provides energy for [b.] ____________ work, such as synthesizing macromolecules; [c.] ____________ work, such as pumping substances across plasma membranes; and [d.] ____________ work, such as the beating of flagella. Because ATP breakdown is [e.] ____________ to endergonic reactions, energy transformation occurs with minimal loss to the cell.

## 6.3 Metabolic Pathways and Enzymes (pp. 108–9)

- Cells have metabolic pathways in which every reaction has a specific enzyme.
- Enzymes speed reactions because they have an active site where a specific reaction occurs.
- Environmental factors like temperature and pH affect the activity of enzymes.
- Inhibition of enzymes is a common way for cells to control enzyme activity.

6. Consider the following diagram of a metabolic pathway:

$$\begin{array}{ccccccccccccc} & E_1 & & E_2 & & E_3 & & E_4 & & E_5 & & E_6 & \\ A & \rightarrow & B & \rightarrow & C & \rightarrow & D & \rightarrow & E & \rightarrow & F & \rightarrow & G \end{array}$$

A–F are a. _____________, and B–G are b. _____________. $E_1$–$E_6$ are c. _____________. A is a d. _____________ for the first enzyme, and B is the product.

7. Label this diagram, using the following alphabetized list of terms.

active site
enzyme (used more than once)
enzyme-substrate complex
products
substrate

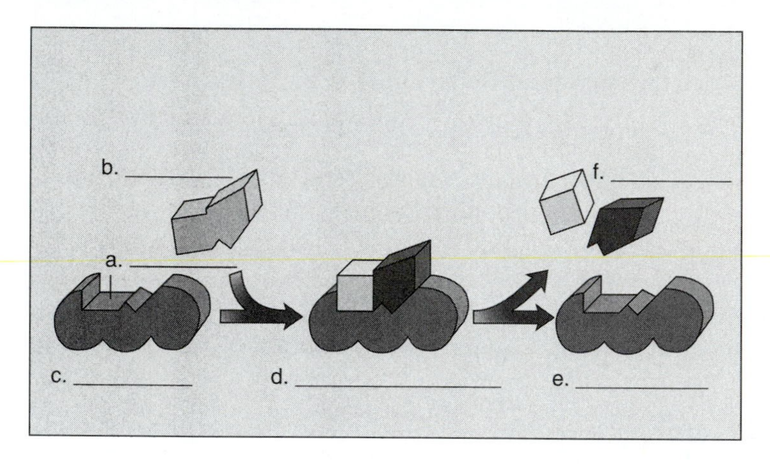


8. Which portion of the diagram in question 7 pertains to enzymes lowering the energy of activation?
a. _____________________________________________

Why? b. _____________________________________________

9. Express the reaction in question 7 in equation form, using E (for enzyme), S (for substrate), and P (for product). a. _____________________________________________

Is the reaction shown in question 7 a synthetic reaction or a degradative reaction? b. _____________

How do you know? c. _____________________________________________

The enzyme-substrate complex and the reaction occur at the d. _____________ site of the enzyme. What is the significance of using the label "enzyme" twice in the diagram in question 7? It shows that e. _____________

_____________________________________________

Why are enzymes named for their substrates (e.g., maltase speeds the breakdown of maltose)? f. _____________

_____________________________________________

10. Complete each statement with the term *increases* or *decreases*.

Raising the temperature generally a.______________ the rate of an enzymatic reaction.

Boiling an enzyme drastically b.______________ the rate of the reaction.

Changing the pH toward the optimum pH for an enzyme c.______________ the rate of the reaction.

Introducing a competitive inhibitor d.______________ the availability of an enzyme for its normal substrate.

Due to feedback inhibition, the affinity of the active site for the substrate e.______________.

11. Enzymes have helpers called a.______________, which b.________________________________________
________________________________________________________________________________

## 6.4 Oxidation-Reduction and the Flow of Energy (pp. 110–11)

- Photosynthesis and cellular respiration are metabolic pathways that include oxidation-reduction reactions. Thereby energy becomes available to living things.

12. Label each of the following as pertaining to cellular respiration (CR) and/or photosynthesis (P):
    a. _____ mitochondria
    b. _____ chloroplasts
    c. _____ breakdown of carbohydrate by oxidation of glucose
    d. _____ synthesis of carbohydrate by reduction of carbon dioxide
    e. _____ high-energy electrons pass down an electron transport system, and ATP is produced
    f. _____ ATP is used to reduce carbon dioxide
    g. _____ NADH carries high-energy electrons to the electron transport system
    h. _____ solar energy energizes electrons
    i. _____ NADPH is used to reduce carbon dioxide

13. Trace the energy content of ATP to the sun by arranging these statements in the proper order. ______________
    a. Carbohydrates are broken down during cellular respiration.
    b. Solar energy is needed for photosynthesis.
    c. Cellular respiration produces ATP molecules.
    d. Carbohydrates are products of photosynthesis.

14. Examine this diagram on chemiosmosis and then complete the sentences.

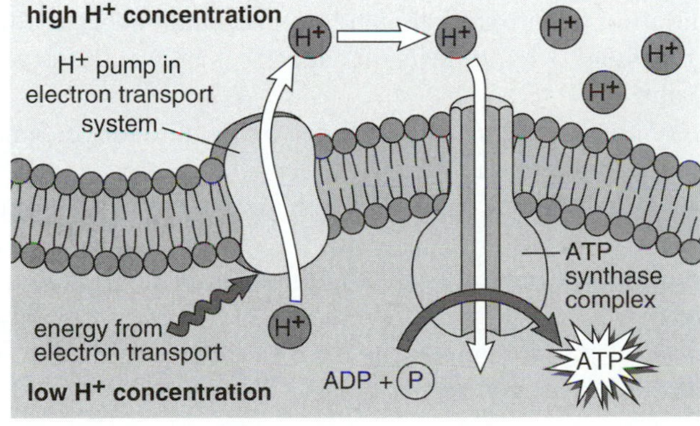


As electrons are passed from carrier to carrier within the electron transport system, energy is released and some of this is used to a.______________ hydrogen ions across a membrane. A high $H^+$ b.______________ builds up. When $H^+$ flows down this gradient through a channel in a protein, the energy released is used to form c.______________ from d.______________.

# Definitions Wordsearch

Review key terms by using the following alphabetized list of terms to fill in the blanks. Then complete the wordsearch.

```
S J E T R E C P T V U I T T H I M K N E W F
N A F D Z C Y T O K I M O D E L B Y E H C I
L O V S O T Y C O D N E I M O P Y C F C L X
T H I R I O T B G T D T N G I H R N Q E J W
G I T T Y C D E R U T A N E D T O N A N E D
R E A T A F R D C P T B C O R H T E M P N Z
A Y M I L Z K A N L C O R O M G T I N X T V
F U I I D A I O T A F L L K E I Y L L O R I
R E N N U G I L N A L I O H T F N M M C O G
L I S C O T N Z C M O S I U A D O J E Y P N
K I E Y A L N T T T S M R O P M C E T T Y P
F L R D G L O M C I R O T R H E Y I A O I T
G O I S I R O T Y C O E A H A L T G R N Y Q
C X I I S M N R T E S L E N S G O O T I N N
O U N L A I R E A C T A N T E M S L S S G U
Q M E O Y B U P L A S M Z L Y S E S B G U U
G I S R W X M P N C E N Y R O M E R U T F M
C V I M O I M H A N I B M R E E H G S G M S
O C S P E R M A T O G E E E S I S T W S I Q
D U C L A I U B N I T O E J U W J K I G U F
```

*cofactors*
*denatured*
*energy*
*entropy*
*enzyme*
*metabolism*
*model*
*oxidation*
*reactant*
*substrate*
*vitamins*

a. ________________ Capacity to do work and bring about change; occurs in a variety of forms.

b. ________________ Measure of disorder or randomness.

c. ________________ All of the chemical changes that occur within a cell.

d. ________________ Substance that participates in a reaction.

e. ________________ Organic catalyst, usually a protein, that speeds up a reaction in cells due to its particular shape.

f. ________________ Reactant in a reaction controlled by an enzyme.

g. ________________ Simulation of a process that aids conceptual understanding until the process can be studied firsthand; a hypothesis that describes how a particular process could possibly be carried out.

h. ________________ Loss of an enzyme's normal shape so that it no longer functions; caused by a less than optimal pH and temperature.

i. ________________ Nonprotein adjunct required by an enzyme to function; many are metal ions, while others are coenzymes.

j. ________________ Essential requirement in the diet, needed in small amounts. They are often part of coenzymes.

k. ________________ Loss of one or more electrons from an atom or molecule; in biological systems, generally the loss of hydrogen atoms.

## Chapter Test

### Objective Questions

Do not refer to the text when taking this test.

____ 1. The useful energy conversion in photosynthesis is
a. chemical to solar.
b. heat to mechanical.
c. mechanical to heat.
d. solar to chemical.

____ 2. Any energy transformation involves the loss of some energy as
a. electricity.
b. heat.
c. light.
d. motion.

____ 3. In the enzymatically controlled chemical reaction $A \rightarrow B + C$, A is the
a. cofactor.
b. enzyme.
c. product.
d. substrate.

____ 4. An enzyme, functioning best at a pH of 3, is in a neutral solution at a temperature of 40°C. Its activity will increase by
a. decreasing the amount of substrate.
b. denaturing the enzyme.
c. increasing the temperature 10 more degrees.
d. making the pH more acidic.

____ 5. In the reaction $A + B \rightarrow C$, the reaction rate may slow down through feedback inhibition by
a. increasing the concentration of A.
b. increasing the concentration of B.
c. increasing the concentration of C.
d. decreasing the concentration of B.

____ 6. The energy laws
a. account for why energy does not cycle.
b. say that some loss of energy always accompanies transformation.
c. say that energy can be made available to living things.
d. All of these are correct.

____ 7. In a metabolic pathway $A \rightarrow B \rightarrow C \rightarrow D \rightarrow E$,
a. A, B, C, and D are substrates.
b. B, C, D, and E are products.
c. each reaction requires its own enzyme.
d. All of these are correct.

____ 8. The enzyme-substrate complex
a. indicates that an enzyme has denatured.
b. accounts for why enzymes lower the energy of activation.
c. is nonspecific.
d. All of these are correct.

____ 9. The tendency for an ordered system to become spontaneously disordered is called
a. thermodynamics.
b. entropy.
c. activation.
d. energy conversion.

____ 10. A coupled reaction occurs when energy released from a(n) ______________ reaction is used to drive a(n) ______________ reaction.
a. endergonic; exergonic
b. breakdown; exergonic
c. exergonic; endergonic
d. chemical; mechanical

____ 11. $NAD^+$ and FAD are
a. dehydrogenases.
b. proteins.
c. coenzymes.
d. Both *a* and *c* are correct.

____ 12. Reduction has occurred when
a. electrons are lost.
b. $C_6H_{12}O_6$ becomes $CO_2$.
c. $O_2$ becomes $H_2O$.
d. heat is given off.
e. ADP becomes ATP.

____ 13. Chloroplasts
a. take in $CO_2$.
b. give off $H_2O$.
c. pass on solar energy.
d. occur in all living things.
e. Both *a* and *c* are correct.

____ 14. ATP is used for
a. chemical work.
b. transport work.
c. mechanical work.
d. All of these are correct.

____ 15. NAD and NADP
a. are found only in plants.
b. do not participate in metabolic reactions.
c. are coenzymes of oxidation-reduction.
d. carry hydrogen atoms.
e. Both *c* and *d* are correct.

____ 16. Which statement is NOT correct about enzymes?
a. They usually end in the suffix "-ase."
b. They catalyze only one reaction.
c. They increase the energy of activation.
d. They bind temporarily with the substrate.

____ 17. Which of these is NOT expected to increase the rate of an enzymatic reaction?
a. add more enzyme
b. remove inhibitions
c. boil rapidly
d. adjust the pH to optimum level

____ 18. Which of these accurately represents a flow of energy from the sun?
a. Plants take in solar energy and use it to oxidize glucose, used by mitochondria to produce ATP.
b. Mitochondria break down glucose to ATP, returned to plants to produce glucose.
c. Plants take in solar energy and use it to transport water up stems so that water is available to all animals.
d. Plants take in solar energy and use it to reduce carbon dioxide so that glucose is made available to animals.
e. Both plants and animals make and use ATP.

____ 19. Since energy does not cycle, animal cells
a. require a continuing source of glucose.
b. are dependent on plant cells.
c. must produce ATP nonstop.
d. All of these are correct.

____ 20. Synthetic reactions
a. require the participation of ATP.
b. do not require enzymes.
c. are represented by $S + E \rightarrow ES \rightarrow P$.
d. are coupled directly to glucose breakdown.

## Critical Thinking Questions

Answer in complete sentences.

21. Why couldn't life exist without a continual supply of solar energy?

22. Why are enzymes absolutely necessary to the continued existence of a cell?

**Test Results:** _____ number correct ÷ 22 = _____ × 100 = _____ %

## Exploring the Internet

The Online Learning Center at *www.mhhe.com/maderbiology8* has additional study material and practice quizzes that can help you master the content of this chapter. You can also find links to websites exploring additional topics in biology. Access to the Online Learning Center is free for those who have purchased a new textbook.

## Answer Key

### Study Exercises

**1. a.** F, The chemical energy of ATP can be transformed into other types of energy such as kinetic energy (muscle contraction). **b.** F, Cells transform the energy of glucose breakdown into ATP molecules, and they do obey the first law of thermodynamics. **c.** T **d.** F, Because our society uses coal as an energy source, it is increasing the entropy of the universe. **2. a.** Ex **b.** En **c.** Ex **d.** Ex **e.** En **3. a.** +P **b.** ATP **c.** –P **d.** ADP **4. a.** L, because during cellular respiration, the chemical energy within a glucose molecule is converted to the chemical energy within ATP. **b.** R, because when muscles contract, the chemical energy within ATP is converted to the kinetic energy of muscle contraction. **c.** R, because when active transport occurs, the energy released by ATP breakdown is used to pump a molecule across the plasma membrane. **5. a.** energy currency **b.** chemical **c.** transport **d.** mechanical **e.** coupled **6. a.** reactants **b.** products **c.** enzymes **d.** substrate **7. a.** active site **b.** substrate **c.** enzyme **d.** enzyme-substrate complex **e.** enzyme **f.** products **8. a.** enzyme-substrate complex **b.** Reactants come together when the enzyme-substrate complex forms. **9. a.** $E + S \rightarrow ES \rightarrow E + P$ **b.** degradative **c.** The reactant is broken down. **d.** active **e.** the enzyme is not broken down and can be used over and over. **f.** Enzymes are specific to their substrates. **10. a.** increases **b.** decreases **c.** increases **d.** decreases **e.** decreases **11. a.** coenzymes **b.** help enzymes function. **12. a.** CR **b.** P **c.** CR **d.** P **e.** CR, P **f.** P **g.** CR **h.** P **i.** P **13.** b d a c **14. a.** pump **b.** concentration **c.** ATP **d.** ADP + P .

## Definitions Wordsearch

**a.** energy **b.** entropy **c.** metabolism **d.** reactant **e.** enzyme **f.** substrate **g.** model **h.** denatured **i.** cofactors **j.** vitamins **k.** oxidation

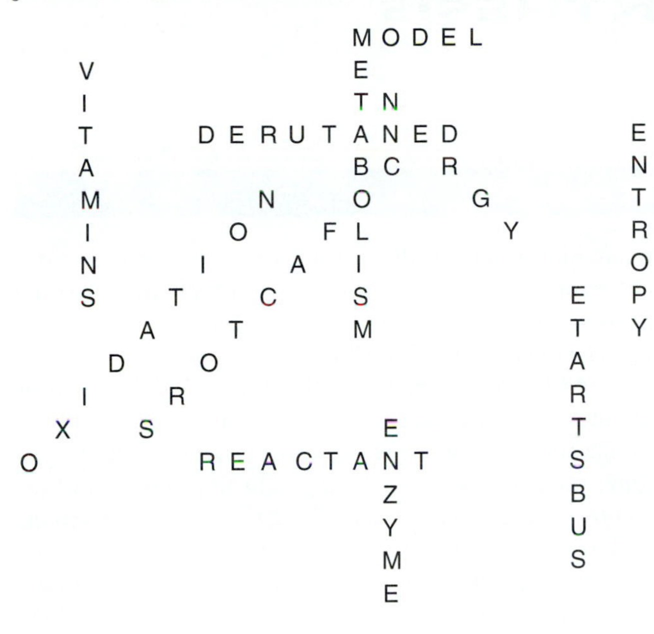

## Chapter Test

**1.** d **2.** b **3.** d **4.** d **5.** c **6.** d **7.** d **8.** b **9.** b **10.** c **11.** d **12.** c **13.** a **14.** d **15.** e **16.** c **17.** c **18.** d **19.** d **20.** a **21.** When chloroplasts carry on photosynthesis, solar energy is converted to the energy of carbohydrates, and when mitochondria complete cellular respiration, the energy stored in carbohydrates is converted to energy temporarily held by ATP. The energy released by ATP breakdown is used by the cell to do various types of work, and eventually it becomes nonusable heat. **22.** Enzymes are absolutely essential for the existence of a cell because they lower the energy of activation of a reaction, thereby requiring less heat to bring about the reaction. At high temperatures, proteins would denature.

# 7

# Photosynthesis

## Chapter Review

Through the process of **photosynthesis,** photosynthetic organisms (i.e., plants) use solar energy to produce carbohydrate, an organic nutrient. Photosynthetic organisms are producers, which provide food for consumers by photosynthesis. Photosynthesis is carried on in the chloroplasts of plants. This process not only produces carbohydrate, but also replenishes oxygen in the atmosphere.

Photosynthesis uses the portion of the electromagnetic spectrum known as visible light. Both **chlorophyll** *a* and chlorophyll *b* absorb violet, blue, and red light better than light of other colors. These pigments are present in the **thylakoid** membrane within the **grana** of **chloroplasts,** and participate in the **light reactions** of photosynthesis.

The light reactions involve a noncyclic electron pathway and a cyclic electron pathway. In the **noncyclic electron pathway,** energized electrons leave chlorophyll *a* of **photosystem II** (PS II), pass down an electron transport system, and enter **photosystem I** (PS I), where they are energized once more before being accepted by $NADP^+$. The overall result from the noncyclic pathway is the production of NADPH and ATP. Oxygen is also liberated when water is split, and electrons enter chlorophyll *a* (PS II) to replace those lost. The NADPH and ATP from the light reactions of photosynthesis are used to build a carbohydrate in the **Calvin cycle reactions,** which occur in the **stroma** of chloroplasts. Only **photosystem I** (PS I) is required for the **cyclic electron pathway,** in which electrons energized by the sun leave the reaction-center chlorophyll *a* and then pass down an **electron transport system** with the concomitant buildup of ATP before returning to chlorophyll *a* (PS I).

ATP production during the light reactions requires **chemiosmosis.** Hydrogen ions are concentrated in the thylakoid space; when water splits, it releases hydrogen ions, and carriers within the cytochrome complex of the electron transport system pump the hydrogen ions to the thylakoid space. The hydrogen ions flow down their concentration gradient through a channel in a protein having an ATP synthase, which forms ATP from ADP and Ⓟ.

During the **Calvin cycle reactions,** carbon dioxide is fixed by **RuBP,** it is reduced to PGAL (this requires the ATP and NADPH from the light reactions), and RuBP is regenerated. One PGAL out of every six joins with another PGAL to form glucose phosphate.

**$C_3$** photosynthesis (the first molecule after fixation is a $C_3$ molecule) occurs when a plant uses the Calvin cycle directly. Plants have also evolved two other types of photosynthesis: **$C_4$** photosynthesis and **CAM** photosynthesis. $C_4$ plants fix carbon dioxide in mesophyll cells (which results in a $C_4$ molecule) and then transport it to bundle sheath cells, where it enters the Calvin cycle. CAM plants fix carbon dioxide at night and then release it during the day to the Calvin cycle. $C_4$ and CAM photosynthesis are adaptations to hot, dry environments; these processes allow the stomata to close to conserve water.

## Study Exercises

Study the text section by section as you answer the questions that follow.

### 7.1 Photosynthetic Organisms (p. 116)

- Plants make use of solar energy in the visible light range when they carry on photosynthesis.
- Photosynthesis takes place in chloroplasts, organelles that contain membranous thylakoids surrounded by a fluid called stroma.

1. Indicate whether the following statements about photosynthesis are true (T) or false (F):
   a. _____ It makes food for animals.
   b. _____ It promotes the breakdown of biodegradable wastes.
   c. _____ It returns carbon dioxide to the atmosphere.
   d. _____ It returns oxygen to the atmosphere.
2. Photosynthesis refers to the ability of plants, algae, and a few kinds of bacteria to make their own a. _____________ in the presence of b. _____________. In plants, photosynthesis is carried on in c. _____________.
3. The green pigment a. _____________ is found within the membrane of the b. _____________, and it is here that c. _____________ energy is captured.
4. A chloroplast contains flattened, membranous sacs called a. _____________ that are stacked like poker chips into b. _____________. The fluid surrounding this is called the c. _____________.

## 7.2 Plants as Solar Energy Converters (p. 118)

- Photosynthesis has two sets of reactions: In the light reactions, solar energy is captured by the pigments in thylakoids; in the Calvin cycle reactions, carbon dioxide is reduced by enzymes in the stroma.

5. Indicate whether the following statements about photosynthetic pigments are true (T) or false (F):
   a. _____ Chlorophylls *a* and *b* absorb violet, blue, and red light best.
   b. _____ Carotenoids absorb light in the yellow-orange range.
   c. _____ Leaves appear green to us because green light is reflected by chlorophyll.
6. Label this diagram of a chloroplast using the following terms:

   stroma — light reactions
   thylakoid membrane — Calvin cycle

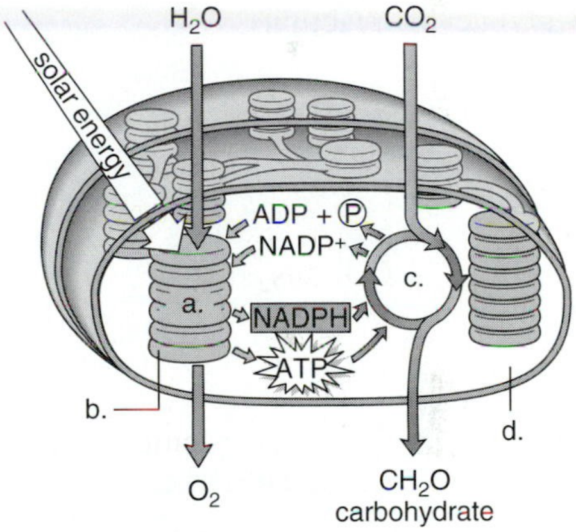

7. From the diagram in question 6, the light reactions would be associated with the a. _____________, and the Calvin cycle reactions would be associated with the b. _____________.
8. The light reactions drive the Calvin cycle reactions. The Calvin cycle reactions use the NADPH and a. _____________ from the light reactions to reduce b. _____________ to a c. _____________.
9. Place the appropriate letters next to each reaction.
   LR—light reaction CC—Calvin cycle reaction:
   a. _____ energy-capturing reaction
   b. _____ carbon dioxide becomes carbohydrate
   c. _____ water gives off oxygen
   d. _____ NADPH and ATP are made
   e. _____ NADPH and ATP are used

## 7.3 Light Reactions (p. 120)

- The light reactions consist of the noncyclic electron pathway and the cyclic electron pathway.

10. In the noncyclic electron pathway the two pigment complexes called a. _____________ and b. _____________ are found within the thylakoid membranes. Within each are green pigments called c. _____________, yellow-orange pigments called d. _____________, and e. _____________ acceptor molecules.

11. Label this diagram of the cyclic electron pathway with the following terms:
    ADP
    ATP
    electron acceptor
    light reactions
    photosystem I
    pigment complex

12. Label this diagram of the noncyclic electron pathway with the following terms:
    electron acceptor (used twice)
    electron transport system
    light reactions
    Calvin cycle reactions
    photosystem I
    photosystem II

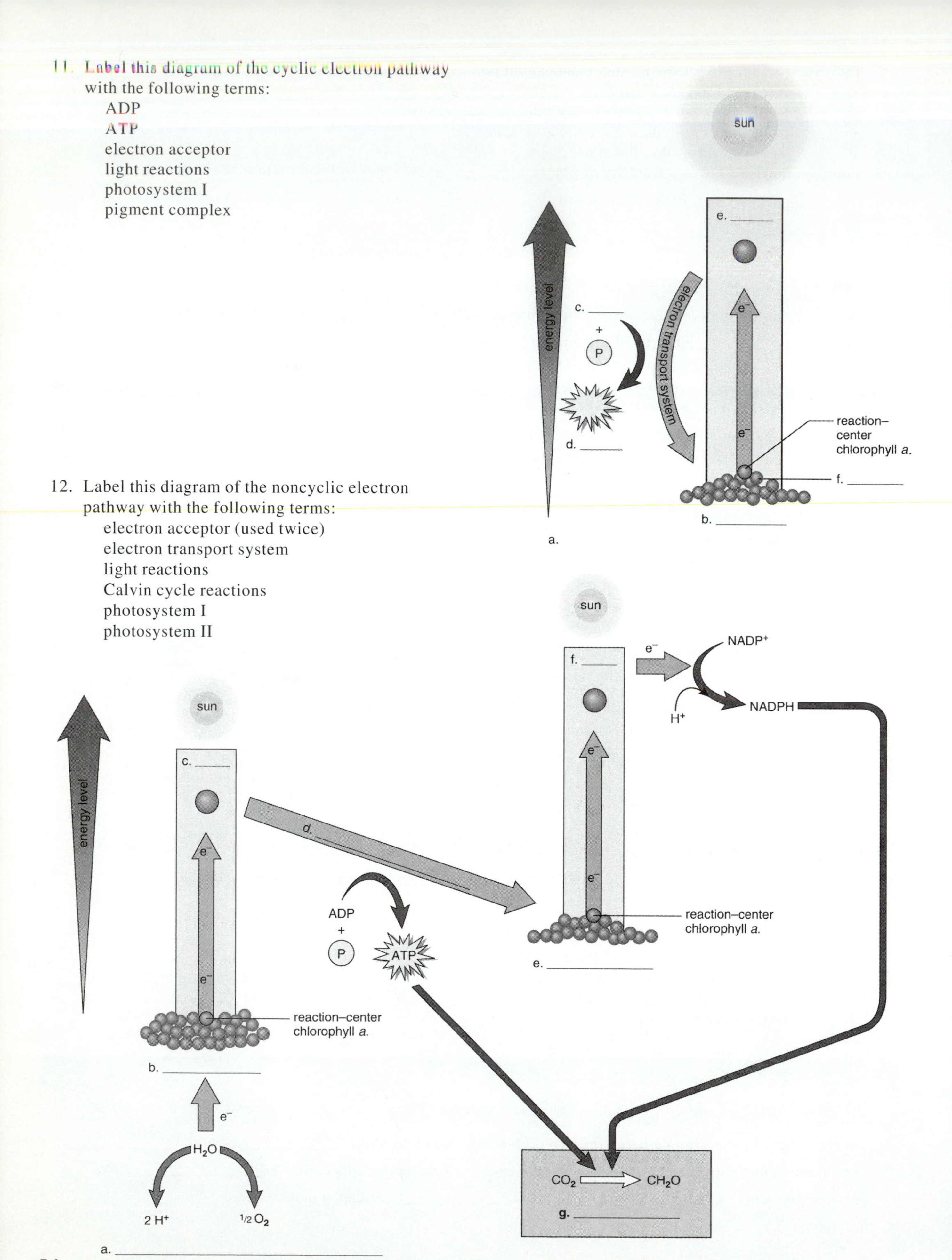

13. What is the role of each of these in the noncyclic electron pathway?

a. reaction-center chlorophyll *a* ______________________________

b. electron acceptors ______________________________

c. $NADP^+$ ______________________________

d. electron transport system ______________________________

e. water ______________________________

## The Organization of the Thylakoid Membrane (p. 122)

14. Match the functions to the complexes in the thylakoid membrane by writing one of the following terms on lines a–d.
photosystem II
cytochrome complex
photosystem I
ATP synthase complex

a. _____ produces ATP from ADP + Ⓟ
b. _____ transports electrons and stores $H^+$ in the thylakoid space
c. _____ captures solar energy; water is split, releasing electrons and oxygen
d. _____ captures solar energy; $NADP^+$ is reduced to NADPH

## ATP Production (p. 122)

- Chemiosmosis depends on an electrochemical gradient that the electron transport system establishes.

15. The concentration of $H^+$ in the a.______________ space is b.______________, compared to the lower $H^+$ concentration in the c.______________. The flow of $H^+$ down its concentration gradient provides the energy for an enzyme called d.______________ to produce e.______________ from ADP + Ⓟ.

## 7.4 Calvin Cycle Reactions (p. 124)

- Carbon dioxide reduction requires energized electrons and energy, supplied by NADPH and ATP.

16. Match the numbers in the diagram to the following descriptions of events in the Calvin cycle. (Some numbers are used more than once.)
a. _____ ATP only required
b. _____ $CO_2$ reduction reaction
c. _____ NADPH and ATP required
d. _____ $CO_2$ taken up, $CO_2$ fixation
e. _____ glucose formed after six turns of cycle
f. _____ five PGAL required to form three molecules of product

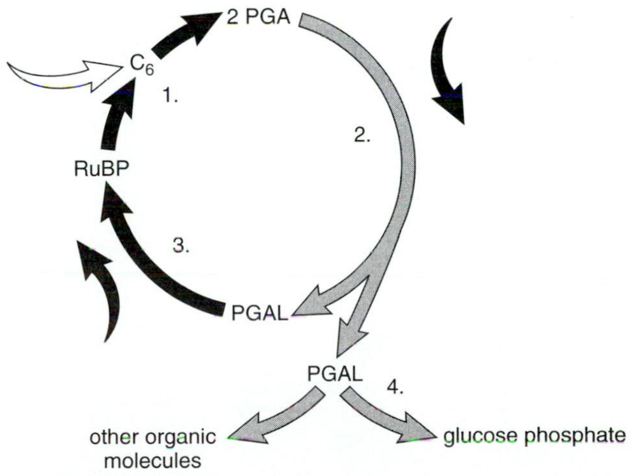

17. Indicate the sequence in which molecules *a–e* appear during or as a result of the Calvin cycle.

____________________

a. $CO_2$
b. glucose phosphate
c. PGA
d. PGAL
e. starch

## 7.5 Other Types of Photosynthesis (p. 126)

- Plants use either $C_3$, $C_4$, or CAM photosynthesis, distinguishable by the manner in which $CO_2$ is fixed.

18. Label the following as describing $C_3$, $C_4$, and/or CAM plants:

a. ____ predominate in spring and cooler summer weather
b. ____ succulent plants, cacti; live in hot, arid regions
c. ____ wheat, rice, oats
d. ____ predominate in hot, dry summer weather
e. ____ product of $CO_2$ fixation is PGA
f. ____ $CO_2$ fixation occurs at night and $C_4$ molecules are stored until daylight
g. ____ photorespiration occurs
h. ____ stomata are closed *during the day* to conserve water
i. ____ product of $CO_2$ fixation is oxaloacetate
j. ____ sugarcane, corn, Bermuda grass
k. ____ chloroplasts only in mesophyll cells
l. ____ chloroplasts in bundle sheath cells and mesophyll

# KEYWORD CROSSWORD

Review key terms by completing this crossword puzzle using the following alphabetized list of terms:

$C_3$ plant
$C_4$ plant
Calvin cycle
CAM plant
chlorophyll
granum
light reaction
photosystem
RuBP
stomata
stroma
thylakoid
visible light

*Across*

3 plant that directly uses the Calvin cycle; the first detected molecule during photosynthesis is PGA, a three-carbon molecule (two words)
5 portion of the electromagnetic spectrum of light that can be seen with the human eye (two words)
7 green pigment that absorbs solar energy and is important in photosynthesis
9 flattened sac within a granum whose membrane contains the photosynthetic pigments (e.g., chlorophyll); where the light reactions occur
10 small openings in leaves through which carbon dioxide enters

*Down*

1 plant that fixes carbon dioxide at night to produce a $C_4$ molecule that releases carbon dioxide to the Calvin cycle during the day (two words)
2 photosynthetic unit where solar energy is absorbed; contains an antenna (photosynthetic pigments) and an electron acceptor
3 plant that fixes carbon dioxide in bundle sheath cells to produce a molecule that releases carbon dioxide to the Calvin cycle in mesophyll (two words)
4 set of photosynthetic reactions that requires solar energy to proceed; it produces ATP and NADPH (two words)
6 stack of chlorophyll-containing thylakoids in a chloroplast
7 uses the products of the light reactions to reduce carbon dioxide to a carbohydrate (two words)
8 five-carbon compound that combines with and fixes carbon dioxide during the Calvin cycle and is later regenerated by the same cycle
10 large, central space in a chloroplast that is fluid filled and contains enzymes used in photosynthesis

# Chapter Test

## Objective Questions

Do not refer to the text when taking this test. In questions 1–6, match the definitions to these terms.

a. chlorophyll
b. oxygen
c. stroma
d. sugar
e. thylakoid membrane
f. water

____ 1. organic product of photosynthesis
____ 2. released by photosynthesis
____ 3. reactant of photosynthesis
____ 4. site of light reactions
____ 5. site of Calvin cycle reactions
____ 6. molecule absorbing solar energy
____ 7. Each of the following is a product of photosynthesis EXCEPT
a. carbon dioxide.
b. organic food.
c. oxygen.
d. carbohydrate.

____ 8. Photosynthesis occurs best at wavelengths that are
a. blue.
b. gamma.
c. infrared.
d. ultraviolet.

____ 9. Each is a product of light-dependent reactions EXCEPT
a. ATP.
b. NADPH.
c. oxygen.
d. sugar.

____ 10. The cyclic pathways of photosynthesis produce
a. ATP only.
b. NADPH only.
c. ATP and NADPH.
d. organic sugars only.

____ 11. Carbon dioxide fixation occurs when $CO_2$ combines with
a. ATP.
b. NADPH.
c. PGAL.
d. RuBP.

____ 12. Which of the following pathways uses the enzyme PEPCase?
a. $C_2$
b. $C_3$
c. CAM
d. CAP

____ 13. The enzyme that produces ATP from ADP + Ⓟ in the thylakoid is
a. RuBP carboxylase.
b. PGAL.
c. ATPase.
d. ATP synthase.
e. coenzyme A.

____ 14. Which statement is NOT true regarding chemiosmosis?
a. $H^+$ concentration is higher in the stroma than in the thylakoid space.
b. The electron transport system pumps $H^+$ from the stroma into the thylakoid space.
c. The ATP synthase complex is present in the thylakoid membrane.
d. All of these are true.

____ 15. Which of the following is NOT a stage in the Calvin cycle?
a. carbon dioxide fixation
b. carbon dioxide oxidation
c. carbon dioxide reduction
d. RuBP regeneration

____ 16. Which of these descriptions is NOT true of photosynthesis?
a. not affected by temperature
b. not affected by solar energy
c. requires a supply of oxygen
d. involves a reduction reaction
e. more likely to occur during the day

____ 17. Which of these descriptions is NOT true of chlorophyll?
a. absorbs solar energy
b. located in the grana
c. located in thylakoid membranes
d. passes electrons directly to $NADP^+$
e. passes electrons to an acceptor molecule

____ 18. The two major sets of reactions involved in photosynthesis are
a. the cyclic and noncyclic electron pathways.
b. glycolysis and the citric acid cycle.
c. the Calvin and citric acid cycles.
d. the Calvin cycle and the electron transport system.
e. the light reaction and the Calvin cycle reaction.

____ 19. Which of the following statements is NOT true of the Calvin cycle?
a. RuBP is regenerated with the use of ATP.
b. Glucose phosphate is synthesized from PGAL.
c. NADPH is used to reduce PGAL to PGA.
d. Five molecules of PGAL are used to reform three molecules of RuBP.

____20. Photosystem II gets replacement electrons from
a. the sun.
b. water molecules.
c. ATP.
d. photosystem I.
e. NADPH.

## Critical Thinking Questions

Answer in complete sentences.

21. How is life dependent on photosynthesis?

22. Why is photosynthesis dependent on the high degree of compartmentalization in cells?

**Test Results:** ______ number correct ÷ 22 = ______ × 100 = ______ %

## Exploring the Internet

The Online Learning Center at *www.mhhe.com/maderbiology8* has additional study material and practice quizzes that can help you master the content of this chapter. You can also find links to websites exploring additional topics in biology. Access to the Online Learning Center is free for those who have purchased a new textbook.

## Answer Key

### Study Exercises

**1. a.** T **b.** F **c.** F **d.** T **2. a.** food **b.** sunlight **c.** chloroplasts **3. a.** chlorophyll **b.** thylakoids **c.** solar **4. a.** thylakoids **b.** grana **c.** stroma **5. a.** T **b.** F **c.** T **6. a.** light reaction **b.** thylakoid membrane **c.** Calvin cycle **d.** stroma **7. a.** thylakoid membrane **b.** stroma **8.** ATP **b.** carbon dioxide **c.** carbohydrate **9. a.** LR **b.** CC **c.** LR **d.** LR **e.** CC **10. a.** photosystem I **b.** photosystem II **c.** chlorophylls **d.** carotenoids **e.** electron **11. a.** light reactions **b.** photosystem I **c.** ADP **d.** ATP **e.** electron acceptor **f.** pigment complex **12. a.** light reactions **b.** photosystem II **c.** electron acceptor **d.** electron transport system **e.** photosystem I **f.** electron acceptor **g.** Calvin cycle reactions. **13. a.** releases electrons that have become excited from solar energy **b.** accepts energized electrons from the reaction-center chlorophyll *a* and sends them to the electron transport system **c.** accepts electrons and hydrogen and becomes NADPH **d.** stores energy for ATP production as the electrons fall to a lower energy level **e.** splits releasing oxygen and hydrogen ions **14. a.** ATP synthase complex **b.** cytochrome complex **c.** photosystem II **d.** photosystem I **15. a.** thylakoid **b.** higher **c.** stroma **d.** ATP synthase **e.** ATP **16. a.** 3 **b.** 2 **c.** 2 **d.** 1 **e.** 4 **f.** 3 **17.** a, c, d, b, e **18. a.** $C_3$ **b.** CAM **c.** $C_3$ **d.** $C_4$ **e.** $C_3$ **f.** CAM **g.** $C_3$ **h.** CAM **i.** $C_4$ **j.** $C_4$ **k.** $C_3$ **l.** $C_4$

## KeyWord CrossWord

| | | | | | | | | | | | | | | | | | | | | |
|---|---|---|---|---|---|---|---|---|---|---|---|---|---|---|---|---|---|---|---|---|
| | | | | | 1 C | | | | | | | | | | | | | | 2 P | |
| 3 C | 3 | | P | L | A | N | T | | | | | | | | | | | | H | |
| 4 | | | | | M | | | | | 4 L | | | | | | | | | O | |
| | | | | | | | 5 V | I | S | I | B | L | E | | L | I | 6 G | H | T | |
| P | | | | | P | | | | | G | | | | | | | R | | O | |
| L | | | 7 C | H | L | O | 8 R | O | P | H | Y | L | L | | | | A | | S | |
| A | | | A | | A | | u | | | T | | | | | | | N | | Y | |
| N | | | L | | N | | B | | | | | | | | | | U | | S | |
| T | | | V | | T | | P | | | R | | | | | | | M | | T | |
| | | | I | | | | | | | E | | | | | | | | | E | |
| | | | N | | | | | | | A | | | | | | | | | M | |
| | | | | | | | | | | C | | | | | | | | | | |
| | | | C | | | | | | | 9 T | H | Y | L | A | K | O | I | D | | |
| | | | Y | | | | | | | I | | | | | | | | | | |
| | | | C | | | | | 10 S | T | O | M | A | T | A | | | | | | |
| | | | L | | | | | T | | N | | | | | | | | | | |
| | | | E | | | | | R | | | | | | | | | | | | |
| | | | | | | | | O | | | | | | | | | | | | |
| | | | | | | | | M | | | | | | | | | | | | |
| | | | | | | | | A | | | | | | | | | | | | |

## Chapter Test

**1.** d **2.** b **3.** f **4.** e **5.** c **6.** a **7.** a **8.** a **9.** d **10.** a **11.** d **12.** c **13.** d **14.** a **15.** b **16.** c **17.** d **18.** e **19.** c **20.** b **21.** Through photosynthesis, plants (and algae) produce food for themselves and all other living things. These organisms are the producers at the start of food chains of all types. Animals feed directly on photosynthesizers or on other animals that have fed on photosynthesizers. **22.** Chemiosmosis cannot occur without a membrane that maintains a difference in hydrogen ion concentration. In plant cells, hydrogen ions build up within the thylakoid space and then they flow down their concentration gradient through an ATP synthase complex located in the thylakoid membrane.

# 8

# Cellular Respiration

## Chapter Review

**Cellular respiration,** the oxidation of glucose to carbon dioxide and water, is an exergonic reaction that drives ATP synthesis, an endergonic reaction. Oxidation involves the removal of hydrogen atoms ($H^+ + e^-$) from substrate molecules, usually by the coenzyme $NAD^+$, but in one case by FAD. Four phases are required: glycolysis, the transition reaction, the citric acid cycle, and the electron transport system.

During **glycolysis,** glucose is converted to pyruvate in the cytoplasm. Glycolysis produces two ATP by substrate-level phosphorylation. When oxygen is available, pyruvate from glycolysis enters mitochondria. In mitochondria, the **transition reaction** and the **citric acid cycle** are located in the matrix, and the **electron transport system** is located on the cristae. Both the transition reaction and the citric acid cycle release carbon dioxide as a result of oxidation of carbohydrate breakdown products.

The electrons carried by NADH and $FADH_2$ enter the electron transport system. The electrons pass down a chain of carriers until they are finally received by oxygen, which combines with $H^+$ forming water. As electrons pass down the electron transport system, energy is released and stored for ATP production. The term oxidative phosphorylation is sometimes used for 32 or 34 ATPs produced as a result of the electron transport system. The protein complexes of the electron transport system pump $H^+$ received from NADH and $FADH_2$ into the intermembrane space, setting up an electrochemical gradient. When $H^+$ flows down this gradient through the ATP synthase complex, energy is released and used to form ATP. This process of producing ATP is called **chemiosmosis.**

Other carbohydrates, as well as protein and fat, can also generate ATP by entering various steps in the degradative paths of glycolysis and the citric acid cycle. These pathways also provide metabolites needed for the synthesis of various important cellular substances.

**Fermentation,** which occurs when oxygen is not available for cellular respiration, involves glycolysis followed by the reduction of pyruvate by NADH. The product can be lactate or alcohol and carbon dioxide. Fermentation produces a net yield of 2 ATP molecules per glucose molecule. Fermentation provides a quick, immediate source of ATP, but lactate buildup is toxic to the cell and creates an **oxygen debt** in the organism.

## Study Exercises

Study the text section by section as you answer the questions that follow.

### 8.1 Cellular Respiration (pp. 132–33)

- During cellular respiration, the breakdown of glucose drives the synthesis of ATP.
- The coenzymes $NAD^+$ and FAD accept electrons from substrates and carry them to the electron transport system in mitochondria.

1. Consider the following equation.

$$C_6H_{12}O_6 + 6\ O_2 \rightarrow 6\ H_2O + 6\ CO_2 + \text{energy}$$

The molecule glucose is (oxidized or reduced) a.______________ while oxygen is (oxidized or reduced) b.______________. This is an (endergonic or exergonic) c.______________ reaction and therefore is used by cells to build up ATP.

2. After completing the diagram, answer these questions:

The (left or right) e. _____________ side of the diagram represents oxidation and the (left or right) f. _____________ side of the diagram represents reduction of NAD. Why is $NAD^+$ called a coenzyme of oxidation-reduction? g. _____________

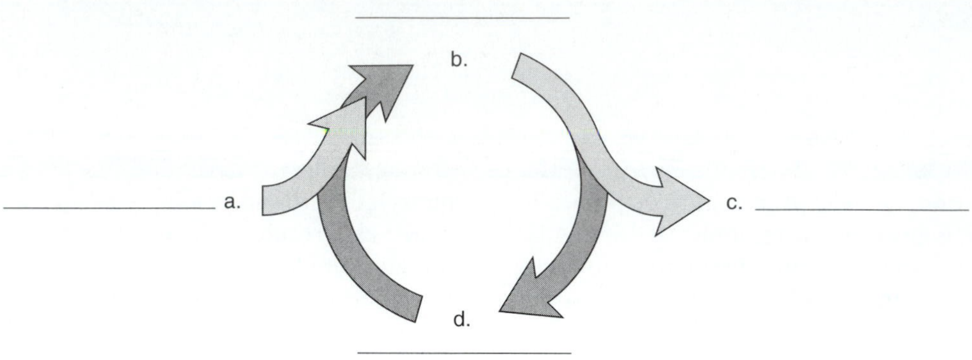

b. _____________

a. _____________

c. _____________

d. _____________

## Phases of Complete Glucose Breakdown (p. 133)

- Complete glucose breakdown involves metabolic pathways, and each reaction requires a specific enzyme.

3. Place the appropriate letters next to each statement.

GL—glycolysis TR—transition reaction CA—citric acid cycle ETS—electron transport system

a. _____ Series of carriers that pass electrons from one to the other.
b. _____ Cyclical series of oxidation reactions that release $CO_2$.
c. _____ Pyruvate is oxidized to an acetyl group.
d. _____ Breakdown of glucose to two molecules of pyruvate.
e. _____ Energy is released and stored for ATP production.
f. _____ Occurs inside mitochondria.
g. _____ Occurs outside the mitochondria in the cytoplasm.
h. _____ Results in only 2 ATP.

## 8.2 Outside the Mitochondria: Glycolysis (pp. 134–35)

- Glycolysis is a metabolic pathway that partially breaks down glucose outside the mitochondria.

4. a. Where does glycolysis occur? _____________
   b. Does it require oxygen? _____________
   c. Glycolysis begins with _____________.
   d. Glycolysis ends with _____________.
   e. How many ATP are produced per glucose molecule as a direct result of glycolysis? _____________
   f. What type of phosphorylation occurs during glycolysis? _____________
   g. What coenzyme carries out oxidation of substrates during glycolysis? _____________
   h. Considering your answers to these questions, what is the output of glycolysis? _____________, _____________, and _____________.

## 8.3 Inside the Mitochondria (pp. 136–40)

- The transition reaction and the citric acid cycle, which occur inside the mitochondria, continue the breakdown of glucose until carbon dioxide and water result.

5. Label this diagram of a mitochondrion using the following terms:
   cristae
   cytoplasm
   inner membrane
   intermembrane space
   matrix
   outer membrane

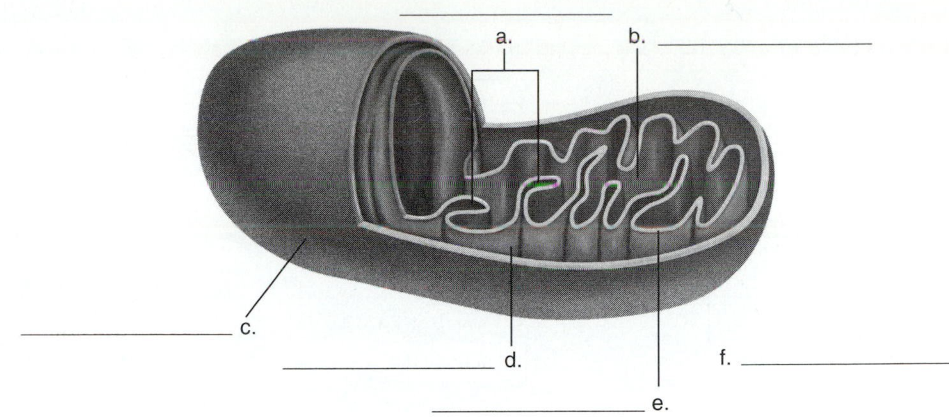


6. Using your labels from question 5, where does each of the following processes occur?
   a. glycolysis ______________
   b. citric acid cycle ______________
   c. electron transport system ______________
7. a. The citric acid cycle begins and ends with what molecule? ______________
   b. A two-carbon molecule acetyl group enters the citric acid cycle. What carbon molecules leave the citric acid cycle? ______________
   c. How many ATP are produced per glucose molecule as a direct result of the citric acid cycle? ______________
   d. What coenzymes carry out oxidation of substrates in the citric acid cycle? ______________
   e. Considering your answers to these questions, what are the outputs of the citric acid cycle? ______________, ______________, ______________, and ______________

## Electron Transport System (p. 138)

8. What coenzymes bring hydrogen atoms ($H^+ + e^-$) to the electron transport system? a.______________
   What happens to the electrons? b.______________
   What happens to the hydrogen ions? c.______________
   What molecule is the final acceptor of electrons from the electron transport system? d.______________
   Each pair of electrons carried by NADH from the citric acid cycle that passes down the electron transport system accounts for the buildup of how many ATP? e.______________
   What type of phosphorylation is associated with the electron transport system? f.______________
9. During chemiosmosis in mitochondria, $H^+$ build up in the a.______________ space.
   When these $H^+$ flow b.______________ their concentration gradient into the matrix, c.______________ is produced from ADP +Ⓟ.
10. Match the complexes with the following functions:

    a. _____ NADH dehydrogenase complex
    b. _____ cytochrome *b-c* complex
    c. _____ cytochrome oxidase complex
    d. _____ ATP synthase complex

    1. passes on electrons and pumps $H^+$ into intermembrane space
    2. carries out ATP synthesis
    3. receives electrons and passes them on to oxygen
    4. oxidizes NADH and pumps $H^+$ into intermembrane space

## Energy Yield from Glucose Metabolism (p. 140)

11. In the following diagram, fill in the blanks with the correct numbers and with the terms NADH, $FADH_2$, and ATP:

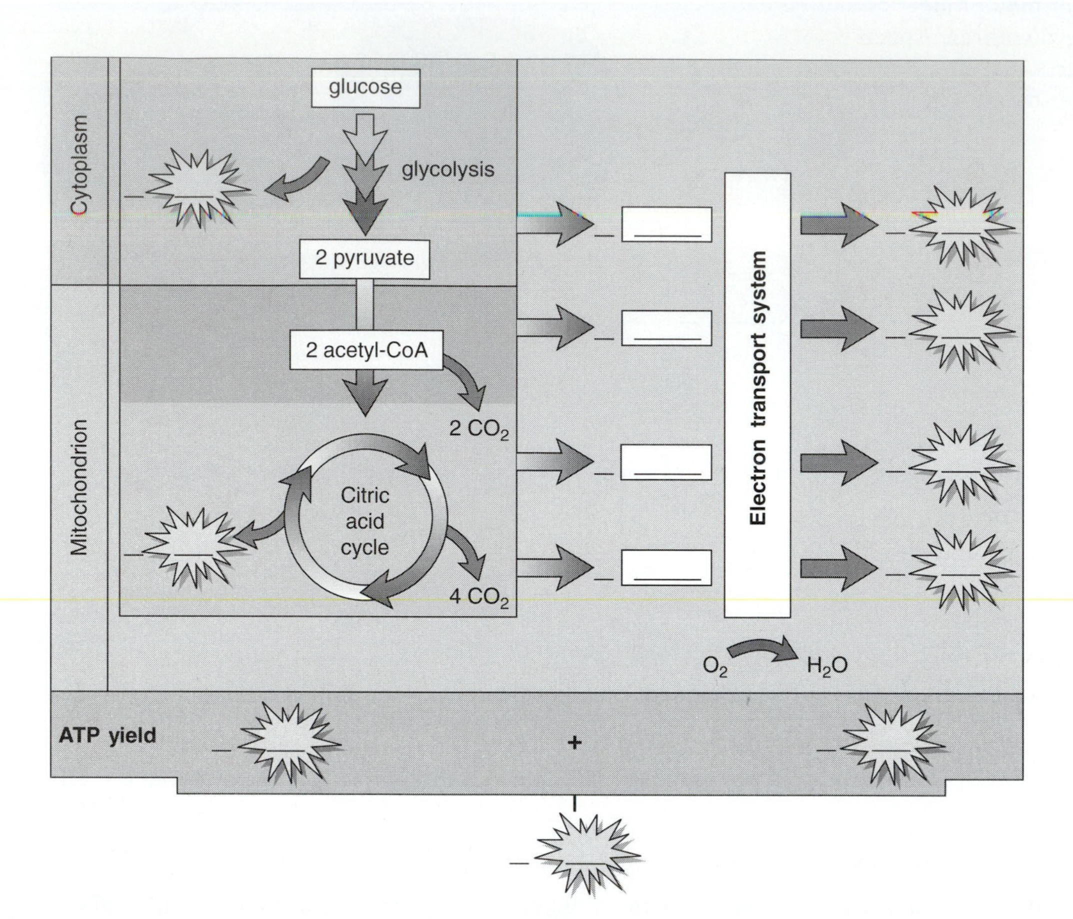


## 8.4 Fermentation (p. 142)

- Fermentation is a metabolic pathway that partially breaks down glucose under anaerobic conditions.

12. a. What happens to pyruvate during fermentation in humans? ____________________
    b. What happens to pyruvate during fermentation in yeast? ____________________
    c. Why is fermentation wasteful? ____________________
    d. What is its advantage? ____________________
    e. What is oxygen debt in humans? ____________________

## 8.5 Metabolic Pool (p. 144)

- A number of metabolites in addition to glucose can be broken down to drive ATP synthesis.

13. The carbon skeleton of amino acids can be respired if the amino acid first undergoes [a] ____________. When fats are respired, glycerol is converted to [b] ____________, fatty acids are converted to the two-carbon molecule [c] ____________, and the acetyl group enters the citric acid cycle. Excess acetyl groups from glucose metabolism can be used to build up fatty acids. Explain why the consumption of carbohydrate makes us fat. [d] ____________________

________________________________________

14. Label the following processes I, II, and/or III, based on this pyruvate diagram:
    a. _____ occurs under anaerobic conditions
    b. _____ fermentation
    c. _____ glycolysis
    d. _____ transition reaction

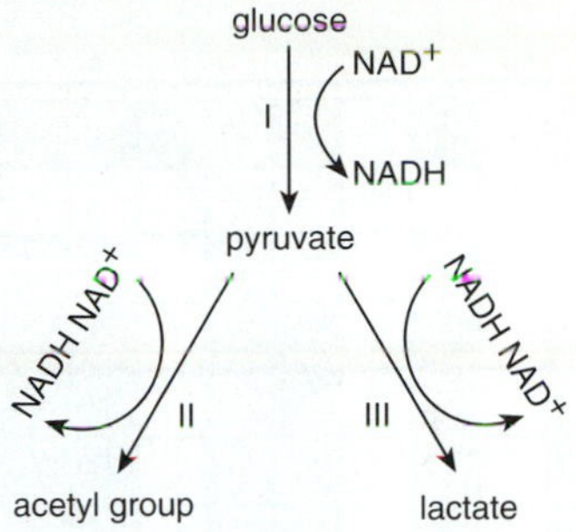

15. Consider III in the diagram for question 14.
    a. Which has more hydrogen atoms, pyruvate or lactate? ____________
    b. In yeast, the product for this reaction is ____________________.
    c. What happens to $NAD^+$ produced by the reaction? ____________
16. Consider II in the diagram for question 14.
    a. What happens to NADH? ______________________________
    b. What happens to the acetyl group? ______________________

# KeyWord CrossWord

Review the key terms by completing this crossword puzzle using the following alphabetized list of terms:

acetyl-CoA
aerobic
cellular respiration
electron transport
FAD
fermentation
glycolysis
metabolic pool
oxygen debt
pyruvate

*Across*

3 metabolic reactions that use the energy from carbohydrate or fatty acid or amino acid oxidation to produce ATP molecules; includes fermentation and aerobic respiration (two words)

6 anaerobic breakdown of glucose that results in two ATP and products such as alcohol and lactate

7 product of glycolysis; its further fate, involving fermentation or entry into a mitochondrion, depends on oxygen availability

8 pathway of metabolism converting glucose to pyruvate; resulting in a net gain of two ATP and two NADH molecules

9 use of oxygen to metabolize lactate, which builds up due to anaerobic conditions (two words)

*Down*

1 metabolites that are the products of and/or the substrates for key reactions in cells, allowing one type of molecule to be changed into another type, such as the conversion of carbohydrates to fats (two words)

2 type of system whereby electrons are passed along a series of carrier molecules, releasing energy for the synthesis of ATP (two words)

4 type of process that requires oxygen

5 molecule made up of a two-carbon acetyl group attached to coenzyme A; the acetyl group enters the citric acid cycle for further oxidation (two words)

6 a coenzyme of oxidation-reduction, sometimes used instead of $NAD^+$

## Chapter Test

### Objective Questions

Do not refer to the text when taking this test.

____ 1. Fermentation is
a. glycolysis and the citric acid cycle.
b. glycolysis and the reduction of pyruvate.
c. glycolysis only.
d. the reduction of pyruvate only.

____ 2. Each of the following is a product of cellular respiration EXCEPT
a. ATP.
b. carbon dioxide.
c. oxygen.
d. water.

____ 3. Per glucose molecule, the net gain of ATP molecules from glycolysis is
a. two.
b. four.
c. six.
d. eight.

____ 4. Fermentation supplies
a. glycolysis with free $NAD^+$.
b. hydrogen to the transition reaction.
c. oxygen as an electron acceptor.
d. the citric acid cycle with oxygen.

____ 5. The process that evolved first was
a. chemiosmosis.
b. glycolysis.
c. the electron transport system.
d. the citric acid cycle.

____ 6. Which is NOT an event of the transition reaction?
a. breaks down pyruvate
b. converts a citrate molecule
c. oxidizes pyruvate
d. transfers an acetyl group

____ 7. Select the incorrect association.
a. electron transport system—cristae
b. fermentation—plasma membrane
c. glycolysis—cytoplasm
d. citric acid cycle—matrix

____ 8. Select the process with the greatest yield of NADH per glucose molecule.
a. glycolysis
b. citric acid cycle
c. substrate-level phosphorylation
d. transition reaction

____ 9. The energy yield by ATP molecules per glucose molecule is closest to
a. 25%.
b. 40%.
c. 50%.
d. 60%.

____ 10. Inside a cell, glycerol is broken down into
a. amino acids.
b. acetyl CoA.
c. fatty acids.
d. PGAL.

____ 11. Which of the following reactions is NOT a part of cellular respiration?
a. glycolysis
b. citric acid cycle
c. electron transport system
d. transition reaction
e. fermentation

____ 12. The coenzyme used in the transition reaction of cellular respiration is
a. ATP.
b. $NAD^+$.
c. NADH.
d. coenzyme A.
e. RuBP.

____ 13. The carbon dioxide given off by cellular respiration is produced by
a. glycolysis.
b. the transition reaction.
c. the citric acid cycle.
d. the electron transport system.
e. Both *b* and *c* are correct.

____ 14. The final acceptor for electrons in cellular respiration is
a. ATP.
b. $NAD^+$.
c. FAD.
d. oxygen.
e. carbon dioxide.

____ 15. Which of the following reactions occurs on the inner membrane of mitochondria?
a. the citric acid cycle
b. the transition reaction
c. the electron transport system
d. glycolysis
e. the Calvin cycle

____ 16. The coenzymes $NAD^+$ and FAD carry hydrogen atoms ($H^+ + e^-$) to the
a. glycolysis reactions.
b. transition reaction.
c. citric acid cycle.
d. Calvin cycle.
e. electron transport system.

____17. Which of the following statements is NOT true about fermentation?
a. It is an anaerobic process.
b. The products are toxic to cells.
c. It results in two ATPs per glucose molecule.
d. In the absence of $O_2$, muscle cells form $CO_2$ and alcohol.
e. It can be used to make bread rise.

____18. Which of the following statements is NOT true regarding fats?
a. Fatty acids are converted to acetyl-CoA.
b. Eighteen-carbon fatty acids are converted to nine acetyl-CoA molecules.
c. Glycerol is converted to PGAL.
d. Fats are the least efficient form of stored energy.
e. Carbohydrates can be converted to fats.

____19. The process directly responsible for most of the ATP formed during cellular respiration is
a. the citric acid cycle.
b. the transition reaction.
c. the electron transport system.
d. chemiosmosis.

____20. A pathway that begins with glucose and ends with pyruvate is
a. glycolysis.
b. the citric acid cycle.
c. the electron transport system.
d. the transition reaction.

## Critical Thinking Questions

Answer in complete sentences.

21. Explain how the human body obtains the reactants for cellular respiration, and explain what happens to the products.

22. In what ways are cellular respiration and photosynthesis similar processes?

**Test Results:** ______ number correct ÷ 22 = ______ × 100 = ______ %

## Exploring the Internet

The Online Learning Center at *www.mhhe.com/maderbiology8* has additional study material and practice quizzes that can help you master the content of this chapter. You can also find links to websites exploring additional topics in biology. Access to the Online Learning Center is free for those who have purchased a new textbook.

## Answer Key

### Study Exercises

**1. a.** oxidized **b.** reduced **c.** exergonic **2. a.** 2H **b.** NADH + $H^+$ **c.** 2H **d.** $NAD^+$ **e.** right **f.** left **g.** It becomes reduced when it accepts electrons from a substrate and becomes oxidized when it passes electrons on to another carrier. **3. a.** ETS **b.** CA **c.** TR **d.** GL **e.** ETS **f.** TR, CA, ETS **g.** GL **h.** GL, CA (two turns per glucose molecule) **4. a.** cytoplasm **b.** no **c.** glucose **d.** pyruvate **e.** two ATP **f.** substrate level **g.** $NAD^+$ **h.** NADH, ATP, pyruvate **5. a.** cristae **b.** matrix **c.** outer membrane **d.** intermembrane space **e.** inner membrane **f.** cytoplasm **6. a.** cytoplasm **b.** matrix **c.** cristae **7. a.** citrate **b.** $CO_2$ **c.** two ATP **d.** $NAD^+$ and FAD **e.** NADH, $FADH_2$, ATP, and $CO_2$ **8. a.** NADH and $FADH_2$ **b.** pass down the system **c.** pumped into intermembrane space **d.** $O_2$ **e.** three ATP **f.** oxidative **9.a.** intermembrane **b.** down **c.** ATP **10. a.** 4 **b.** 1 **c.** 3 **d.** 2 **11.** see Figure 8.9, page 140, in text **12. a.** reduced to lactate **b.** reduced to alcohol and $CO_2$ **c.** produces only two ATP **d.** does not require oxygen **e.** $O_2$ needed to metabolize lactate **13. a.** deamination **b.** PGAL **c.** acetyl-CoA **d.** The PGAL is converted to glycerol and acetyl groups that result from carbohydrate breakdown can be joined to form fatty acids. Glycerol + 3 fatty acids forms fat molecules. **14. a.** I, III **b.** III **c.** I **d.** II **15. a.** lactate **b.** alcohol and $CO_2$ **c.** returns to glycolysis **16. a.** goes to the electron transport system **b.** enters the citric acid cycle

## KeyWord CrossWord

| | | | | | | | | | | | | | | | | | | | | |
|---|---|---|---|---|---|---|---|---|---|---|---|---|---|---|---|---|---|---|---|---|
| | 1 M | | 2 E | | | | | | | | | | | | | | | | | |
| 3 C | E | L | L | U | L | 4 A | R | | R | E | S | P | I | R | 5 A | T | I | O | N | |
| | T | | E | | | E | | | | | | | | | C | | | | | |
| | A | | C | | | R | | | | | | | | | E | | | | | |
| | B | | T | | | O | | | 6 F | E | R | M | E | N | T | A | T | I | O | N |
| | O | | R | | | B | | | A | | | | | | Y | | | | | |
| | L | | O | | | I | | | D | | | | | | L | | | | | |
| | I | | N | | | C | | | | | | | | | — | | | | | |
| | C | | | | | | | | | | | | | | C | | | | | |
| | | | T | | | | | | | | | | | | o | | | | | |
| | P | | R | | | | | | | 7 P | Y | R | U | V | A | T | E | | | |
| | O | | A | | | | | | | | | | | | | | | | | |
| | O | | N | | | | | | | 8 G | L | Y | C | O | L | Y | S | I | S | |
| | L | | S | | | | | | | | | | | | | | | | | |
| | | | P | | | | | | | | | | | | | | | | | |
| | | | 9 O | X | Y | G | E | N | | D | E | B | T | | | | | | | |
| | | | R | | | | | | | | | | | | | | | | | |
| | | | T | | | | | | | | | | | | | | | | | |

## Chapter Test

**1.** b **2.** c **3.** a **4.** a **5.** b **6.** b **7.** b **8.** b **9.** b **10.** d **11.** e **12.** d **13.** e **14.** d **15.** c **16.** e **17.** d **18.** d **19.** d **20.** a **21.** Glucose enters the body at the digestive tract and oxygen enters at the lungs. Glucose and oxygen are delivered to cells by the cardiovascular system. Water from cellular respiration enters the blood and is used by the body or excreted; we breathe out the carbon dioxide. **22.** Both photosynthesis and cellular respiration consist of a series of reactions that the overall reaction does not indicate. Both pathways make use of an electron transport system located in membrane to build up an electrochemical gradient of $H^+$. When $H^+$ flows down this gradient through an ATP synthase complex, ATP is produced. Both pathways use a coenzyme of oxidation/reduction—photosynthesis uses NADP and cellular respiration uses NAD. The same molecules (PGA and PGAL) occur in the Calvin cycle and the citric acid cycle, but in photosynthesis, PGA is reduced to PGAL, and in cellular respiration, PGAL is oxidized to PGA.

# Part II Genetic Basis of Life

# 9

# The Cell Cycle and Cellular Reproduction

## Chapter Review

The **cell cycle** is a repeating sequence of growth, replication of DNA, and cell division. Most of the cell cycle is spent in interphase. Interphase includes the four stages $G_1$, S, $G_2$, and M. During the $G_1$ stage, the organelles increase in number; during the S stage, DNA replication occurs; and during the $G_2$ stage, various proteins are synthesized. During the M stage, mitosis occurs. The cell cycle is controlled by internal and external signals. Cell division is balanced by apoptosis.

Each species has a characteristic number of chromosomes. The total number is the **diploid** number, and half this number is the **haploid** number.

Replication of DNA precedes cell division. The duplicated chromosome is composed of two sister chromatids attached at a centromere. During mitosis, the centromeres split, and daughter chromosomes go into each new cell.

Mitosis has the following phases: prophase, when the chromosomes have no particular arrangement; prometaphase, when the spindle forms and kinetochores develop; metaphase, when the chromosomes are aligned at the metaphase plate; anaphase, when daughter chromosomes move toward opposite poles; and telophase, when new nuclear envelopes form around the daughter chromosomes and cytokinesis begins. **Cytokinesis** (division of the cytoplasm) in animal cells occurs by furrowing, while in plant cells, a **cell plate** forms.

Control of the cell cycle is an area of intense investigation, particularly because cancer cells divide uncontrollably. **Cancer** is characterized by this lack of control. Cancer cells are nondifferentiated, have abnormal nuclei, do not require growth factors, and are not constrained by their neighbors. After forming a tumor, cancer cells metastasize and start new tumors elsewhere in the body. Proto-oncogenes and tumor-suppressor genes are normal genes that bring on cancer when they mutate because they code for factors involved in cell growth.

Avoiding unnecessary radiation, exposure to toxic chemicals, and adopting a diet rich in fruits and vegetables is protective against the development of cancer.

Prokaryotes divide by **binary fission**—replication of the single chromosome and elongation of the cell that pulls the chromosomes apart. Inward growth of the plasma membrane and formation of new cell wall divide the cell in two.

Binary fission (in prokaryotes) and mitosis (in unicellular eukaryotic protists and fungi) allow these organisms to reproduce asexually. Mitosis in multicellular eukaryotes is primarily for the purpose of development, growth, and repair of tissues.

## Study Exercises

Study the text section by section as you answer the questions that follow.

### 9.1 The Cell Cycle (pp. 150–51)

- Cell division in eukaryotes is a part of the cell cycle. First, cells get ready to divide and then they divide.

1. Cell cycle stages are labelled $G_1$, S, $G_2$, or M. What do these letters stand for?

a. $G_1$ ________________________________

b. S ________________________________

c. $G_2$ ________________________________

d. M ________________________________

2. Match the letters in the following diagram to these phrases.

_______ Mitosis and cytokinesis occur.

_______ DNA replication occurs as chromosomes duplicate.

_______ Growth occurs as cell prepares to divide.

_______ Growth occurs as organelles double.

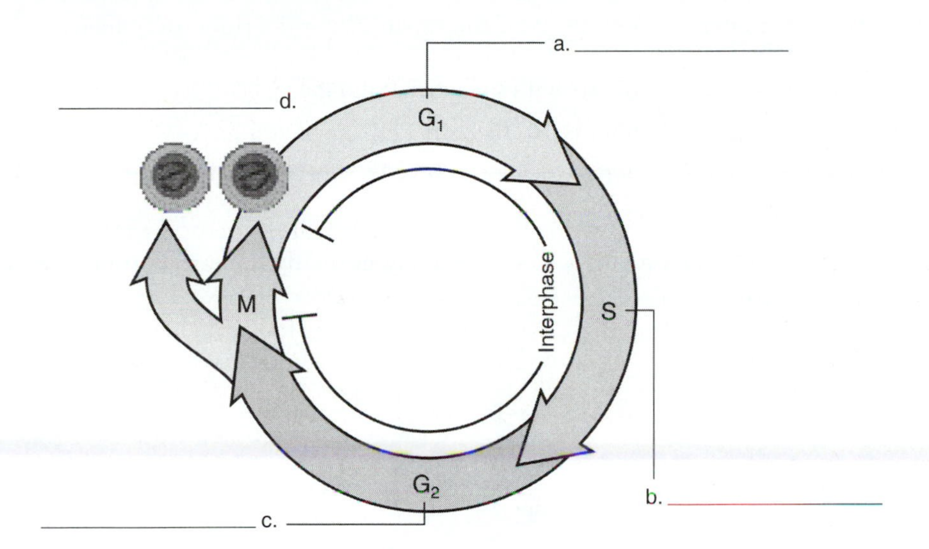


3. Match the checkpoints in the following diagram to these phrases.

_______ Apoptosis can occur if DNA is damaged.

_______ Mitosis will not occur if DNA is damaged or not replicated.

_______ Mitosis stops if chromosomes are not properly aligned.

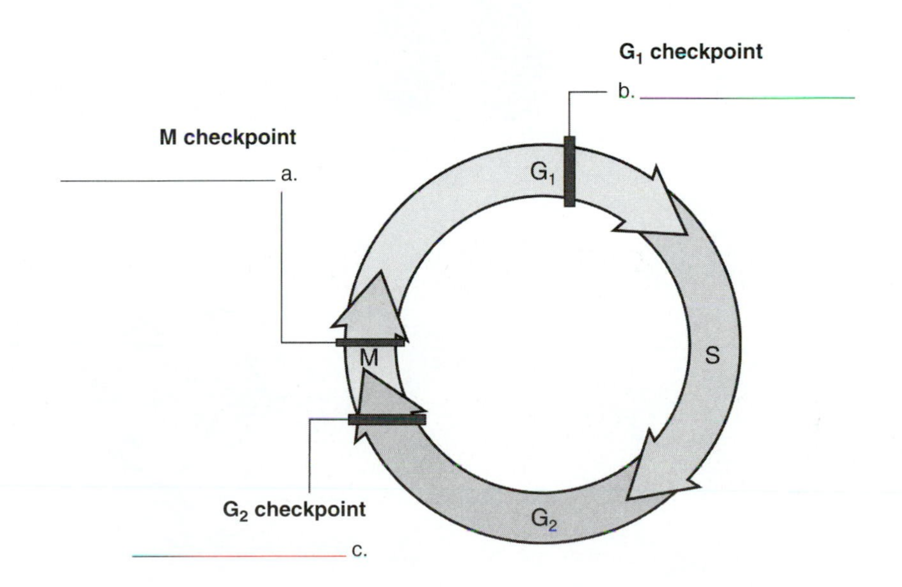

4. Place an X on the line beside each statement that is true of apoptosis.

a. _____ Some cells in the body are always undergoing apoptosis.
b. _____ Apoptosis is abnormal and occurs rarely.
c. _____ Cells are destroyed during apoptosis by external enzymes.
d. _____ Cells are destroyed during apoptosis by internal enzymes.
e. _____ At the end of apoptosis, the cell looks like it did before.
f. _____ At the end of apoptosis, the cell is fragmented.

## 9.2 Mitosis and Cytokinesis (pp. 153–57)

- Each eukaryotic species has a characteristic number of chromosomes.
- Mitosis is a type of nuclear division that ensures each new eukaryotic cell has a full set of chromosomes.
- Mitosis is necessary to the development, growth, and repair of multicellular organisms.

5. Complete each of the following statements with the correct number:

In corn, the haploid chromosome number is 10. Its body cells normally have a.______________ chromosomes. The diploid chromosome number in the domestic cat is 38. Normally, its sperm and egg cells have b.______________ chromosomes. The horse has a haploid chromosome number of 32. In this animal, 2n = c.______________. The sperm and egg cells of a dog normally have 39 chromosomes. In this animal, n = d.______________.

6. Label the following diagram:

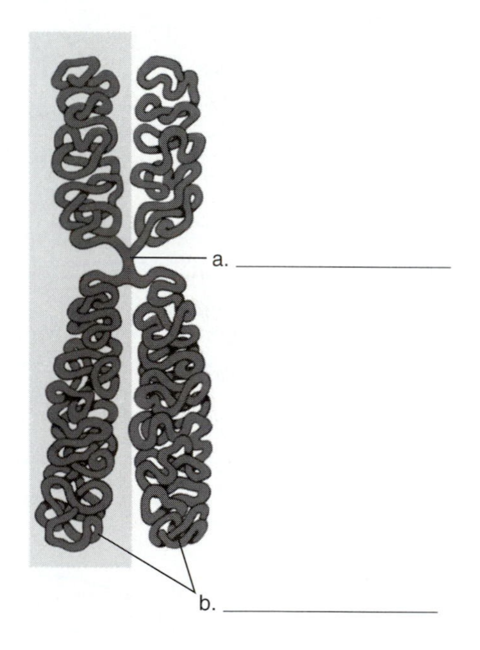


7. Label this diagram, using the following alphabetized list of terms.

aster
centriole
centromere
centrosome
chromosome
kinetochore
nuclear membrane fragment

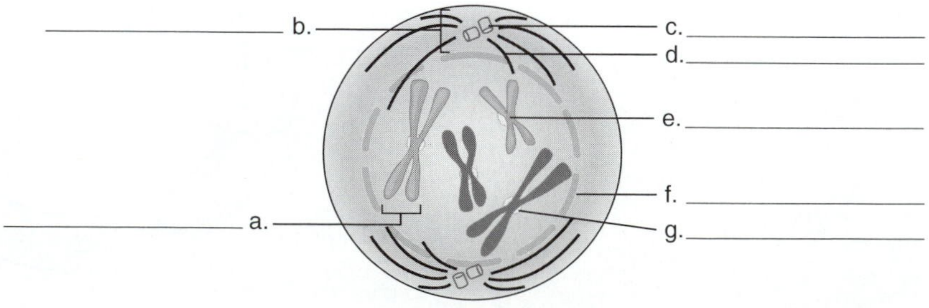

a. ______________
b. ______________
c. ______________
d. ______________
e. ______________
f. ______________
g. ______________

8. Complete the following diagrams to show the arrangement and movement of chromosomes during animal cell mitosis. Briefly describe the events of each phase on the lines provided.

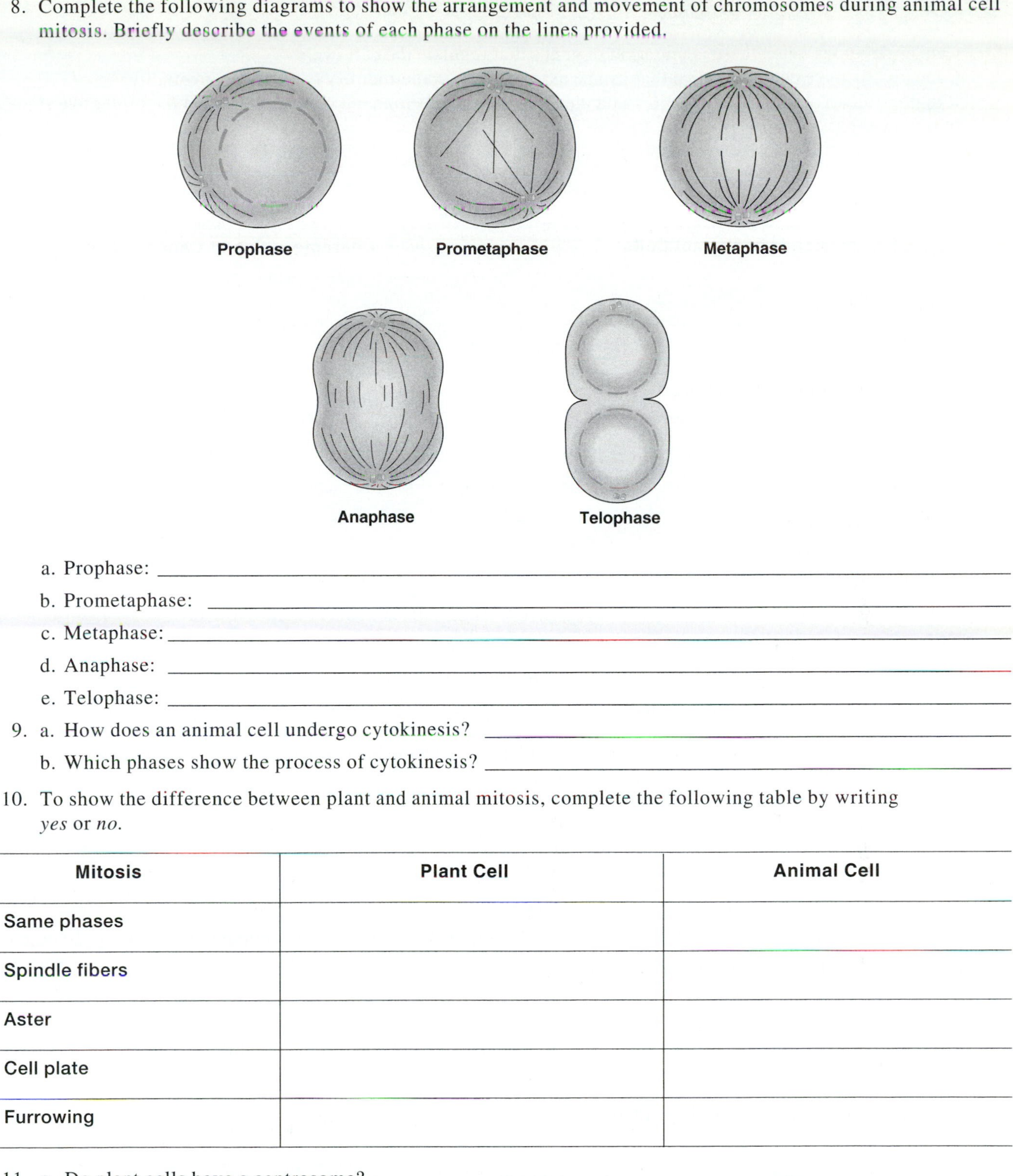


a. Prophase: ____________________

b. Prometaphase: ____________________

c. Metaphase: ____________________

d. Anaphase: ____________________

e. Telophase: ____________________

9. a. How does an animal cell undergo cytokinesis? ____________________

b. Which phases show the process of cytokinesis? ____________________

10. To show the difference between plant and animal mitosis, complete the following table by writing *yes* or *no*.

| Mitosis | Plant Cell | Animal Cell |
|---|---|---|
| Same phases | | |
| Spindle fibers | | |
| Aster | | |
| Cell plate | | |
| Furrowing | | |

11. a. Do plant cells have a centrosome? ____________________

____________________

b. Do plant cells have centrioles? ____________________

____________________

c. Are centrioles necessary to spindle formation? Explain. ____________________

____________________

## 9.3 The Cell Cycle and Cancer (pp. 158–61)

- Cancer develops when there is mutation of genes that regulate the cell cycle.
- Cancer cells have characteristics that can be associated with their ability to divide controllably.
- It is possible to avoid certain agents that contribute to the development of cancer and to take protective steps to reduce the risk of cancer.

12. Complete the following table:

| Characteristics of Normal Cells | Characteristics of Cancer Cells |
|---|---|
| controlled growth | a. |
| contact inhibition | b. |
| one organized layer in tissue culture | c. |
| differentiated cells | d. |
| normal nuclei | e. |

In question 13, fill in the blanks.

13. Instead of growing in a.______________ layer(s), as normal cells do, cancer cells grow in b.______________ layer(s), losing the property of c.______________ inhibition. Cancer cells divide to form a growth, or d.______________. The cells of e.______________ tumors remain in one place. The cells of f.______________ tumors wander, a characteristic called g.______________.

14. Put a check next to the items that pertain to a regulatory pathway that controls the cell cycle in cells.
    a. _____ growth factor receptors in plasma membrane
    b. _____ signaling proteins within cytoplasm
    c. _____ various genes in nucleus
    d. _____ proteins that directly control the cell cycle
    e. _____ oncogenes and tumor-suppressor genes

15. Place the appropriate letter(s) next to each statement.

    P—proto-oncogenes O—oncogenes T—tumor-suppressor genes MT—mutated tumor-suppressor gene

    a. _____ cell division is always promoted (requires two)
    b. _____ normal genes in cells that regulate cell division (requires two)
    c. _____ mutated genes that cause cancer (requires two)

16. The right diet can help reduce the chances of cancer. What would you advise about the following?
    a. obesity ______________
    b. intake of salt-cured, smoked, or nitrite-cured foods ______________
    c. fat intake ______________
    d. intake of fruits and vegetables containing vitamins A and C ______________
    e. intake of high-fiber foods ______________

## 9.4 Prokaryotic Cell Division (pp. 162–63)

- Binary fission allows prokaryotes to reproduce and ensures that each new cell has a chromosome.

17. Indicate whether the following statements about binary fission are true (T) or false (F):

_____ a. A unicellular organism reproduces by this process.
_____ b. DNA replicates before the onset of this process.
_____ c. Elongation of the cell requires that the two circular chromosomes combine.
_____ d. Eukaryotic cells also divide through this process.
_____ e. The plasma membrane grows inward when the cell reaches about twice its original length.
_____ f. Two daughter cells genetically identical to the parent cell are produced.

## Definitions Wordsearch

Review key terms by using the following alphabetized list of terms to fill in the blanks. Then complete the wordsearch.

```
S J E T R E C P T V U I T T H I M K N E W F L M
N A F D Z C Y T O K I N E S I S B Y E H C I G X
L O I S O T Y C O D N T I M O P Y C F C L X Y K
T H I R I O T B G T D E A G I H R N Q E J W K U
G I G T Y C O P R O T R I N S T O N A N U D L P
R E N T A F R D C P T P I O U H T E M P R Z N A
A Y L I L Z K A Q L C H R O M A T I N X S V W O
F U I I D A I Y T A Y A L K E I I L L O N I N N
R E N N U G G L N S L S O H T F N M M C G G S G
L I T P O E N Z I M O E I U A D O J J Y N N S N
K I E Y R L N U T T S U R O P M C E E T N P I N
F L R G G O T M C I R O T R H E Y I I O I T S T
G O K S I S P T Y C O E A H A L T G G N Y Q O R
C I I I U M N H T E S L F N S G O O D I N N T N
N U N L A I U B A I I O S J E M S L G S G U I W
Q M E O Y B U P L S S M O L Y S E S R G U U M U
G I S R W X M P N C E N T R O M E R E T F M Z M
C V I M O I M H A N I B U R E E H G U G M S Q M
O C S P E R M A T O G E N E S I S T W S I Q F I
D U C L A I U B N I T O E J U W J K I G U F X U
```

centromere
chromatin
cytokinesis
interkinesis
interphase
mitosis
prophase
spindle
synapsis

a. ________________ Type of cell division in which daughter cells receive the exact chromosome and genetic makeup of the parent cell; occurs during growth and repair.
b. ________________ Division of the cytoplasm following mitosis and meiosis.
c. ________________ Stages of the cell cycle ($G_1$, S, $G_2$) during which growth and DNA synthesis occur when the nucleus is not actively dividing.
d. ________________ Network of fibrils consisting of DNA and associated proteins observed within a nucleus that is not dividing.
e. ________________ Microtubule structure that brings about chromosomal movement during nuclear division.
f. ________________ Mitosis phase during which chromatin condenses so that chromosomes appear.
g. ________________ Constricted region of a chromosome where sister chromatids are attached to one another and where the chromosome attaches to a spindle fiber.
h. ________________ Period between meiosis I and meiosis II during which no DNA replication takes place.
i. ________________ Pairing of homologous chromosomes during meiosis I.

# Chapter Test

## Objective Questions

Do not refer to the text when taking this test. In questions 1–4, match the description to the phase.

a. anaphase
b. metaphase
c. prophase
d. telophase

____ 1. Chromosomes first become visible.
____ 2. Chromatids separate at centromere.
____ 3. Chromosomes are aligned at the metaphase plate.
____ 4. Last phase of nuclear division.

____ 5. The diploid chromosome number in an organism is 42. The number of chromosomes in its sperm and egg cells is normally
a. 21.
b. 42.
c. 63.
d. 84.

____ 6. Which statement about mitosis is NOT correct?
a. does not affect the nuclear envelope
b. forms four daughter cells
c. makes diploid nuclei
d. prophase is the first active phase

____ 7. How does mitosis in plant cells differ from that in animal cells?
a. Animal cells do not form a spindle.
b. Animal cells lack cytokinesis.
c. Plant cells lack a cell plate.
d. Plant cells lack centrioles.

____ 8. Select the incorrect association.
a. $G_1$—cell grows in size
b. $G_2$—protein synthesis occurs
c. mitosis—nuclear division
d. S—DNA fails to duplicate

____ 9. The phase of cell division in which the nuclear envelope and nucleolus are disappearing as the spindle fibers are appearing is called
a. anaphase.
b. prophase.
c. telophase.
d. metaphase.

____ 10. In animal cells, cytokinesis takes place by
a. membrane fusion.
b. a furrowing process.
c. formation of a cell plate.
d. cytoplasmic contraction.

____ 11. If a cell is to divide, DNA replication must occur during
a. prophase.
b. metaphase.
c. anaphase.
d. telophase.
e. interphase.

____ 12. If a cell had 18 chromosomes, how many chromosomes would each daughter cell have after mitosis?
a. 9
b. 36
c. 18
d. cannot be determined

____ 13. Normal growth and repair of the human body requires
a. mitosis.
b. binary fission.
c. Both *a* and *b* are correct.
d. Neither *a* nor *b* are correct.

____ 14. The cell cycle
a. includes mitosis as an event.
b. includes only the stages $G_1$, S, and $G_2$.
c. is under cellular, but not under genetic, control.
d. involves proteins but not chromosomes.

____ 15. When do chromosomes move to opposite poles?
a. prophase
b. metaphase
c. anaphase
d. telophase

____ 16. Which statement about binary fission is NOT correct?
a. Unicellular prokaryotes reproduce by binary fission.
b. Like mitosis, binary fission uses a spindle.
c. Binary fission is a form of asexual reproduction.
d. Binary fission ensures that daughter cells receive the same genetic material as the parent cell.

____ 17. Which of the following does NOT describe the behavior of cells in a malignant tumor?
a. carry out metastasis
b. lose the ability of contact inhibition
c. multiply rapidly
d. remain in one site

____ 18. An agent that contributes to the development of cancer is a(n)
a. carcinogen.
b. oncogene.
c. promoter.
d. tumor-suppressor gene.

____ 19. Which of the following is NOT a suggested measure to prevent cancer?
a. Avoid foods of the cabbage family.
b. Cut down on salt-cured foods.
c. Eat more high-fiber foods.
d. Increase the intake of vitamins A and C.

____20. Which of the following will NOT help you to prevent cancer?
a. Avoid carcinogenic chemicals.
b. Stop smoking.
c. Lower total fat intake.
d. Eat fewer high-fiber foods.
e. Eat more broccoli and cauliflower.

## Critical Thinking Questions

Answer in complete sentences.

21. Why might mitosis and multicellularity be considered adaptive for organisms?

22. Why are mutations sometimes particularly beneficial to organisms that reproduce asexually?

**Test Results:** ______ number correct ÷ 22 = ______ × 100 = ______ %

## Exploring the Internet

The Online Learning Center at *www.mhhe.com/maderbiology8* has additional study material and practice quizzes that can help you master the content of this chapter. You can also find links to websites exploring additional topics in biology. Access to the Online Learning Center is free for those who have purchased a new textbook.

## Answer Key

### Study Exercises

**1. a.** growth **b.** synthesis **c.** growth **d.** mitosis **2. a.** Growth occurs as organelles **b.** DNA replication **c.** Growth occurs as cell prepares **d.** Mitosis **3. a.** Mitosis stops **b.** Apoptosis can occur **c.** Mitosis will not occur **4.** a, d, f are true. **5. a.** 20 **b.** 19 **c.** 64 **d.** 39 **6. a.** centromere **b.** sister chromatids **7. a.** chromosome **b.** centrosome **c.** centriole **d.** aster **e.** centromere **f.** nuclear membrane fragment **g.** kinetochore

**8.** Diagrams:

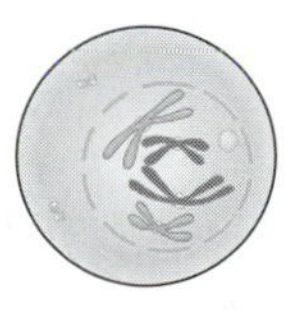
Prophase

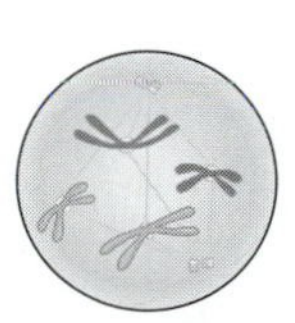
Prometaphase

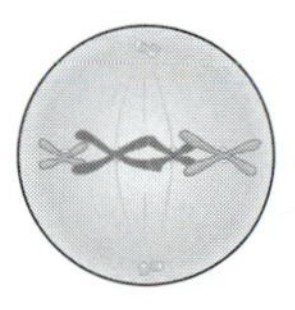
Metaphase

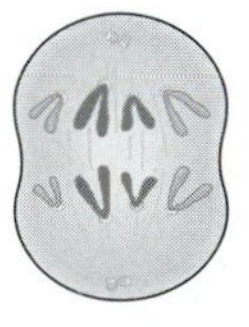
Anaphase

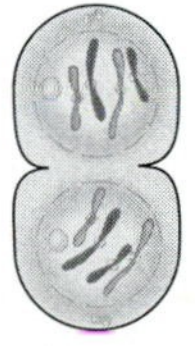
Telophase

**8. a.** Chromosomes are now distinct, nucleolus is disappearing, centrosomes begin moving apart, and nuclear envelope is fragmenting. **b.** Spindle is in process of forming, and kinetochores of chromosomes are attaching to kinetochore spindle fibers. **c.** Chromosomes are at the metaphase plate. **d.** Daughter chromosomes are at the poles of the spindle. **e.** Daughter cells are forming as nuclear envelopes and nucleoli appear. **9. a.** cleavage furrowing **b.** anaphase and telophase

**10.**

| Plant Cell | Animal Cell |
|---|---|
| yes | yes |
| yes | yes |
| no | yes |
| yes | no |
| no | yes |

**11. a.** Yes and they form a spindle. **b.** no **c.** It seems not, because plant cells have no centrioles, yet they have a spindle. **12. a.** uncontrolled growth **b.** no contact inhibition **c.** disorganized, multilayered **d.** nondifferentiated cells **e.** abnormal nuclei **13. a.** one **b.** multiple **c.** contact **d.** tumor **e.** benign **f.** malignant **g.** metastasis **14.** a, b, c, d, e **15. a.** O, MT **b.** P, T **c.** O, MT **16. a.** avoid **b.** avoid **c.** reduce **d.** increase **e.** increase **17. a.** T **b.** T **c.** F **d.** F **e.** T **f.** T

## Definitions Wordsearch

**a.** mitosis **b.** cytokinesis **c.** interphase **d.** chromatin **e.** spindle **f.** prophase **g.** centromere **h.** interkinesis **i.** synapsis

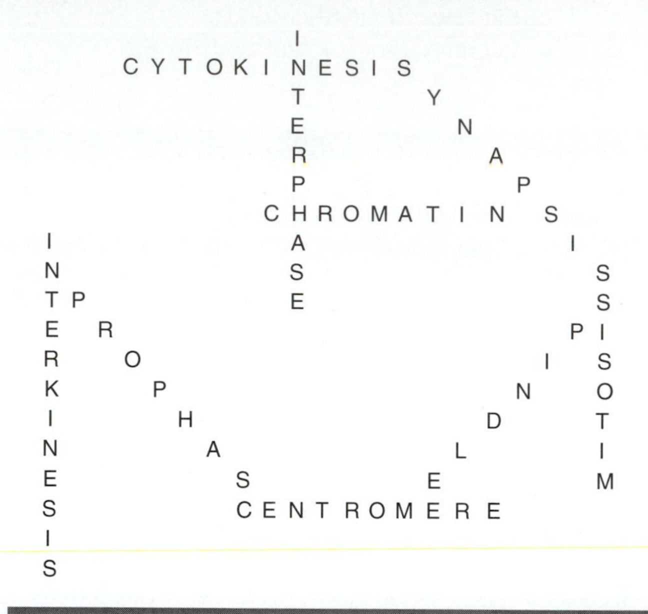


## Chapter Test

**1.** c **2.** a **3.** b **4.** d **5.** a **6.** b **7.** d **8.** d **9.** b **10.** b **11.** e **12.** c **13.** a **14.** a **15.** c **16.** b **17.** d **18.** a **19.** a **20.** d **21.** Complex organisms are multicellular with specialized tissues and organs. In addition, mitosis allows growth and repair of the organism. **22.** Without mutations, asexual reproduction produces offspring that are identical to the parent. With mutations, variations may occur that will be adaptive, particularly if the environment is changing.

# 10

# Meiosis and Sexual Reproduction

## Chapter Review

Working in conjunction with fertilization, **meiosis** ensures the constancy of the chromosome number from generation to generation, by reducing the chromosome number from diploid to haploid. Prior to meiosis, the parent cell is **diploid.** Replication has occurred, and the chromosomes are duplicated. Meiosis involves two consecutive cell divisions that produce four **haploid** daughter cells.

**Genetic recombination** provides a means of insuring variation among offspring. It occurs during sexual reproduction due to **crossing-over,** which often occurs between nonsister chromatids in prophase I; due to **independent assortment,** where gametes receive various recombined chromosomes; and subsequently **fertilization,** where the gamete receives recombined chromosomes from both parents.

**Meiosis I** and **meiosis II** have these phases: prophase, metaphase, anaphase, and telophase. During meiosis I, homologous chromosomes separate. Following meiosis I, daughter cells are haploid. During meiosis II, sister chromatids of each duplicated chromosome separate. Following meiosis II, the chromosomes are no longer duplicated.

Comparing meiosis with mitosis: Meiosis occurs only during the production of gametes and reduces the chromosome number. Mitosis occurs during growth and repair of tissues.

In the animal life cycle, including the human life cycle, meiosis is a part of gamete formation. The diploid adult produces gametes by meiosis, and the gametes are the only haploid stage of the life cycle. The human life cycle includes mitosis and meiosis (part of **spermatogenesis** in males and **oogenesis** in females). In the two-generation life cycle of plants, meiosis produces spores, which are haploid, and fertilization results in a zygote, which is diploid. In fungi and some algae, only the zygote is diploid, and it undergoes meiosis to produce haploid spores.

## Study Exercises

Study the text section by section as you answer the questions that follow.

### 10.1 Halving the Chromosome Number (pp. 168–69)

- Due to meiosis, reproductive cells contain half the total number of chromosomes.
- Meiosis requires two cell divisions and results in four daughter cells.

1. The nuclear division that reduces the chromosome number from the a.______________ number to the b.______________ number is called meiosis.
2. In a life cycle, the zygote always has the a.______________ number of chromosomes and gametes have the b.______________ number.
3. Indicate whether the following statements regarding the role of meiosis are true (T) or false (F):
   a. _____ In animals, meiosis forms gametes that fuse to form a zygote.
   b. _____ Meiosis forms haploid cells in the life cycle of animals.
   c. _____ Meiosis produces four diploid cells over two divisions.

4. Label this summary diagram of meiosis using the following alphabetized list of terms.

centromere
DNA replication
meiosis I
meiosis II
nucleolus
sister chromatids
synapsis

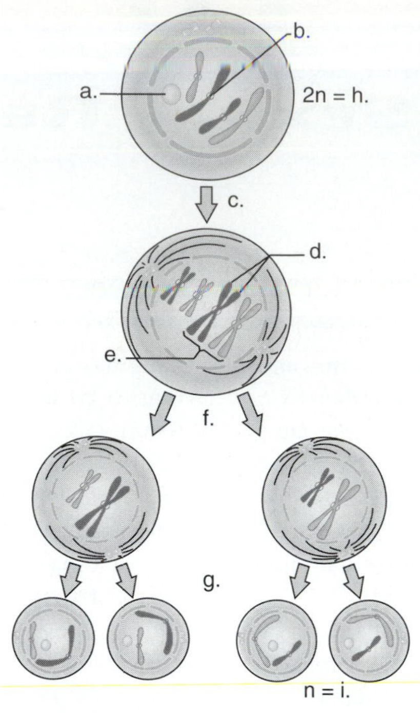


h. Solve 2n = ______________

i. Solve n = ______________

j. Why is it correct to symbolize meiosis as 2n → n? ____________________________________________

____________________________________________________________________________________

5. a. What is the diploid number of chromosomes in the diagram in question 4? ____________________

b. Which structures separate during meiosis I? ____________________________________________

c. Which structures separate during meiosis II? ___________________________________________

d. Does chromosome duplication occur between meiosis I and meiosis II? ______________________

e. Why or why not? _______________________________________________________________

## 10.2 Genetic Recombination (pp. 170–71)

- The process of meiosis ensures that the daughter cells will not have the same number and types of chromosomes as the parental cell.

6. Match the descriptions to these terms.

bivalents    crossing-over    genetic recombination    synapsis

a. ______________________ Homologues line up side by side.

b. ______________________ Nonsister chromatids exchange genetic material.

c. ______________________ Two chromosomes stay in close association with each other.

d. ______________________ The arrangement of genetic material is new due to crossing-over.

7. Using ink for one duplicated chromosome and pencil for the other, color these bivalents before and after crossing-over has occurred.

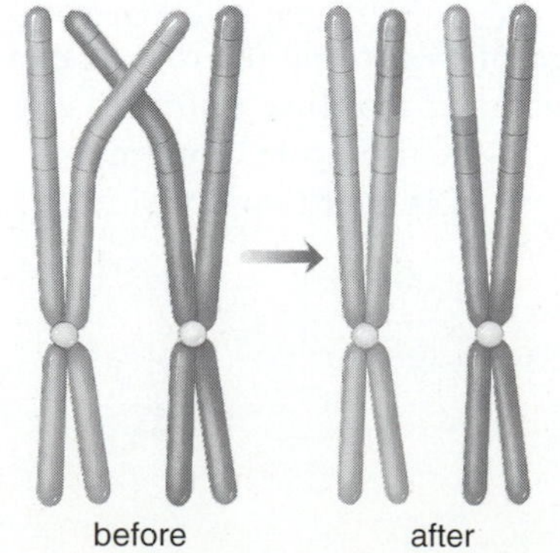

8. Show how genetic recombination occurs as a result of sexual reproduction by matching the following statements:
   1. Zygote carries a unique combination of chromosomes and genes.
   2. Gametes carry different combinations of chromosomes.
   3. Daughter chromosomes carry different combinations of genes.

a. _____ Homologues separate independently.
b. _____ Crossing-over occurs.
c. _____ Gametes fuse during fertilization.

## 10.3 The Phases of Meiosis (pp. 172–73)

- Meiosis I and meiosis II each have four phases.

9. Label the stage and complete these diagrams to show the arrangement and movement of chromosomes during meiosis I and meiosis II. (The diagram for meiosis II pertains to only one daughter cell from meiosis I.)

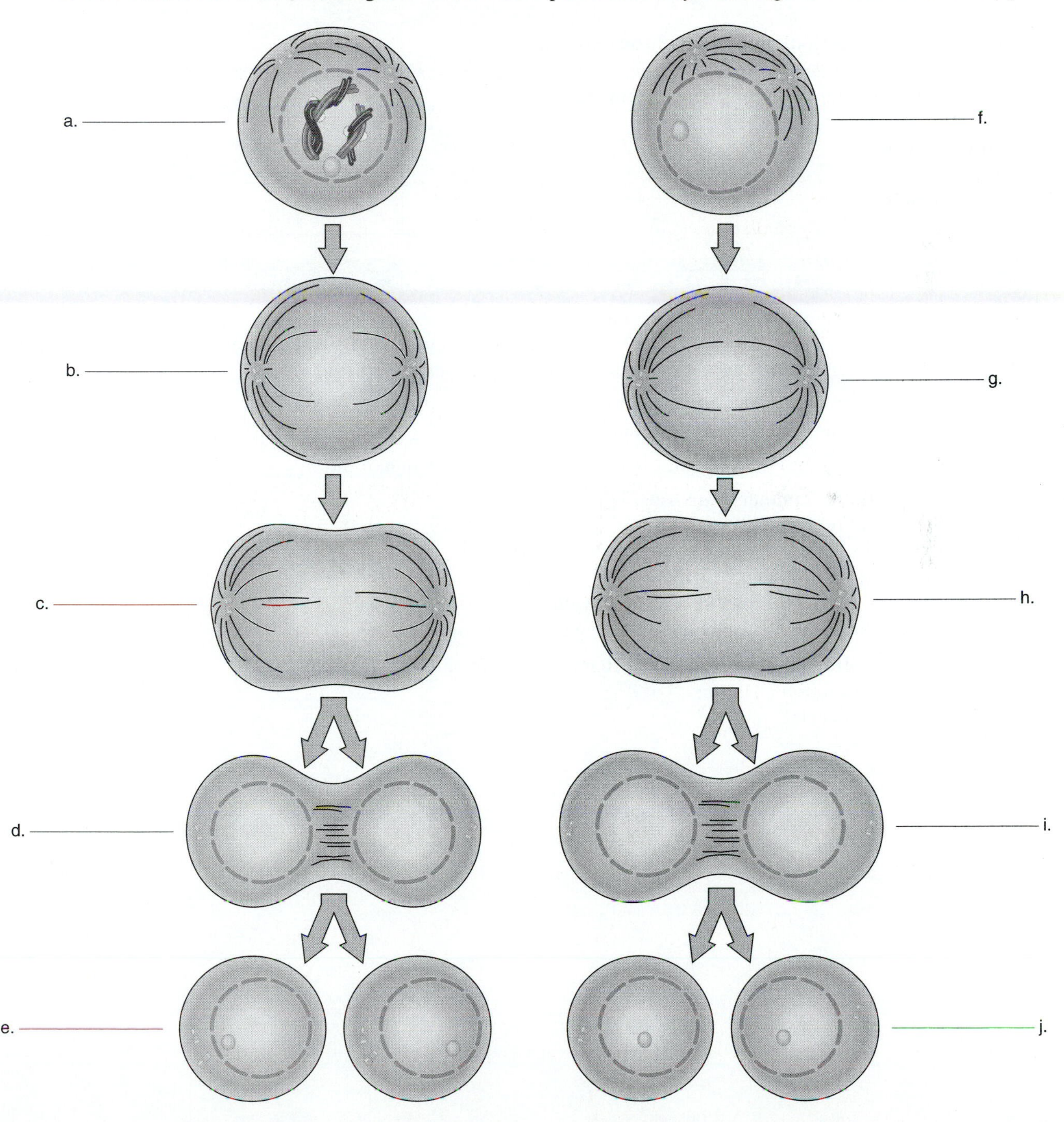

## 10.4 Comparison of Meiosis with Mitosis (pp. 174–75)

- Meiosis differs from mitosis both in occurrence and in process.

10. Complete the table by writing *yes* or *no* to distinguish meiosis from mitosis.

| Event | Meiosis | Mitosis |
|---|---|---|
| one division | | |
| two successive divisions | | |
| daughter chromosomes separate | | |
| homologous chromosomes (homologues) separate | | |
| daughter cells with the diploid number of chromosomes | | |
| daughter cells with the haploid number of chromosomes | | |
| somatic (body) cells | | |
| formation of gametes | | |

## 10.5 The Human Life Cycle (pp. 176–77)

- The human life cycle includes both mitosis and meiosis.
- In humans, and many other animals, meiosis is a part of the production of sperm in males and eggs in females.
- When the sperm fertilizes the egg, the full number of chromosomes is restored in offspring.

11. In the animal life cycle, meiosis results in haploid a.______________; in the plant life cycle, meiosis results in haploid b.______________; in the fungal life cycle, meiosis results in haploid c.______________.

12. Indicate whether these statements are true (T) or false (F).
    a. _____ Meiosis in human males is a part of spermatogenesis.
    b. _____ Mitosis in human females is a part of oogenesis.
    c. _____ Oogenesis occurs in the testis.
    d. _____ A zygote undergoes mitosis during development of the embryo.
    e. _____ Oogenesis produces four functional egg cells from one cell.

13. State whether the following processes occur in males (M), females (F), or both (B):
    a. _____ gamete formation
    b. _____ spermatogenesis
    c. _____ oogenesis
    d. _____ polar body formation

## KeyWord CrossWord

Review key terms by completing this crossword puzzle using the following alphabetized list of terms:

bivalent
crossing-over
diploid
gamete
haploid
homologous chromosome
homologue
meiosis
oogenesis
polar body
secondary oocyte
spermatogenesis

*Across*

2 member of a pair of chromosomes that carry genes for the same traits and synapse during prophase of the first meiotic division (two words)

9 homologous chromosomes, each having sister chromatids, joined by a nucleoprotein lattice during meiosis

10 cell condition in which only one of each type of chromosome is present

*Down*

1 type of nuclear division that occurs as part of sexual reproduction and in which the daughter cells receive the haploid number of chromosomes

3 egg production in females by the process of meiosis and maturation

4 haploid sex cell

5 in oogenesis, the functional product of meiosis I; becomes the egg (two words)

6 exchange of segments between nonsister chromatids of a bivalent during meiosis (two words)

7 sperm production in males by the process of meiosis and maturation

8 cell condition in which two of each type of chromosome is present

10 member of a homologous pair of chromosomes

11 in oogenesis, a nonfunctional product; three of the four meiotic products are of this type (two words)

## Chapter Test

### Objective Questions

Do not refer to the text when taking this test. In questions 1–8, match each of the descriptions with the following processes:

a. mitosis only
b. meiosis only
c. mitosis and meiosis
d. neither mitosis nor meiosis

_____ 1. Begins with a diploid nucleus
_____ 2. Complete after one division
_____ 3. Requires two successive divisions
_____ 4. Homologous chromosomes separate
_____ 5. Produces haploid nuclei
_____ 6. Produces diploid nuclei
_____ 7. Normally produces triploid cells
_____ 8. Occurs during growth and repair of tissues

In questions 9–15, label each statement with one of the following choices:

a. meiosis I
b. meiosis II

____ 9. Synapsis of homologues occurs
____10. Separation of homologues occurs
____11. Results in one oocyte and/or two to three polar bodies in human females
____12. Results in four sperm cells in human males
____13. Daughter cells have double-stranded chromosomes
____14. Daughter nuclei produced have single-stranded chromosomes
____15. Crossing-over occurs
____16. Which of the following is NOT a valid contrast between mitosis and meiosis:

| | Mitosis | Meiosis |
|---|---|---|
| a. | Requires one set of phases | Requires two sets of phases |
| b. | Occurs when body cells divide | Occurs during gamete production |
| c. | Results in four daughter nuclei | Results in two daughter nuclei |
| d. | Results in daughter nuclei with diploid number of chromosomes | Results in daughter nuclei with haploid number of chromosomes |

____17. Polar bodies are formed during

a. meiosis.
b. mitosis.
c. oogenesis.
d. spermatogenesis.
e. Both *a* and *c* are correct.

____18. During anaphase of meiosis II,

a. homologues separate.
b. daughter chromosomes separate.
c. daughter centrioles separate.
d. duplicated chromosomes separate.

____19. During interkinesis

a. chromosome duplication occurs.
b. chromosomes consist of two chromatids.
c. meiosis I is complete.
d. Both *b* and *c* are correct.

____20. By the end of meiosis I,

a. crossing-over has occurred.
b. daughter chromosomes have separated.
c. synapsis of homologues has occurred.
d. each daughter nucleus is genetically identical to the original cell.
e. Both *a* and *c* are correct.

## Critical Thinking Questions

Answer in complete sentences.

21. How might sexual reproduction be advantageous to the organism?

22. Based on the behavior of chromosomes during meiosis, how does a species with six chromosome pairs have an evolutionary advantage over a species with two chromosome pairs?

**Test Results:** ______ number correct ÷ 22 = ______ × 100 = ______ %

## Exploring the Internet

The Online Learning Center at *www.mhhe.com/maderbiology8* has additional study material and practice quizzes that can help you master the content of this chapter. You can also find links to websites exploring additional topics in biology. Access to the Online Learning Center is free for those who have purchased a new textbook.

# Answer Key

## Study Exercises

**1. a.** diploid (2n) **b.** haploid (n) **2. a.** diploid (2n) (or full) **b.** haploid **3. a.** T **b.** T **c.** F **4. a.** nucleolus **b.** centromere **c.** DNA replication **d.** sister chromatids **e.** synapsis **f.** meiosis I **g.** meiosis II **h.** 4 **i.** 2 **j.** A diploid cell becomes haploid. The parent cell is diploid and undergoes meiosis, which results in four daughter cells, each of which is haploid. **5. a.** 4 **b.** homologous chromosomes **c.** daughter chromosomes **d.** no **e.** The chromosomes are already duplicated. **6. a.** synapsis **b.** crossing-over **c.** bivalents **d.** genetic recombination **7.** See Fig. 10.3, page 170, in text **8. a.** 2 **b.** 3 **c.** 1 **9. a.** prophase I **b.** metaphase I **c.** anaphase I **d.** telophase I **e.** interkinesis **f.** prophase II **g.** metaphase II **h.** anaphase II **i.** telophase II **j.** daughter cells. See Figs. 10.6 and 10.7, pp. 172–73.

**10.**

| Meiosis | Mitosis |
|---|---|
| no | yes |
| yes | no |
| yes, meiosis II | yes |
| yes, meiosis I | no |
| no | yes |
| yes | no |
| no | yes |
| yes | no |

**11. a.** gametes **b.** spores **c.** spores **12. a.** T **b.** F **c.** F **d.** T **e.** F **13. a.** B **b.** M **c.** F **d.** F

## KeyWord CrossWord

| | | | | | | | | | | | | | | | | | | | | 1 M |
|---|---|---|---|---|---|---|---|---|---|---|---|---|---|---|---|---|---|---|---|---|
| 2 H | 3 O | M | O | L | O | 4 G | O | U | 5 S | | 6 C | H | R | O | M | O | 7 S | O | M | E |
| | O | | | | | A | | | E | | R | | | | | | P | | | I |
| | G | | | | | M | | | C | | O | | | | | | E | | | O |
| | E | | | | | E | | | O | | S | | | | | | R | | | S |
| | N | | | 8 D | | T | | | N | | S | | | | | | M | | | I |
| | E | | | I | | E | | | D | | I | | | | | | A | | | S |
| | S | | | P | | | | | A | | N | | | | | | T | | | |
| 9 B | I | V | A | L | E | N | T | | R | | G | | 10 H | A | 11 P | L | O | I | D | |
| | S | | | O | | | | | Y | | - | | O | | O | | G | | | |
| | | | | I | | | | | | | O | | M | | L | | E | | | |
| | | | | D | | | | | O | | V | | O | | A | | N | | | |
| | | | | | | | | | O | | E | | L | | R | | E | | | |
| | | | | | | | | | C | | R | | O | | | | S | | | |
| | | | | | | | | | Y | | | | G | | B | | I | | | |
| | | | | | | | | | T | | | | U | | O | | S | | | |
| | | | | | | | | | E | | | | E | | D | | | | | |
| | | | | | | | | | | | | | | | Y | | | | | |

## Chapter Test

**1.** c **2.** a **3.** b **4.** b **5.** b **6.** a **7.** d **8.** a **9.** a **10.** a **11.** a **12.** b **13.** a **14.** b **15.** a **16.** c **17.** e **18.** b **19.** d **20.** e **21.** Sexual reproduction results in genetic recombinations among offspring due to (1) crossing-over of nonsister chromatids, (2) independent assortment of homologues, and (3) fertilization. Certain recombinations may result in a variation that makes the organism more suited to the environment. **22.** By independent assortment of homologues, the species with six chromosome pairs can produce 64 different combinations of chromosomes by meiosis. The species with only two pairs can produce only four, and therefore lacks the same potential for variety.

# 11

# Mendelian Patterns of Inheritance

## Chapter Review

At the time Mendel started his work, the blending concept of inheritance was prevalent. Mendel disproved this concept through well-designed experiments that offered statistical evidence. By analyzing the 3:1 results among the $F_2$ generation of a monohybrid cross, Mendel arrived at the law of segregation: factors (alleles) for a trait occur in pairs in an organism; they separate into different sex cells during gamete formation. This explains why the recessive phenotype—absent in the $F_1$ generation—reappeared in the $F_2$ generation.

Solving genetics problems requires distinguishing between the **genotype** (genetic makeup) and **phenotype** (appearance) of an individual. For any pair of **alleles,** the **dominant allele** is given as an uppercase letter, and the **recessive allele** is given as a lowercase letter. A cross is done by using the laws of probability, most often by employing a **Punnett square,** which offers a mechanism whereby all possible types of sperm fertilize all possible types of eggs. The results can be expressed as the phenotypic ratio or can be used to state the chances of an individual having a particular genotype.

Mendel used a one-trait **testcross** to verify his law of segregation. A cross represented by $Aa \times aa$ offers the best chance of producing a recessive offspring. Today, the testcross is used to determine whether an individual is **heterozygous** or **homozygous** dominant.

The $F_2$ results of a dihybrid cross allowed Mendel to formulate his law of independent assortment: during gamete formation, the factors of one pair separate independently from the factors of other pairs. This law explains why the $F_2$ generation contained four types of genotypes—that is, all possible combinations of dominant and recessive characteristics. The Punnett square can also be used to solve two-trait problems, including the two-trait testcross.

Studies of human genetics have shown that many autosomal genetic disorders can be explained on the basis of simple Mendelian inheritance. When studying human genes, biologists often construct pedigree charts to show the pattern of inheritance of a characteristic within a family. The particular pattern indicates the manner in which a characteristic is inherited.

Tay-Sachs disease, cystic fibrosis, and PKU are autosomal recessive disorders that have been studied in detail. Neurofibromatosis and Huntington disease are autosomal dominant disorders that have been well studied.

There are many exceptions to Mendel's laws. The phenotypes of individuals arise from many kinds of genotypes and inheritance patterns. **Dominance,** for example, can be complete or **incomplete. Codominance** can also exist among the offspring of a genetic cross. Genes for some traits have **multiple alleles,** whereas, sometimes, genes interact through pleiotropy or epistasis. On the other hand, **polygenic inheritance** controls some traits. The relative effect of a genotype varies, depending on the influence of the environment.

## Study Exercises

Study the text section by section as you answer the questions that follow.

### 11.1 Gregor Mendel (pp. 182–83)

- Mendel discovered certain laws of heredity after doing experiments with garden peas during the mid-1800s.

1. When Mendel began breeding experiments, other breeders had different ideas about heredity. Place a check next to the statements that represent the ideas at that time:
   a. _____ A cross between a red flower and a white flower results in all offspring having red flowers.
   b. _____ A cross between a red flower and white flower results in some offspring having white flowers.
   c. _____ In a genetic cross, both parents contribute equally to the offspring.
   d. _____ Parents of contrasting appearance will produce offspring of intermediate appearance.
   e. _____ The blending concept of inheritance explained the results of genetic crosses.

2. Mendel's work reflected several methods and advantages that contributed to his success. Place a check next to the statements that represent those methods and advantages:
   a. _____ Each trait studied (e.g., seed shape, flower color) displayed many different phenotypes.
   b. _____ Breeding experiments have a statistical basis.
   c. _____ The garden pea plants used had a long generation time.
   d. _____ The plants used were easy to cultivate.
   e. _____ The plants used could not self-pollinate.

## 11.2 One-Trait Inheritance (pp. 184–88)

- A one-trait cross told Mendel that each organism contains two factors for each trait and the factors segregate during formation of gametes.
- Today, it is known that alleles located on chromosomes control the traits of individuals and homologous pairs of chromosomes separate during meiosis I.

3. Mendel arrived at the law of segregation by interpreting the results of his one-trait crosses. Place a check next to the interpretations that he used.
   a. _____ $F_1$ organisms contain one copy for each hereditary factor.
   b. _____ Factors segregate when gametes form.
   c. _____ Gametes fuse randomly during fertilization.
   d. _____ Allelic pairs assort in a dependent manner.
4. The length of stem in the plants that Mendel studied had two alleles: *T* (tall) and *t* (short). Using these letters, write the alleles for the heterozygous genotype [a]______________, the homozygous dominant genotype [b]______________, and the homozygous recessive genotype [c]______________.
5. When Mendel crossed true breeding tall plants with true breeding short plants, the $F_1$ plants were [a]______________. When he crossed $F_1 \times F_1$, the offspring included [b]______________ plants for every [c]______________ plant. Because some of the $F_2$ plants were short, he concluded that the $F_1$ generation was *Tt;* therefore, each original parent plant had passed on only one factor. Mendel's law of segregation states: [d] ____

________________________________________________________________________________

________________________________________________________________________________

6. a. Complete the following table to show the difference between genotype and phenotype:

| Genotype | Genotype | Phenotype |
|---|---|---|
| *TT* | ______________ | ______________ |
| _____ | heterozygous | ______________ |
| *tt* | ______________ | ______________ |

   If a plant's phenotype is short, its genotype(s) can be [b]______________.

   If a plant's phenotype is tall, its genotype(s) can be [c]______________.
7. Among humans, the allele for dark hair (*D*) is dominant to the allele for blonde hair (*d*). Consider the cross *Dd* × *Dd*. To answer these questions, use fractions except when asked for a percentage.
   a. What is the chance that either parent will produce a gamete with a dominant allele? ______________
   b. Using the multiplicative law of probability, calculate the chance of a homozygous dominant offspring (dark hair). Show your work. ______________
   c. What is the chance this couple will have a homozygous dominant offspring? ______________
   d. What is the chance either parent will produce a gamete with a recessive allele? ______________
   e. Using the multiplicative law of probability, calculate the chance of a homozygous recessive offspring (blonde hair). Show your work. ______________________________
   f. What is the chance this couple will have a homozygous recessive offspring? ______________

g. Using the multiplicative law and additive law, calculate the chance of a heterozygous offspring (dark hair).

______________________________________________

h. What is the chance this couple will have a heterozygous offspring? ______________

i. Using the additive law, calculate the chance of an offspring with the dominant phenotype (dark hair).

______________________________________________

j. Your calculations indicate that the phenotypic ratio for this cross is ¾ dark hair to ¼ blonde hair, or a phenotypic ratio of ______________________.

8. a. In peas, yellow seed color is dominant to green. The key is: *Y* = ______________, *y* = ______________.

b. Fill in this Punnett square for the cross *Yy* × *Yy*.

| | | |
|---|---|---|
| | | |
| | | |

c. The genotypic ratio among the offspring from this cross is: ______________________.

d. The phenotypic ratio among the offspring from this cross is: ______________________ (yellow:green).

9. The gametes combine at [a]______________, and usually, a [b]______________ number must be counted before a 3:1 ratio can be verified.

## One-Trait Testcross (p. 188)

- A testcross can determine the genotype of an individual with the dominant phenotype.

10. Researchers do a [a]______________ (i.e., the dominant phenotype is mated to the recessive phenotype) to determine if the dominant phenotype is homozygous or heterozygous. If the individual is homozygous dominant, the $F_1$ generation is expected to be [b]______________. If the individual is heterozygous, a phenotypic ratio of [c]______________________ is expected.

11. In humans, widow's peak (*W*) is dominant to straight hairline (*w*). Consider the cross *Ww* × *Ww*. The chance of a child with widow's peak is [a]______________ %, and the chance of a child with straight hairline is [b]______________ %.

12. Consider the cross *Ww* × *ww*. The chance of a child with widow's peak is [a]______________ %, and the chance of a child with straight hairline is [b]______________ %.

13. Among humans, dark eyes (*B*) is dominant to blue eyes (*b*). In a family, one parent has dark eyes and the other has blue eyes. Among their offspring, two develop dark eyes and two develop blue eyes. Most likely, the genotypes of the parents are [a]______________ (dark-eyed parent) and [b]______________ (blue-eyed parent).

14. In fruit flies, a cross between long-winged (*L*) flies and short-winged (*l*) flies produces only long-winged flies. Most likely, the genotypes of the parental flies are [a]______________ (long-winged parents) and [b]______________ (short-winged parents).

## 11.3 Two-Trait Inheritance (p. 189)

- A two-trait cross told Mendel that every possible combination of factors is present in the gametes.
- Today, it is known that homologous pairs of chromosomes separate independently during meiosis I, producing all possible combinations of alleles in the gametes.

15. In pea plants, $T$ = tall and $t$ = short, $G$ = green pods, and $g$ = yellow pods. When Mendel crossed homozygous tall plants having green pods with pure, short plants having yellow pods, the $F_1$ plants all had the genotype a.______________ and the phenotype b.______________. If $T$ always stayed with $G$ and $t$ always stayed with $g$ in the gametes, then how many different phenotypes would be among the $F_2$ plants? c.______________ Mendel observed four different phenotypes and formulated his law of independent assortment, which states: d. ________________________________________

16. The process of meiosis explains the law of segregation and the law of independent assortment. Considering the movement of chromosomes, why is only one allele for each trait present in the gametes? a. ______________ Why are all combinations of alleles present in the gametes? b. ______________

### Two-Trait Genetics Problems (p. 191)

- A testcross can determine the genotype of an individual who is dominant in two traits.

17. In horses, black ($B$) is dominant to brown ($b$) and a trotter ($T$) is dominant to a pacer ($t$). Use fractions in your answers.

    Consider the cross $Bb \times bb$. Among the offspring, the chance of black coat is a.______________ and the chance of brown coat is b.______________.

    Consider the cross $Tt \times Tt$. Among the offspring, the chance of a trotter is c.______________ and the chance of a pacer is d.______________.

    Consider the cross $BbTt \times bbTt$, and use the multiplicative law of probability to determine the chances of the following:

    black trotter e.______________

    black pacer f.______________

    brown trotter g.______________

    brown pacer h.______________

    What is the phenotypic ratio expected for the preceding cross?

    i.______________

    Check your answer by doing a Punnett square. j.

| | | |
|---|---|---|
| | | |
| | | |
| | | |
| | | |

18. Given the cross $BbTt \times BbTt$, determine the ratio and phenotypes expected.____________ ____________:
____________ ____________:____________ ____________:____________ ____________

19. a. Do a Punnett square for the cross $BbTt \times Bbtt$.

|  |  |  |
|---|---|---|
|  |  |  |
|  |  |  |
|  |  |  |
|  |  |  |

b. What is the phenotypic ratio among offspring? ________________________:____________________
________________________:________________________

20. In rabbits, black (*B*) is dominant to brown (*b*), and spotted coat (*S*) is dominant to solid coat (*s*). A black, spotted rabbit is mated to a brown, solid one, and all ten of their offspring are black and spotted. The genotypes of the parents are [a]____________ and [b]____________.

21. In humans, widow's peak (*W*) is dominant to straight hairline (*w*), and short fingers (*S*) are dominant to long fingers (*s*).

If the two parents are heterozygous in both regards, what is the chance of any offspring having widow's peak and short fingers? [a]____________

If one parent is heterozygous in both regards and the other is homozygous recessive, what is the chance of an offspring with widow's peak and short fingers? [b]____________ or [c]____________%.

## 11.4 Human Genetic Disorders (p. 193)

- Many genetic disorders are inherited according to the laws first established by Gregor Mendel.
- The pattern of inheritance indicates whether the disorder is a simple recessive disorder or an autosomal dominant disorder.

22. A man who is heterozygous for neurofibromatosis reproduces with a woman who is normal. The chances (percent) of the offspring having neurofibromatosis are [a]____________ and of the offspring being normal are [b]____________.

23. An unaffected man carrying the allele for PKU reproduces with an unaffected woman carrying the same allele. The chances (percent) of the offspring having PKU are [a]____________ and of the offspring being normal are [b]____________.

24. Match the descriptions to these types of disorders.
    1. cystic fibrosis
    2. Huntington disease
    3. neurofibromatosis
    4. phenylketonuria (PKU)
    5. Tay-Sachs disease

a. ____ lysosomal storage disease
b. ____ benign tumors in skin or deeper
c. ____ progressive nervous system degeneration
d. ____ disorder affecting function of mucous and sweat glands
e. ____ essential liver enzyme deficiency

25. Answer the questions for the following pedigree:

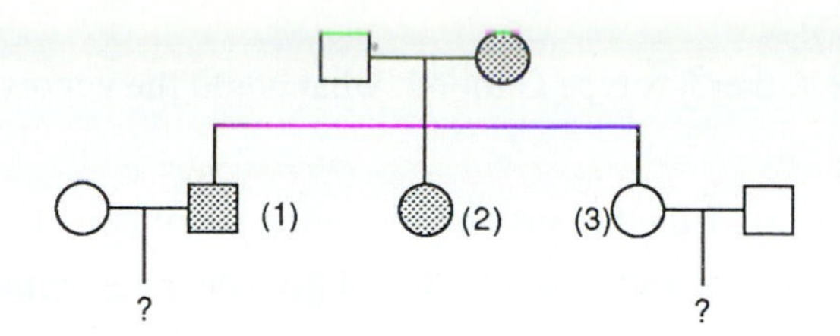


a. What is the mode of inheritance shown in this pedigree? _____________

b. What is the genotype of person 1? _____________

c. What are the chances of person 1 having normal children? _____________

d. What are the chances of person 3 having normal children? _____________

26. Answer the questions for the following pedigree:

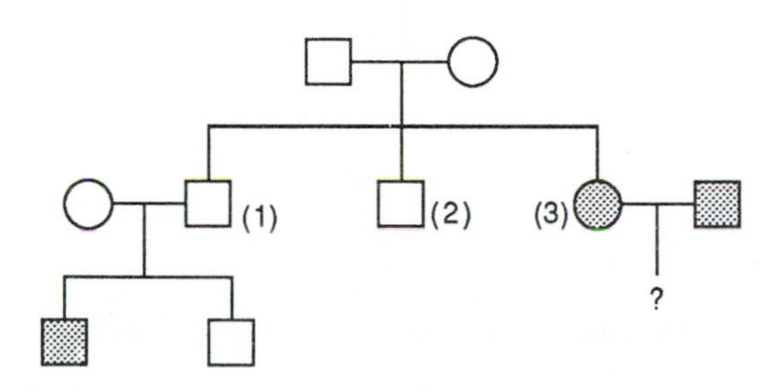


a. What is the mode of inheritance in this pedigree? _____________

b. For person 2, the genotype is _____________.

c. For person 2, the phenotype is _____________.

d. For person 1, the genotype is _____________.

e. How did you determine this? _____________________________________________

f. What are the chances that person 3's children will be normal? _____________

## 11.5 Beyond Mendelian Genetics (pp. 196–99)

- The genotype should be expanded to include all the genes because genes often work together to control the phenotype.
- There are forms of inheritance that involve degrees of dominance, interactions, multiple alleles, and polygenes.
- Environmental conditions can influence gene expression.

27. When a curly-haired person reproduces with a straight-haired person and their children have wavy hair, this is an example of _________________________.

28. Both a man and a woman have sickle-cell trait. List all phenotypes among the offspring, as well as the chance (percent) of each occurring. a. _________________________, b. _________________________, c. _________________________.

29. A gene that affects more than one characteristic is an example of _____________.

30. A man with blood type A reproduces with a woman who has blood type B. Their child has blood type O. Give the genotype of all persons involved: man a. _____________, woman b. _____________, and child c. _____________.

31. If a child has AB blood and the father has type B blood, what could the genotype of the mother be?

______________

32. If a child has BO blood and the father has type O blood, what could the genotype of the mother be?

______________

33. Roan cattle have red hairs and white hairs because of codominance of two alleles. Using the key in the diagram that follows, cross a roan cow and a roan bull, and give the phenotypic ratio. [a.]______________ What ratio would have resulted if red hair was dominant over white hair? [b.]______________

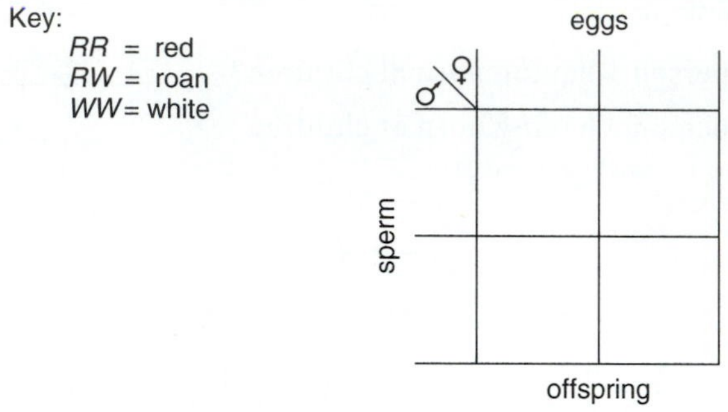

Key:
*RR* = red
*RW* = roan
*WW* = white

eggs
♂ ♀
sperm
offspring

34. An investigator notes that a population contains a range of phenotypes that fits a bell-shaped curve. What type of inheritance pattern is this? [a.]______________ The investigator decides that three pairs of alleles are involved. List all the possible genotypes for an intermediate phenotype. [b.] ______________________________
Explain your answer. [c.] ______________________________________________________________

______________________________________________________________

35. Consider a model in which there are three gene pairs of alleles; a dominant allele in any pair adds pigment to the skin. Use the letters *A, B, C* to indicate pigment formation and *a, b, c* to indicate lack of pigment formation.

a. What is the genotype for the darkest individual? ______________

b. What is the genotype for the lightest individual? ______________

c. What is the genotype for the offspring from a cross of the individuals from *a* and *b*? ______________

d. How does the skin color of this person compare to either of the parents? ______________

36. In humans, *A* = a normal amount of the pigment melanin and *a* = albinism (absence of pigment) and *B* = brown eyes and *b* = blue eyes.

a. What is the eye color for the genotype *BBAa*? ______________

b. What is the eye color for the genotype *bbaa*? ______________

c. If a man with the genotype *BBAa* reproduces with a woman having the genotype *bbaa,* what is the chance (percent) of producing a child with normal amounts of melanin? ______________

d. What is the chance of producing a child with albinism? ______________

37. In rabbits there are four alleles for coat color, but each rabbit has only two of these. What type of inheritance pattern is this? ______________

# CHAPTER TEST

## OBJECTIVE QUESTIONS

Do not refer to the text when taking this test.

____ 1. When Gregor Mendel began crossing plants, most breeders of organisms believed that
a. dominance was complete.
b. many genes affected one trait.
c. red and white flowers produced pink offspring.
d. the genetic material was always stable.

____ 2. Two different phenotypes result among the offspring from a genetic cross. The genotypes of the parents are
a. *TT* and *TT*.
b. *TT* and *Tt*.
c. *Tt* and *tt*.
d. *tt* and *tt*.

____ 3. The phenotypic ratio from a genetic cross is 1:1:1:1. The genotypes of the parents are
a. *TTGG* × *TtGg*.
b. *TtGG* × *Ttgg*.
c. *TtGg* × *ttgg*.
d. *Ttgg* × *ttgg*.

____ 4. What were the genotypes of the parents in the $F_1$ generation of Mendel's one-trait cross for stem length?
a. both *TT*
b. both *Tt*
c. *TT* and *Tt*
d. *Tt* and *tt*

____ 5. In guinea pigs, the allele for dark color (*B*) is dominant to the allele for light color (*b*). In a cross between two heterozygous organisms, the chance for producing a light-colored offspring is
a. 4/5.
b. 3/4.
c. 1/2.
d. 1/4.

____ 6. The two factors for each trait separate when gametes form, so that each gamete contains only one factor for each trait. This statement is part of Mendel's law of
a. dominance.
b. independent assortment.
c. random recombination.
d. segregation.

____ 7. Select the incorrect association.
a. alleles—*A* and *p*
b. heterozygous—*Aa*
c. homozygous—*DD*
d. homozygous—*pp*

____ 8. From the cross *Aa* × *Aa*, the probability of producing a homozygous dominant offspring is
a. 25%.
b. 33%.
c. 50%.
d. 75%.

____ 9. From the cross *Dd* × *Dd*, the probability of producing the dominant phenotype is
a. 25%.
b. 50%.
c. 75%.
d. 100%.

____ 10. How many kinds of gametes can an organism with genotype *AaBB* produce?
a. one
b. two
c. three
d. four

____ 11. Members of one pair of factors separate independently of the members of another pair of factors. This is a statement of Mendel's law of
a. dominant factors.
b. independent assortment.
c. random recombination.
d. segregation.

____ 12. Two organisms, each with the genotype *TtGg*, mate. The chance of producing an offspring that has the dominant phenotype for height and the recessive phenotype for color is
a. 9/16.
b. 7/16.
c. 6/16.
d. 3/16.

____ 13. In humans, *B* = short fingers and *b* = long fingers, *W* = widow's peak and *w* = straight hairline. From the cross *BbWW* × *Bbww*, the chance of an offspring having both dominant traits is
a. 9/16.
b. 3/4.
c. 3/8.
d. 1/4.

____ 14. Why is it that two normal parents could have a child with PKU?
a. PKU is a dominant inherited disorder.
b. PKU is a recessive inherited disorder.
c. PKU results due to an error in gamete formation.
d. There is no known explanation.

____ 15. If only one parent is a carrier for Huntington disease, what is the chance a child will have the condition?
a. 25%
b. 50%
c. 75%
d. no chance

____ 16. Polygenic inheritance can explain
a. a range in phenotypes among the offspring.
b. the occurrence of degrees of dominance.
c. the inheritance of behavioral traits.
d. Both *a* and *c* are correct.

___17. Sickle-cell disease illustrates
a. dominance.
b. recessiveness.
c. incomplete dominance.
d. multiple pairing.

___18. A female with light brown skin will be able to have a child with very dark skin if she reproduces with a very dark-skinned male, or a dark child if she reproduces with a light-skinned male.
a. true
b. false

___19. Children with which of the following blood types could not have parents who both have type A blood?
a. type A
b. type O
c. type AB
d. type B
e. Both *c* and *d* are correct.

___20. Inheritance by multiple alleles is illustrated by the inheritance of
a. skin color.
b. blood type.
c. sickle-cell disease.
d. Both *b* and *c* are correct.

For questions 21–26, match the types of genetic crosses to each of these examples:

a. multiple alleles
b. incomplete dominance
c. disorder dominant
d. recessive disorder
e. polygenic
f. two-trait cross

___21. $I^A i \times I^A I^B$
___22. neurofibromatosis
___23. skin color
___24. Tay-Sachs disease
___25. $YySs \times YySs$
___26. wavy hair × straight hair

## Critical Thinking Questions

Answer in complete sentences.

27. Discuss the concept that chance has no memory.

28. What evidence suggests that Huntington disease is not inherited as a simple autosomal dominant disorder?

**Test Results:** ______ number correct ÷ 28 = ______ × 100 = ______ %

## Exploring the Internet

The Online Learning Center at *www.mhhe.com/maderbiology8* has additional study material and practice quizzes that can help you master the content of this chapter. You can also find links to websites exploring additional topics in biology. Access to the Online Learning Center is free for those who have purchased a new textbook.

## Answer Key

### Study Exercises

**1.** c, d, e **2.** b, d **3.** b, c **4. a.** *Tt* **b.** *TT* **c.** *tt* **5. a.** tall **b.** three tall **c.** one short **d.** See page 184 in text **6. a.** See Table 11.1, page 185, in text **b.** *tt* **c.** *TT* or *Tt* **7. a.** ½ **b.** ½ × ½ = ¼ **c.** 1 out of 4, or 25% **d.** ½ **e.** ½ × ½ = ¼ **f.** 1 out of 4, or 25% **g.** chance of *Dd* = ½ × ½ = ¼; chance of *dD* = ½ × ½ = ¼; chance of heterozygous offspring is ¼ + ¼ = ½ **h.** 2 out of 4, or 50% **i.** ¼ + ½ = ¾ **j.** 3:1 **8. a.** yellow seed, green seed

**8. b.**

| | Y | y |
|---|---|---|
| Y | YY | Yy |
| y | Yy | yy |

**8. c.** 1:2:1 **d.** 3:1 **9. a.** random **b.** large **10. a.** testcross **b.** all dominant phenotypes **c.** 1:1 **11. a.** 75 **b.** 25 **12. a.** 50 **b.** 50 **13. a.** *Bb* **b.** *bb* **14. a.** *LL* **b.** *ll* **15. a.** *TtGg* **b.** tall with green pods **c.** two **d.** see page 189 in text **16. a.** homologous pairs separate **b.** homologous pairs separate independently **17. a.** ½ (50%) **b.** ½ (50%) **c.** ¾ (75%) **d.** ¼ (25%) **e.** ½ × ¾ = ⅜ **f.** ½ × ¼ = ⅛ **g.** ½ × ¾ = ⅜ **h.** ½ × ¼ = ⅛ **i.** 3:1:3:1
**j.**

| | bT | bt |
|---|---|---|
| BT | BbTT | BbTt |
| Bt | BbTt | Bbtt |
| bT | bbTT | bbTt |
| bt | bbTt | bbtt |

**18.** 9 black trotters: 3 black pacers: 3 brown trotters: 1 brown pacer
**19. a.**

| | Bt | bt |
|---|---|---|
| BT | BBTt | BbTt |
| Bt | BBtt | Bbtt |
| bT | BbTt | bbTt |
| bt | Bbtt | bbtt |

**b.** 3 black trotters: 3 black pacers: 1 brown pacer: 1 brown trotter **20. a.** *BBSS* **b.** *bbss* **21. a.** 9/16 **b.** ¼ **c.** 25 **22. a.** 50% **b.** 50% **23. a.** 25% **b.** 75% **24. a.** 5 **b.** 3 **c.** 2 **d.** 1 **e.** 4 **25. a.** autosomal dominant **b.** *Aa* **c.** 50% **d.** 100% **26. a.** autosomal recessive **b.** *AA* or *Aa* **c.** normal **d.** *Aa* **e.** because he has an affected child **f.** 0% **27.** incomplete dominance **28. a.** sickle-cell disease, 25 % **b.** sickle-cell trait, 50% **c.** normal, 25% **29.** pleiotropy **30. a.** $I^A i$ **b.** $I^B i$ **c.** *ii* **31.** $I^A I^A$ *or* $I^A i$ **32.** $I^B I^B$ *or* $I^B i$ **33. a.** 1:2:1 **b.** 3:1 **34. a.** polygenic **b.** *AaBbCc, AABbcc, aABBcc, aaBBCc, aabBCC, AAbbCc* **c.** All capitals have the same quantitative effect. **35. a.** *AABBCC* **b.** *aabbcc* **c.** A genotype that has any three capitals and any three lowercase letters. **d.** intermediate between the two **36. a.** brown eyes **b.** pink eyes (albinism) **c.** 50% **d.** 50% **37.** multiple alleles

## Chapter Test

**1.** c **2.** c **3.** c **4.** b **5.** d **6.** d **7.** a **8.** a **9.** c **10.** b **11.** b **12.** d **13.** b **14.** b **15.** b **16.** d **17.** c **18.** b **19.** e **20.** b **21.** a **22.** c **23.** e **24.** d **25.** f **26.** b **27.** The concept that chance has no memory refers to the idea that each pregnancy has the same probability as the previous one. If two heterozygous parents already have three children with a widow's peak and they are expecting a fourth child, this child still has a 75% chance of having a widow's peak and a 25% chance of having a straight hairline. **28.** The severity and time of onset can vary, and it appears that persons most at risk have inherited the disorder from their fathers. Analysis has revealed that Huntington disease is due to many repeats of the base triplet CAG. The more repeats, the earlier the onset and the more severe the symptoms.

# 12

# Chromosomal Patterns of Inheritance

## Chapter Review

Genes are located on the chromosomes, which explains the similarity of gene behavior during sexual reproduction. All chromosomes except one pair are called **autosomes—**the nonsex chromosomes. The other pair are the **sex chromosomes.** This pair determines the sex of the new individual. The father can contribute an **X chromosome** or a **Y chromosome** to his offspring, while the mother can only contribute an X chromosome. Therefore, the sex of the offspring is determined by the genetic contribution of the father.

The inheritance patterns of chromosomes explain sex determination in animals, as supported by Morgan's experiments. **X-linked** inheritance refers to genes located on the X chromosomes.

Because males normally receive only one X chromosome, they are subject to disorders caused by the inheritance of a recessive allele on the X chromosome. Well-known X-linked disorders are color blindness, muscular dystrophy, and hemophilia. Another X-linked disorder is fragile X syndrome.

All the genes on one chromosome form a **linkage group,** broken only when crossing-over occurs. Genes that are linked tend to go together into the same gamete. If crossing-over occurs, a two-trait cross involving dihybrids gives all possible phenotypes among the offspring, but the expected ratio is greatly changed. The frequency of recombinant gametes that occurs due to crossing-over has been used to map chromosomes.

Chromosomal mutations include changes in chromosome number. Changes in chromosome number include polyploidy, monosomy, and trisomy. The usual cause of monosomy and trisomy is nondisjunction during meiosis, which can result in an abnormal number of autosomes to be inherited, as in Down syndrome. Nondisjunction of the sex chromosomes can cause abnormal sex chromosome numbers in offspring, as in Turner syndrome, Klinefelter syndrome, poly-X, and Jacobs syndrome.

Chromosomal mutations also include changes in chromosome structure, such as deletions, translocations, duplications, and inversions. Examples of deletion syndromes are Williams and cri du chat syndromes, and an example of a translocation syndrome is Alagille syndrome.

Amniocentesis and chorionic villi sampling can provide fetal cells for the karyotyping of chromosomes.

## Study Exercises

Study the text section by section as you answer the questions that follow.

### 12.1 Chromosomal Inheritance (pp. 204–8)

- Certain chromosomes are called the sex chromosomes because they determine the sex of the individual.
- Certain traits, unrelated to sex, are controlled by genes located on the sex chromosomes.
- Sex-linked traits are usually carried on the X chromosome. Males, with only one X, are more likely to exhibit X-linked disorders.
- The pattern of inheritance can indicate whether the disorder is an X-linked disorder.

1. Match the descriptions with these terms:

   X and Y chromosomes  XX  homologous chromosomes  autosomes  XY

   a. ______________ sex chromosomes
   b. ______________ all chromosomes but the sex chromosomes
   c. ______________ pairs of chromosomes
   d. ______________ female
   e. ______________ male

2. Indicate whether the following statements about Morgan's findings with *Drosophila* are true (T) or false (F):

_____ a. An allele on the X chromosome was not found on the Y chromosome.

_____ b. *Drosophila* has the same sex chromosome pattern as humans.

_____ c. Carriers are $X^Rx^r$.

_____ d. White eyes are dominant to red eyes.

3. Bar eye in *Drosophila* is an X-linked characteristic in which bar eye (*B*) is dominant over normal eye (*b*). The genotype of a normal-eyed female is a. _____________, and the genotype of a bar-eyed male is b. _____________.

4. a. Draw a Punnett square to show the genotypes among the offspring from a cross of the flies in question 3.

Determine the chance (percent) of each of the following phenotypes:

b. bar-eyed males _____________

c. bar-eyed females _____________

d. normal-eyed males _____________

e. normal-eyed females _____________

5. If $X^B$ = normal vision and $X^b$ = color blindness, state the sex and the phenotype of each of these genotypes:

$X^BX^B$ a. _____________________

$X^BX^b$ b. _____________________

$X^bX^b$ c. _____________________

$X^BY$ d. _____________________

$X^bY$ e. _____________________

6. a. Why do more males than females have X-linked genetic disorders? _____________________

b. Why do sons inherit the disorder from their mothers? _____________________

7. a. Indicate the genotype of each person in the following pedigree chart. Use alleles *A* or *a* attached to an X chromosome in each case.

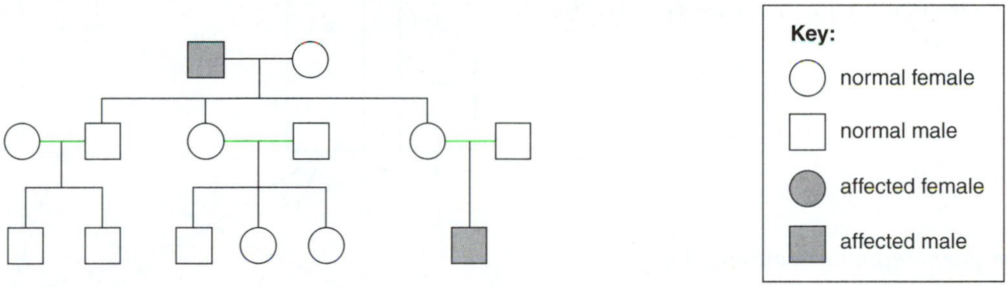

Key:
normal female
normal male
affected female
affected male

b. How do you know that this is a pedigree chart for an X-linked recessive trait? _____________________

8. Match the descriptions to these disorders:

hemophilia    muscular dystrophy    color blindness

a. _____________ muscle weakness

b. _____________ can't see reds and greens

c. _____________ bleeder's disease

9. Use the Punnett square at right to show the expected outcome if a color-blind woman reproduces with a man who has normal vision.

a. What are the chances of a color-blind daughter? ______________

b. What are the chances of a color-blind son? ______________

10. A son is color blind, but his mother and father are not color blind. Give the genotype of all persons involved.

son ____________________

mother ____________________

father ____________________

11. Hemophilia is an X-linked recessive disorder. A woman who is a hemophiliac reproduces with a man who is not. Use $h$ = recessive allele, $H$ = dominant allele:

a. The genotype of the woman is ______________.

b. The genotype of the man is ______________.

c. What is the genotype of sons from this cross? ______________

d. What is the genotype of daughters from this cross? ______________

e. What is the chance (percent) of the daughters' sons having hemophilia? ______________

f. What is the chance of the sons' sons (assume noncarrier wife) having hemophilia? ______________

12. Fragile X syndrome is due to the inheritance of a(n) [a]__________________________. The fragile location seems to be due to a multiple [b]______________ of the bases CGG.

## 12.2 Gene Linkage (pp. 209–15)

- Alleles on the same chromosome are said to be linked, and crossing-over makes it possible to determine the order of genes on the chromosome.

13. All the genes on one chromosome form a linkage group and tend to be inherited together. Mendel's law of independent assortment ______________ (does/does not) hold for linked genes.

In questions 14–15, consider that, in humans, arched eyebrow ($E$) is dominant over curved eyebrow ($e$) and hitchhiker thumb ($T$) is dominant over normal thumb ($t$). Imagine that these two genes are linked and that two dihybrids having these gametes reproduce.

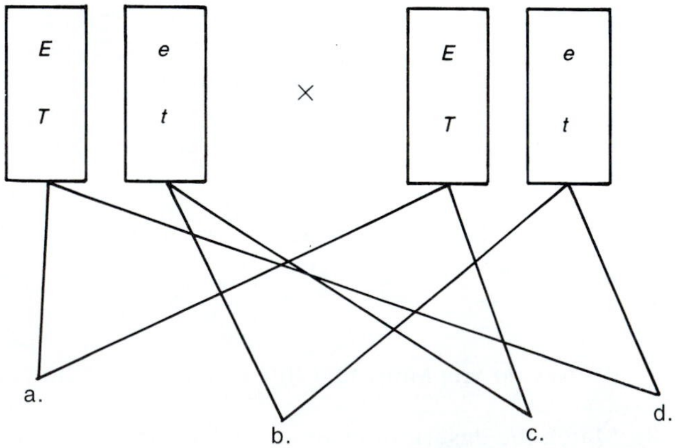

E T | e t × E T | e t

a. b. c. d.

14. From the diagram, indicate the phenotype for the following offspring:

a. ______________

b. ______________

c. ______________

d. ______________

15. What is the phenotypic ratio among the offspring? [a]______________ What would the ratio have been if the genes were on nonhomologous chromosomes? [b]______________

16. The percentage of crossing-over indicates the distance between two gene loci. For example, *A* and *B* recombine with *a* and *b* in 10% of the offspring, *A* and *C* recombine with *a* and *c* in 20% of the offspring, and *B* and *C* recombine with *a* and *c* in 10% of the offspring.

    How many map units are between the following?

    a. *A* and *C* _____________

    b. *A* and *B* _____________

    c. *B* and *C* _____________

    d. Draw a line and map the positions of *A*, *B*, and *C*.

## 12.3 Changes in Chromosome Number (pp. 212–13)

- Chromosomal mutations can be caused by a change in chromosome number.

17. The diploid chromosome number of a species is 24. Give the chromosome number if the species has the following conditions:

    a. pentaploid condition _____________

    b. tetraploid condition _____________

    c. condition with one trisomy _____________

    d. condition with one monosomy _____________

18. Indicate whether the following statements about Down syndrome are true (T) or false (F):

    a. _____ Characteristics include a wide, flat face and slanting eyelids.
    b. _____ In some cases, an extra copy of chromosome 21 is attached to chromosome 16.
    c. _____ Children of younger women are more likely to be affected than children of older women.
    d. _____ It most often occurs due to a nondisjunction.
    e. _____ The *Gart* gene may be involved in the mental retardation that accompanies Down syndrome.
    f. _____ Persons with the defect usually have three copies of chromosome 18.

19. Match the following conditions to each of the descriptions. Terms can be used more than once.

    1. Turner syndrome
    2. Klinefelter syndrome
    3. poly-X individual
    4. Jacobs syndrome
    5. fragile X syndrome

    a. _____ female with no apparent physical abnormalities
    b. _____ hyperactive or autistic as children, protruding ears as adults, mentally retarded
    c. _____ male with some breast development, large hands
    d. _____ XYY male
    e. _____ XXY male
    f. _____ XXX female
    g. _____ XO female
    h. _____ X chromosome is nearly broken, leaving its tip hanging

20. Identify the types of chromosomal mutations shown in the following illustration.

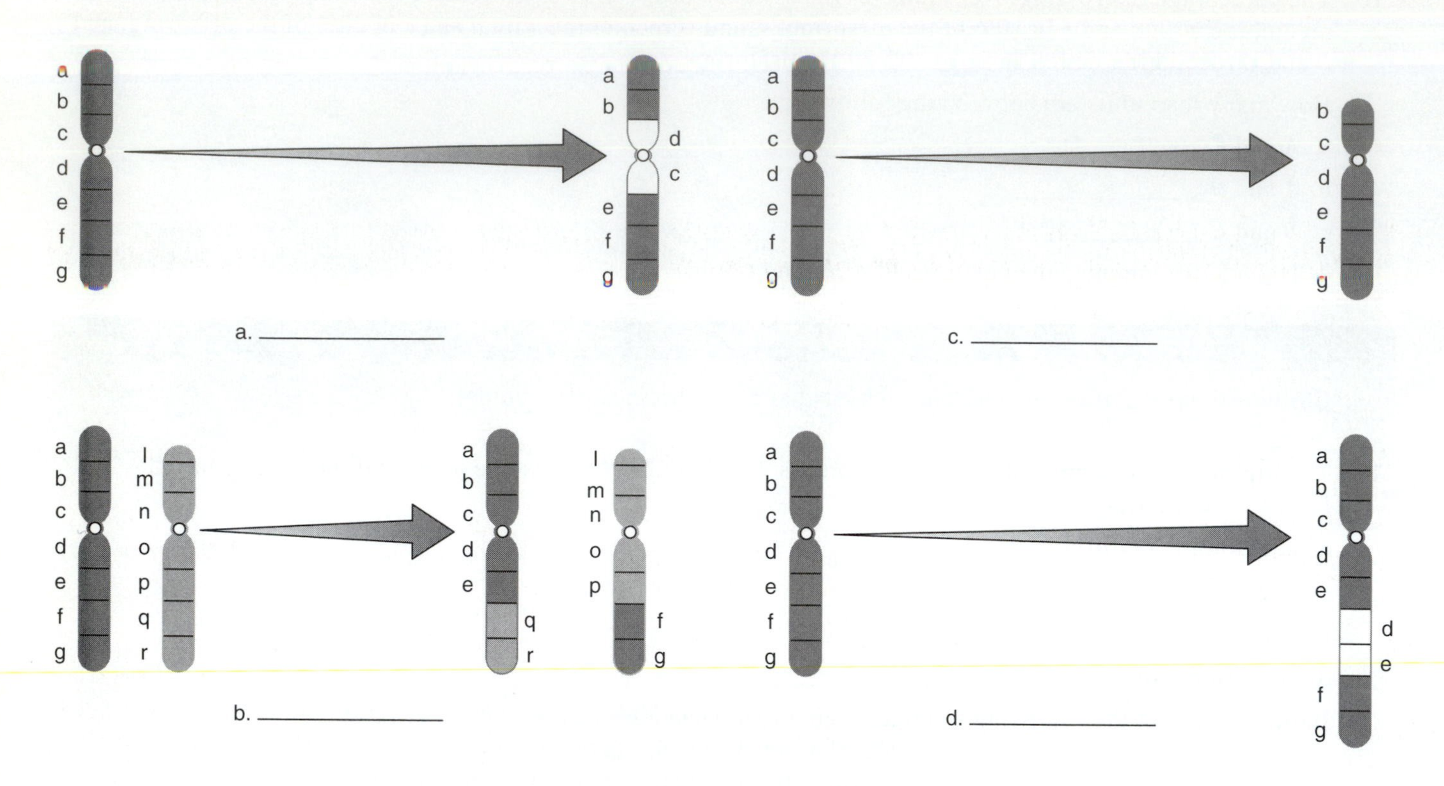


21. Label the photograph, using these terms:

autosomes   homologous pair   karyotype of a male   sex chromosomes

1 2 3 4 5
6 7 8 9 10
11 12 13 14 15
16 17 18 X Y
19 20 21 22

a. ______________

b. ______________
pair xy

c. ______________

d. ______________
pairs 1–22

22. Place A for amniocentesis and C for chorionic villi sampling on the appropriate lines.
   a. _____ 14 to 17 weeks of pregnancy
   b. _____ 5 weeks of pregnancy
   c. _____ Cells are obtained by suction.
   d. _____ Cells are obtained by needle.
   e. _____ Cells are from a cavity about the embryo/fetus.
   f. _____ Cells are from the extraembryonic membrane.

## DEFINITIONS WORDMATCH

Review key terms by completing this matching exercise, selecting from the following alphabetized list of terms:

amniocentesis
chromosomal mutation
karyotype
linkage group
nondisjunction
sex chromosomes
X-linked trait

a. _______________ Alleles on the same chromosome are linked in the sense that they tend to move together to the same gamete; crossing-over interferes with linkage.

b. _______________ Failure of homologues or sister chromatids to separate during the formation of gametes.

c. _______________ Variation in regard to the normal number of chromosomes inherited or in regard to the normal sequence of alleles on a chromosome; the sequence can be inverted, translocated from a nonhomologous chromosome, deleted, or duplicated.

d. _______________ Arrangement of all the chromosomes within a cell by pairs in a fixed order.

e. _______________ Procedure for removing amniotic fluid surrounding the developing fetus for the testing of the fluid or cells within the fluid.

f. _______________ Chromosome that determines the sex of an individual; in humans, females have two X chromosomes and males have an X and Y chromosome.

# Chapter Test

## Objective Questions

Do not refer to the text when taking this test.

____ 1. Which of the following statements is NOT true?
   a. Genes are on the chromosomes and behave similarly.
   b. The chromosomes, but not the alleles, occur in pairs in diploid cells.
   c. Both the chromosomes and the alleles of each pair separate independently during meiosis.
   d. Fertilization restores the diploid chromosome number and the paired condition for alleles in the zygote.

____ 2. Who determines the sex of offspring?
   a. male
   b. female
   c. both male and female
   d. alternately male, then female

____ 3. If $B$ = normal vision and $b$ = color blindness, then the genotype $X^BX^b$ is a
   a. male with normal vision.
   b. male with color blindness.
   c. female with normal color vision.
   d. carrier female with normal color vision.
   e. female who is color blind.

____ 4. Which represents the genotype of a carrier female?
   a. $X^AX^A$
   b. $X^AX^a$
   c. $X^aX^a$
   d. Both *b* and *c* are correct.

____ 5. If the female in question 4 reproduces with a normal male, then in the $F_1$ generation,
   a. all males will have the recessive phenotype, and all females will be normal.
   b. a male will have a 50% chance of the recessive phenotype, and a female will have a 50% chance of being a carrier.
   c. 50% of males will have the recessive phenotype, and 50% of females will have the recessive phenotype.
   d. The answer depends on whether the father is heterozygous.

____ 6. A male in the P generation has the recessive X-linked condition, but all $F_1$ offspring are normal. Why would you expect some of the $F_2$ offspring to have the condition?
   a. The $F_1$ males and females are heterozygous and carriers.
   b. All of the $F_1$ females are carriers.
   c. Half of the $F_1$ females are carriers.
   d. $F_2$ males always have the condition like their grandfather.

Questions 7 and 8 pertain to this pedigree chart.

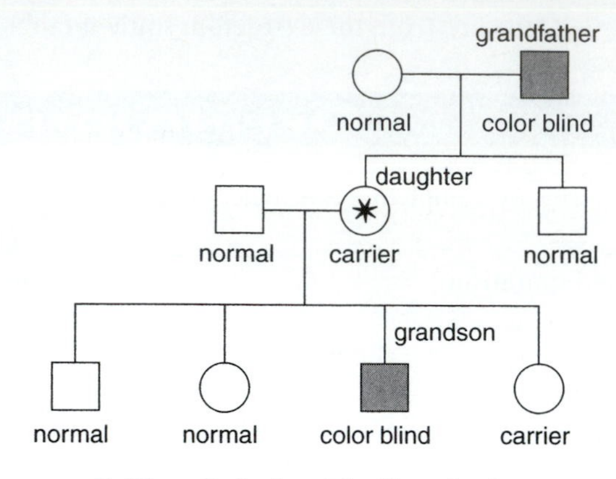


____ 7. The allele for this disorder is
   a. dominant.
   b. recessive.
   c. X-linked and recessive.
   d. None of these is correct.

____ 8. The genotype of the starred individual is
   a. *Aa*.
   b. *aa*.
   c. $X^AX^a$.
   d. $X^AY^a$.

____ 9. The X-linked disorder prevalent among royal families of Europe at the turn of the century was
   a. muscular dystrophy.
   b. color blindness.
   c. hemophilia.
   d. fragile X syndrome.
   e. cri du chat syndrome.

____ 10. Which of the following is NOT considered an X-linked recessive disorder?
   a. fragile X syndrome
   b. color blindness
   c. hemophilia
   d. muscular dystrophy (some forms)
   e. Jacob syndrome

____ 11. Which of the following is NOT a characteristic of an X-linked recessive disorder?
   a. More males than females are affected.
   b. An affected son can have parents who have the normal phenotype.
   c. For a female to have the characteristic, her father must also have it.
   d. The characteristic often skips a generation from the grandmother to the granddaughter.
   e. If a woman has the characteristic, all of her sons will have it.

____ 12. Which of the following statements is NOT true about a linkage group?
a. It includes all alleles on one chromosome.
b. Traits controlled by linked genes tend to be inherited together.
c. If linkage is complete, a dihybrid produces only two types of gametes in equal proportions.
d. Incomplete linkage can be due to crossing-over between nonsister chromatids.

____ 13. Genes *A* and *B* are 8 chromosome map units apart. Genes *B* and *C* are 10 units apart. Genes *A* and *C* are 2 map units apart. The order of these genes on the chromosome is
a. *ABC*.
b. *ACB*.
c. *BAC*.
d. *CBA*.

____ 14. The diploid number of an organism is 36. Its pentaploid chromosome number is
a. 5.
b. 48.
c. 72.
d. 90.

____ 15. The haploid number of an organism is 16. The number of chromosomes in its body cells, if they have one trisomy, is
a. 3.
b. 17.
c. 33.
d. 46.

____ 16. Which phrase best describes the human karyotype?
a. 46 pairs of autosomes
b. one pair of sex chromosomes and 23 pairs of autosomes
c. X and Y chromosomes and 22 pairs of autosomes
d. one pair of sex chromosomes and 22 pairs of autosomes

____ 17. The gene arrangement on a chromosome changes from *ABCDEFG* to *ABCDEDEFG*. This is an example of
a. deletion.
b. duplication.
c. inversion.
d. linkage.

____ 18. Which of the following conditions is NOT an example of a chromosomal mutation?
a. inversion
b. translocation
c. deletion
d. duplication
e. linkage

____ 19. Which chromosomal mutation does NOT require the presence of another chromosome?
a. translocation
b. duplication
c. inversion
d. All of these are correct.

____ 20. Which type of chromosomal mutation occurs when two simultaneous breaks in a chromosome lead to the loss of a segment?
a. inversion
b. translocation
c. deletion
d. duplication

## Critical Thinking Questions

Answer in complete sentences.

21. Why is it evident that a gene for maleness exists on the Y chromosome among humans?

22. Why are color-blind women rare?

**Test Results:** ______ number correct ÷ 22 = ______ × 100 = ______ %

## EXPLORING THE INTERNET

The Online Learning Center at *www.mhhe.com/maderbiology8* has additional study material and practice quizzes that can help you master the content of this chapter. You can also find links to websites exploring additional topics in biology. Access to the Online Learning Center is free for those who have purchased a new textbook.

## ANSWER KEY

### STUDY EXERCISES

**1. a.** X and Y chromosomes **b.** autosomes **c.** homologous chromosomes **d.** XX **e.** XY **2. a.** T **b.** T **c.** T **d.** F **3. a.** $X^bX^b$ **b.** $X^BY$
**4. a.**

| | $X^B$ | Y |
|---|---|---|
| $X^b$ | $X^BX^b$ | $X^bY$ |
| $X^b$ | $X^BX^b$ | $X^bY$ |

**b.** 0% **c.** 100% **d.** 100% **e.** 0% **5. a.** female with normal vision **b.** female who is a carrier **c.** female who is color blind **d.** male with normal vision **e.** male who is color blind **6. a.** If a male inherits the recessive gene, he always has the disorder. **b.** Only mothers pass on an X chromosome to their sons. **7. a. top:** $X^aY$, $X^AX^A$ **middle:** $X^AX^A$, $X^AY$, $X^AX^A$, $X^AY$, $X^AX^a$, $X^AY$ **bottom:** $X^AY$, $X^AY$, $X^AY$, $X^AX^A$, $X^AX^A$, $X^aY$ **b.** Only males have the disorder, and it passes from grandfather to grandson by way of a female. **8. a.** muscular dystrophy **b.** color blindness **c.** hemophilia **9. a.** none **b.** 100% **10.** son $X^bY$, mother $X^BX^b$, father $X^BY$ **11. a.** $X^hX^h$ **b.** $X^HY$ **c.** $X^hY$ **d.** $X^HX^h$ **e.** 50% **f.** 0% **12. a.** abnormal X chromosome **b.** repeat **13.** does not **14. a.** arched, hitchhiker **b.** curved, normal **c.** arched, hitchhiker **d.** arched, hitchhiker **15. a.** 3 arched, hitchhiker: 1 curved, normal **b.** 9:3:3:1 **16. a.** 20 units **b.** 10 units **c.** 10 units
**16. d.**

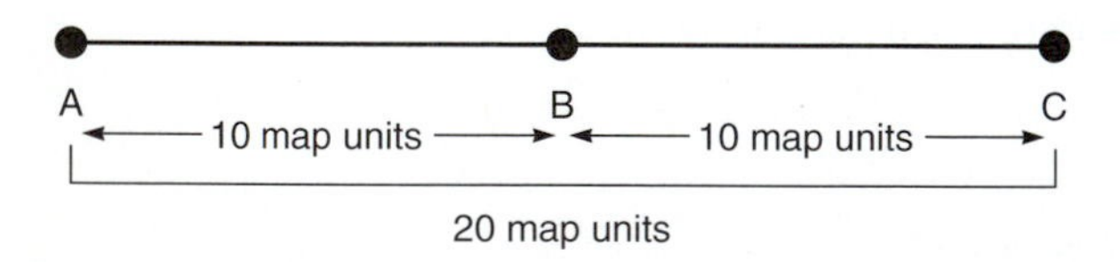


**17. a.** 60 **b.** 36 **c.** 25 **d.** 23 **18. a.** T **b.** F **c.** F **d.** T **e.** T **f.** F **19. a.** 3 **b.** 5 **c.** 2 **d.** 4 **e.** 2 **f.** 3 **g.** 1 **h.** 5 **20. a .** inversion **b.** translocation **c.** deletion **d.** duplication **21. a.** homologous pair **b.** sex chromosomes **c.** autosomes **d.** karyotype of a male **22. a.** A **b.** C **c.** C **d.** A **e.** A **f.** C

### DEFINITIONS WORDMATCH

**a.** linkage group **b.** nondisjunction **c.** chromosomal mutation **d.** karyotype **e.** amniocentesis **f.** sex chromosomes

### CHAPTER TEST

**1.** b **2.** a **3.** d **4.** b **5.** b **6.** b **7.** c **8.** c **9.** c **10.** e **11.** d **12.** d **13.** c **14.** d **15.** c **16.** c **17.** b **18.** e **19.** c **20.** c **21.** The presence of the Y chromosome in humans produces maleness and the absence of this chromosome leads to the absence of maleness. **22.** A color-blind woman has to receive an allele for color blindness from both parents. If she receives only one allele for color blindness and one normal allele, she will not be color blind.

# 13

# DNA Structure and Functions

## Chapter Review

Several experiments proved that **DNA** is the genetic material. Griffith's work revealed the presence of a transforming substance in the pneumococcus infecting mice. Avery and associates reported that the transforming substance was DNA. The Hershey and Chase experiments offered more convincing evidence for the genetic role of DNA.

Several lines of investigation contributed to a knowledge of DNA structure. Chargaff showed that the amount of **purine** (**adenine, guanine**) equals the amount of **pyrimidine** (**cytosine, thymine**). Franklin's X-ray diffraction analysis revealed the helical shape of the molecule.

Watson and Crick used the information gained from the experiments of others to build a model of DNA. Alternating sugar-phosphate molecules compose the sides of a ladder, with hydrogen-bonded base pairs composing the rungs. This ladder is twisted into a helix. This model also accurately predicted the model of DNA replication. As the helix unzips, each parental strand serves as the template for the synthesis of a new daughter strand. Through replication, each duplex produced is identical to the original double helix. This **semiconservative** mode of replication was demonstrated through the experiments of Meselson and Stahl. Replication in prokaryotes and eukaryotes is bidirectional along the chromosome, although the details of the process differ. Some **genetic mutations** due to DNA damage can be reversed by **DNA repair enzymes.**

## Study Exercises

Study the text section by section as you answer the questions that follow.

### 13.1 The Genetic Material (pp. 224–26)

- Early experiments showed that DNA is the genetic material.
- DNA is the genetic material, and therefore its structure and functions constitute the molecular basis of inheritance.
- DNA stores information that controls both the development and metabolism of a cell.

1. Check the descriptions that are requirements for a substance that is genetic material.
   a. _____ has the constancy to store information, and therefore serve as a blueprint for each generation
   b. _____ can be replicated and each new cell has a copy
   c. _____ can undergo mutations resulting in variability between species
   d. _____ cannot be transmitted from generation to generation so that veracity exists
   e. _____ conducts photosynthesis ensuring that each organism gets the energy it needs

2. Indicate whether these statements about Griffith's transformation experiments are true (T) or false (F).
   a. _____ The normal S strain was virulent.
   b. _____ The normal R strain was not virulent.
   c. _____ The heat-killed S strain was not virulent.
   d. _____ A mixture of heat-killed S strain and live R strain was virulent.

3. Based on your answers to question 2, what was Griffith's conclusion? ______________________________

   ______________________________________________________________________

4. Indicate whether these statements about Avery's transformation experiments are true (T) or false (F).
   a. _____ Action of a DNase on the transforming substance prevented transformation.
   b. _____ The transforming substance had many nucleotides.
   c. _____ Protein from S strain bacteria transformed R strain bacteria.
   d. _____ RNA was the transforming substance.
5. Based on your answers to question 4, what was Avery's conclusion?

______________________________________________________________________

In question 6, fill in the blanks.

6. The following diagram shows that the two separate experiments used $^{32}P$ to label a. ______________ and $^{35}S$ to label b. ______________ of viruses. In each experiment, the viruses were allowed to infect bacteria, and then a blender was used to separate the capsids from the bacteria. Radioactivity was found inside the cell only when c. ______________ was labeled. Replication of viruses followed, so the hypothesis that d. ______________ is the genetic material was supported.

   For *e–i,* tell the location of radioactivity in the diagram on the lines provided.

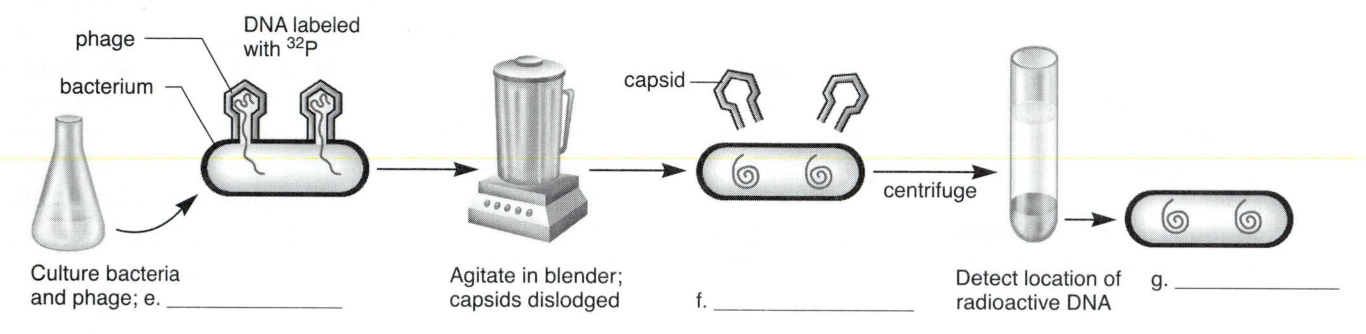


**a. Viral DNA is labeled**

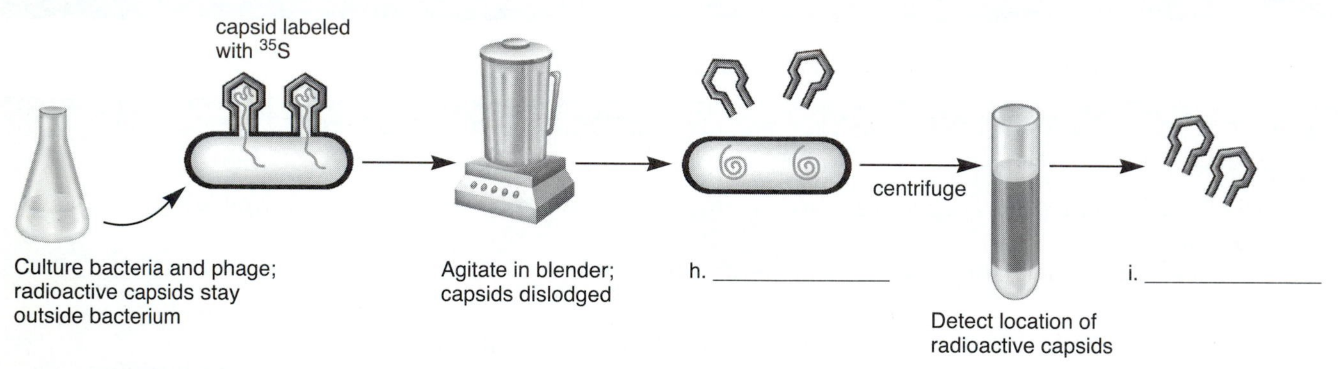


**b. Viral capsid is labeled**

## 13.2 The Structure of DNA (pp. 227–29)

- DNA is double helix; each of the two strands is a polymer of four different nucleotides.
- Four nucleotides permit a great deal of variability, and the sequence of base pairs in DNA varies from gene to gene.
- Hydrogen bonding between complementary bases joins the two strands.

7. Four different nucleotides are found in DNA. Check the way(s) these nucleotides differ.
   a. _____ They differ in their sugar content.
   b. _____ They differ in their phosphate content.
   c. _____ They differ in their base content.
8. What are the four different nucleotide bases in DNA? ______________________________

9. Study the following table:

| Species | A | T | G | C |
|---|---|---|---|---|
| *Homo sapiens* | 31.0 | 31.5 | 19.1 | 18.4 |
| *Drosophila melanogaster* | 27.3 | 27.6 | 22.5 | 22.5 |
| *Zea mays* | 25.6 | 25.3 | 24.5 | 24.6 |
| *Neurospora crassa* | 23.0 | 23.3 | 27.1 | 26.6 |
| *Escherichia coli* | 24.6 | 24.3 | 25.5 | 25.6 |
| *Bacillus subtilis* | 28.4 | 29.0 | 21.0 | 21.6 |

a. Is the amount of each base constant between species? ______________

b. What is Chargaff's first rule? ____________________________________________

____________________________________________

c. With which requirement for the genetic material in question 1 (p. 107) do you associate this rule? _______

____________________________________________

d. What is constant within each species, as stated in Chargaff's second rule? ______________

____________________________________________

e. With which requirement for the genetic material in question 1 (p. 107) do you associate this rule? _______

____________________________________________

10. Examine this diagram, which shows the ladder structure of DNA.

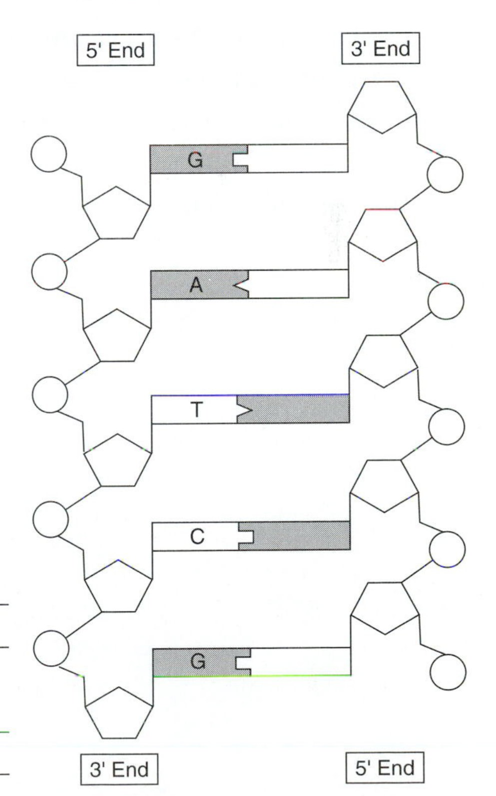


a. DNA is a polymer of ______________.

b. Draw a box around one nucleotide.

c. The molecules making up the sides of the ladder are ______________.

d. Label a sugar and a phosphate.

e. What is meant by the phrase *complementary base pairing?*

____________________________________________

f. Add the bases complementary to those on the left.

g. What do you have to do to the ladder structure to have it match the Watson and Crick model? ______________

h. Explain what is meant by *double-stranded helix.* ______________

____________________________________________

i. Explain what is meant by *antiparallel strands.* ______________

____________________________________________

## 13.3 Replication of DNA (pp. 230–33)

- DNA is able to replicate, and in this way genetic information is passed from one cell generation to the next.

11. Study this diagram of replication.

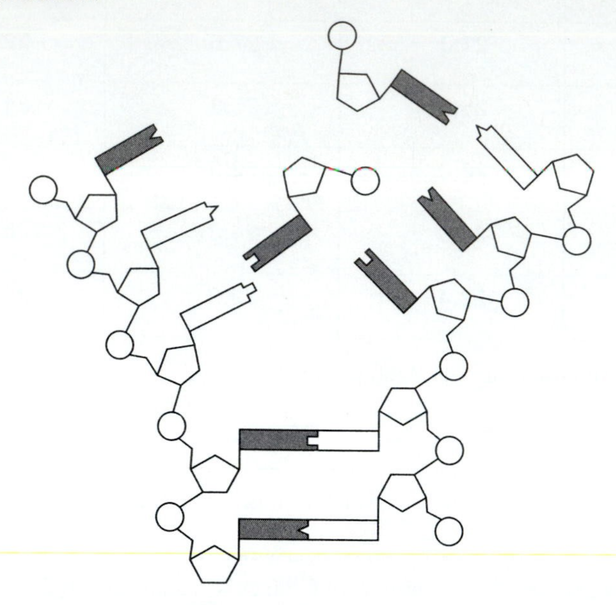

a. The bases in parental DNA are held together by what type of bond (not shown)? _____________

b. What happens to these bonds for replication to take place? _____________

c. During replication, new nucleotides move into proper position by what methodology? _____________

d. Elongation of DNA is catalyzed by an enzyme called _____________. When replication is finished there will be two DNA molecules.

e. Each double helix consists of an _____________ strand and a _____________ strand. Therefore, the process is called _____________.

f. Each double helix has _____________ (the same, a different) sequence of complementary paired bases.

12. Use the diagram to help you fill in the blanks, and complete the explanation of the Meselson and Stahl experiment.

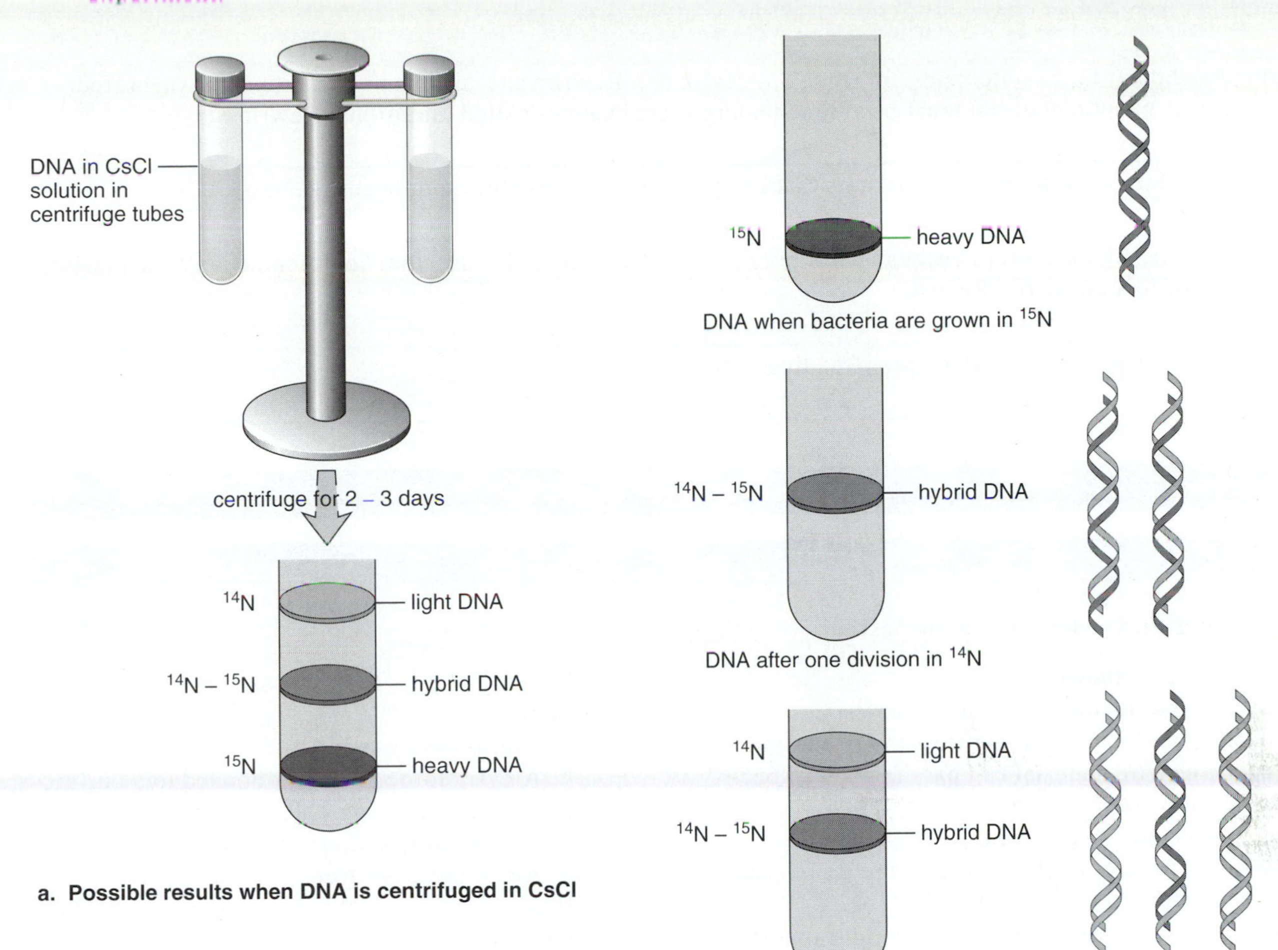


**a. Possible results when DNA is centrifuged in CsCl**

**b. Steps in Meselson and Stahl experiment**

First, Meselson and Stahl grew bacteria in a medium with a.______________ (see upper right). By this step, only b.______________ molecules of DNA were in the cells. When they switched the medium to c.______________, then d.______________ DNA molecules were in the cell after one cell division. After two divisions, half of the DNA molecules were e.______________, and the other half were f.______________.

g. Color the DNA molecules to be consistent with the description in the paragraph.

13. Meselson and Stahl concluded that DNA replication is a.______________, meaning that each daughter molecule contains a(n) b.______________ strand and a(n) c.______________ strand.

## Prokaryotic Versus Eukaryotic Replication (p. 223)

14. Place the appropriate letter(s) next to each statement to indicate DNA replication in:
P—prokaryotes E—eukaryotes P,E—both

a. _____ Cells complete replication in a matter of hours.
b. _____ Cells complete replication in a matter of minutes.
c. _____ It begins at numerous origins.
d. _____ It begins at one origin.
e. _____ It is bidirectional.
f. _____ A replication fork is involved.

## Replication Errors (p. 233)

- Mutations occur when there are errors during the replication process.

15. Indicate whether these statements are true (T) or false (F). Rewrite any false statements to make them true.
    a. _____ A genetic material must be able to undergo rare changes called mutations. Rewrite:
    ______________________________
    b. _____ Mutations introduce variations that can possibly cause evolution to occur. Rewrite:
    ______________________________
    c. _____ Mutations are rare because RNA polymerase checks to make sure that complementary base pairing occurs correctly. Rewrite:
    ______________________________
    d. _____ This process is called "proofreading." Rewrite:
    ______________________________

# Chapter Test

## Objective Questions

Do not refer to the text when taking this test.

_____ 1. Griffith concluded that
   a. bacteria do not have genetic material.
   b. the genetic material controls the phenotype.
   c. the genetic material can pass from dead bacteria to live bacteria.
   d. Both *b* and *c* are correct.

_____ 2. Which of these would be true of the transforming substance that Avery isolated?
   a. After isolation, it no longer could transform R strain bacteria into S strain bacteria.
   b. Its effect could be destroyed if subjected to ribonuclease, which degrades RNA.
   c. Its effect could be destroyed if subjected to digestion by trypsin, an enzyme that digests protein.
   d. Its effect could be destroyed if exposed to DNase.

_____ 3. Hershey and Chase found that
   a. the entire virus enters bacteria, so determining whether the capsid or the DNA controls replication of viruses is difficult.
   b. just the capsid enters bacteria and controls replication of viruses.
   c. just the DNA enters bacteria and controls replication of viruses.
   d. the capsid must be digested for DNA to control replication of viruses.

_____ 4. In a DNA molecule, the sugar
   a. bonds covalently to phosphate groups.
   b. bonds covalently to nitrogen-containing bases.
   c. is deoxyribose.
   d. All of these are correct.

_____ 5. Which of these is NOT true of complementary base pairing?
   a. A is always bonded to T.
   b. A pyrimidine is always bonded to a purine.
   c. The amount of A + T is always equal to the amount of G + C.
   d. All of these are true.

_____ 6. If the structure of DNA is compared to a ladder, then the
   a. sides of the ladder consist of phosphate and sugar.
   b. rungs of the ladder are hydrogen-bonded bases.
   c. ladder is twisted.
   d. All of these are correct.

_____ 7. During DNA replication,
   a. the nucleotides separate and reassemble, allowing genetic variability.
   b. the daughter molecules are just like the parental molecule so that constancy is maintained.
   c. one daughter molecule resembles the parental molecule, and one does not, so that variability and constancy are achieved at the same time.
   d. All of these are correct.

_____ 8. Semiconservative replication means that
   a. sometimes DNA can replicate and sometimes it cannot—this accounts for aging.
   b. sometimes daughter DNA molecules are exact copies of parental molecules and sometimes they are not, so that genetic variability may occur.
   c. a new DNA molecule consists of an old strand and a new strand.
   d. All of these are correct.

____ 9. X-ray diffraction data suggested that, in the ladder structure of DNA,
a. the sides are composed of bases, and the rungs are composed of phosphate and sugar molecules.
b. nucleotides have a different composition than previously thought.
c. DNA has a center from which the ladders project.
d. the ladder is twisted.

____10. Which is(are) correct regarding DNA?
a. C is paired with G.
b. The sugar is deoxyribose.
c. Hydrogen bonds exist between the bases.
d. All of these are correct.

____11. Before replication begins,
a. enzymes must be present.
b. the parental strands must unzip.
c. "free" nucleotides must be present.
d. All of these are correct.

____12. It is NOT required that the genetic material
a. be replicated.
b. handle energy.
c. store information.
d. undergo mutation.

____13. Griffith found that heat-killed S strains are
a. mobile.
b. not mobile.
c. not virulent.
d. virulent.

____14. Hershey and Chase used radioactive
a. C to label nucleic acids.
b. C to label protein.
c. N to label protein.
d. S to label protein.

____15. By Chargaff's rule,
a. G = A.
b. A = T.
c. C = T.
d. G = T.

____16. In the DNA double helix, if 20% of the bases are A, then ____________ of the bases are G.
a. 10%
b. 20%
c. 30%
d. 80%

____17. Franklin offered information about the DNA molecule's
a. base content.
b. length.
c. shape.
d. sugar content.

____18. Replication of DNA cannot begin until the helix
a. joins.
b. transcribes.
c. transposes.
d. unwinds.

____19. Each is true of replication in prokaryotes EXCEPT that it
a. does not produce replication forks.
b. is unidirectional.
c. proceeds from a single loop of DNA.
d. is relatively rapid.

____20. Each is true of replication in eukaryotes EXCEPT that it
a. begins at numerous origins of replication.
b. begins at replication forks.
c. is faster than that of prokaryotes.
d. precedes cell division in eukaryotes.

## Critical Thinking Questions

Answer in complete sentences.

21. What characteristics of proteins do you think might impede their use as the genetic material? Base your answer on the three requirements of the genetic material listed on page 224 of the text.

22. Why are mutations necessary to the process of evolution?

**Test Results:** ______ number correct ÷ 22 = ______ × 100 = ______ %

## Exploring the Internet

The Online Learning Center at *www.mhhe.com/maderbiology8* has additional study material and practice quizzes that can help you master the content of this chapter. You can also find links to websites exploring additional topics in biology. Access to the Online Learning Center is free for those who have purchased a new textbook.

## Answer Key

### Study Exercises

**1.** a, b, c **2. a.** T **b.** T **c.** T **d.** T **3.** Somehow, the virulence of the S strain was transferred to the R strain because the R strain had been transformed. **4. a.** T **b.** T **c.** F **d.** F **5.** DNA is the transforming substance and, therefore, the hereditary material. **6. a.** DNA **b.** capsids **c.** DNA **d.** DNA **e.** Radioactive DNA enters bacterium. **f.** Radioactivity stays within bacteria. **g.** radioactivity in precipitate **h.** Radioactivity stays within capsids **i.** Radioactivity in liquid medium (See also Figure 13.3, p. 226 in text.) **7.** c **8.** adenine (A), guanine (G), cytosine (C), thymine (T) **9. a.** no **b.** The quantity of A, T, C, and G varies from species to species. **c.** The genetic material can undergo mutations resulting in variability between species. **d.** The amount of A = T and the amount of G = C **e.** The genetic material has the constancy to store information so it can serve as a blueprint for each generation. **10. a.** nucleotides **b.** See figure at right. **c.** sugar (deoxyribose) and phosphate **d.** See figure at right. **e.** A binds with T, and G binds with C. **f.** See figure at right. **g.** twist it **h.** Each nucleotide polymer is a strand; when the ladder twists, a helix results. **i.** The strands run opposite to one another. **11. a.** hydrogen bond **b.** They become unzipped. **c.** complementary base pairing **d.** DNA polymerase **e.** old, new, semiconservative **f.** the same **12. a.** $^{15}N$ **b.** heavy **c.** $^{14}N$ **d.** hybrid **e.** light **f.** hybrid **g.** See Figure 13.8, page 231, in text **13. a.** semiconservative **b.** old **c.** new **14. a.** E **b.** P **c.** E **d.** P **e.** P, E **f.** E **15. a.** T **b.** T **c.** F Mutations are rare because DNA polymerase . . . **d.** T

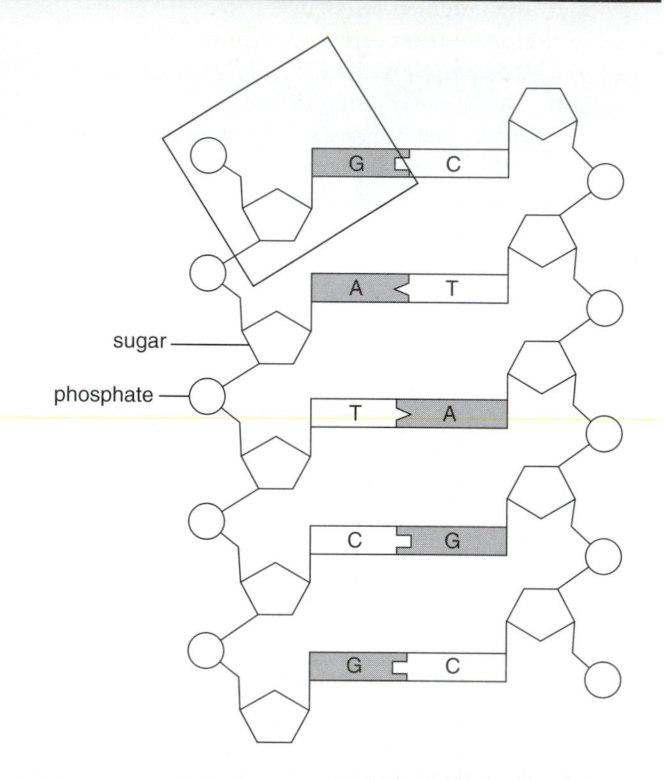


### Chapter Test

**1.** d **2.** d **3.** c **4.** d **5.** c **6.** d **7.** b **8.** c **9.** d **10.** d **11.** d **12.** b **13.** c **14.** d **15.** b **16.** c **17.** c **18.** d **19.** b **20.** c **21.** Proteins have numerous functions in cells; some have a structural role and some have a metabolic role. It seems unlikely, then, that they would also store genetic information. Genetic material should have a means of self-replicating and proteins have no such ability. A change in amino acid sequence could be a mutation, but how would such mutations be passed on to offspring? **22.** Mutations are the ultimate means of introducing variation into a population. Recombinations that occur during sexual reproduction can produce a different phenotype, but without mutations there could be no changes in the genetic material.

# 14

# Gene Activity: How Genes Work

## Chapter Review

The knowledge of gene activity arose from the experiments of several investigators. Garrod reasoned the basis for inborn errors of metabolism. Beadle and Tatum suggested the *one gene—one enzyme hypothesis*. Pauling and Itano refined this to the *one gene—one polypeptide hypothesis*.

RNA differs from DNA in several ways: (1) the pentose sugar is ribose, not deoxyribose; (2) the base uracil replaces thymine; and (3) RNA is single stranded. According to the central dogma of molecular biology, DNA is the template for its own replication and also for RNA formation. The sequence of bases in DNA specifies the proper sequence of amino acids in a polypeptide. The genetic code is a triplet code, and each codon (code word) consists of three bases. The code is almost universal among organisms.

Polypeptide (protein) synthesis requires transcription and translation. **Transcription,** the synthesis of an RNA off a DNA template in the nucleus, begins when RNA polymerase attaches to a promoter. Elongation of an RNA molecule occurs through the process of complementary base pairing until there is a stop DNA sequence. **Messenger RNA (mRNA),** which now carries codons, is processed before it leaves the nucleus; in particular, introns are removed.

**Translation,** the making of a polypeptide in the cytoplasm, requires several types of RNA. **Ribosomal RNA (rRNA)** and various proteins make up a ribosome where a polypeptide is formed. As a ribosome moves down an mRNA strand, the codons pair with the anticodons of **transfer RNAs (tRNA),** which bring amino acids to the ribosomes. Because of this process, the amino acids are joined according to the sequence of bases in DNA.

## Study Exercises

Study the text section by section as you answer the questions that follow.

### 14.1 The Function of Genes (pp. 238–40)

- Each gene specifies the amino acid sequence of one polypeptide of a protein, molecules that are essential to the structure and function of a cell.

1. Indicate whether these statements are true (T) or false (F).
   a. _____ Beadle and Tatum induced mutations in asexual haploid spores.
   b. _____ Beadle and Tatum proposed the one gene–one enzyme hypothesis.
   c. _____ Garrod was the first to suggest an association between genes and proteins.
   d. _____ Pauling and Itano showed that a mutation leads to a change in the structure of protein in hemoglobin.
   e. _____ Pauling and Itano proposed the one gene–one enzyme hypothesis.
2. The work of these investigators made it possible to conclude that ____________________________________.

## FROM DNA TO RNA TO PROTEIN (P. 240)

- The expression of genes leading to a protein product involves two steps, called transcription and translation.
- Three different types of RNA molecules are involved in transcription and translation.

3. Because genes (DNA) reside in the a.______________ of the cell and polypeptide (protein) synthesis occurs in the b.______________, they must have a go-between. The most likely molecule to fill this role is c.______________.

4. Indicate whether these statements about differences between DNA and RNA are true (T) or false (F).
   a. _____ DNA is double stranded; RNA is single stranded.
   b. _____ DNA is a polymer; RNA is a building block of that polymer.
   c. _____ DNA occurs in three forms; RNA occurs in only one form.
   d. _____ The sugar of DNA is ribose, which is absent in RNA.
   e. _____ Uracil, in RNA, replaces the base thymine, found in DNA.

5. Complete this table to describe the function of the various types of RNA involved in protein synthesis.

| RNA | Function |
|---|---|
| messenger RNA (mRNA) | a. ______________ |
| ribosomal RNA (rRNA) | b. ______________ |
| transfer RNA (tRNA) | c. ______________ |

6. Label this diagram, which pertains to the central concept of genetics.

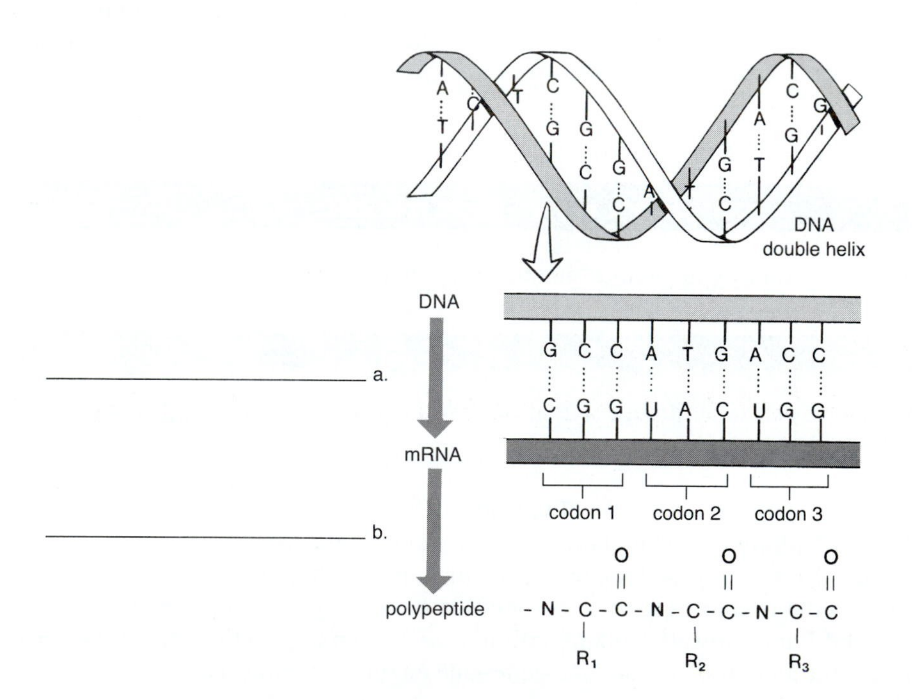

## 14.2 The Genetic Code (p. 241)

- The genetic code is a triplet code; each code word, called a codon, consisting of three nucleotide bases, stands for a particular amino acid of a polypeptide.

7. Study this figure, which lists the mRNA codons.

| First Base | Second Base | | | | Third Base |
|---|---|---|---|---|---|
| | U | C | A | G | |
| U | UUU phenylalanine | UCU serine | UAU tyrosine | UGU cysteine | U |
| U | UUC phenylalanine | UCC serine | UAC tyrosine | UGC cysteine | C |
| U | UUA leucine | UCA serine | UAA *stop* | UGA *stop* | A |
| U | UUG leucine | UCG serine | UAG *stop* | UGG tryptophan | G |
| C | CUU leucine | CCU proline | CAU histidine | CGU arginine | U |
| C | CUC leucine | CCC proline | CAC histidine | CGC arginine | C |
| C | CUA leucine | CCA proline | CAA glutamine | CGA arginine | A |
| C | CUG leucine | CCG proline | CAG glutamine | CGG arginine | G |
| A | AUU isoleucine | ACU threonine | AAU asparagine | AGU serine | U |
| A | AUC isoleucine | ACC threonine | AAC asparagine | AGC serine | C |
| A | AUA isoleucine | ACA threonine | AAA lysine | AGA arginine | A |
| A | AUG (*start*) methionine | ACG threonine | AAG lysine | AGG arginine | G |
| G | GUU valine | GCU alanine | GAU aspartic acid | GGU glycine | U |
| G | GUC valine | GCC alanine | GAC aspartic acid | GGC glycine | C |
| G | GUA valine | GCA alanine | GAA glutamic acid | GGA glycine | A |
| G | GUG valine | GCG alanine | GAG glutamic acid | GGG glycine | G |

a. What does it mean to say that the genetic code is a triplet code? ______________________

b. What are the mRNA codons for leucine? ______________________

______________________

c. What does it mean to say that the genetic code is degenerate? ______________________

d. What does it mean to say that the genetic code is unambiguous? ______________________

______________________

## 14.3 The First Step: Transcription (pp. 242–43)

- During transcription, a DNA strand serves as a template for the formation of an RNA molecule.

8. During transcription, an RNA molecule is formed that has a sequence of bases a.______________ to a portion of one DNA strand. The bases pair in this manner: A in DNA pairs with b.______________, and G pairs with c.______________ (and vice versa) in the mRNA being formed. If the sequence of bases in DNA is CGA AGC TCT, then the sequence in mRNA is d.______________________
Why is there a space between every three bases? e.______________________

______________________

______________________

9. a. Which one, exons or introns, is spliced out when primary RNA is processed? ______________

b. Which one, the cap at the 5′ end or the poly-A tail at the 3′ end, tells a ribosome where to attach? ______________

c. Which one, a spliceosome or a ribozyme, is an intron and also an enzyme? ______________

d. How is it possible for mRNA processing to produce different products in different cells? ______________

______________________________________________________________

## 14.4 The Second Step: Translation (pp. 244–48)

- During translation, the amino acids of a specific polypeptide are joined in the sequence directed by a type of RNA called messenger RNA.
- Transfer RNA (tRNA) molecules transfer amino acids to the ribosomes.

10. Features of tRNA structure include: At one end an a. ______________ attaches, and at the other end there is a(n) b. ______________, complementary to a codon in mRNA.

11. Features of rRNA structure include: Each ribosome is composed of a a. ______________ subunit and a b. ______________ subunit. Ribosomes have a binding site for c. ______________ and two d. ______________ at a time. Several ribosomes moving down the same mRNA plus the mRNA is called a e. ______________.

12. Label this diagram using the alphabetized list of terms.

amino acid
A site
codon
E site
large ribosomal subunit
mRNA
P site
small ribosomal subunit

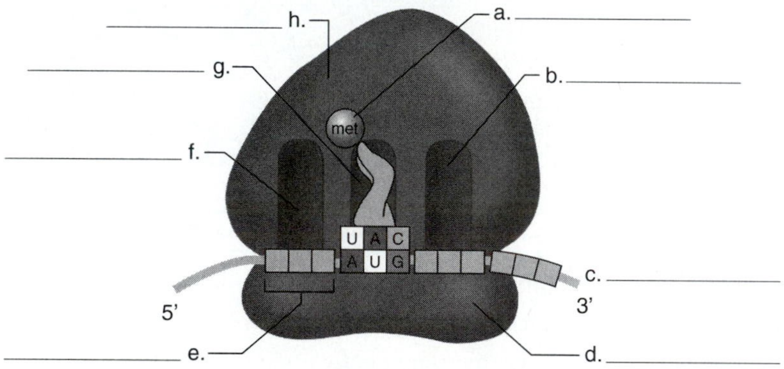

h. ______________
a. ______________
g. ______________
b. ______________
f. ______________
met
U A C
A U G
5'
3'
c. ______________
e. ______________
d. ______________

13. Use the diagram in question 12 to help you answer questions with regard to initiation.
During initiation, two ribosomal subunits come together to form a a. ______________. An initiator tRNA is at the P site of the ribosome. According to the diagram above, the A site is ready for a tRNA that has the anticodon b. ______________.

14. What are the three steps of translation? a. ______________ b. ______________ c. ______________

15. Use this diagram to help you answer questions with regard to elongation.

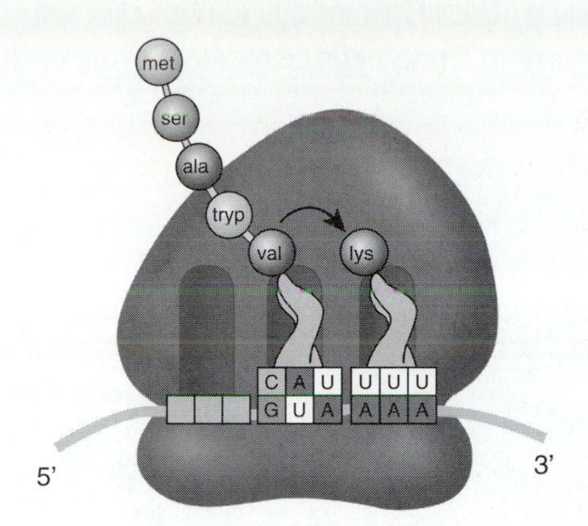


15. During elongation, the tRNA at the a._____________ site passes its peptide to the tRNA-amino acid at the b._____________ site. Then c._____________ occurs, and the spent tRNA leaves from the d._____________ site. The tRNA bearing the peptide moves to the P site and the A site is now empty. The next tRNA-amino acid arrives and this tRNA pairs with its e._____________. This sequence of events reoccurs until the polypeptide is complete.

16. Use this diagram to help you answer questions with regard to termination.

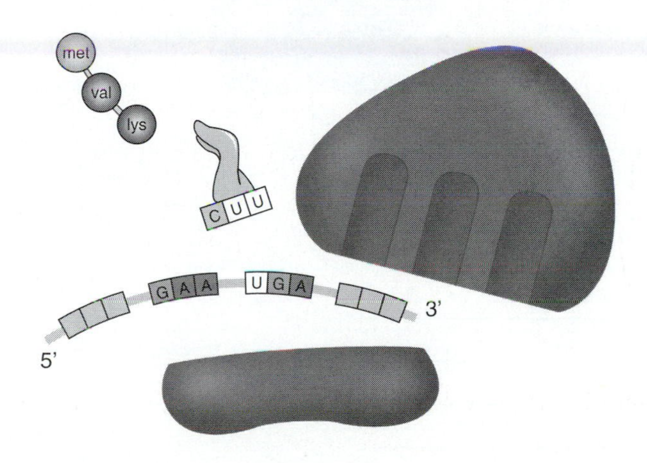


During termination, the ribosomal a._____________ separate, and the b._____________ is released from the last tRNA before the stop codon UGA.

# Polypeptide Synthesis Maze

Can you find your way through the maze to a polypeptide by identifying each of the components depicted?

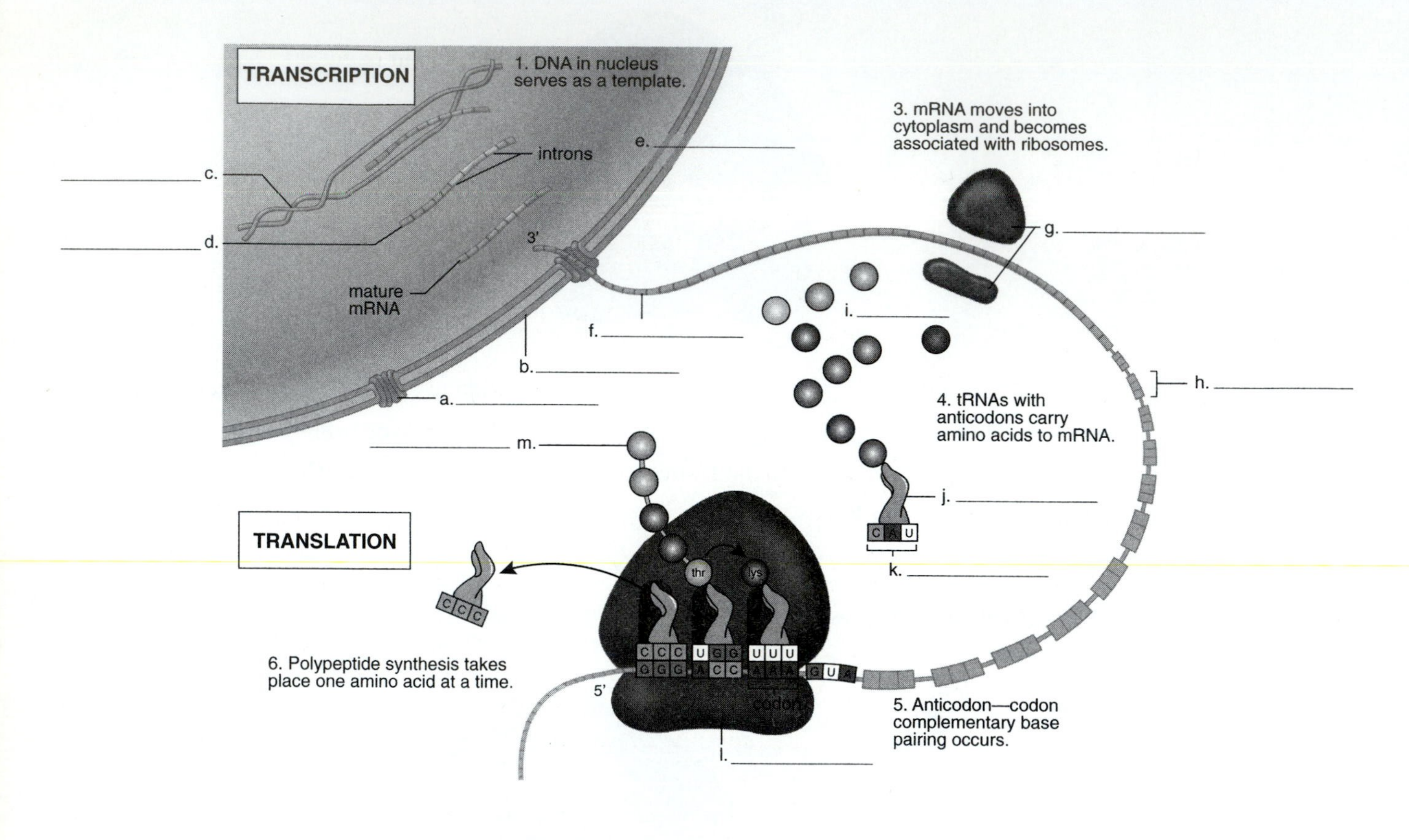


a. ______________________________

b. ______________________________

c. ______________________________

d. ______________________________

e. ______________________________

f. ______________________________

g. ______________________________

h. ______________________________

i. ______________________________

j. ______________________________

k. ______________________________

l. ______________________________

m. ______________________________

If you identified all correctly, you have found your way out.

# Chapter Test

## Objective Questions

Do not refer to the text when taking this test.

____ 1. Select the one characteristic that is NOT different between DNA and RNA.
- a. identity of the nucleotide sugar
- b. identity of one of the bases
- c. number of strands in the molecule
- d. solubility in water

____ 2. Select the incorrect association.
- a. mRNA—takes DNA message to the ribosome
- b. mRNA—takes amino acids to the ribosome
- c. rRNA—combines with protein in ribosomal subunits
- d. tRNA—has an anticodon

____ 3. Select the incorrect association.
- a. transcription—DNA synthesized
- b. transcription—RNA synthesized
- c. translation—occurs at the ribosome
- d. transition—polypeptide is made

____ 4. If each codon consisted of two bases, there would be ______________ different codons.
- a. 4
- b. 16
- c. 64
- d. 128

____ 5. The base sequence of DNA is ATAGCATCC. The sequence of RNA transcribed from this strand is
- a. ATAGCATCC.
- b. CCTACGATA.
- c. CCUACGAUA.
- d. UAUCGUAGG.

____ 6. An mRNA base sequence is UUAGCA. The two anticodons complementary to this are
- a. AAT CGT.
- b. AAU CGU.
- c. TTA GCA.
- d. UUA GCA.

____ 7. A DNA base sequence is 90 bases long. How many codons can this sequence order?
- a. 270
- b. 180
- c. 90
- d. 30

____ 8. An RNA base sequence is 120 bases long. How many anticodons can it order?
- a. 240
- b. 180
- c. 120
- d. 40

____ 9. During elongation,
- a. the P site, but not the A site, of a ribosome is required.
- b. the tRNA at the P site transfers the growing polypeptide to the tRNA–amino acid at the A site.
- c. several ribosomes are involved per polypeptide made.
- d. the amino acids line up and then are joined by peptide bonds.

____ 10. Which of the following pairs is NOT a valid comparison of DNA and RNA?

| | DNA | RNA |
|---|---|---|
| a. | double helix | single stranded |
| b. | replicates | replicates |
| c. | deoxyribose | ribose |
| d. | thymine | uracil |

____ 11. Which of these is true of an anticodon but is not true of a codon?
- a. part of an RNA molecule
- b. sequence of three bases
- c. part of a tRNA molecule
- d. part of a mRNA molecule

____ 12. RNA nucleotides are joined during transcription by
- a. helicase.
- b. DNA polymerase.
- c. RNA polymerase.
- d. ribozymes.

____ 13. Which statement is NOT true?
- a. Transcription in eukaryotes occurs in the nucleus.
- b. Introns are DNA segments found within a gene but not expressed.
- c. Exons are portions of a gene that are ultimately expressed.
- d. Ribozymes are protein enzymes that remove introns during RNA processing.

____ 14. Which of these is happening when translation takes place?
- a. mRNA is still in the nucleus.
- b. tRNAs are bringing amino acids to the ribosomes.
- c. rRNA is exposing its anticodons.
- d. DNA is being replicated.
- e. All of these are correct.

_____15. Which of these is true concerning translation?
a. Each polypeptide is synthesized one amino acid at a time.
b. The amino acids are joined by RNA polymerase at the same time.
c. Each ribosome is responsible for adding a single amino acid to each polypeptide.
d. The same type of polypeptide often contains a different sequence of amino acids.
e. All of these are true.

_____16. Which of the following is NOT correct?
a. mRNA is produced in the nucleus and processed in the cytoplasm.
b. Several ribosomes move along mRNA at a time.
c. DNA has a triplet code, and each triplet stands for an amino acid.
d. tRNA brings amino acids to ribosomes, where they contribute to polypeptide formation.

_____17. If the triplet code in DNA is TAG, what is the anticodon?
a. UTC
b. AUG
c. UAG
d. ATG

_____18. The substitution of histidine for tyrosine will have little effect if
a. the shape of the protein does not change.
b. the active site does not change.
c. histidine and tyrosine have similar properties.
d. All of these are correct.

## Critical Thinking Questions

Answer in complete sentences.

19. What is the significance of a universal genetic code throughout the domains of life?

20. Why is control of polypeptide (protein) synthesis advantageous to the cell, compared to other kinds of molecules?

**Test Results:** _____ number correct ÷ 20 = _____ × 100 = _____ %

## Exploring the Internet

The Online Learning Center at *www.mhhe.com/maderbiology8* has additional study material and practice quizzes that can help you master the content of this chapter. You can also find links to websites exploring additional topics in biology. Access to the Online Learning Center is free for those who have purchased a new textbook.

## Answer Key

### Study Exercises

**1. a.** T **b.** T **c.** T **d.** T **e.** F **2.** the genes determine the proteins of the cell. **3. a.** nucleus **b.** cytoplasm **c.** RNA. **4. a.** T **b.** F **c.** F **d.** F **e.** T **5. a.** takes a message from DNA in the nucleus to the ribosomes in the cytoplasm. **b.** is found in ribosomes, where polypeptides are synthesized. **c.** transfers amino acids to the ribosomes. **6. a.** transcription **b.** translation **7. a.** Every three bases stands for an amino acid. **b.** UUA, UUG, CUU, CUC, CUA, CUG **c.** There can be more than one codon for each amino acid. **d.** Each codon has only one meaning. **8. a.** complementary **b.** U **c.** C **d.** GCU UCG ACA **e.** The code is a triplet code and each codon contains three bases. **9. a.** introns **b.** cap **c.** ribozyme **d.** The product depends on which exons remain. **10. a.** amino acid **b.** anticodon **11. a.** small **b.** large **c.** mRNA **d.** tRNAs **e.** polyribosome **12. a.** amino acid **b.** A site **c.** mRNA **d.** small ribosomal subunit **e.** codon **f.** E site **g.** P site **h.** large ribosomal subunit **13. a.** ribosome **b.** CAU **14. a.** initiation **b.** elongation **c.** termination **15. a.** P **b.** A **c.** translocation **d.** E **e.** codon **16. a.** subunits **b.** polypeptide

## Polypeptide Synthesis Maze

**a.** nuclear pore **b.** nuclear envelope **c.** DNA **d.** primary mRNA **e.** mRNA processing **f.** mRNA **g.** large and small ribosomal subunits **h.** codon **i.** amino acids **j.** tRNA **k.** anticodon **l.** ribosome **m.** peptide

## Chapter Test

**1.** d **2.** b **3.** a **4.** b **5.** d **6.** b **7.** d **8.** d **9.** b **10.** b **11.** c **12.** c **13.** d **14.** b **15.** a **16.** a **17.** c **18.** d **19.** It shows that all organisms have a common origin. Related organisms share genetic characteristics. **20.** Polypeptides have a wide variety of functions in cells, ranging from structural roles to enzymatic activity. Other molecules are not this varied in their abilities.

# 15

# Regulation of Gene Activity and Gene Mutations

## Chapter Review

In prokaryotic cells, gene regulation usually occurs at the level of transcription. Examples are **operons**—the *trp* operon and the *lac* operon. The *trp* operon is a **repressible operon** because the repressor must bind with a **corepressor** (i.e., tryptophan) before the complex can bind to the operator and stop protein synthesis.In the *lac* **operon,** a repressor protein coded by a **regulator gene** ordinarily binds the **operator** so that RNA polymerase is unable to bind and transcription is therefore unable to take place. When lactose is present, it binds to the **repressor,** and then this combination is unable to bind to the operator. The *lac* operon is an **inducible operon.**

Eukaryotic cells have four levels of control (gene regulation): transcriptional, posttranscriptional, translational, and posttranslational. Transcriptional control includes the organization of the chromatin and the use of transcription factors. Posttranscriptional control includes differences in mRNA processing and the speed with which mRNA leaves the nucleus. Translational control pertains to the life expectancy of mRNA molecules, which can vary; some mRNAs may need modification before they can be translated. Posttranslational control includes feedback control of enzymes, as well as the possible need for additional changes before a protein is functional.

Mutations are changes in DNA nucleotide base sequences. Types of mutations include frameshrift and point mutations. Mutations can be spontaneous or due to an environmental mutagen. **Carcinogens** are mutagens that cause cancer.

**Cancer** is due to a series of genetic mutations among regulatory genes that control the cell cycle. These mutations activate oncogenes or deactivate tumor-suppressor genes, resulting in uncontrolled cell division that leads to a tumor.

## Study Exercises

Study the text section by section as you answer the questions that follow.

### 15.1 Prokaryotic Regulation (pp. 252–54)

- Regulator genes control the expression of genes that code for a protein product.

1. a. Label this diagram of a *lac* operon, using the alphabetized list of terms.

   mRNA
   operator
   promoter
   regulator gene
   repressor protein
   structural genes
   transcription is prevented

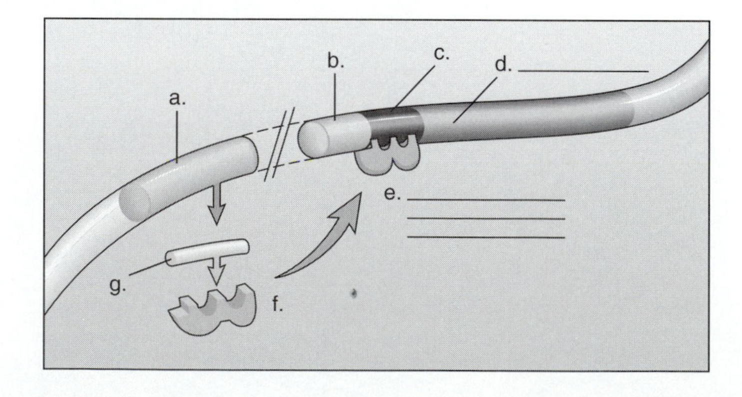

b. Which of these codes for a repressor? ____________

c. To which of these does RNA polymerase bind? ____________

d. To which of these does the repressor bind? ____________

e. Which of these codes for enzymes of the pathway? ____________

2. Cross out all portions of the diagram that are not in use if the *lac* operon is turned off.

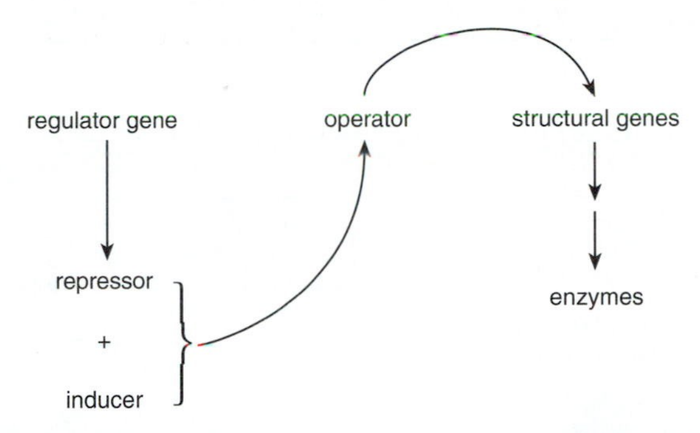


3. Cross out all portions of the diagram that are not in use if the *lac* operon is turned on.

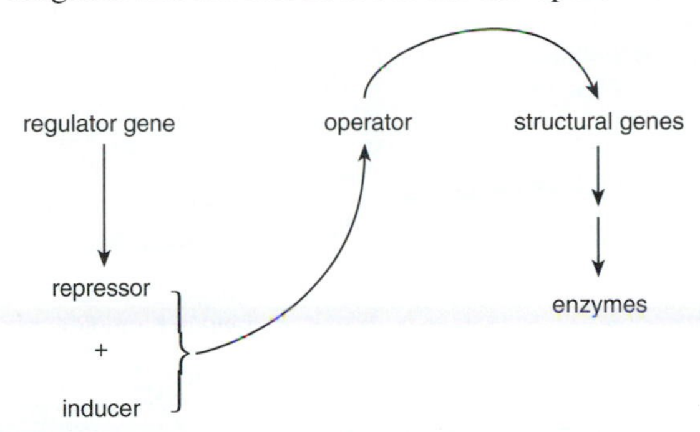


4. Put this sequence of events in order to describe how *E. coli* ensures that the *lac* operon is maximally turned on when glucose is absent.
   a. _____ Cyclic AMP builds up.
   b. _____ Now RNA polymerase is better able to bind to the promoter.
   c. _____ Cyclic AMP binds to catabolite activator protein (CAP).
   d. _____ The complex attaches to a CAP binding site next to the *lac* promoter.

5. a. With the *trp* operon, are the structural genes ordinarily turned on or off? ____________

   b. Why is the *trp* operon a repressible operon? ________________________________

## 15.2 Eukaryotic Regulation (pp. 255–60)

- The control of gene expression can occur at all stages from transcription to the activity of proteins in the eukaryotic cell.
- The structural organization of chromatin helps control gene expression in eukaryotes.

6. Complete the following table:

| Levels of Control of Gene Activity | Affects the Activity of |
|---|---|
| | |
| | |
| | |
| | |

7. Indicate whether these statements about euchromatin and heterochromatin are true (T) or false (F).
   a. _____ Decompacted euchromatin is inactive.
   b. _____ Looped euchromatin is actively being transcribed.
   c. _____ Highly compacted and condensed heterochromatin is inactive.
   d. _____ Heterochromatin is actively being transcribed.
8. a. Which statement in question 7 is supported by knowledge of Barr bodies? _____
   b. Why? _____
   _____
9. a. Which statement in question 7 is supported by knowledge of lampbrush chromosomes? _____
   b. Why? _____
10. Consider this diagram:

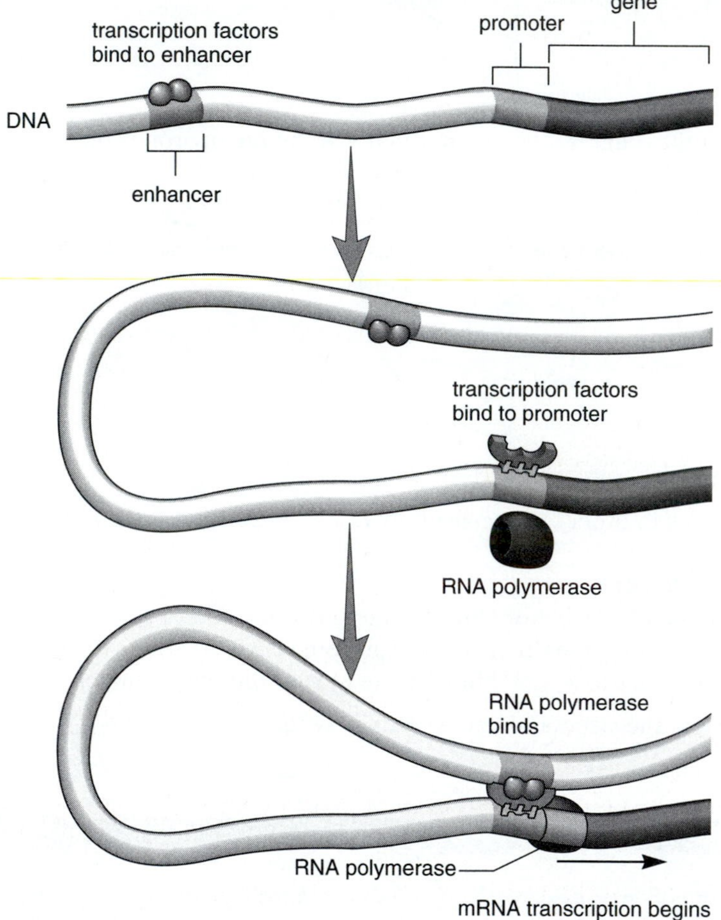


   a. What are transcription factors? _____
   _____
   _____
   b. Where do transcription factors bind? _____
   c. What conformational change occurs before transcription begins? _____
11. Questions 7–9 pertain to what level of genetic control in eukaryotes? _____
12. Place the appropriate letters next to each statement.

   PTC—posttranscriptional control  TL—translational control  PTL—posttranslational control

   a. _____ The product of a metabolic pathway binds to enzyme that speeds the first reaction of the pathway.
   b. _____ mRNA persists for different lengths of times in cells.
   c. _____ Estrogen interferes with ribonuclease activity in certain cells. Ribonuclease destroys mRNA.
   d. _____ Different patterns of mRNA splicing.

## 15.3 Genetic Mutations (pp. 260–62)

- Mutations occur when the nucleotide base sequence of DNA changes.
- Mutations can lead to proteins that do not function or do not function properly.
- Mutations of regulatory genes are now known to cause cancer.

13. The original base sequence is UACUACUAC.
    a. Name the mutation that reads UAUACUACU. ____________
    b. Name the mutation that reads UACUAGUAC. ____________
    c. Which of these two types of mutations causes sickle-cell disease? ____________
14. a. Aside from replication errors, what affects the rate of mutation? ____________
    b. How is DNA protected against mutations due to mutagens? ____________
15. a. What are carcinogens? ____________
    b. Name two environmental factors that are carcinogens. ____________
16. Match the statements to these terms:
    frameshift   point   mutagen   transposons
    a. ____________ type of mutation (requires two answers)
    b. ____________ environmental influence that causes mutations
    c. ____________ DNA sequence that can move between chromosomes
17. The mutation of a a.____________ to an oncogene can lead to a b.____________. When a tumor is at its place of origin, it is said to be c.____________. The development of cancer usually leads from a localized tumor to a d.____________ tumor, found at some distance from the e.____________ tumor.

    The *p53* gene is a f.____________ gene. The p53 protein acts as a g.____________ factor and can turn on expression of genes whose products are h.____________ inhibitors. i.____________, programmed cell death, can also be stimulated by the *p53* gene.
18. Place the appropriate letter(s) next to each statement. Two answers are required for each statement.
    P—proto-oncogenes   O—oncogenes   T—tumor-suppressor genes   MT—mutated tumor-suppressor gene
    a. _____ cell division is always promoted
    b. _____ normal genes in cells that regulate cell division
    c. _____ mutated genes that cause cancer

# Chapter Test

## Objective Questions

Do not refer to the text when taking this test.

_____ 1. A drug prevents the exit of mRNA from the nucleus. This control is
   a. transcriptional.
   b. posttranscriptional.
   c. translational.
   d. posttranslational.

_____ 2. Select the incorrect association.
   a. promoter—accepts RNA polymerase
   b. regulator—makes inducer
   c. repressor—binds to operator
   d. structural gene—transcriptional unit

_____ 3. Select the correct statement about the *lac* operon.
   a. It involves at least five enzymes.
   b. It is an inducible operon.
   c. It was first studied in *B. subtilis*.
   d. Tryptophan is the inducer in this operon.

_____ 4. Select the correct statement about the *trp* operon.
   a. It is an inducible operon.
   b. It was first studied in *B. subtilis*.
   c. Lactose is the inducer.
   d. Tryptophan is a corepressor.

_____ 5. Select the incorrect statement about heterochromatin.
a. It is found in Barr bodies.
b. It is genetically inactive.
c. Its pattern is highly diffuse.
d. Polytene chromosomes are not in this state.

_____ 6. Chromosome puffs indicate that the
a. DNA is being destroyed.
b. DNA is very active.
c. genetic material is radioactive.
d. RNA remains bound to DNA.

_____ 7. An enhancer affects what level of genetic control?
a. transcriptional
b. posttranscriptional
c. translational
d. posttranslational

_____ 8. A DNA base sequence changes from ATGCGG to ATCGG. This type of mutation is
a. metastasis.
b. frameshift.
c. point.
d. translocation.

_____ 9. If a mutation occurs, then
a. the code changes.
b. some particular codon or codons changes.
c. some particular anticodon or anticodons changes.
d. Both *a* and *b* are correct.
e. All of these are correct.

_____ 10. A promoter
a. turns on and off the transcription of a set of structural genes.
b. binds to RNA polymerase.
c. codes for the enzymes necessary for the transcription of polypeptides.
d. is an intron that breaks up a structural gene.

For questions 11–14, match the descriptions to these levels of control.
a. transcriptional
b. posttranscriptional
c. translational
d. posttranslational

_____ 11. Antisense RNA controls the expression of a gene in bacteria.

_____ 12. Given lactose, *E. coli* begins to make three enzymes that metabolize this sugar.

_____ 13. The mRNA leaving the nucleus of the hypothalamus and thyroid gland is different.

_____ 14. Feedback controls the metabolic activity of genes.

_____ 15. Which statement is descriptive of euchromatin?
a. A Barr body is in this form.
b. It is highly compacted and visible with the light microscope.
c. Lampbrush chromosomes are an example.
d. It is diffuse in its pattern when viewed under the microscope.
e. Both *c* and *d* are correct.

For questions 16–18, match the descriptions to these parts of an operon.
a. regulator
b. promoter
c. structural genes
d. operator

_____ 16. A group of genes that code for enzymes active in a particular metabolic pathway.

_____ 17. A segment of DNA that acts as an on/off switch for transcription of the structural gene.

_____ 18. A gene that codes for a protein that either directly combines with the operator or else must first join with a metabolite before joining with the operator.

## Critical Thinking Questions

Answer in complete sentences.

19. How can you reason that a human muscle cell contains a gene to make the polypeptides in hemoglobin?

20. How might cancer be cured through a study of gene control systems?

**Test Results:** _____ number correct ÷ 20 = _____ × 100 = _____ %

## Exploring the Internet

The Online Learning Center at *www.mhhe.com/maderbiology8* has additional study material and practice quizzes that can help you master the content of this chapter. You can also find links to websites exploring additional topics in biology. Access to the Online Learning Center is free for those who have purchased a new textbook.

## Answer Key

### Study Exercises

**1. a.** see page 252 in text **b.** regulator gene **c.** promoter **d.** operator **e.** structural genes **2.** cross out *inducer, structural gene, enzymes* **3.** cross out arrow to operator **4.** a, c, d, b **5. a.** turned on **b.** because the operon is ordinarily turned on, and the corepressor combines with the repressor to turn it off

**6.**

| Levels of Control of Gene Activity | Affects the Activity of |
|---|---|
| transcriptional | DNA |
| posttranscriptional | mRNA during formation and processing |
| translational | mRNA life expectancy during protein synthesis |
| posttranslational | protein |

**7. a.** F **b.** T **c.** T **d.** F **8. a.** a **b.** because a Barr body is a highly condensed, inactive X chromosome **9. a.** d **b.** because lampbrush chromosomes are looped, and transcription is actively going on **10. a.** factors that must be in place for transcription to begin **b.** enhancer and promoter **c.** A looping occurs to bring the transcription factors attached to the enhancer and promoter next to one another. **11.** transcriptional **12. a.** PTL **b.** TL **c.** TL **d.** PTC **13. a.** frameshift **b.** point **c.** point **14. a.** exposure to mutagens **b.** through the action of repair enzymes **15. a.** environmental factors that can cause mutations leading to cancer **b.** organic chemicals, radiation **16. a.** frameshift, point **b.** mutagen **c.** transposon **17. a.** proto-oncogene **b.** tumor **c.** in situ **d.** metastatic **e.** primary **f.** tumor-suppressor **g.** transcription **h.** cell-cycle **i.** Apoptosis **18. a.** O, MT **b.** P, T **c.** O, MT

### Chapter Test

**1.** b **2.** b **3.** b **4.** d **5.** c **6.** b **7.** a **8.** b **9.** e **10.** b **11.** b **12.** a **13.** b **14.** d **15.** e **16.** c **17.** d **18.** a **19.** The muscle cell is derived from the first cell, the zygote, by mitosis. Mitosis assures that all daughter cells receive all chromosomes with copy of each gene. **20.** Cancer might be cured by learning how to suppress the effects of oncogenes, preventing them from promoting uncontrolled cell division. Learning how to activate tumor-suppressor genes can also stabilize cell growth.

# 16

# Biotechnology and Genomics

## Chapter Review

To clone a gene, a vector is first prepared. To genetically engineer a **plasmid** or virus, **restriction enzymes** are used to cleave plasmid DNA and to cleave foreign DNA. The "sticky ends" produced facilitate the insertion of foreign DNA into **vector** DNA. The foreign gene is sealed into the vector DNA by **DNA ligase.** When the plasmid replicates or the virus reproduces, the foreign gene is **cloned.**

The **polymerase chain reaction (PCR)** uses the enzyme DNA polymerase to make multiple copies of target DNA. Then the base sequence of this DNA can be determined, or it can be subjected to DNA fingerprinting. During **DNA fingerprinting,** restriction enzymes are used to fragment DNA. Gel electrophoresis separates the fragments according to size and charge. Analysis of the resulting pattern identifies the DNA as belonging to a particular human or other organism.

**Transgenic organisms** have also been made. Genetically engineered bacteria, agricultural plants, and farm animals now produce commercial products of interest to humans, such as hormones and vaccines. Transgenic bacteria also perform oil spill cleanup, extract minerals, and produce chemicals. Transgenic agricultural plants have been engineered to resist herbicides and pests. Transgenic animals have been given bovine growth hormone. Pigs have been genetically altered to serve as a source of organs for transplant patients. Cloning of animals is now possible.

Constructing a genetic map of the chromosomes is still in progress. Sequencing of all the bases in the human genome is now complete. This information is expected to facilitate testing for genetic disorders, developing treatments that lead to a longer life and making gene therapy commonplace.

Gene therapy is used to correct the genotype of humans and to cure various human ills. Ex vivo gene therapy involves withdrawing cells from the patient; inserting a functioning gene, usually via a retrovirus; and then returning the treated cells to the patient. Many investigators are trying to develop in vivo gene therapy, in which viruses, laboratory-grown cells, or synthetic chemicals will be used to carry healthy genes into the patient.

## Study Exercises

Study the text section by section as you answer the questions that follow.

### 16.1 DNA Cloning (pp. 268–69)

- Using recombinant DNA technology, bacteria and viruses can be genetically engineered to clone a gene.
- The polymerase chain reaction (PCR) makes multiple copies of DNA segments. Analysis of the DNA usually follows.

1. In the diagram, write the numbers of the following descriptions in the appropriate blanks.
   1. Cloning occurs when host cell reproduces.
   2. Host cell takes up recombined plasmid.
   3. DNA ligase seals human gene and plasmid.
   4. Restriction enzyme cleaves DNA.

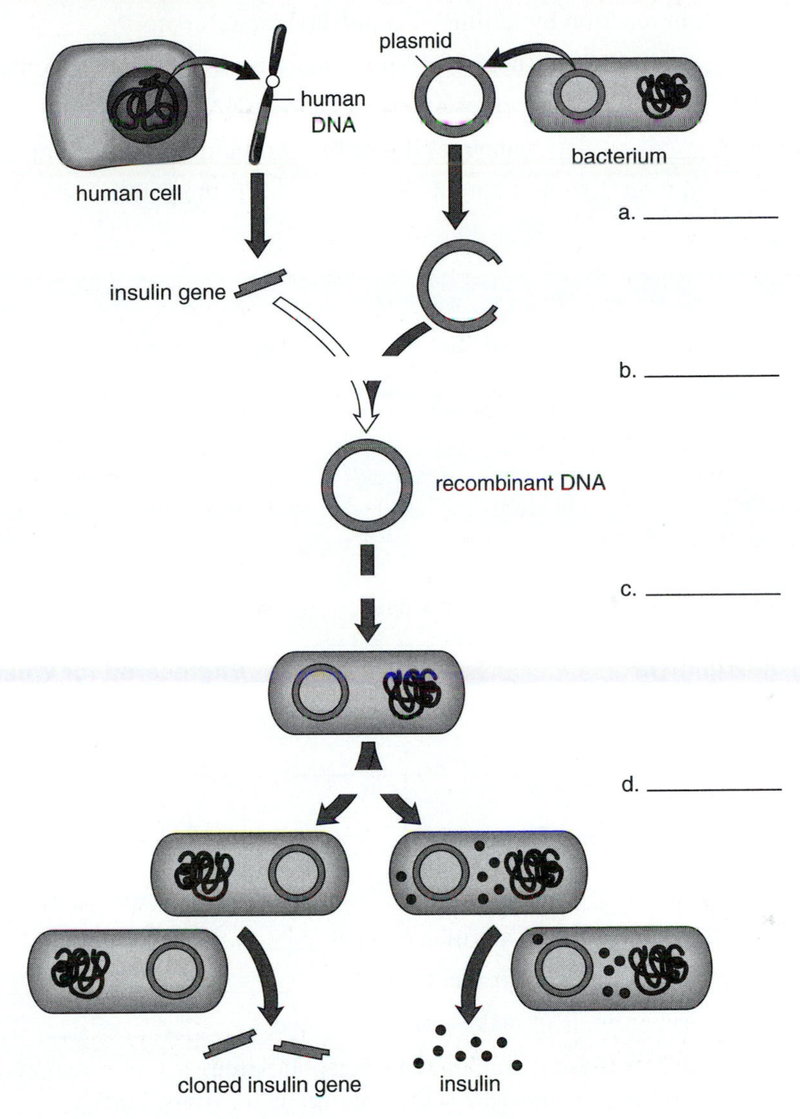


2. What is meant by the expression that restriction enzymes produce "sticky ends"? ___________________________

_______________________________________________

3. Change the following false statements to true statements:
   a. Plasmids are used as vectors in genetic engineering experiments involving humans. Rewrite: ___________

   _______________________________________________

   b. Recombinant DNA contains two types of bacterial DNA recombined together. Rewrite: ___________

   _______________________________________________

   c. Genetic engineering usually means that an organism receives genes from a member of its own species.
   Rewrite: _______________________________________________

   _______________________________________________

   d. Gene cloning occurs when a gene produces many copies of various genes. Rewrite: ___________

   _______________________________________________

In questions 4–6, fill in the blanks.

4. For human gene expression to take place in bacteria, the genes must be accompanied by a. ______________ regions, necessary for the expression of mammalian genes. If genes have been made using reverse transcriptase, then the cDNA contains the b. ______________ but not the introns.
5. Explain the polymerase chain reaction by telling what *polymerase* refers to: a. ______________; and what chain reaction means: b. ______________. At the beginning of the reaction, very little DNA may be available, but at the end of the reaction, c. ______________ copies of a segment of DNA are available.
6. In DNA fingerprinting, a. ______________ enzymes digest the two samples to be compared. b. ______________ separates the fragments, and their different lengths are compared. If the pattern is similar, the samples are from c. ______________.

## 16.2 Biotechnology Products (pp. 270–71)

- Bacteria, agricultural plants, and farm animals have been genetically engineered to produce commercially available products.
- Human tissues and farm animals are being genetically engineered to serve as a source of organs for human transplant patients.
- Agricultural plants and farm animals have been genetically engineered to improve their yield.
- It is now possible to clone animals, and cloning is used to produce multiple copies of farm animals that have been genetically engineered.

7. Complete the following table on transgenic organisms:

| Type of Organism | Engineered for What Purpose |
| --- | --- |
| | |
| | |
| | |

8. a. The advantage of using bacteria to make a product is that ______________.
   b. The advantage of using plants to make a product is that ______________.
   c. The advantage of using farm animals to make a product is that ______________.
9. Place a check to indicate which of these is true of xenotransplantation.
   a. _____ Xenotransplantation uses humans as a source of organs for transplants.
   b. _____ Xenotransplantation uses other species, such as the pig, as a source of organs for transplants.
   c. _____ Pigs can be genetically altered to prevent rejection of their organs by humans.
   d. _____ Other species can pass new and different viruses to humans.
10. Put these statements in the proper sequence (1–5) to describe the making of a transgenic female goat that will produce a medicine needed by humans in its milk.
   a. _____ Development within a host animal
   b. _____ Remove egg from donor animal
   c. _____ Isolate a human gene
   d. _____ Microinject the human gene into the egg of the donor animal
   e. _____ Transgenic goat is born
11. Put these statements in the proper sequence (1–6) to describe the cloning of the transgenic goat produced in question 10.
   a. _____ Birth of cloned transgenic goats
   b. _____ Remove nuclei from adult cells of transgenic goat
   c. _____ Collect the milk that contains the medicine of interest
   d. _____ Development within host goats
   e. _____ Microinject 2n nuclei into the donor eggs
   f. _____ Remove eggs from donor animal

## 16.3 The Human Genome Project (p. 273)

- The human genome project has two goals: to genetically map each chromosome and to sequence the DNA bases of each chromosome.

12. Place a check beside those statements that are true.
    a. _____ We now know the sequence of base pairs in the human genome.
    b. _____ We now know the sequence of all the genes on all the chromosomes.
    c. _____ Humans have hundreds of thousands more genes than do the other animals.
    d. _____ The sequence of the bases in our DNA differs greatly from that of bacteria.
13. Place a check beside the statements that are true. Knowing the sequence of bases in our DNA holds great promise for which of these?
    a. _____ Helping locate genes on the chromosomes
    b. _____ Helping determine why some people get sick
    c. _____ Helping develop specific drugs for the various causes of the same type of disorder

## 16.4 Gene Therapy (p. 276)

- Gene therapy is now being used to replace defective genes with healthy genes and to help cure various human ills.

14. Change these false statements to true statements.
    a. Ex vivo methods of gene therapy require that the therapeutic gene be placed in the body either directly or by using a viral vector. Rewrite: ________________________________________________

    ____________________________________________________________________

    b. A common ex vivo method is to microinject normal genes into bone marrow stem cells removed from the patient. Then the stem cells are returned to the patient. Rewrite: ______________________________

    ____________________________________________________________________

    c. Gene therapy is currently restricted to curing genetic diseases and is not used to treat illnesses like cystic fibrosis or cardiovascular diseases. Rewrite: ________________________________________

    ____________________________________________________________________

## Definitions Wordmatch

Review key terms by completing this matching exercise, selecting from the following alphabetized list of terms:

clone
DNA ligase
gene cloning
plasmid
polymerase chain reaction
recombinant DNA
restriction enzyme
transgenic organism
vector
xenotransplantation

a. ________________ Bacterial enzyme that stops viral reproduction by cleaving viral DNA; used to cut DNA at specific points during production of recombinant DNA.

b. ________________ Free-living organisms in the environment that have had a foreign gene inserted into them.

c. ________________ Production of identical copies; in genetic engineering, the production of many identical copies of a gene.

d. ________________ Self-duplicating ring of accessory DNA in the cytoplasm of bacteria.

e. ________________ Technique that uses the enzyme DNA polymerase to produce copies of a particular piece of DNA within a test tube.

f. ________________ DNA that contains genes from more than one source.

# Chapter Test

## Objective Questions

Do not refer to the text when taking this test.

____ 1. Select the incorrect description of a plasmid.
  a. used as vector
  b. consists of chromosomal DNA
  c. found in some bacteria
  d. small, ringlike structure

____ 2. Restriction enzymes
  a. cleave DNA into small fragments.
  b. restrict the growth of eukaryotic cells.
  c. seal pieces of DNA together.
  d. serve as introns in cells.

____ 3. The problem with using pigs as a source of human organs is
  a. it's not possible to genetically alter pig cells to avoid rejection.
  b. their organs are too small in comparison to human organs.
  c. pig organs might carry viruses that would be new to humans.
  d. transplant patients would rather die than receive organs from a pig.

____ 4. To clone a transgenic animal, you have to have
  a. an egg donor.
  b. special food to feed them.
  c. a way to protect human beings from coming in contact with transgenic animals.
  d. a transgenic animal.
  e. Both *a* and *d* are correct.

____ 5. A final step in the use of a plasmid to clone a gene is
  a. to insert a foreign gene into a bacterium.
  b. to introduce a plasmid into a treated cell.
  c. to remove a plasmid from a bacterium.
  d. reproduction of plasmid in a cell.

____ 6. The polymerase chain reaction does NOT typically
  a. take place in vats called bioreactors.
  b. occur at a high temperature.
  c. copy the entire human genome at one time.
  d. work if the target DNA is limited in size.
  e. Both *a* and *c* are not typical.

____ 7. A transgenic organism is
  a. free-living and receives a foreign gene.
  b. free-living and transmits a foreign gene.
  c. parasitic and receives a foreign gene.
  d. parasitic and transmits a foreign gene.

____ 8. The human genome
  a. will probably be completely sequenced in 50 years.
  b. has no usefulness to humans.
  c. contains only about 30,000 genes.
  d. is known only to government employees because it is top secret.

____ 9. Gene therapy is
  a. on the back burner because human genes are so different from one another.
  b. going full speed ahead despite its being still investigative.
  c. the use of foreign genes to cure a human ill.
  d. Both *a* and *c* are correct.
  e. Both *b* and *c* are correct.

____ 10. Genetically engineered plants have been or will be used to
  a. resist insects.
  b. resist herbicides.
  c. produce protein-enhanced beans, corn, and wheat.
  d. produce animal neuropeptides, blood factors, and growth hormones.
  e. All of these are correct.

____ 11. A DNA fingerprint is the type of
  a. plasmid chosen for recombinant DNA.
  b. restriction enzyme used to fragment DNA.
  c. fragment pattern that results following gel electrophoresis.
  d. sequence pattern discovered in a section of DNA.

____ 12. Genetically engineered bacteria can be used to
  a. protect plants from frost.
  b. clean oil spills on beaches.
  c. produce organic chemicals.
  d. extract copper and gold from low-grade sources.
  e. All of these are correct.

____ 13. Which statement(s) are true?
  a. A biotechnology product is a protein made by a transgenic organism desired by humans.
  b. Transgenic organisms have had a foreign gene inserted into their cells.
  c. If a transgenic organism can be cloned, it would only be necessary to make the first organism transgenic.
  d. Most likely, it will never be possible to insert foreign genes into the human genome, and therefore there can never be transgenic humans.
  e. All but *d* are true statements.

____ 14. Which of these would you NOT expect to be a biotechnology product produced by a bacterium?
  a. steroid sex hormones
  b. hormones of any kind
  c. nucleic acids instead of proteins
  d. Both *b* and *c* are correct.

____ 15. If a cell is altered while outside the human body for gene therapy, it is considered ____________ therapy.
   a. ex vivo
   b. in vivo
   c. in vitro
   d. extraneous
   e. intravenous

____ 16. What does the Human Genome Project have to do with gene therapy?
   a. Once we know the location of all the genes on the chromosomes, it will make it easier to isolate particular genes to cure humans.
   b. Both the Human Genome Project and gene therapy are unethical, and therefore will never be completed.
   c. Both the Human Genome Project and gene therapy require the use of restriction enzymes and gel electrophoresis.
   d. Both the Human Genome Project and gene therapy require the prior use of the polymerase chain reaction.
   e. All but *b* are correct.

____ 17. The polymerase chain reaction
   a. produces many copies of different segments of DNA.
   b. produces many copies of the same segment of DNA.
   c. requires denaturing double-stranded DNA into single strands by heating.
   d. Both *b* and *c* are correct.

____ 18. Which of these is NOT a step to prepare recombinant DNA?
   a. remove plasmid from bacterial cell
   b. use restriction enzyme to acquire foreign gene and cut open vector
   c. use ligase to seal foreign gene into vector
   d. use a virus to carry recombinant DNA into a plasmid

____ 19. Which of these is a benefit to having insulin produced by biotechnology?
   a. It can be mass-produced.
   b. It is nonallergenic.
   c. It is less expensive.
   d. All of these are correct.

____ 20. Vaccines produced by biotechnology could be
   a. pathogens treated to be nonvirulent.
   b. proteins produced by a pathogen's gene.
   c. only enzymes taken from a pathogen.
   d. Both *a* and *b* are correct.

## Critical Thinking

Answer in complete sentences.

21. How do studies of genetic engineering prove that the genetic code is nearly universal?

22. What do you think are some objections our society may have regarding genetic engineering?

**Test Results:** ______ number correct ÷ 22 = ______ × 100 = ______ %

## Exploring the Internet

The Online Learning Center at *www.mhhe.com/maderbiology8* has additional study material and practice quizzes that can help you master the content of this chapter. You can also find links to websites exploring additional topics in biology. Access to the Online Learning Center is free for those who have purchased a new textbook.

## Answer Key

### Study Exercises

**1. a.** 4 **b.** 3 **c.** 2 **d.** 1 **2.** Cleavage results in unpaired bases. **3. a.** . . . involving bacteria **b.** contains DNA from two different sources **c.** . . . from a member of a different species **d.** . . . many copies of the same gene **4. a.** regulatory **b.** exons **5. a.** DNA polymerase, the enzyme involved in DNA replication **b.** the reaction occurs over and over **c.** many **6. a.** restriction **b.** Gel electrophoresis **c.** the same individual
**7.**

| Type of Organism | Engineered for What Purpose |
| --- | --- |
| bacteria | to make products to protect plants, for oil spill cleanup, to produce chemicals, and to mine metals |
| plants | to resist insects, pesticides, and herbicides, and to make products |
| animals | to have improved qualities and to make products |

**8. a.** they will take up plasmids. **b.** they will grow from single cells (protoplasts). **c.** the product is easily obtainable in milk. **9.** b, c, d **10.** c, b, d, a, e **11.** b, f, e, d, a, c **12.** a **13.** a, b, c **14. a.** In vivo **b.** . . . use a viral vector to carry normal genes . . . **c.** . . . is not restricted to curing genetic disease and is used to treat illnesses . . .

### Definitions Wordmatch

**a.** restriction enzyme **b.** transgenic organism **c.** clone **d.** plasmid **e.** polymerase chain reaction **f.** recombinant DNA

### Chapter Test

**1.** b **2.** a **3.** c **4.** e **5.** d **6.** e **7.** a **8.** c **9.** b **10.** e **11.** c **12.** e **13.** e **14.** a **15.** a **16.** a **17.** d **18.** d **19.** d **20.** b **21.** Genes transmitted to new cells through vectors and other means are still transcribed and translated by the same process and with the same accuracy. **22.** Some people may object to genetic engineering on religious grounds because we are now able to change the inherited characteristics of organisms, including human beings. Some scientists have concerns because a disease-causing transgenic bacterium may be produced for which humans have no immunity or a transgenic bacterium, plant, or animal may be produced that could wreak havoc in the environment.

# Part III Evolution

# 17

# Darwin and Evolution

## Chapter Review

In the mid-nineteenth century, Charles Darwin's studies led to his hypothesis on **evolution.** Darwin was a student of biology and geology. His observations on the life-forms of the Galápagos Islands, including their **biogeography,** influenced the formation of his hypothesis.

Darwin's view of evolution, in stark contrast to the pre-Darwinian outlook, supported the common descent of organisms. During this process, members of a species evolve **adaptations** through **natural selection.** Inherited variations in the members of a population establish the raw material for these adaptations. Through potential overpopulation, and the inevitable competition among population members, certain organisms are more likely to survive and reproduce. (These organisms exhibit **fitness.**) Over generations, this process results in adaptation of the population to the environment. This change in the population over time is called evolution.

Cuvier and Lamarck, contemporaries of Darwin, expressed ideas about evolution that differed from Darwin's. Cuvier, the founder of the science of **paleontology,** proposed the hypothesis of **catastrophism.** Lamarck correctly recognized the process of descent with modification among organisms in a population, but he explained the mechanism for that process incorrectly as the **inheritance of acquired characteristics.**

Numerous lines of evidence currently support Darwin's theory of common descent and evolution through natural selection. Evidence includes studies from fossils, biogeography, comparative anatomy, and comparative biochemistry.

## Study Exercises

Study the text section by section as you answer the questions that follow.

### 17.1 History of Evolutionary Thought (pp. 282–84)

- Evolution has two aspects: descent from a common ancestor and adaptation to the environment.
- In the eighteenth century, scientists became especially interested in classifying and understanding the relationships between the many forms of present and past life.
- Gradually in the eighteenth century, scientists began to accept the view that the earth is very old and that life-forms evolve or change over time.

1. Before each of the following statements, write *pre* for pre-Darwinian or *post* for post-Darwinian view of evolution:
   a. _____ Adaptation to the environment occurs through the work of a creator.
   b. _____ Hypotheses are tested through observation and experimentation.
   c. _____ Species are related by a common descent.
   d. _____ The earth is relatively young, with an age measured in thousands of years.

2. Indicate whether these statements, related to the principles of mid-eighteenth-century taxonomy, are true (T) or false (F).
   a. _____ A binomial system of nomenclature can be used to classify organisms.
   b. _____ Different organisms can be arranged by increasing order of complexity.
   c. _____ Each species has an ideal structure and function fixed in the sequential ladder of life.
   d. _____ Gradations occur naturally between species.

3. Label each of the following as reflecting the thinking of these scientists:

   C—Cuvier L—Lamarck

   a. _____ A new stratum or mix of fossils in a region signals that a local catastrophe occurred.
   b. _____ Members of a population change over time through the inheritance of acquired characteristics.
   c. _____ The structure of an animal can be deduced by studying its fossil bones.
   d. _____ The increasing complexity of organisms through evolutionary descent is the result of a natural force.

## 17.2 Darwin's Theory of Evolution (pp. 285–91)

- Charles Darwin's trip around the Southern Hemisphere aboard the HMS *Beagle* provided him with evidence that the earth is very old and that evolution does occur.

4. Indicate whether these statements are true (T) or false (F).
   a. _____ Darwin had no suitable background to be the naturalist on board the HMS *Beagle*.
   b. _____ Darwin had taken various science courses and had worked with people who were experts in their fields.
   c. _____ The HMS *Beagle* took a trip to South America and then returned.
   d. _____ The HMS *Beagle* went around the world in the Southern Hemisphere.

### Occurrence of Descent (p. 285)

5. **Fossils:** Darwin noticed a close a._____________ between modern forms and extinct species known only through fossils. He began to think that these fossil forms might be b._____________ to modern species. If so, the implication is that new species appear on earth as a result of biological change.

   **Biogeography:** Darwin noticed that whenever the environment changed, the types of species c._____________. He also observed that similar environments have different but d._____________ adapted species. This indicates that species are suited to the environment.

   **Darwin's finches:** Darwin speculated that a mainland finch was the e._____________ ancestor for all the different species of finches on the Galápagos Islands. This shows that speciation occurs.

   **Conclusion:** Based on his observations, Darwin came to accept the idea of f._____________; that is, life-forms change over time.

### Natural Selection and Adaptation (p. 288)

- Both Darwin and Alfred Wallace proposed natural selection as a mechanism by which adaptation to the environment occurs. This mechanism is consistent with our present-day knowledge of genetics.

6. What is the proper sequence for these statements describing Darwin's theory of natural selection?

   _____________

   a. The result of organic evolution is many different species, each adapted to specific environments.
   b. Many more individuals are produced each generation than can survive and reproduce.
   c. Gradually, over long periods, a population can become well adapted to a specific environment.
   d. The members of a population have heritable variations.
   e. Some individuals have adaptive characteristics that enable them to survive and reproduce better than others in the environment.

7. In each of the following pairs of situations, place a check beside the members that are better adapted:

a. In a forest, certain ground plants
_____ (1) are able to grow in the shade.
_____ (2) require full sunlight.

b. In the depths of the ocean, certain fishes
_____ (1) need to eat only infrequently.
_____ (2) must eat continuously.

c. In a mountain village, some inhabitants
_____ (1) get dizzy when the oxygen level falls below normal.
_____ (2) do not get dizzy when the oxygen level falls below normal.

## 17.3 Evidence of Evolution (pp. 292–97)

- The fossil record, biogeography, comparative anatomy, and comparative biochemistry support a hypothesis of common descent.

In questions 8–11, match the statements to the type of evidence supporting evolution.

a. The succession of life-forms is revealed through preserved remnants.
b. Related organisms share homologous structures.
c. Closely related organisms have high correlations in DNA base sequences.
d. Organisms arise and disperse from place of origin.
e. Classification includes at least seven categories.

8. _____ fossil record
9. _____ biogeography
10. _____ comparative anatomy
11. _____ comparative biochemistry

# Chapter Test

## Objective Questions

Do not refer to the text when taking this test.

_____ 1. Each is a pre-Darwinian view of evolution EXCEPT
a. adaptation to the environment comes from a creator.
b. each species is specially created.
c. hypotheses regarding species can be tested by experimentation.
d. the earth is relatively young.

_____ 2. Each is an idea from taxonomy in the mid-eighteenth century EXCEPT
a. a fixity of species exists.
b. humans occupy the last rung of a ladder of life.
c. natural gradations exist between species.
d. species have a special creation.

_____ 3. The science of paleontology was founded by
a. Cuvier.
b. Darwin.
c. Lamarck.
d. Lyell.

_____ 4. The Galápagos Islands are off the western coast of
a. Africa.
b. Asia.
c. North America.
d. South America.

_____ 5. Darwin claimed that the beak size of finch species was related to their
a. body size.
b. flight pattern.
c. food source.
d. time of reproduction.

_____ 6. Select the statement that is NOT a tenet of Darwin's theory of natural selection.
a. Members of a population have heritable variations.
b. Members of a population will compete.
c. Populations tend to reproduce in small numbers.
d. Some population members have adaptive characteristics.

____ 7. Each could be an example of adaptation EXCEPT a
a. plant that has the broadest leaves.
b. plant that has the greatest height.
c. predator that has the keenest eyesight.
d. prey species that runs the slowest.

____ 8. An adaptation promotes
a. only the chance to reproduce.
b. survival only.
c. the chance to survive and reproduce.
d. neither the chance to reproduce nor the chance to survive.

____ 9. Darwin's studies closely matched the independent work of
a. Cuvier.
b. Lamarck.
c. Lyell.
d. Wallace.

____ 10. Vertebrate forelimbs are most likely to be studied in
a. biogeography.
b. comparative anatomy.
c. comparative biochemistry.
d. ecological physiology.

____ 11. Biochemical evidence supporting evolution would show that
a. there are more base differences between yeasts and humans than between pigs and humans.
b. there are more base differences between monkeys and humans than between pigs and humans.
c. monkeys and humans have almost the same sequence of bases.
d. Both *a* and *c* are correct.

____ 12. Comparative anatomy demonstrates that
a. each species has its own structures, indicating no relationship with any other species.
b. different vertebrates have widely different body plans.
c. different species can have similar structures traceable to a common ancestor.
d. fossils bear no anatomical similarities to modern-day species.

____ 13. The study of biogeography shows that
a. the same species of plants and animals are found on different continents whenever the environment is the same.
b. one species can spread out and give rise to many species, each adapted to varying environments.
c. the structure and function of organisms bear no relationship to the environment.
d. barriers do not prevent the same species from spreading around the world.

____ 14. Which is NOT true of fossils?
a. They are evidences of life in the past.
b. They look exactly like modern-day species, regardless of their age.
c. In general, the older the fossil, the less it resembles modern-day species.
d. They indicate that life has a history.

____ 15. Darwin reasoned that, if the world is very old, then
a. taxonomy will have to give up the binomial system of nomenclature.
b. evolution could not have occurred.
c. geological changes occur in a relatively short period of time.
d. there was time for evolution to occur.

## Critical Thinking Questions

Answer in complete sentences.

16. A line of talented pianists, each practicing diligently, is found over five generations in a family. Offer a modern-day explanation as opposed to a Lamarckian explanation.

17. In a sample of geological strata, where are the oldest life-forms most likely to be found? Where are the most recent life-forms?

**Test Results:** ______ number correct ÷ 17 = ______ × 100 = ______ %

## Exploring the Internet

The Online Learning Center at *www.mhhe.com/maderbiology8* has additional study material and practice quizzes that can help you master the content of this chapter. You can also find links to websites exploring additional topics in biology. Access to the Online Learning Center is free for those who have purchased a new textbook.

## Answer Key

### Study Exercises

**1. a.** pre **b.** post **c.** post **d.** pre **2. a.** T **b.** T **c.** T **d.** F **3. a.** C **b.** L **c.** C **d.** L **4. a.** F **b.** T **c.** F **d.** T **5. a.** resemblance or similarity **b.** related **c.** changed **d.** similarly **e.** common **f.** evolution **6.** d, b, e, c, a **7. a.** 1 **b.** 1 **c.** 2 **8.** a **9.** d **10.** b **11.** c

### Chapter Test

**1.** c **2.** c **3.** a **4.** d **5.** c **6.** c **7.** d **8.** c **9.** d **10.** b **11.** d **12.** c **13.** b **14.** b **15.** d **16.** An ability to play the piano due to practice, cannot be passed on. Traits are passed on by way of the gametes. It is possible, however, that the genes in a particular family endow recipients with a musical ability, including anatomical characteristics that facilitate playing the piano. **17.** The oldest forms are in the deepest strata. The most recent life-forms are in the more recently added strata, which are not as deep.

# 18

# Process of Evolution

## Chapter Review

Gene and chromosomal **mutations** and also recombination, which may produce a more favorable combination of alleles, are sources of new gene types. The **Hardy-Weinberg principle** refers to a constancy of the **gene pool** as long as there are no mutations, no gene flow, random mating, no genetic drift, and no selection. The reverse of these conditions causes evolution to occur. **Gene flow** will cause the gene pool of two populations to become similar, and **genetic drift** will cause them to become dissimilar. Today it is possible to see that natural selection occurs when certain alleles become more frequent in a gene pool. **Directional selection** occurs when an extreme phenotype is favored and shifts in one direction; **stabilizing selection** occurs when an intermediate phenotype is favored; and **disruptive selection** occurs when more than one extreme phenotype is favored over any intermediate phenotype.

Mutation and gene flow maintain variation within a population despite natural selection. Balanced polymorphism exists due to a heterozygote that hides the recessive allele from selection.

**Speciation** occurs when populations become isolated from one another, most often due to a geographic barrier. **Allopatric speciation** requires a geographic barrier; **sympatric speciation** does not. **Species** remain reproductively isolated due to prezygotic and postzygotic isolating mechanisms.

## Study Exercises

Study the text section by section as you answer the following questions.

1. An investigator determines, by inspection, that 4% of a population is albino. Answer the following questions about this population:
   a. $q^2$ = _____________.
   b. This represents the percentage of the population that is _____________.
   c. What is the frequency of the recessive allele in this population? $q$ = _____________
   d. Considering the frequency of the recessive allele, what is the frequency of the dominant allele?
   $p$ = _____________
   e. If $p$ = this value, then $p^2$ = _____________.
   f. This is the frequency of the population that is _____________.
   g. The value of $2pq$ = _____________.
   h. This represents the frequency of the population that is _____________.
   i. What percentage of the population has normal pigmentation? _____________
2. Forty-nine percent of a population cannot taste a chemical called PTC. Presence of a dominant allele is necessary to taste this substance. Complete the following information about the gene pool of the population:
   a. $q^2$ = _____________
   b. $q$ = _____________
   c. $p$ = _____________
   d. $p^2$ = _____________
   e. $2pq$ = _____________

3. a. Using this Punnett square, show that the next generation of the population in question 2 will have exactly the same composition, assuming a Hardy-Weinberg equilibrium.

| | (     )*T* | (     )*t* |
|---|---|---|
| (     )*T* | (     )*TT* | (     )*Tt* |
| (     )*t* | (     )*Tt* | (     )*tt* |

b. The frequency of *T* = ______________.

c. The frequency of *t* = ______________.

d. Describe the gene pool of the next generation. ________________________________

e. What does this prove? ________________________________

f. How would we know when evolution occurs? ________________________________

- Mutations, gene flow, nonrandom mating, genetic drift, and natural selection can cause allele frequency changes in a population over time.

4. Label the statements with the correct agents of evolutionary change:

gene flow    genetic drift    mutations    natural selection    nonrandom mating

a. ______________ Investigators have discovered that multiple alleles are common in a population.

b. ______________ Populations are subject to new alleles entering by the migration of organisms between populations.

c. ______________ Female birds of paradise choose mates with the most splendid feathers.

d. ______________ Investigators discovered that if they randomly picked out a few flies from each generation to start the next generation, gene pool frequency changes appeared.

e. ______________ Giraffes with longer necks get a larger share of resources and tend to have more offspring.

5. Match these descriptions to one of the agents of evolutionary change listed in question 4. (Some agents are used more than once.)

a. ______________ Dwarfism is common among the Amish of Lancaster County, Pennsylvania.

b. ______________ Cheetahs are homozygous for a larger proportion of their genes.

c. ______________ This agent of change tends to make the members of a population dissimilar to one another.

d. ______________ This agent of change tends to make the members of a population similar to one another.

e. ______________ Certain members of a population are more fit than other members.

f. ______________ Bacteria and insects become resistant to agents that formerly killed them.

## 18.1 Evolution in a Genetic Context (pp. 302–6)

- The Hardy-Weinberg principle defines evolution in terms of allele frequency changes in a population over time.
- The raw material for evolutionary change is mutations, both genetic and chromosomal. Recombination of genes is another source in sexually reproducing organisms.

6. Indicate whether these statements, related to sources of variation among diploid members in a population, are true (T) or false (F).
   a. ______ Genetic mutations occur at random.
   b. ______ The only mutations that occur are those that make organisms more fit.
   c. ______ Some chromosomal mutations are simply a change in chromosomal number.
   d. ______ An offspring receives recombined genes because of the events of gametogenesis and fertilization.
   e. ______ Recombination is a significant source of variation because many traits are polygenic.

## 18.2 Natural Selection (pp. 306–9)

- Natural selection is changes in allele frequencies in a population due to the differential ability of certain phenotypes to reproduce.
- Natural selection results in adaptation to the environment. The three types of natural selection are: directional selection, stabilizing selection, and disruptive selection.

7. Natural selection can now be understood in terms of genetics. Many of the variations that exist between members of a population are due to differences in a.______________. Some of these genotypes result in b.______________ better adapted to the environment. Individuals better adapted to the environment reproduce to a(n) c.______________ extent, and therefore, these genotypes and phenotypes become more prevalent in the population.

8. Match the observations to the correct type of natural selection at work:
   directional    disruptive    stabilizing
   a. ________________ Trees in a windy area tend to remain the same size each year.
   b. ________________ The brain size of hominids steadily increases.
   c. ________________ The same species of moths tends to have blue stripes in open areas and orange stripes in forested areas.

9. Match the types of natural selection listed in question 8 with the following diagrams:

a.

b.

c.

10. Variation is maintained in a population despite directional and stabilizing selection when members have a.______________ alleles for every trait. That's because the b.______________ allele is hidden by the c.______________ allele. In instances such as sickle-cell disease, in which the d.______________ genotype is more fit, the two e.______________ genotypes are maintained due to the reproductive process, which involves meiosis and fertilization.

## 18.3 Speciation (pp. 310–13)

- New species come about when populations are reproductively isolated from other, similar populations.
- Adaptive radiation is the rapid development of several species from a single species; each species is adapted in a unique way.

11. Bush babies (a type of primate) living higher in the tropical canopy are a different species from those living lower in the canopy. Answer the following *yes* or *no:*

   a. Would the two species of bush babies reproduce with each other? _____________

   b. Would a premating mechanism separate the two species of bush babies? _____________

   c. Could the two species of bush babies eat the same food? _____________

   d. Would the two species of bush babies have to look dissimilar enough to be distinguishable by the naked eye? _____________

12. Match the numbered statements to the letters in the diagram. (Some answers use more than one number and also numbers are used more than once.)

   1. A newly formed barrier comes between the populations.
   2. The barrier is removed.
   3. A species contains one population or more than one interbreeding populations.
   4. Reproductive isolation has occurred.
   5. Two species now exist.
   6. Reproductive isolation has occurred without benefit of a barrier.
   7. Drift alone can cause gene pools to become dissimilar.
   8. Divergent evolution occurs.
   9. Allopatric speciation occurs.
   10. Sympatric speciation occurs.

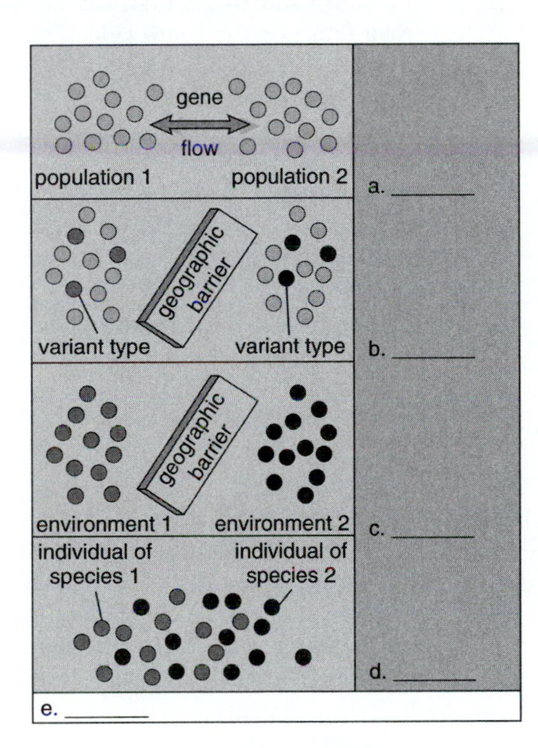


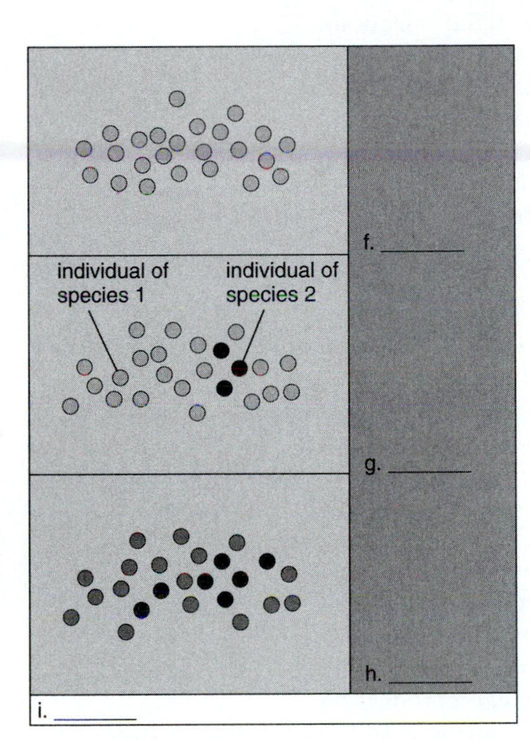


13. Label each of these statements with *pre* for premating isolating mechanism or *post* for postmating isolating mechanism.

   a. _____ Zygote mortality exists.
   b. _____ Species reproduce at different times.
   c. _____ Species have genitalia unsuitable to each other.

# Chapter Test

## Objective Questions

Do not refer to the text when taking this test.

In questions 1–5, assume that 16% of the organisms in a population are homozygous recessive. Describe the current gene pool.

____ 1. frequency of *aa*

____ 2. frequency of *a*

____ 3. frequency of *A*

____ 4. frequency of *AA*

____ 5. frequency of *Aa*

____ 6. Each is a condition of the Hardy-Weinberg principle EXCEPT that
   a. gene flow is absent.
   b. genetic drift does not occur.
   c. mutations are lacking.
   d. random mating does not occur.

____ 7. Establishment of polydactylism among the Amish is an example of
   a. artificial selection.
   b. natural selection.
   c. the bottleneck effect.
   d. the founder effect.

____ 8. Industrial melanism is an example of selection that is
   a. directional.
   b. disruptive.
   c. sexual.
   d. stabilizing.

____ 9. In a population of mature trees, it is NOT true that disruptive selection
   a. favors the shortest trees.
   b. favors the tallest trees.
   c. will not occur if the environment is diverse.
   d. will occur in nondiverse environment.

____ 10. Each factor contributes to the maintenance of variation from a recessive allele EXCEPT
   a. diploidy.
   b. heterozygosity.
   c. homozygosity.
   d. sexual reproduction.

____ 11. The criterion used to distinguish between two species is based on
   a. geography.
   b. physical traits.
   c. reproduction.
   d. time.

____ 12. Select the prezygotic isolating mechanism.
   a. $F_2$ fitness
   b. gamete mortality
   c. habitat type
   d. hybrid sterility

____ 13. Select the postzygotic isolating mechanism.
   a. behavior
   b. mechanical differences in genitalia
   c. temporal factors
   d. zygote mortality

____ 14. Which type of speciation requires a geographical barrier?
   a. allopatric
   b. sympatric

____ 15. Which type of speciation occurs from polyploidy in plants?
   a. allopatric
   b. sympatric

____ 16. Two populations of field mice become separated by an interstate highway so that gene flow is impossible. After ten years, a researcher detects an evolutionary change in the populations: mice on the north side of the highway have longer nails than those on the south side. The two habitats, however, appear identical in every respect. The probable reason for the difference is ____________ each group.
   a. natural selection within
   b. nonrandom mating among
   c. genetic drift within
   d. mutations in
   e. adaptation of

____ 17. During the usual process of speciation, a species is first isolated
   a. behaviorally.
   b. geographically.
   c. reproductively.
   d. mechanically.
   e. genetically.

____ 18. Two species have been observed in nature to copulate successfully with each other, yet they are unable to produce a hybrid line. This might be due to all of the following EXCEPT
   a. gamete mortality.
   b. hybrid sterility.
   c. zygote mortality.
   d. courtship behavioral differences.

____ 19. The frequency of the dark form of the peppered moth increased in industrial areas of England during the nineteenth century. This is because
   a. predatory birds changed their preference from the light to the dark form.
   b. the light form was more sensitive to deteriorating environmental conditions.
   c. dark-colored moths had a better survival rate as the Industrial Revolution progressed.
   d. environmental pollutants became responsible for increased mutation rates.
   e. the allele for dark color is dominant to that for light color.

____20. Evolution by natural selection requires
a. variation.
b. heritable genetic differences.
c. differential adaptedness.
d. differential reproduction.
e. All of these are correct.

## Critical Thinking Questions

Answer in complete sentences.

21. One percent of a population consists of albinos. A student studies the population and claims that albinism can be removed from the population by preventing the mating of all albinos. Will this approach work?

22. A dominant allele produces a desirable coloration pattern in a fish species. A pond owner stocks a pond with a small number of heterozygous members showing this desirable trait, hoping to maintain it. Will this approach work?

**Test Results:** _____ number correct ÷ 22 = _____ × 100 = _____ %

## Exploring the Internet

The Online Learning Center at *www.mhhe.com/maderbiology8* has additional study material and practice quizzes that can help you master the content of this chapter. You can also find links to websites exploring additional topics in biology. Access to the Online Learning Center is free for those who have purchased a new textbook.

## Answer Key

### Study Exercises

**1. a.** 0.04 or 4% **b.** homozygous recessive **c.** 0.2 **d.** 0.8 **e.** 0.64 or 64% **f.** homozygous dominant **g.** 0.32 or 32% **h.** heterozygous **i.** 0.96 or 96% **2. a.** 0.49 **b.** 0.7 **c.** 0.3 **d.** 0.09 **e.** 0.42
**3.**

| | (0.3)*T* | (0.7)*t* |
|---|---|---|
| (0.3)*T* | (0.09)*TT* | (0.21)*Tt* |
| (0.7)*t* | (0.21)*Tt* | (0.49)*tt* |

**a.** In the generation shown in the Punnett square, *TT* = 0.09, *Tt* = 0.42, *tt* = 0.49; this is exactly the same as the parental generation in question 2. **b.** 0.3 **c.** 0.7 **d.** exactly the same as the previous generation **e.** that sexual reproduction alone does not change allele frequencies **f.** when gene pool frequencies change **4. a.** mutations **b.** gene flow **c.** nonrandom mating **d.** genetic drift **e.** natural selection **5. a.** genetic drift (founder effect) **b.** genetic drift (bottleneck effect) **c.** mutations **d.** natural selection **e.** mutations **f.** natural selection **6. a.** T **b.** F **c.** T **d.** T **e.** T **7. a.** genotype **b.** phenotypes **c.** greater **8. a.** stabilizing **b.** directional **c.** disruptive **9. a.** directional **b.** stabilizing **c.** disruptive **10. a.** two **b.** recessive **c.** dominant **d.** heterozygous **e.** homozygous **11. a.** no **b.** yes **c.** yes **d.** no **12. a.** 3 **b.** 1 **c.** 8 **d.** 2, 4, 5 **e.** 9 **f.** 3 **g.** 6, 5 **h.** 4, 7 **i.** 10 **13. a.** post **b.** pre **c.** pre

### Chapter Test

**1.** 0.16 = 16% **2.** 0.4 **3.** 0.6 **4.** 0.36 = 36% **5.** 0.48 = 48% **6.** d **7.** d **8.** a **9.** d **10.** c **11.** c **12.** c **13.** d **14.** a **15.** b **16.** c **17.** b **18.** d **19.** c **20.** e. **21.** It will not work because a recessive allele causes albinism. Eighteen percent of the population members are heterozygotes and will protect the recessive allele, even though it is hidden from expression phenotypically. **22.** It may not work because of genetic drift—the population may diverge to homozygous forms through random events.

# 19

# Origin and History of Life

## Chapter Review

The primitive atmosphere contained mainly water vapor, nitrogen, and carbon dioxide; because it contained no oxygen, it is known as a reducing atmosphere. As the Earth cooled, rain began to fall, and the first atmospheric gases reacted with one another in the original ocean. First, small organic molecules and then macromolecules formed. The **RNA-first hypothesis** states that RNA could have been the first genetic material because it may have had enzymatic properties that allowed it to reproduce itself and form proteins. Alternately, the first macromolecules could have been proteins that carried on metabolism, allowing growth to occur; only later did DNA genes come into being. A plasma membrane is required for the **protocell** to exist. The protocell was a heterotrophic fermenter. Later, photosynthesis and then cellular respiration occurred.

**Fossils** are the remains of past life. The fossil record sketches the history of life in broad terms. Prokaryotes were alone for about 1.5 billion years, and they diversified metabolically. Eukaryotes, which may have come about by **endosymbiosis,** arose at about the time the atmosphere became an oxidizing one because of cyanobacterial release of oxygen. Multicellularity and sexual reproduction began about 600 million years ago. The Cambrian period of the Paleozoic era began with an explosion of animals whose abundance in the fossil record is attributed to their having skeletons.

During the Paleozoic era, the swamp forests of the Carboniferous period contained seedless vascular plants, insects, and amphibians—all of whom lived on land. Cycads and reptiles were prevalent in the Mesozoic era, and twice during this era, dinosaurs of enormous size evolved. Mammals evolved earlier but did not diversify until the Cenozoic era, after the dinosaurs were extinct. The Neogene period of the Cenozoic era is associated with the evolution of primates—first prosimians evolved, then monkeys, apes, and humans. The Cenozoic era is the present era.

Many environmental factors influence evolution. **Continental drift** has affected biogeography and helps explain the distribution pattern of today's land organisms. Mass extinctions have played a dramatic role in the history of life. While some mass extinctions may have been caused by meteorite impact, others may be climatic fluctuations due to continental drift.

## Study Exercises

Study the text section by section as you answer the questions that follow.

### 19.1 Origin of Life (pp. 320–23)

- A chemical evolution proceeded from atmospheric gases to small organic molecules to macromolecules to protocells.
- The primitive atmosphere contained no oxygen, and the first cell was a heterotrophic fermenter.
- The first cell was bounded by a membrane and contained a replication system—that is, DNA, RNA, and proteins.

1. In the top half of the diagram, place these labels next to the correct arrows:

   cooling    energy capture    polymerization

   Then place these labels in the boxes:

   inorganic chemicals    macromolecules    plasma membrane    primitive earth    small organic molecules

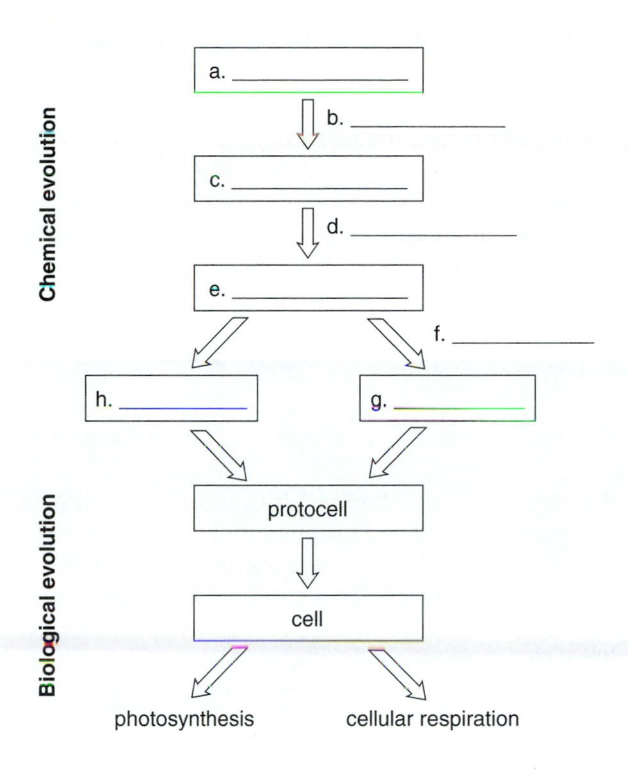


2. A student decides to reproduce Miller's experiment, so she assembles all the necessary equipment and adds the proper gases. What is still needed and why? ______________________________

   ______________________________________________________________

3. A student decides to reproduce Fox's experiment, so he puts a solution of amino acids on heated rocks. What is still needed and why? ______________________________

   ______________________________________________________________

4. What is the evidence for suggesting that it was an "RNA world" some 4 billion years ago? __________

   ______________________________________________________________

5. Indicate whether these statements about the protocell are true (T) or false (F).

   a. _____ carried on cellular respiration
   b. _____ was a heterotrophic fermenter
   c. _____ contained a self-replication system that allowed it to reproduce
   d. _____ had a plasma membrane

6. a. To be a true cell, which statement, labeled as false in question 5, must be fulfilled? __________
   b. Add the label *self-replicating system* to the diagram in question 1.

7. a. Why does it seem logical that the protocell was heterotrophic? ______________________
   b. Why would the protocell have been a fermenter? ______________________________

8. What is the proper sequence for the order of events in the evolution of a self-replicating system, assuming that it was an "RNA world" at the time? ______________________

a. replication of RNA
b. reverse transcription of DNA
c. RNA → proteins
d. DNA → RNA → proteins

9. Place the appropriate letter next to each statement about the Earth's atmosphere to describe it as either:

P—primitive  C—current

a. _____ It exists without the ozone shield.
b. _____ It favors the polymerization of organic molecules.
c. _____ It is a reducing atmosphere.
d. _____ It is an oxidizing atmosphere.
e. _____ It tends to break down organic molecules.
f. _____ Oxygen-producing autotrophs made it.

## 19.2 History of Life (pp. 324–33)

- The fossil record allows us to trace the history of life, divided into the Precambrian, and the Paleozoic, Mesozoic, and Cenozoic eras.
- The first fossils are of prokaryotes and date from about 3.5 billion years ago. Prokaryotes diversified for about 1.5 billion years before the eukaryotic cell and multicellular forms evolved during the Precambrian.
- Fossils of complex multicellular marine invertebrates and vertebrates appeared during the Cambrian period of the Paleozoic era. During the Carboniferous period, swamp forests on land contained seedless vascular plants, insects, and amphibians.
- Cycads and reptiles, including dinosaurs, were prevalent in the Mesozoic era. Mammals and flowering plants evolved during the Cenozoic era.

10. Why are fossils either shells, bones, or teeth? ______________________

11. Indicate whether these comparisons between relative dating and absolute dating of fossils are true (T) or false (F).

| | Relative Dating | Absolute Dating |
|---|---|---|
| a. _____ | Date such as 3.5 MYA is known. | Date such as 3.5 MYA is not known. |
| b. _____ | Strata location must be known. | Strata location does not need to be known. |
| c. _____ | Only comparative age is known. | Age independent of other fossils is known. |

12. In the Precambrian, the first cells were prokaryotes that carried on a. __________ because the atmosphere contained no b. __________. Some of the earliest cells dated 3.5 billion years ago are found in c. __________, of which living examples are found in shallow seas today. One of the main events of the Precambrian is the evolution of photosynthesizers, which added d. __________ to the atmosphere. The eukaryotic cell, which appeared in the fossil record about 2.0 billion years ago, probably acquired its e. __________ gradually. For example, the f. __________ were added by the process called g. __________. Finally, h. __________, which may have been preceded by i. __________ reproduction, came into being.

13. Write a sentence based on the text material that includes the terms listed.

a. Cambrian period, invertebrates ______________________

______________________

b. Carboniferous forests, insects, seedless vascular plants, amphibians ______________________

______________________

14. Rewrite these false statements concerning the Mesozoic era to make them true.

a. The dominant plants of this era were flowering plants. ______________________________

______________________________________________________________________

b. The dominant animals on land were mammals. ______________________________

______________________________________________________________________

c. The dinosaurs lived on to become the dominant animals of the Cenozoic era. ______________

______________________________________________________________________

15. a. What type of vertebrate diversified during the Cenozoic era? ______________

b. What type of plant diversified during the Cenozoic era? ______________

## 19.3 Factors That Influence Evolution (pp. 334–37)

- The position of the continents changes over time because the Earth's crust consists of moving plates.
- Continental drift can explain patterns of past and present distributions of life-forms. Continental drift and meteorite impacts may have contributed to several mass extinction episodes during the history of life.

16. Indicate whether these statements about continental drift and the evolution of life are true (T) or false (F).

a. _____ The Earth existed for some time before Pangaea formed.
b. _____ The first continent was Pangaea.
c. _____ Laurasia and Gondwanaland resulted from the breakup of Pangaea.
d. _____ South America and Africa have matching coastlines.
e. _____ Widely separated continents have similar fossils that date from before separation occurred.

17. a. According to this diagram, the Earth's plates move like a(n) ______________________________.

b. What happens at ocean ridges? ______________________________

c. What happens at ocean trenches? ______________________________

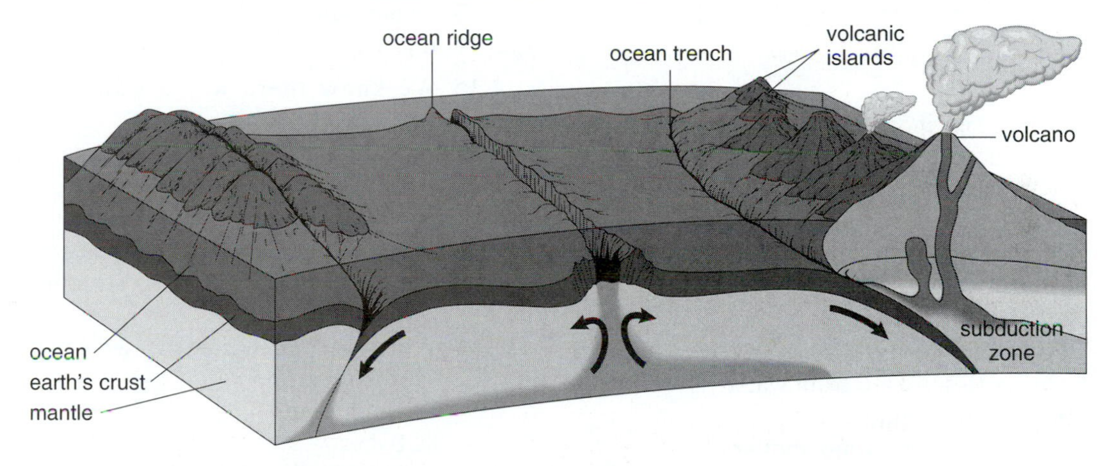


### Mass Extinctions (p. 336)

18. Place a check in front of the statement(s) that contributed to mass extinctions.

a. _____ The wrath of God showed its power.
b. _____ Asteroid impact spewed dust into the atmosphere.
c. _____ Climates changed due to continental drift.
d. _____ Species have a finite life existence.

# Chapter Test

## Objective Questions

Do not refer to the text when taking this test.

____ 1. Each of the following was present in the primitive atmosphere of the Earth EXCEPT
a. carbon dioxide.
b. carbon monoxide.
c. molecular nitrogen.
d. molecular oxygen.

____ 2. Miller's experiments produced
a. coacervate droplets from macromolecules.
b. inorganic substances from organic molecules.
c. organic molecules from inorganic substances.
d. protocells from macromolecules.

____ 3. Select the correct sequence that occurred on the primitive Earth.
a. gases, small molecules, macromolecules, protocells
b. macromolecules, small molecules, protocells, gases
c. protocells, macromolecules, small molecules, gases
d. small molecules, gases, macromolecules, protocells

____ 4. Microspheres formed from the polymerization of
a. amino acids.
b. nucleotides.
c. sugars.
d. water.

____ 5. Clay may have promoted the formation of macromolecules because it
a. attracts small organic molecules.
b. is dry.
c. lacks zinc and iron.
d. wards off energy.

____ 6. Protocells exhibited each of the following EXCEPT
a. the ability to separate from water.
b. a lipid-protein membrane.
c. conduction of energy metabolism.
d. means of self-replication.

____ 7. The most primitive life-forms were
a. anaerobic photosynthesizers.
b. eukaryotic plants.
c. eukaryotic protists.
d. prokaryotic cells.

____ 8. Currently, the ozone in the atmosphere
a. enhances photosynthesis.
b. promotes the origin of life today.
c. protects organisms from the effect of ultraviolet rays.
d. reacts with and destroys pollutants.

____ 9. Which is NOT a characteristic of the Cenozoic era?
a. appearance of first hominids
b. extinction of dinosaurs
c. dominance on land by flowering plants
d. increase in number of herbaceous plants

____ 10. Hydrothermal vents
a. could be a site for the origination and polymerization of monomers.
b. could be a site where nitrogen gas could have been reduced to $NH_3$.
c. are a site where chemosynthetic bacteria proliferate.
d. occur at ocean ridges where waters are deep and pressure is great.
e. All of these are correct.

In questions 11–14, match the descriptions with these time periods:
a. Precambrian
b. Paleozoic
c. Mesozoic
d. Cenozoic

____ 11. single cells
____ 12. mammals
____ 13. amphibians
____ 14. reptiles

____ 15. We know there was a Cambrian explosion because
a. a mass extinction took place due to all the dust in the air.
b. the fossil record is very rich.
c. so many fish were in the seas.
d. prokaryotes became so structurally diversified.

____ 16. Which of these organisms were prevalent in the Mesozoic era?
a. amphibians
b. fishes
c. cycads, dinosaurs
d. primates

____ 17. What happens when continents collide?
a. Mountain ranges develop.
b. The Earth moves slightly in the solar system.
c. Subduction zones appear.
d. All of these are correct.

____ 18. Mass extinctions
a. refer to loss of mass by many species.
b. may be due to asteroid impacts.
c. may be due to climatic changes.
d. Both *b* and *c* are correct.

## Critical Thinking Questions

Answer in complete sentences.

19. How do the currently existing RNA viruses lend support to the hypothesis that RNA could have been the first genetic material?

20. Some scientists hypothesize that birds are dinosaurs. How might biotechnological techniques help settle the question?

**Test Results:** ______ number correct ÷ 20 = ______ × 100 = ______ %

## Exploring the Internet

The Online Learning Center at *www.mhhe.com/maderbiology8* has additional study material and practice quizzes that can help you master the content of this chapter. You can also find links to websites exploring additional topics in biology. Access to the Online Learning Center is free for those who have purchased a new textbook.

## Answer Key

### Study Exercises

**1. a.** primitive earth **b.** cooling **c.** inorganic chemicals **d.** energy capture **e.** small organic molecules **f.** polymerization **g.** plasma membrane **h.** macromolecules. See also Figure 19.4, page 323, in text. **2.** an energy source, because amino acids do not react unless energy is provided **3.** To obtain microspheres, proteinoids must be placed in water. **4.** discovery of ribozymes, nucleotides with enzymatic properties **5. a.** F **b.** T **c.** F **d.** T **6. a.** statement c **b.** label *self-replicating system* between protocell and cell **7. a.** The ocean contained organic molecules that could serve as food. **b.** The atmosphere did not contain any oxygen. **8.** a, c, b, d **9. a.** P **b.** P **c.** P **d.** C **e.** C **f.** C **10.** These parts do not decompose. **11. a.** F **b.** T **c.** T **12. a.** anaerobic fermentation **b.** oxygen **c.** stromatolites **d.** oxygen **e.** organelles **f.** mitochondria (chloroplasts) **g.** endosymbiosis **h.** multicellularity **i.** sexual **13. a.** During the Cambrian period, many and various shelled invertebrates appeared. **b.** The seedless vascular plants that characterized the Carboniferous forests are of minor importance today, but at that time, they provided a home for insects and various amphibians—some of whom were very large. **14. a.** . . . were nonflowering plants (gymnosperms) **b.** . . . were reptiles, including dinosaurs **c.** The mammals lived on . . . **15. a.** mammals **b.** flowering plants (angiosperms) **16. a.** T **b.** F **c.** T **d.** T **e.** T **17. a.** conveyor belt **b.** seafloor spreading occurs **c.** subduction occurs **18.** b, c

### Chapter Test

**1.** d **2.** c **3.** a **4.** a **5.** a **6.** d **7.** d **8.** c **9.** b **10.** d **11.** a **12.** d **13.** b **14.** c **15.** b **16.** c **17.** a **18.** d **19.** Some RNA viruses exist that manage to replicate without need of DNA. The RNA genome of these viruses can be used as an mRNA to produce protein and additional RNA genomes. This suggests that the first cell would have been able to function without a DNA genome. **20.** The polymerase chain reaction could be used to make multiple copies of dinosaur genes, and then DNA fingerprinting could be used to compare the dinosaur genome to genome of birds. If the two patterns are quite similar, it would lend support to the hypothesis that birds are dinosaurs.

# 20

# Classification of Living Things

## Chapter Review

**Taxonomy** deals with the naming of organisms; each species is given a binomial name consisting of the genus and the specific epithet. Members of the same species share anatomical similarities and reproduce only with each other.

**Systematics** includes taxonomy and **classification.** The classification categories are: **species, genus, family, order, class, phylum,** and **kingdom.** Recently, a higher category, **domain,** has been added.

Classification should reflect **phylogeny,** which can be described in terms of **phylogenetic trees. Homology**—determined by similarity in structure due to a common ancestor, molecular data, and the fossil record—helps decipher phylogenies. **Cladistic systematics** uses shared, derived characters to construct **cladograms.** However, convergent evolution can make it seem as if groups share derived characters when they actually do not. In **phenetic systematics,** species are classified on the basis of shared similarities, regardless of whether the similarities are due to convergent evolution. **Traditional systematics** uses common ancestry and the degree of structural difference to construct phylogenetic trees.

The five-kingdom classification system recognizes Plantae, Animalia, Fungi, Protista, and Monera.

The three-domain system used by this text is based on molecular data and recognizes three domains: **Bacteria, Archaea,** and **Eukarya.** The domain Eukarya contains the kingdoms **Plantae, Animalia, Fungi,** and **Protista.**

## Study Exercises

Study the text section by section as you answer the questions that follow.

### 20.1 Taxonomy (pp. 342–45)

- Each known species has been given a binomial name consisting of the genus and specific epithet.
- Species are distinguished on the basis of structure and reproductive isolation. This chapter stresses reproductive isolation.

1. The branch of biology concerned with identifying and naming organisms is called __________________.
2. The [a] __________________ system of naming species contains two parts. The first part of the name is the [b] __________________. The second part of the name is the [c] __________________.
   Both names together are the [d] __________________ name. The [e] __________________ name can be used alone to refer to a group of related species.
3. A species can be distinguished by its distinctive structural [a] __________________, even though members of a species show [b] __________________. The biological definition of species states that members of a species [c] __________________ and share the same gene pool. But we know that [d] __________________ occurs between members of different species. In this chapter, we define a species as a(n) [e] __________________ category below the rank of genus. All species within a genus share a recent [f] __________________.

## Classification Categories (pp. 344–45)

- Classification usually involves the assignment of species to a genus, family, order, class, phylum, kingdom, and domain (the largest classification category).

4. a. Create a mnemonic device that will help you remember the order of the classification categories from domain to species. ______________________________

   b. Which category is just below family? ______________________________

   c. Which category is just above class? ______________________________

   d. Which category is just below class? ______________________________

   e. Which two categories are used in a binomial name? ______________, ______________

5. a. What is a structural, chromosomal, or molecular feature that distinguishes one group of organisms from another? ________

   b. Which taxonomic category has the most general characters in common? ________

   c. Which category has the most specific characters in common? ________

## 20.2 Phylogenetic Trees (pp. 346–50)

- Systematics encompasses both taxonomy (the naming of organisms) and classification (placing species in the proper categories).
- The fossil record, homology, and molecular data are used to decide the evolutionary relatedness of species and the construction of phylogenetic trees, diagrams that show their relatedness.

6. The goal of systematics is to determine [a]______________, the evolutionary history of a group of organisms. Common ancestors and lines of descent are found in diagrams called [b]______________.
7. A(n) [a]______________ character is one that is present in ancestral forms; a(n) [b]______________ character is one that is found in descendants.
8. To determine phylogeny, systematists rely on [a]______________, [b]______________, and the [c]______________ to tell primitive from derived characters.
9. Sometimes, the fossil record reveals how [a]______________ a particular group is. Some fossils are intermediate enough to show possible [b]______________ between two groups.
10. Similarity in structure due to descent from a common ancestor defines [a]______________ structures. Analogous structures have the same [b]______________, but there is no recent [c]______________ ancestor for the groups being studied. [d]______________ structures reveal that groups are closely related. [e]______________ structures appear in groups not closely related. In regard to molecular data, the more closely related the two groups of animals, the [f]______________ the differences between their genes.

## 20.3 Systematics Today (pp. 351–53)

- There are three main schools of systematics: cladistic, phenetic, and traditional systematics.

11. The three primary schools of systematics are [a]______________, [b]______________, and [c]______________. In the traditional school, the [d]______________ of structural difference is important. Therefore, a traditionalist does not group dinosaurs and birds together. In the cladistic school, the common ancestor is included in the group; therefore, dinosaurs are [e]______________ with birds.
12. To construct a cladogram, a cladist determines which characters are [a]______________ and which are [b]______________. A cladogram is composed of [c]______________, each of which contains a common ancestor and species derived from that ancestor.

13. In the diagram that follows, *a* is a(n) a. ______________________________ and *b* is a(n) b. ______________.

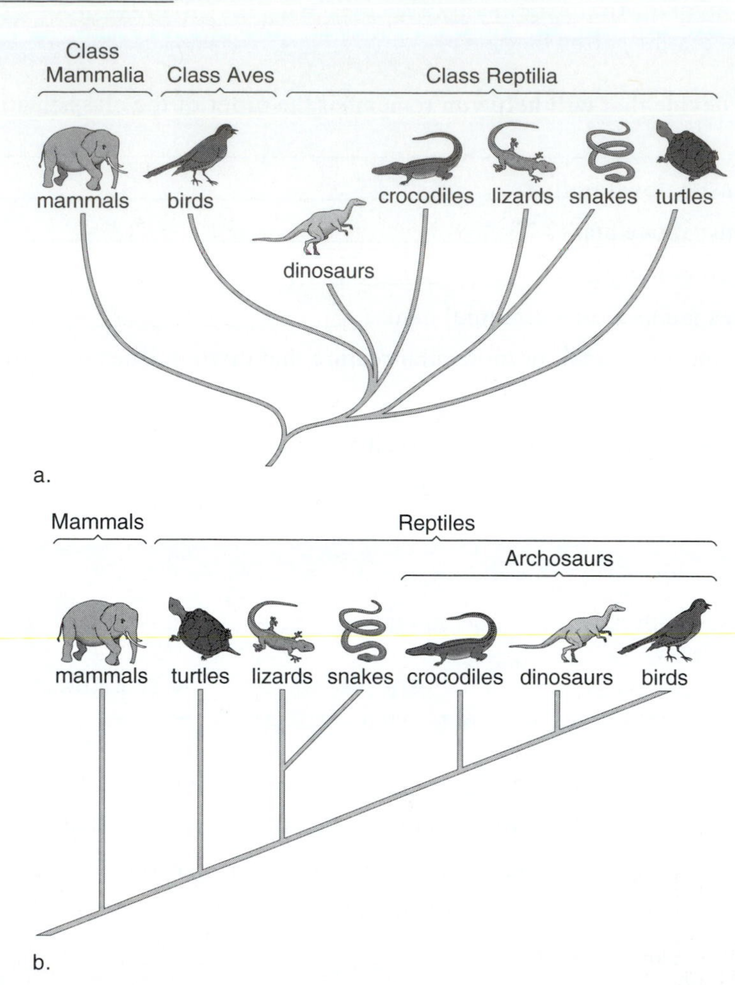


## 20.4 Classification Systems (pp. 354–56)

- The five-kingdom system contains these kingdoms: Plantae, Animalia, Fungi, Protista, and Monera.
- The three-domain system recognizes three domains: Bacteria, Archaea, and Eukarya. The domain Eukarya contains the kingdoms Protista, Fungi, Plantae, Animalia.

14. a. The three domains of life are ____________, ____________, ____________.

    b. Why are Archaea placed in their own domain? ______________________________

15. Match the description to these classification categories.
    1. domain Archaea and domain Bacteria
    2. domain Eukarya, kingdom Protista
    3. domain Eukarya, kingdom Fungi
    4. domain Eukarya, kingdom Plantae
    5. domain Eukarya, kingdom Animalia

    a. _____ eukaryotic, multicellular, motile, and ingest their food.
    b. _____ eukaryotic, unicellular or multicellular; absorb, ingest, or photosynthesize their food.
    c. _____ prokaryotic, unicellular, usually absorb or photosynthesize their food, motile or nonmotile.
    d. _____ eukaryotic, multicellular, absorb food, nonmotile.

16. Using the same listing of kingdoms as in question 15 match the five kingdoms to these organisms.
    a. _____ includes algae, protozoans, water molds, and slime molds
    b. _____ includes humans, clams, insects
    c. _____ includes yeasts, mushrooms, and molds
    d. _____ includes trees, grasses, and vines

# Chapter Test

## Objective Questions

Do not refer to the text when taking this test.

____ 1. Which of the following sequences is in correct order, starting from the most specific but fewest in number of species?
a. class, family, kingdom, order, phylum
b. family, order, class, phylum, kingdom
c. order, family, kingdom, class, phylum
d. phylum, order, kingdom, family, class

____ 2. In the scientific name *Elaphe obsoleta,* which is the genus name?
a. *Elaphe*
b. *obsoleta*
c. *bairdi*
d. *Elaphe obsoleta*

____ 3. In the scientific name *Elaphe obsoleta,* which name is the specific epithet?
a. *Elaphe*
b. *obsoleta*
c. *bairdi*
d. *Elaphe obsoleta*

____ 4. Structures having the same makeup but different functions are _______; structures having different makeup but similar functions are _______.
a. analogous; homologous
b. homologous; analogous

____ 5. Members of the same _______ are most similar to each other.
a. species
b. genus
c. class
d. family

____ 6. Organisms that can interbreed and bear fertile offspring are in the same
a. order.
b. family.
c. class.
d. species.

____ 7. Insect wings and bird wings are examples of _______ structures.
a. homologous
b. analogous
c. vestigial
d. None of these are correct.

____ 8. The presence of _______ structures strongly indicates that organisms are related.
a. homologous
b. analogous
c. vestigial
d. None of these are correct.

____ 9. Taxonomy is the branch of biology concerned with the
a. interaction of living organisms.
b. identifying, naming, and classifying living organisms.
c. history of humans.
d. history of dinosaurs.
e. interaction of living organisms with the inorganic environment.

____ 10. According to traditionalists, which of the following are descended from amphibians?
a. birds
b. reptiles
c. mammals
d. fish
e. plants

____ 11. Bacteria, including cyanobacteria, belong to the domain
a. Bacteria.
b. Protista.
c. Fungi.
d. Plantae.
e. Animalia.

____ 12. Molds and mushrooms belong to the kingdom
a. Monera.
b. Protista.
c. Fungi.
d. Plantae.
e. Animalia.

____ 13. Which of the following statements is false?
a. Taxonomists are biologists who classify living things.
b. Moving from genus to domain, more different types of species are included in each higher category.
c. Species in the same genus share very specific characteristics.
d. Organisms placed in the same genus are least closely related.

____ 14. In which domain(s) are the members unicellular and without a nucleus?
a. Bacteria
b. Protista
c. Fungi
d. Archaea
e. Both *b* and *c* are correct.
f. Both *a* and *d* are correct.

____ 15. Whenever a phylogenetic tree branches, there is assumed to be
  a. no living member of that group.
  b. always a living member of that group.
  c. a common ancestor.
  d. an embryo only.

____ 16. Those who believe that the ancestor for all mammals is a mammal, not a reptile, are in the ______ school.
  a. traditional
  b. cladistic
  c. phylogenetic

____ 17. When determining phylogeny, systematists use
  a. homology.
  b. the fossil record.
  c. molecular data.
  d. All of these are correct.
  e. None of these are correct.

____ 18. Acquisition of the same or similar characteristics in distantly related lines of descent is called
  a. parallel evolution.
  b. convergent evolution.
  c. evolution by natural selection.
  d. evolution by mutation.

____ 19. A similar banding pattern found in almost all species of moths is an example of
  a. parallel evolution.
  b. convergent evolution.
  c. evolution by natural selection.
  d. evolution by mutation.

____ 20. Analyzing molecular data to determine phylogeny involves which of the following techniques?
  a. comparing the base sequences of DNA and RNA
  b. immunological studies
  c. DNA-DNA hybridization
  d. Both *a* and *b* are correct.
  e. All of these are correct.

____ 21. The degree of relatedness of species can be indicated by
  a. the fossil record.
  b. homology.
  c. analyzing molecular data.
  d. All of these are correct.

____ 22. The fossil record is incomplete because
  a. not every organism becomes a fossil.
  b. most organisms decay before they become buried.
  c. fossils must survive intense geological processes.
  d. All of these are correct.

____ 23. The traditional school of systematics
  a. came after the cladistic school.
  b. stresses common ancestry.
  c. stresses the degree of structural difference among divergent groups.
  d. Both *b* and *c* are correct.
  e. All of these are correct.

____ 24. Cladists
  a. agree with the traditional approach to systematics.
  b. believe that any group must contain the ancestor to that group.
  c. believe that the common ancestor for mammals is a reptile.
  d. All of these are correct.

____ 25. A clade is a
  a. tool used to excavate archeological sites.
  b. piece of laboratory equipment used to sterilize glassware.
  c. division of a cladogram.
  d. common ancestor, with its descendent species, on a cladogram.
  e. Both *c* and *d* are correct.
  f. None of these are correct.

## Critical Thinking Questions

Answer in complete sentences.

26. Why is it imperative that organisms be given a scientific name in Latin, rather than simply using common names?

27. Why do you think the species concept can often be difficult to test in wild populations?

**Test Results:** ______ number correct ÷ 27 = ______ × 100 = ______ %

## Exploring the Internet

The Online Learning Center at *www.mhhe.com/maderbiology8* has additional study material and practice quizzes that can help you master the content of this chapter. You can also find links to websites exploring additional topics in biology. Access to the Online Learning Center is free for those who have purchased a new textbook.

## Answer Key

### Study Exercises

**1.** taxonomy **2. a.** binomial **b.** genus **c.** specific epithet **d.** species **e.** genus **3. a.** characteristics **b.** variations **c.** interbreed **d.** hybridization **e.** taxonomic **f.** common ancestor **4. a.** Example: Daring Karen Pushed Cans Off Friendly Grandmother's Stove (domain, kingdom, phylum, class, order, family, genus, species) **b.** genus **c.** phylum **d.** order **e.** genus, species **5. a.** character **b.** domain **c.** species **6. a.** phylogeny **b.** phylogenetic trees **7. a.** primitive **b.** derived **8. a.** homology **b.** molecular data **c.** fossil record **9. a.** old **b.** relationships **10. a.** homologous **b.** function **c.** common **d.** Homologous **e.** Analogous **f.** fewer **11. a.** cladistic **b.** phenetic **c.** traditional **d.** degree **e.** classified **12. a.** primitive **b.** derived **c.** clades **13. a.** phylogenetic tree **b.** cladogram **14. a.** Bacteria, Archaea, Eukarya **b.** The Archaea have nucleotide sequences that are unique and do not match those of Eukarya or Bacteria. **15. a.** 5 **b.** 2 **c.** 1 **d.** 3 **16. a.** 2 **b.** 5 **c.** 3 **d.** 4

### Chapter Test

**1.** b **2.** a **3.** b **4.** b **5.** a **6.** d **7.** b **8.** a **9.** b **10.** b **11.** a **12.** c **13.** d **14.** f **15.** c **16.** b **17.** d **18.** b **19.** a **20.** e **21.** d **22.** d **23.** d **24.** b **25.** e **26.** The common name for organisms will vary from country to country because of language differences. Even among those speaking the same language, different common names are sometimes used for the same organism. However, with the scientific Latin name, we know we are referring to the same organism. **27.** Two species may never meet to test whether they can effectively interbreed.

# PART IV MICROBIOLOGY AND EVOLUTION

# 21

# VIRUSES, BACTERIA, AND ARCHAEA

## CHAPTER REVIEW

**Viruses** are noncellular entities consisting of an outer capsid and an inner core of nucleic acid. They are obligate intracellular parasites that reproduce inside cells. **Bacteriophages** undergo either a **lytic cycle,** in which they break out of the host cell, or a **lysogenic cycle,** in which viral DNA is integrated into bacterial DNA. Animal viruses have a membranous envelope they acquire when they bud from the host cell; some RNA viruses are **retroviruses** that transcribe RNA into DNA, which then becomes incorporated into the host genome.

The **prokaryotes** include Bacteria and Archaea. Prokaryotes lack a nucleus and most other cytoplasmic organelles found in eukaryotic cells. They reproduce asexually by **binary fission,** but genetic recombination occurs by **conjugation, transformation,** and **transduction.** Prokaryotes differ in their need for oxygen, but most bacteria are aerobic heterotrophs that act as decomposers. Some bacteria are **symbiotic,** including several that are parasitic. Other prokaryotes acquire energy and nutrients by using photosynthetic or **chemoautotrophic** processes.

**Bacteria** (domain Bacteria) are the most common type of prokaryote. Bacterial cell walls contain either thick layers (Gram-positive bacteria) or thin layers (Gram-negative bacteria) of the molecule peptidoglycan. Classification of bacteria was previously made on the basis of Gram-positive, Gram-negative, or other criteria. However, bacterial taxonomy is now based primarily on comparisons of bacterial ribosomal RNA sequences. Cyanobacteria are ecologically important photosynthesizers.

The **archaea** (domain Archaea) are prokaryotes with biochemical differences that distinguish them from bacteria and eukaryotes. Archaea are more closely related to eukarya than bacteria. Types of archaea are methanogens, halophiles, and thermoacidophiles.

## STUDY EXERCISES

Study the text section by section as you answer the questions that follow.

### 21.1 THE VIRUSES (PP. 362–67)

- Viruses are noncellular, while prokaryotes are fully functioning cellular organisms.
- All viruses have an outer capsid composed of protein and an inner core of nucleic acid. Some have an outer membranous envelope.
- Viruses are obligate intracellular parasites, including bacteriophages (reproduce inside bacteria), and plant and animal viruses.

1. Fill in the following table to contrast viruses and bacteria:

| Viruses | Bacteria |
|---|---|
| Structure | |
| Life cycle occurs where? | |
| Parasitic? | |

## Viral Reproduction (pp. 364–66)

2. a. Label the diagram, which describes how viruses such as bacteriophages replicate, using the alphabetized list of terms:

   attachment
   biosynthesis
   integration
   lysogenic cycle
   lytic cycle
   maturation
   penetration
   release

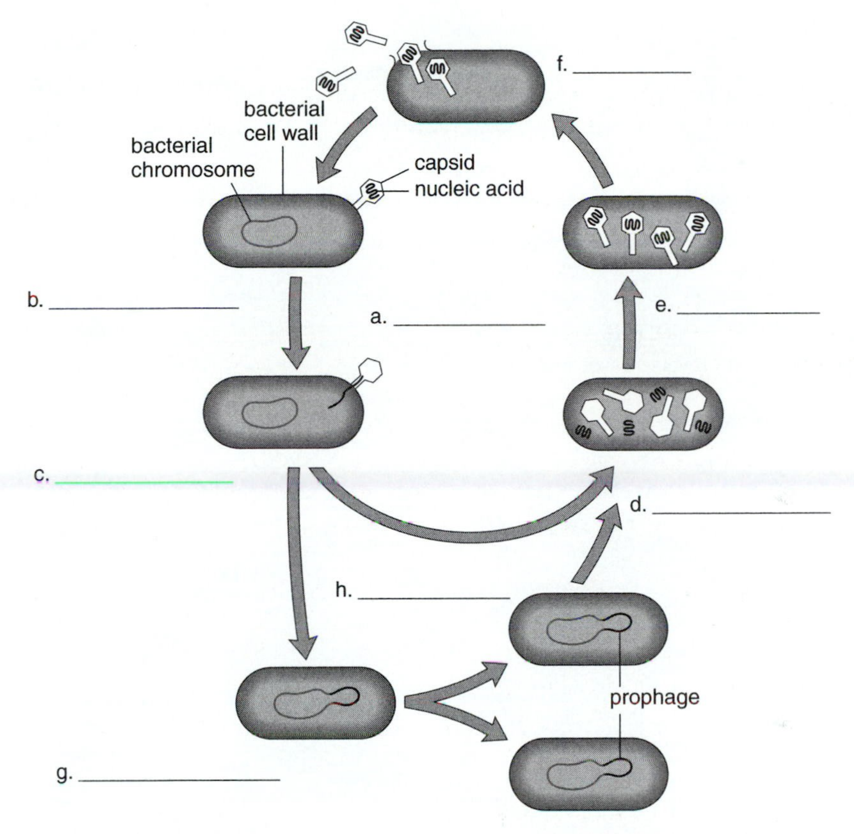


b. Which cycle produces viruses? ___________________

c. Which cycle is dormant? ___________________

d. Which cycle kills, or lyses, the host? ___________________

3. Place the correct number from the following diagram next to its description:
   a. _____ reverse transcription
   b. _____ integration
   c. _____ biosynthesis
   d. _____ attachment
   e. _____ fusion
   f. _____ maturation
   g. _____ release
   h. _____ replication
   i. _____ uncoating

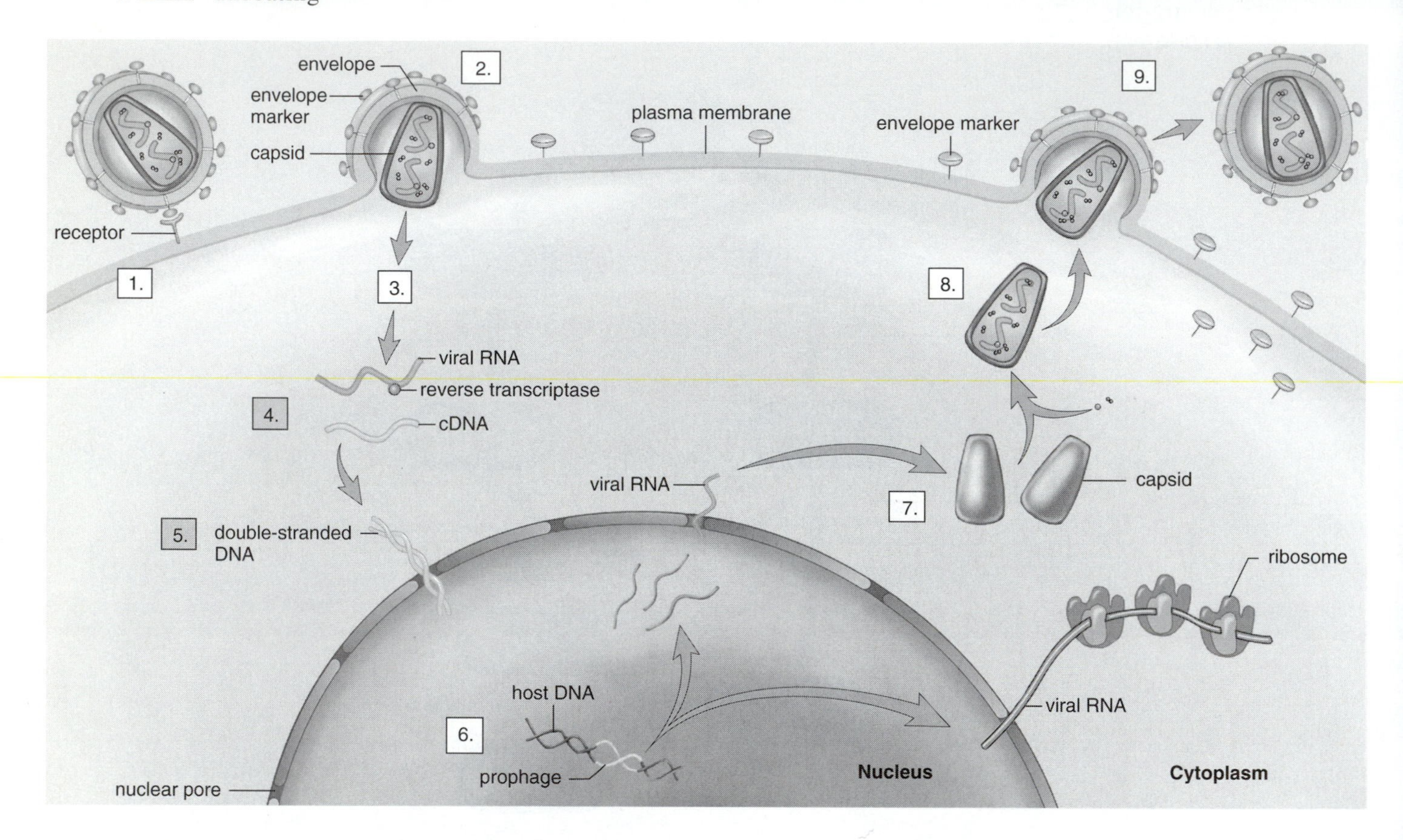


4. The life cycle shown in question 3 is that of a(n) [a]__________________. In this life cycle, [b]__________________ integrates into the host genome until transcription occurs. Then new viruses are produced by: [c]__________________, [d]__________________, and [e]__________________. When the viruses leave the cell, they are surrounded by a(n) [f]__________________.

5. Indicate whether these statements are true (T) or false (F).
   a. _____ Antibiotics are helpful for viral infections.
   b. _____ Antiviral drugs act by interfering with viral replication.
   c. _____ There are no vaccines for viral infections.
   d. _____ Prions are neither viruses nor bacteria; they are protein particles.

- Prokaryotes, the archaea and bacteria, lack a nucleus and most other cytoplasmic organelles found in eukaryotic cells.

6. Label this diagram of a bacterial cell using the alphabetized list of terms.
   capsule
   cell wall
   cytoplasm
   fimbriae
   flagellum
   nucleoid
   plasma membrane
   plasmid
   ribosome

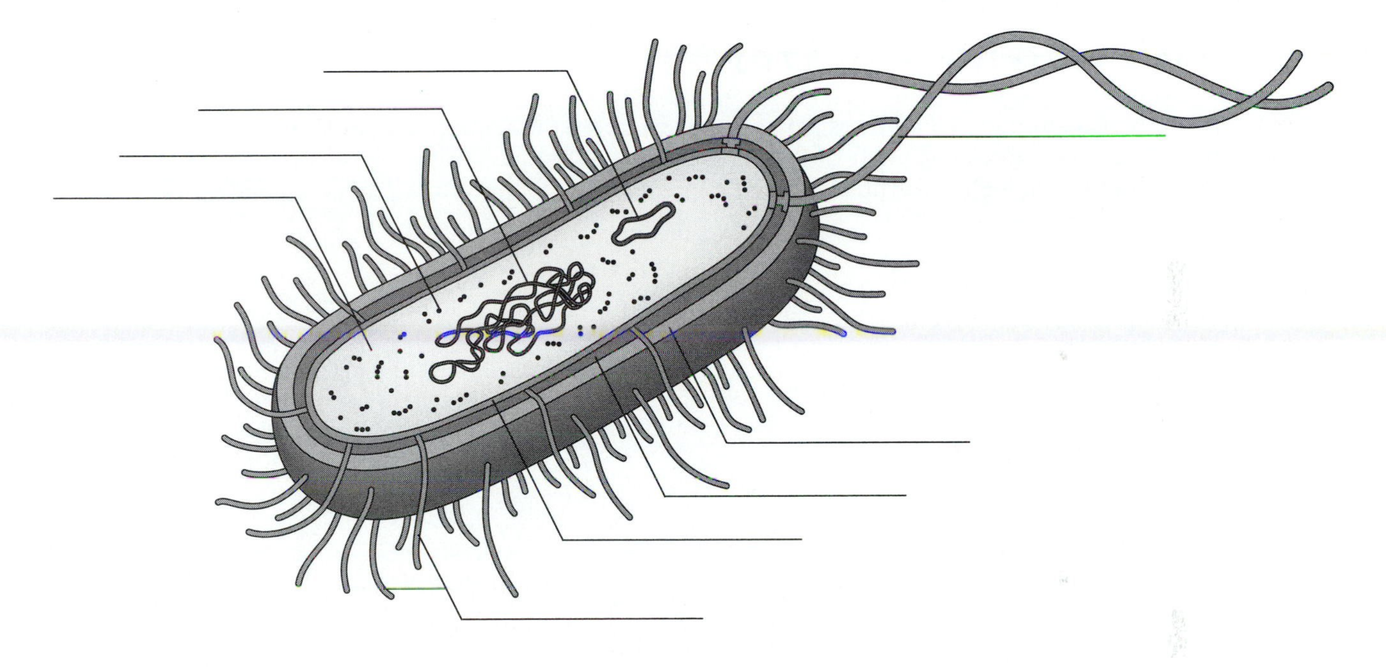

7. Based on the diagram you labeled in question 6, which of these structures are present in eukaryotic animal/plant cells but are not in a prokaryotic cell?
   a. _____ plasma membrane
   b. _____ nucleus
   c. _____ ribosomes
   d. _____ mitochondria
   e. _____ cell wall
   f. _____ chloroplasts
   g. _____ flagella

8. Based on the diagram you labeled in question 6, what four structures are present in a prokaryotic cell but absent from a eukaryotic cell? What are their functions?

| | Structure | Function |
|---|---|---|
| a. | ______________________ | ______________________ |
| b. | ______________________ | ______________________ |
| c. | ______________________ | ______________________ |
| d. | ______________________ | ______________________ |

## Reproduction in Prokaryotes (p. 369)

- Prokaryotes reproduce asexually by binary fission. Mutations introduce variations; genetic recombinations have been observed among bacteria.

9. Match the descriptions to these terms, which pertain to the reproduction or survival of bacteria:
   1. binary fission
   2. conjugation
   3. transformation
   4. transduction
   5. endospores

   a. _____ bacteria picks up free pieces of DNA
   b. _____ a means of survival
   c. _____ asexual division
   d. _____ donor cell passes DNA to recipient cell
   e. _____ bacteriophages carry bacterial DNA from one cell to the next

## Prokaryotic Nutrition (p. 370)

- Some prokaryotes require oxygen; others are obligate anaerobes or facultative anaerobes.
- Some prokaryotes are autotrophs, being either photoautotrophs or chemoautotrophs.
- Most prokaryotes are chemoheterotrophs. Many chemoheterotrophs are symbiotic, being mutualistic, commensalistic, or parasitic.

10. Match the descriptions to these organisms:
    1. chemoautotrophic prokaryotes
    2. cyanobacteria
    3. parasitic bacteria
    4. saprotrophic bacteria

    a. _____ decomposers
    b. _____ $O_2$ given off
    c. _____ $NH_3 \rightarrow NO_2^-$
    d. _____ disease
11. Match the relationships to these terms:
    1. commensalistic
    2. mutualistic
    3. parasitic
    4. symbiosis

    a. _____ includes all the others
    b. _____ bacteria living in nodules of legumes
    c. _____ bacteria living in intestines
    d. _____ bacteria that cause strep throat

## 21.3 The Bacteria (pp. 371–73)

- Gram staining, shape of cell, type of nutrition, and other biochemical characteristics are used to differentiate groups of bacteria.

12. Label the three shapes of bacteria in this diagram:

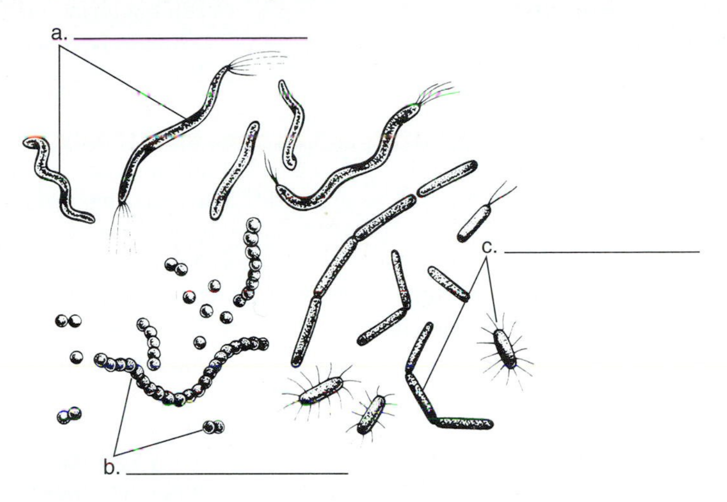


13. Place a check next to all characteristics typical of cyanobacteria.
    a. _____ many forms of nutrition
    b. _____ always photosynthetic
    c. _____ have flagella
    d. _____ form lichens
    e. _____ associated with algal bloom
    f. _____ nitrogen fixing

## 21.4 The Archaea (pp. 373–75)

- The archaea have biochemical characteristics that distinguish them from the bacteria.
- The archaea are specialized and live in extreme habitats.

14. Archaea are able to live in extreme environments. Name the type of archaea that live in the following habitats:
    a. swamps and marshes; produce methane ____________________
    b. salty environments (Great Salt Lake in Utah) ____________________
    c. hot and acidic environments (hot sulfur springs of Yellowstone National Park) ____________________

## Chapter Test

### Objective Questions

Do not refer to the text when taking this test.

____ 1. Which is NOT generally true of viruses?
a. have a nucleic acid core
b. have a capsid
c. have a specific host range
d. reproduce independently

____ 2. Which viral life cycle does NOT immediately rupture the bacterial cell?
a. lysogenic
b. lytic

____ 3. Which is true concerning animal viruses?
a. lack an envelope when they leave the host cell
b. attack the host cell by exocytosis
c. have an outer coat of nucleic acid
d. some have RNA genomes

____ 4. In the lytic cycle, the term *maturation* refers to the
a. translation of RNA.
b. integration of cDNA.
c. assembly of parts into new viruses.
d. All of these are correct.

____ 5. Which shape is NOT representative of bacteria?
a. bacillus
b. coccus
c. flagellar
d. spirillum

____ 6. Which of the following is NOT a form of genetic recombination in bacteria?
a. binary fission
b. conjugation
c. transduction
d. transformation

____ 7. The function of the bacterial endospore is to
a. increase the rate of anaerobic respiration.
b. promote asexual reproduction.
c. protect against attack from immune systems.
d. withstand harsh environmental conditions.

____ 8. A bacterium that can exist in the presence or absence of oxygen is a(n)
a. autotroph.
b. facultative anaerobe.
c. obligate anaerobe.
d. saprotroph.

____ 9. A bacterium produces vitamins for a host while gaining a habitat. This relationship is
a. commensalism.
b. mutualism.
c. parasitism.
d. predation.

____ 10. Which of these is NOT a correct contrast between bacteria and eukaryotes?

| | **Bacteria** | **Eukaryotic** |
|---|---|---|
| a. | binary fission | mitotic cell division |
| b. | nucleoid | nucleus |
| c. | asexual only | asexual and sexual |
| d. | nonmotile | motile |

____ 11. The cyanobacteria differ from other bacteria in which of the following ways?
a. They are unicellular, and other bacteria are filamentous.
b. They are able to form spores, whereas other bacteria cannot.
c. They are autotrophic, whereas other bacteria never are.
d. They release oxygen during photosynthesis, whereas other bacteria do not.

____ 12. A virus infecting a bacterium injects its ______ and leaves behind its ______.
a. protein coat; outer capsule
b. nucleic acid; capsid
c. nucleus; nucleoplasm
d. genes; metabolic enzymes

____ 13. Bacteria have
a. a cell wall with a construction similar to that of plants.
b. flagella with a construction different from that of eukaryotes.
c. mitochondria but not chloroplasts.
d. All of these are correct.

____ 14. Viruses are not in the classification system because they
a. are obligate parasites.
b. are noncellular.
c. can integrate into the host genome.
d. All of these are correct.

____ 15. Chemoautotrophic bacteria
a. give off oxygen just like plants do.
b. are exemplified by the nitrifying bacteria that oxidize ammonia ($NH_3$) to nitrites ($NO_2^-$).
c. are decomposers like all bacteria.
d. Both *b* and *c* are correct.

____ 16. Why can't cyanobacteria be classified with the eukaryotic algae?
a. They fix atmospheric nitrogen.
b. They form a symbiotic relationship with fungi.
c. They cause disease.
d. They do not have a nucleus.

____17. Which of these is (are) a true statement(s)?
a. Archaea are in a separate kingdom.
b. Archaea are in their own domain.
c. Archaea are found in most every habitat.
d. Archaea are found in extreme habitats like swamps, salty lakes, hot, acidic aquatic habitats.
e. Both *b* and *d* are correct.

____18. How are archaea different from bacteria?
a. Archaea have a nucleus and bacteria do not.
b. Archaea live in extreme habitats and bacteria do not.
c. Archaea have nucleotide sequences not found in bacteria.
d. Archaea have a cell wall and bacteria do not.
e. Both *b* and *c* are correct.

## Critical Thinking Questions

Answer in complete sentences.

19. Taking into consideration the specificity of viruses, explain how HIV (the AIDS virus) may have come into being.

20. Bacteria have diverse lifestyles. Give examples.

**Test Results:** ______ number correct ÷ 20 = ______ × 100 = ______ %

## Exploring the Internet

The Online Learning Center at *www.mhhe.com/maderbiology8* has additional study material and practice quizzes that can help you master the content of this chapter. You can also find links to websites exploring additional topics in biology. Access to the Online Learning Center is free for those who have purchased a new textbook.

## Answer Key

### Study Exercises

**1.**

| Viruses | Bacteria |
|---|---|
| capsid plus nucleic acid core | prokaryotic cell |
| in host cell | independently |
| always | sometimes |

**2. a.** See Figure 21.3, page 365, in text. **b.** lytic **c.** lysogenic **d.** lytic **3. a.** 4 **b.** 6 **c.** 7 **d.** 1 **e.** 2 **f.** 8 **g.** 9 **h.** 5 **i.** 3 **4. a.** retrovirus **b.** cDNA **c.** biosynthesis **d.** maturation **e.** release **f.** envelope **5. a.** F **b.** T **c.** F **d.** T **6. a.** cytoplasm **b.** ribosome **c.** nucleoid **d.** plasmid **e.** flagellum **f.** capsule **g.** cell wall **h.** plasma membrane **i.** fimbriae **7.** b, d, f **8. a.** plasmid, accessory ring of DNA **b.** fimbriae, adhere to surfaces **c.** glycocalyx/slime layer, protection **d.** nucleoid, location of DNA **9. a.** 3 **b.** 5 **c.** 1 **d.** 2 **e.** 4 **10. a.** 4 **b.** 2 **c.** 1 **d.** 3 **11. a.** 4 **b.** 2 **c.** 1 **d.** 3 **12. a.** spirillum **b.** coccus **c.** bacillus **13.** b, d, e, f **14. a.** methanogens **b.** halophiles **c.** thermoacidophiles

### Chapter Test

**1.** d **2.** a **3.** d **4.** c **5.** c **6.** a **7.** d **8.** b **9.** b **10.** d **11.** d **12.** b **13.** b **14.** b **15.** b **16.** d **17.** e **18.** c **19.** HIV attacks a particular immune cell. Therefore, the RNA found in these viruses must be derived from this type of cell. **20.** Bacteria carry on various means of nutrition: saprotrophic, chemosynthetic, and photosynthetic. Bacteria are symbiotic: mutualistic, commensalistic, or parasitic. Bacteria vary in their need for oxygen and can be anaerobic, facultative, or aerobic. Saprotrophic bacteria can digest almost any type of material and can live and survive under all sorts of conditions.

# 22

# The Protists

## Chapter Review

**Protists** are members of the domain Eukarya and the kingdom Protista. Most protists are unicellular organisms, but there are also some multicellular forms.

**Algae** are aquatic photosynthesizers. Algae are classified according to their pigments (colors). Green algae are photoautotrophs and are diverse: Some are unicellular or colonial flagellates, some are filamentous, and some are multicellular sheets. **Diatoms** and **dinoflagellates** are unicellular producers in oceans; red and brown algae are seaweeds; **euglenoids** are unicellular and some have chloroplasts in addition to flagella. Every type of life cycle is seen among the algae.

**Protozoans** are unicellular, aquatic heterotrophs that ingest their food. They are classified according to their type of locomotor organelle. The **amoeboids, radiolarians,** and **foraminiferans** locomote by pseudopods; the **ciliates** are very complex and locomote by cilia. The **sporozoans** have no locomotor organelles. Sporozoans are all animal parasites. Malaria is a significant disease caused by a sporozoan.

**Slime molds,** which are terrestrial, and **water molds,** which are aquatic, have some characteristics in common with fungi and some characteristics that separate them from fungi. They have an amoeboid stage and then form fruiting bodies, which produce spores dispersed by the wind. Water molds have filamentous bodies.

## Study Exercises

Study the text section by section as you answer the questions that follow.

### 22.1 General Biology of Protists (pp. 380–81)

- Endosymbiosis played a role in the origin of the eukaryotic cell.
- The protists are largely unicellular, but are varied and complex in structure and life cycle.

1. Match the organisms with these phyla:
   1. phylum Chlorophyta
   2. phylum Phaeophyta
   3. phylum Bacillariophyta
   4. phylum Pyrrophyta
   5. phylum Euglenophyta
   6. phylum Rhodophyta

   a. _____ dinoflagellates
   b. _____ euglenoids
   c. _____ brown algae
   d. _____ red algae
   e. _____ green algae
   f. _____ golden-brown algae

## 22.2 Diversity of Protists (pp. 382–94)

- The green algae, the red algae, and the brown algae are protists. All of these have chlorophyll *a* and photosynthesize.
- Diatoms are unicellular with a two-piece construction of the cell wall.
- The dinoflagellates, the euglenoids, and the zooflagellates use flagella for locomotion.
- The amoeboids, the foraminiferans, and the radiolarians use pseudopods for locomotion.
- The ciliates are a diverse group that use cilia for locomotion.
- The sporozoans are nonmotile spore-forming parasites.
- Slime molds include the plasmodial slime molds and the cellular slime molds.
- Water molds are filamentous decomposers.

2. Green algae are believed to be related to plants because they have a cell wall that contains a. ____________________, they possess chlorophylls b. ______ and c. ______, and they store reserve food as d. ____________________.

3. Complete the table describing the algae by placing the following terms in the appropriate columns (some terms are used more than once).

| I | II |
|---|---|
| unicellular | zoospores |
| filamentous | daughter colonies |
| colonial | conjugation |
| multicellular | alternation of generations |

| Algae | I. | II. |
|---|---|---|
| *Chlamydomonas* | | |
| *Spirogyra* | | |
| *Ulva* | | |
| *Volvox* | | |

4. Label this diagram of the life cycle of *Chlamydomonas*, using these terms:

fertilization   meiosis   zoospores   zygote

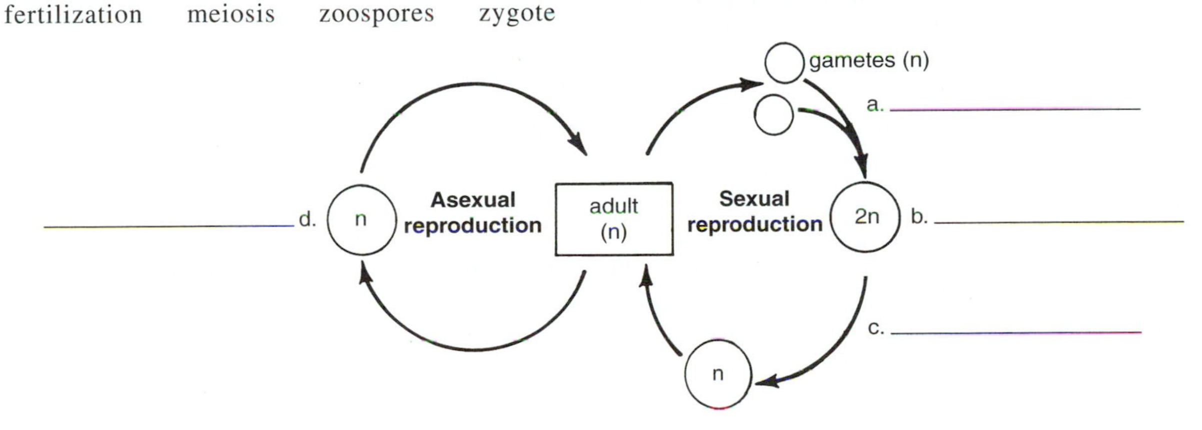


e. Which portions of this life cycle are haploid? ____________________

f. Which portion is diploid? ____________________

g. What type of life cycle is this? ____________________

5. Label this diagram of the life cycle of *Ulva*, using the following alphabetized list of terms.

fertilization
gametophyte
meiosis
sporophyte

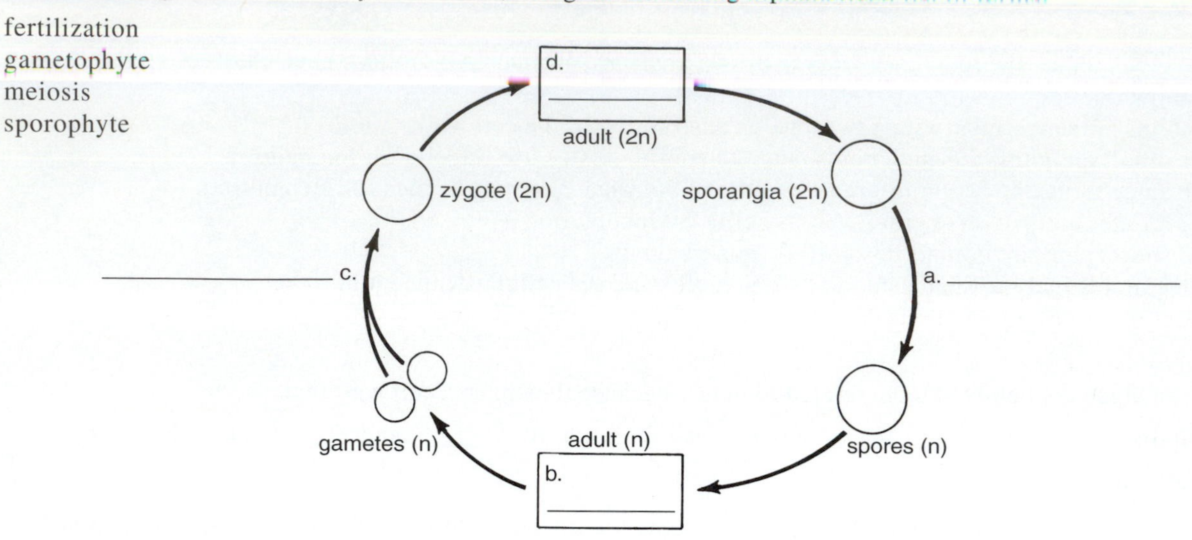


6. Match the traits with these protists (some numbers are used more than once):
   1. brown algae
   2. diatoms
   3. euglenoids
   4. dinoflagellates

   a. _____ are numerous photosynthesizers in the ocean
   b. _____ have animal-like and plantlike characteristics
   c. _____ have chlorophylls *a* and *c*, and carotenoid pigment
   d. _____ have silica-impregnated valves
   e. _____ are used as filtering agents and scouring powders
   f. _____ cause red tide
   g. _____ have a symbiotic relationship with corals
   h. _____ are seaweeds

7. Place the appropriate letter next to each description.

   B—brown algae R—red algae

   a. _____ *Fucus,* a rockweed
   b. _____ *Laminaria,* a kelp
   c. _____ adapted to cold, rough water
   d. _____ adapted to warm, gentle water
   e. _____ economically important as source of agar

8. Protozoans are typically a. _______________, b. _______________, and c. _______________ organisms. Some protozoans d. _______________ and engulf their food; others are e. _______________ and absorb nutrients; others are f. _______________ and cause disease.

9. Complete this table, classifying the protozoans by means of locomotion and giving an example organism of each:

| Classes | Organelle of Locomotion | Example |
|---|---|---|
| Zooflagellates | | |
| Amoeboids | | |
| Ciliates | | |
| Sporozoans | | |

10. Label this diagram of *Paramecium,* using the following alphabetized list of terms.

anal pore
contractile vacuole
food vacuole
macronucleus
micronucleus
oral groove
pellicle

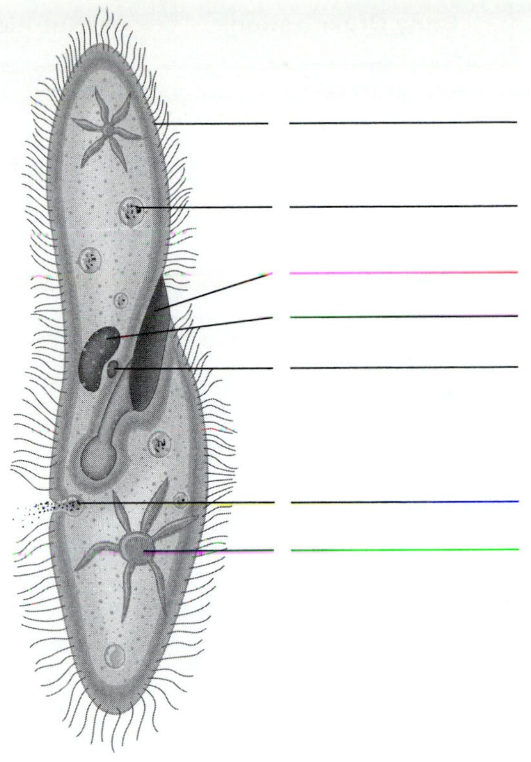

11. Ciliates, such as *Paramecium,* have hundreds of a.________________ that extend through a pellicle. Beneath the pellicle are numerous oval capsules that contain b.________________, used for defense. Food is swept down a(n) c.________________, at the end of which food vacuoles form. Ciliates have two nuclei: a large d.________________ that controls normal metabolism and one or more micronuclei used during conjugation.

12. Fill in the blanks about the life cycle of *Plasmodium vivax,* the organism that causes malaria.

In the gut of a a.________________ (male/female) mosquito, gametes of the malarial parasite *Plasmodium vivax* fuse. This is the b.________________ (sexual/asexual) phase of the *Plasmodium vivax* life cycle. The sporozoites then migrate to the c.________________ of the mosquito. When an infected mosquito bites a human, the protist exits the mosquito and enters the human through the salivary duct, where the spores enter d.________________, which rupture, releasing spores and toxins into the bloodstream. This causes the symptoms of e.________________________________________________. Some spores become f.________________ if taken up by a female mosquito. Blood taken up by a mosquito enters its food canal. From there, the gametes travel to the g.________________ of the mosquito, where the cycle begins again.

13. Complete this table to describe slime molds and water molds.

| Type of Mold | Body Organization | Nutrition | Reproduction |
|---|---|---|---|
| Plasmodial slime molds | | | |
| Cellular slime molds | | | |
| Water molds | | | |

## Objective Questions

Do not refer to the text when taking this test.

____ 1. Green algae
- a. store reserve food as lipid.
- b. do not give off oxygen.
- c. possess chlorophylls *a* and *b*.
- d. have a cell wall that contains pectin.

____ 2. Classification of algae according to color
- a. can no longer be justified.
- b. is based on the type of pigments they contain.
- c. suggests that they do not have chlorophyll.
- d. means that some algae are colorless.

____ 3. Which is NOT true of *Chlamydomonas?*
- a. has an eyespot
- b. produces zoospores
- c. is multicellular
- d. All of these are true.

____ 4. *Volvox*
- a. does not reproduce.
- b. is a colonial alga.
- c. produces daughter colonies.
- d. Both *b* and *c* are correct.

____ 5. Which of these is NOT true of *Spirogyra?*
- a. has a spiral chloroplast
- b. is filamentous
- c. carries out conjugation
- d. reproduces asexually by forming spores

____ 6. Some brown algae
- a. live at sea.
- b. are quite large.
- c. produce algin.
- d. All of these are correct.

____ 7. Diatoms
- a. reproduce sexually.
- b. have a cell wall impregnated with cellulose.
- c. are flagellated.
- d. resemble a pill box.

____ 8. Which of these is NOT true of dinoflagellates?
- a. are numerous in the ocean
- b. have the same pigments as green algae
- c. protected by cellulose plates
- d. have two flagella

____ 9. Which is (are) true of euglenoids?
- a. They have flagella.
- b. Some have chloroplasts.
- c. They reproduce asexually.
- d. All of these are correct.

____ 10. Both red algae and brown algae
- a. have the same pigments.
- b. are delicate in appearance.
- c. are seaweeds.
- d. are economically unimportant.

____ 11. Protozoans are not animals because they are
- a. pigmented.
- b. motile.
- c. unicellular.
- d. All of these are correct.

____ 12. Amoeboids
- a. have pseudopods.
- b. never have a shell.
- c. always live in fresh water.
- d. All of these are correct.

____ 13. Ciliates
- a. have a macronucleus and a micronucleus.
- b. do not move.
- c. are parasitic.
- d. are usually saprotrophic.

____ 14. A trypanosome causes
- a. malaria.
- b. trichinosis.
- c. an intestinal infection.
- d. African sleeping sickness.

____15. Which one of these is NOT an alga?
a. *Chlamydomonas*
b. *Volvox*
c. *Paramecium*
d. All of these are algae.

____16. In the life cycle of *Plasmodium vivax,* the cause of one type of malaria,
a. sexual reproduction occurs in a mosquito.
b. red blood cells burst, causing chills and fever.
c. spores and gametes form.
d. All of these are correct.

____17. Slime molds
a. are exactly like fungi.
b. have a body composed of hyphae.
c. produce spores.
d. All of these are correct.

## Critical Thinking Questions

18. How does the versatility of the ciliate *Paramecium* compare to that of a multicellular organism cell?

19. Algae and protozoans are in the same kingdom. Do they seem closely related? Why or why not?

**Test Results:** ______ number correct ÷ 19 = ______ × 100 = ______ %

## Exploring the Internet

The Online Learning Center at *www.mhhe.com/maderbiology8* has additional study material and practice quizzes that can help you master the content of this chapter. You can also find links to websites exploring additional topics in biology. Access to the Online Learning Center is free for those who have purchased a new textbook.

## Answer Key

### Study Exercises

**1. a.** 4 **b.** 5 **c.** 2 **d.** 6 **e.** 1 **f.** 3 **2. a.** cellulose **b.** *a* **c.** *b* **d.** starch
**3.**

| I. | II. |
|---|---|
| unicellular | zoospores |
| filamentous | conjugation |
| multicellular | alternation of generations |
| colonial | daughter colonies |

**4. a.** fertilization **b.** zygote **c.** meiosis **d.** zoospores **e.** zoospores, adult, gametes **f.** zygote **g.** haploid **5. a.** meiosis **b.** gametophyte **c.** fertilization **d.** sporophyte **6. a.** 2, 4 **b.** 3 **c.** 1, 2 **d.** 2 **e.** 2 **f.** 4 **g.** 4 **h.** 1 **7. a.** B **b.** B **c.** B **d.** R **e.** R **8. a.** heterotrophic **b.** unicellular **c.** motile **d.** capture **e.** saprotrophic **f.** parasitic **9.** See Table 22.1, page 389, in text. **10.** See Figure 22.16*b,* page 391, in text. **11. a.** cilia **b.** trichocysts **c.** gullet **d.** macronucleus **12. a.** female **b.** sexual **c.** salivary gland **d.** red blood cells **e.** recurring chills and fever **f.** gametes **g.** gut. See Figure 22.17, page 392, in text.

13.

| Body Organization | Nutrition | Reproduction |
|---|---|---|
| Plasmodium | Phagocytosis | Sporangium produces spores, which release haploid flagellated cells that fuse, forming zygote |
| Individual amoeboid cells | Phagocytosis | Sporangium produces spores |
| Filamentous, cell walls are cellulose | Parasitism, decomposition | Meiosis produces gametes; otherwise asexual by zoospores |

## Chapter Test

**1.** c **2.** b **3.** c **4.** d **5.** d **6.** d **7.** d **8.** b **9.** d **10.** c **11.** c **12.** a **13.** a **14.** d **15.** c **16.** d **17.** c **18.** The versatility of a ciliate cell is greater, because it's capable of many more functions. In multicellular organisms, cells are generally specialized and carry out a specific function. **19.** They do not seem related in that algae are photosynthetic and protozoans are heterotrophic. They do seem related in that some of the algae are motile in the same way protozoans are.

# 23

# The Fungi

## Chapter Review

**Fungi** are saprotrophic, multicellular eukaryotes. The body of a fungus is composed of **hyphae,** collectively called a **mycelium.** Hyphae produce nonmotile and often windblown **spores** during both asexual and sexual reproduction.

During sexual reproduction, hyphae tips fuse; **dikaryotic** hyphae result before zygote formation and zygotic meiosis occur. The **zygospore** fungi are nonseptate (hyphae do not have cross walls), and during sexual reproduction, they form a thick-walled **zygospore.** The sac fungi are septate (hyphae have cross walls), and during sexual reproduction, dikaryotic hyphae end in saclike cells (**asci**) within a **fruiting body.** Each ascus produces eight ascospores. **Yeasts** are unicellular fungi; most reproduce by budding. The club fungi are septate, and during sexual reproduction, dikaryotic hyphae end in club-shaped structures called **basidia** that produce basidiospores. The imperfect fungi reproduce asexually by forming conidiospores. Sexual reproduction has not been observed in imperfect fungi.

Fungi form symbiotic relationships with algae in **lichens** and with seed plants in **mycorrhizas.**

## Study Exercises

Study the text section by section as you answer the questions that follow.

### 23.1 Characteristics of Fungi (pp. 398–99)

- Fungi are saprotrophic detritivores that aid the cycling of inorganic nutrients in ecosystems.
- The body of a fungus is multicellular; it is composed of thin filaments called hyphae.
- As an adaptation to life on land, fungi produce nonmotile and often windblown spores during asexual and sexual reproduction.

1. Indicate whether these statements about fungi are true (T) or false (F).
   a. _____ usually multicellular
   b. _____ usually unicellular
   c. _____ composed of hyphae
   d. _____ saprotrophic
   e. _____ can be parasitic
   f. _____ can be photosynthetic
   g. _____ cell wall contains cellulose
   h. _____ cell wall contains chitin
   i. _____ have flagella at some time in their life cycle
   j. _____ do not have flagella at any time in their life cycle
   k. _____ form spores only during asexual reproduction
   l. _____ form spores during both asexual and sexual reproduction
2. Fungi are mostly [a]________________ detritivores that assist in the recycling of nutrients in ecosystems. The bodies of most fungi are made up of filaments called [b]________________, a collection of which are called a(n) [c]________________. If hyphae have cross walls, they are [d]________________; if they do not have cross walls, they are [e]________________.

## 23.2 Evolution of Fungi (pp. 400–7)

- Fungi are classified according to differences in their life cycle and the type of structure that produces spores.

3. Match the types of fungi with the following phyla:
   1. phylum Zygomycota
   2. phylum Ascomycota
   3. phylum Basidiomycota
   4. phylum Deuteromycota

   a. _____ club fungi
   b. _____ zygospore fungi
   c. _____ sac fungi
   d. _____ imperfect fungi
   e. _____ mushrooms
   f. _____ cup fungi
   g. _____ bread mold
   h. _____ *Penicillium*

### Zygospore Fungi (p. 400)

- Zygospore fungi have a dormant stage consisting of a thick-walled zygospore.

4. Label the diagram of the life cycle of black bread mold on the next page using the alphabetized list of terms (some are used more than once).

   asexual reproduction
   gametangia
   meiosis
   mycelium
   nuclear fusion
   sexual reproduction
   sporangiophore
   sporangium
   spores
   zygospore
   zygospore germination
   zygote

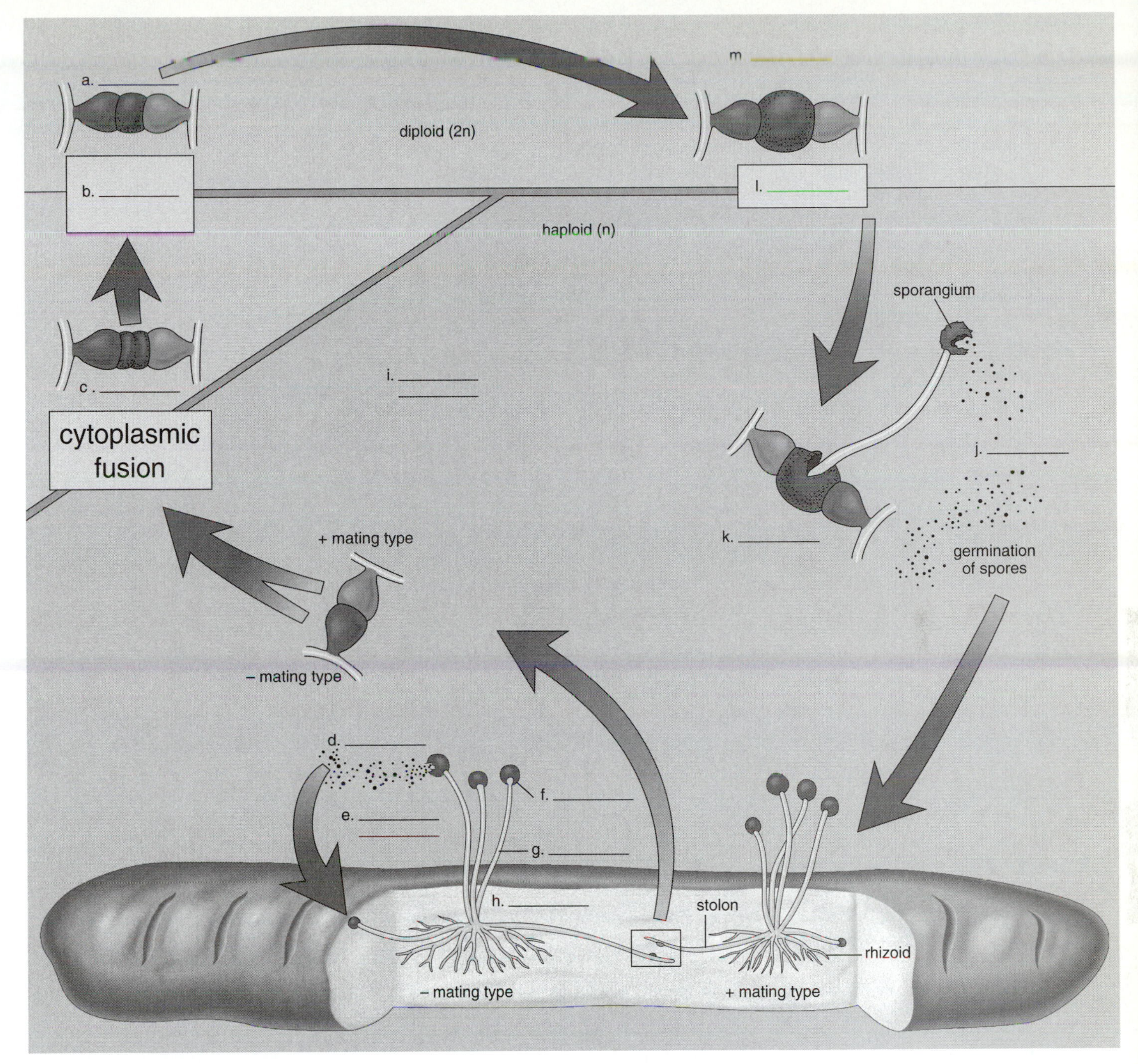


5. Answer these questions based on the life cycle of the black bread mold.

a. Is the adult diploid or haploid? ________________

b. In which cycle (asexual or sexual) are haploid spores produced? ________________

c. Where are the spores produced? ________________

d. What is the name of the enlarged diploid zygote formed in sexual reproduction? ________________

e. How are the spores dispersed from the sporangium? ________________

## Sac Fungi (p. 403)

- During sexual reproduction of sac fungi, the fingerlike sac (ascus) produce spores. Asci are located in fruiting bodies.

6. Place the appropriate letter next to each description.
   F—free-living sac fungi P—parasitic sac fungi
   a. _____ powdery mildew that grows on leaves
   b. _____ red mold that grows on bread
   c. _____ cup fungi that grows on the forest floor
   d. _____ chestnut blight that grows on chestnut trees
   e. _____ ergot that grows on rye plants
   f. _____ unicellular yeasts
7. Why are all these fungi classified as sac fungi? ______________________________

   ______________________________
8. Explain what is happening in each of the following sequential drawings of asci:

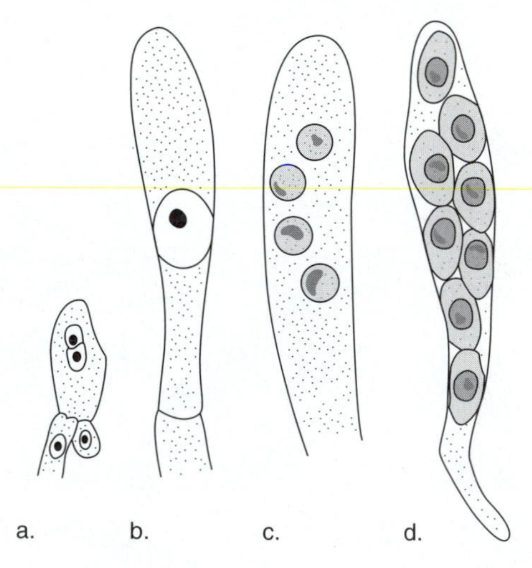


## Club Fungi (p. 405)

- During sexual reproduction of club fungi, club-shaped structures (basidia) produce spores. Basidia are located in fruiting bodies.

9. What do mushrooms, puffballs, bird's nest fungi, stinkhorn fungi, and bracket fungi have in common?

   ______________________________
10. Name two well-known parasites of cereal crops. ______________________________
11. Label the diagram of the life cycle of a mushroom on the next page, using the alphabetized list of terms.

| | |
|---|---|
| basidiospores | gill (portion of) |
| basidium | meiosis |
| cap | monokaryotic (n) |
| cytoplasmic fusion | nuclear fusion |
| dikaryotic (n+n) | nuclei |
| dikaryotic mycelium | spore germination |
| diploid (2n) | spore release |
| fruiting body | stalk |
| gill | zygote |

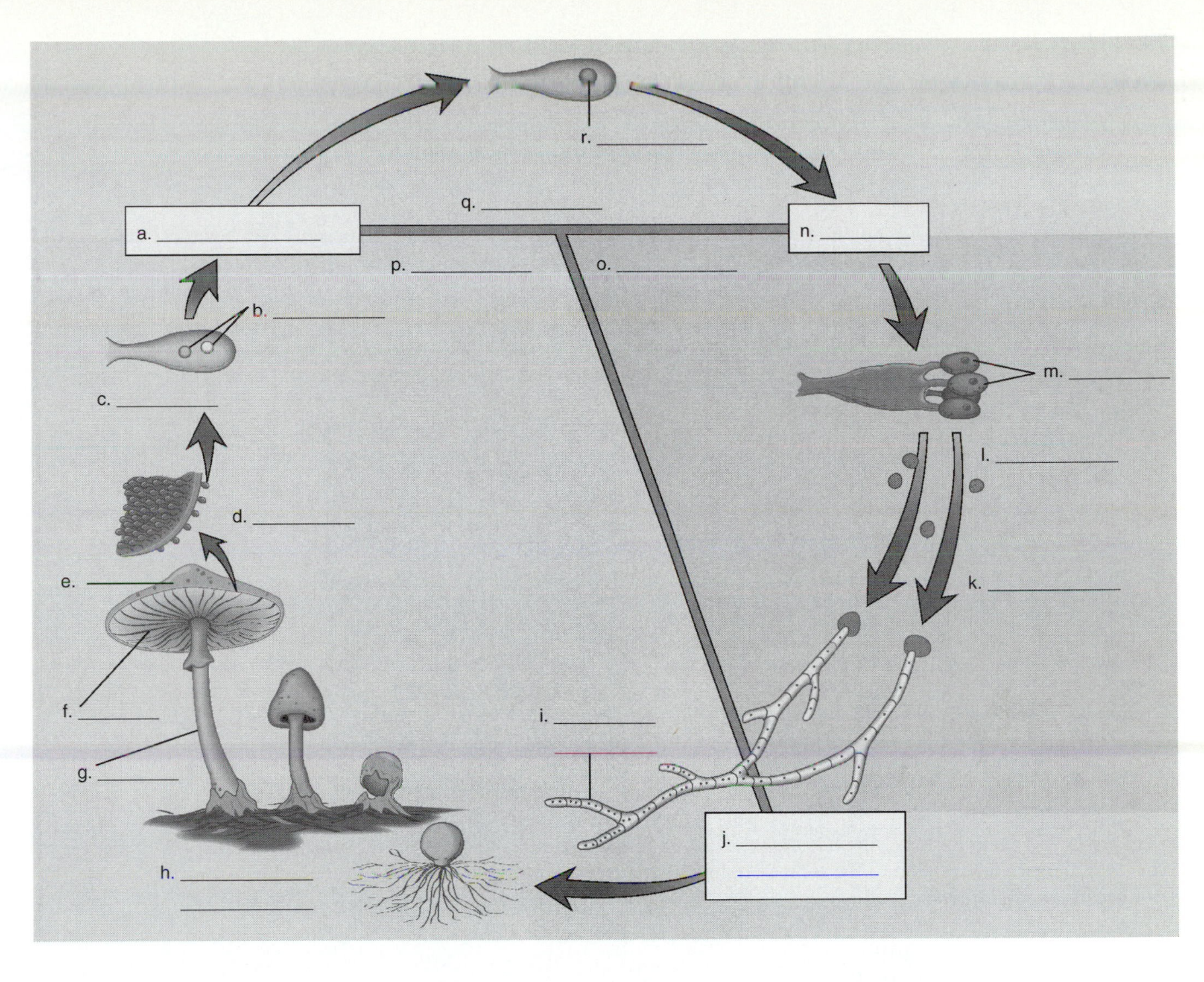


12. In the life cycle of a mushroom, the basidiocarp is a(n) a._________________, which contains club-shaped structures called b._________________, where c._________________ are produced. Beneath each mushroom is a dikaryotic d._________________ that exists for years.

## Imperfect Fungi (p. 407)

- The fungi imperfecti always reproduce asexually by conidiospores; sexual reproduction has not yet been observed in these organisms.

13. Complete the following table:

| Fungus | Significance | Associated Disease |
|---|---|---|
| *Penicillium* | | |
| *Aspergillus* | | |
| *Candida albicans* | | |

14. Like sac and club fungi, imperfect fungi reproduce asexually by producing spores called a.__________________. Unlike sac and club fungi, however, sexual reproduction b.__________________ in imperfect fungi.

## 23.3 Symbiotic Relationships of Fungi (pp. 408–9)

- Lichens, which can live in extreme environmental conditions, are an association between a fungus and a cyanobacterium or a green alga. The fungus may be somewhat parasitic on the alga.
- Mycorrhizas are a mutualistic association between a fungus and the roots of a plant, such that the fungus helps the plant absorb minerals and the plant supplies the fungus with carbohydrates.

15. Label this diagram of a lichen, using these terms:
    algal cells
    fungal hyphae

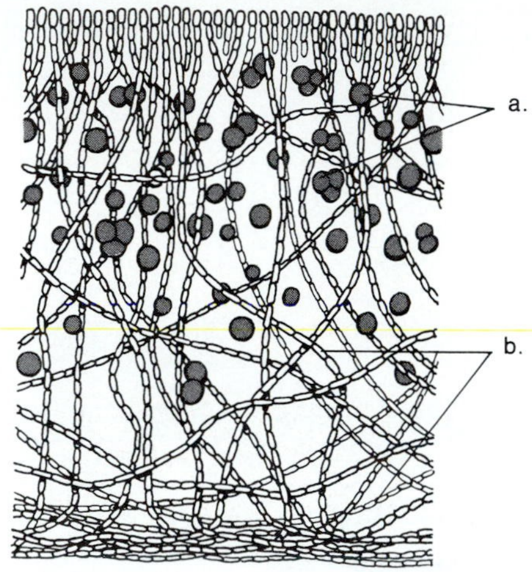


16. Match each description to the type of lichen, using these terms:
    crustose fruticose foliose
    a. ______________ compact
    b. ______________ leaflike
    c. ______________ shrublike
17. a.__________________ (fungus roots), b.__________________ relationships between a(n) c.__________________ and d.__________________ roots, help plants acquire e.__________________ nutrients.

# CHAPTER TEST

## OBJECTIVE QUESTIONS

Do not refer to the text when taking this test.

____ 1. Most fungi
- a. are plant parasites.
- b. form mycorrhizas.
- c. are saprotrophic.
- d. are deuteromycetes.

____ 2. Which terms are mismatched?
- a. hyphae—mycelium
- b. ascocarp—fruiting body
- c. basidiospore—sporangium
- d. zygomycete—bread mold

____ 3. Yeasts
- a. usually reproduce by budding.
- b. never form spores.
- c. are dikaryotic.
- d. All of these are correct.

____ 4. Nonseptate describes a fungus
- a. whose means of sexual reproduction is unknown.
- b. whose hyphae do not have cross walls.
- c. whose hyphae are dikaryotic.
- d. that is parasitic.

____ 5. Club fungi
- a. include the mushrooms.
- b. have a basidiocarp that looks like a cup.
- c. include more parasites than all the other types of fungi.
- d. All of these are correct.

____ 6. Which of these is mismatched?
- a. red bread molds—zygomycete
- b. bracket fungi—basidiomycete
- c. yeasts—ascomycete
- d. rust and smuts—basidiomycetes

____ 7. Conidiospores are
- a. formed asexually.
- b. formed sexually.
- c. formed either asexually or sexually.
- d. never formed.

____ 8. In fungi, the gametes are
- a. heterogametes.
- b. flagellated.
- c. the ends of hyphae.
- d. produced by meiosis.

____ 9. Rusts and smuts
- a. parasitize cereal crops.
- b. are found infrequently.
- c. do not form basidiocarps.
- d. Both *a* and *c* are correct.

For questions 10–15 match the terms with these types of fungi (fungi types are used more than once; each question can have more than one answer):
- a. bread mold
- b. sac fungi
- c. club fungi
- d. other fungi

____ 10. conidiospores

____ 11. athlete's foot

____ 12. sporangia

____ 13. fruiting bodies

____ 14. cup fungi

____ 15. bracket fungi

____ 16. Which of the following is NOT characteristic of lichens?
- a. soil formers
- b. algal cells and fungal hyphae
- c. form a type of moss
- d. can live in extreme conditions

____ 17. Which of the following is NOT true of fungi?
- a. saprotrophic nutrition
- b. eukaryotic cells
- c. reproduce by means of spores
- d. are always multicellular

____ 18. A fruiting body is
- a. a special type of vacuole found in fungi.
- b. a symbiotic relationship between algae and bacteria.
- c. a reproductive structure found in fungi.
- d. always the same shape.

____ 19. Sexual reproduction in a bread mold involves the production of
- a. a sperm and an egg.
- b. flagellated zoospores.
- c. zygospores.
- d. fruiting bodies.

____ 20. In a mushroom, the ______ is (are) analogous to the asci of a sac fungus.
- a. stalk
- b. cap
- c. basidia
- d. spores

## Critical Thinking Questions

Answer in complete sentences.

21. How have yeasts proven invaluable to human civilization?

22. How do you think the Earth would change ecologically if fungi were not present?

**Test Results:** _____ number correct ÷ 22 = ______ × 100 = ______ %

## Exploring the Internet

The Online Learning Center at *www.mhhe.com/maderbiology8* has additional study material and practice quizzes that can help you master the content of this chapter. You can also find links to websites exploring additional topics in biology. Access to the Online Learning Center is free for those who have purchased a new textbook.

## Answer Key

### Study Exercises

**1. a.** T **b.** F **c.** T **d.** T **e.** T **f.** F **g.** F **h.** T **i.** F **j.** T **k.** F **l.** T **2. a.** saprotrophic **b.** hyphae **c.** mycelium **d.** septate **e.** nonseptate **3. a.** 3 **b.** 1 **c.** 2 **d.** 4 **e.** 3 **f.** 2 **g.** 1 **h.** 4 **4.** See Figure 23.3, page 401, in text. **5. a.** haploid **b.** both **c.** in the sporangia **d.** zygospore **e.** by the wind **6. a.** P **b.** F **c.** F **d.** P **e.** P **f.** F **7.** because they form asci during sexual reproduction **8. a.** Two nuclei are fusing. **b.** Zygote (2n) has formed. **c.** Meiosis has occurred. **d.** A mitotic division has resulted in eight ascospores **9.** They are all basidiomycetes. **10.** rust and smuts **11.** See Figure 23.6, page 404, in text. **12. a.** fruiting body **b.** basidia **c.** basidiospores **d.** mycelium

**13.**

| Significance | Associated Disease |
|---|---|
| Makes penicillin | None |
| Makes various additives and causes disease | *Aspergillosis* |
| Yeast | Vaginal infections, thrush |

**14. a.** conidiospores **b.** has never been observed **15. a.** algal cells **b.** fungal hyphae **16. a.** crustose **b.** foliose **c.** fruticose **17. a.** Mycorrhizas **b.** mutualistic **c.** fungus **d.** plant **e.** mineral

### Chapter Test

**1.** c **2.** c **3.** a **4.** b **5.** a **6.** a **7.** a **8.** c **9.** a **10.** b, c **11.** d **12.** a **13.** b, c **14.** b **15.** c **16.** c **17.** d **18.** c **19.** c **20.** c **21.** Yeast makes bread rise, and bread is the staff of life in many parts of the world. Yeast is also used to make wine, and in some countries where the purity of the drinking water is in question, wine is the preferred drink at meals. Today, yeast is used for recombinant DNA experiments that require a eukaryote as the experimental material. **22.** Recycling would falter, and organic waste would accumulate. Without efficient recycling, the carrying capacities of ecosystems would diminish.

# PART V PLANT EVOLUTION AND BIOLOGY

# 24

# EVOLUTION AND DIVERSITY OF PLANTS

## CHAPTER REVIEW

Plants are multicellular photosynthetic eukaryotes that may have evolved from a freshwater green alga. The four evolutionary events in plants are (1) development of an internal embryo (embryo is protected from desiccation); (2) development of **vascular tissue**; (3) production of **seeds**; and (4) advent of the **flower**. These evolutionary events adapted plants to life on land. All plants have a life cycle that includes an alternation of generations.

The **nonvascular plants** include hornworts, liverworts, and mosses, in which the **gametophyte** is dominant. Flagellated sperm swim from the **antheridia** to the **archegonia,** and in mosses the **sporophyte** is a stalk plus a capsule where spores form.

The extinct vascular plants of phylum Rhyniophyta may be ancestral to today's seedless vascular plants, such as **club mosses, horsetails, whisk ferns,** and **ferns.** Ferns and other seedless vascular plants have a large, conspicuous sporophyte with vascular tissue, and a small, but independent gametophyte. In ferns, the zygote develops into the dominant sporophyte with large **fronds.**

Seed plants are heterosporous—that is, they produce separate male and female gametophytes. The male gametophyte is the mature **pollen grain** that produces sperm, and the female gametophyte develops within an ovule, which eventually becomes a seed. Fertilization results in an embryo enclosed within a seed. **Gymnosperms** have exposed ovules and seeds. **Angiosperms** are the flowering plants. They produce seeds covered by **fruit** derived from an **ovary,** part of the flower. Development of the flower, which attracts insects and other animals that carry out cross-fertilization, allowed angiosperms to spread and become the dominant plant today.

## STUDY EXERCISES

Study the text section by section as you answer the questions that follow.

### 24.1 EVOLUTIONARY HISTORY OF PLANTS (PP. 414–17)

- Plants are multicellular photosynthetic organisms adapted to living on land. Among various adaptations, all plants have an alternation of generations life cycle.
- The presence of vascular tissue and reproductive strategy are used to compare and classify plants.

1. Match the evolutionary events to the following major plant groups (the groups can be used more than once):
   1. nonvascular plants
   2. seedless vascular plants
   3. gymnosperms
   4. angiosperms

   a. _____ production of seeds
   b. _____ development of internal embryo
   c. _____ advent of flower
   d. _____ development of vascular tissue

2. Match the evolutionary events to the following adaptations for a land existence. Place the appropriate letter next to each statement.

   S—seeds, IE—internal embryo, F—flower, VT— vascular tissue

   a. _____ dispersal of offspring
   b. _____ protection of desiccation of the embryo
   c. _____ conduction of water and solutes through the plant
   d. _____ reproductive structure that attracts pollinators

3. Indicate whether these statements about plants are true (T) or false (F).

a. _____ adapted to living on land
b. _____ diploid sporophyte produces diploid spores
c. _____ haploid gametophyte, which produces sex cells
d. _____ photosynthetic organisms

4. Match the plants to these major plant groups (the groups can be used more than once):

1. nonvascular plants
2. seedless vascular plants
3. gymnosperms
4. angiosperms

a. _____ club mosses
b. _____ conifers
c. _____ mosses
d. _____ phylum Anthophyta
e. _____ ferns
f. _____ cycads

## 24.2 Nonvascular Plants (pp. 417–19)

- The nonvascular plants are low-growing and contain little or no vascular tissue. The gametophyte (haploid generation) is dominant and produces windblown spores.

5. Label this diagram of part of a moss life cycle using the alphabetized list of terms.

antheridium
archegonium
egg
sperm

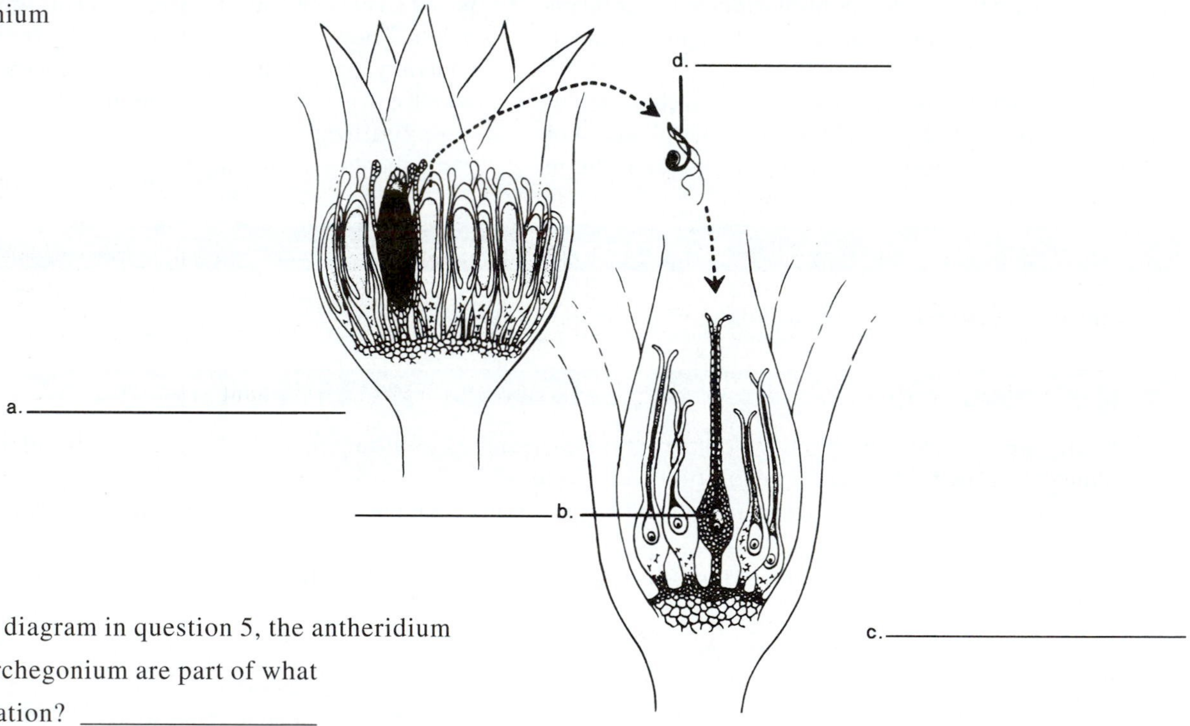

6. a. In the diagram in question 5, the antheridium and archegonium are part of what generation? __________

b. Do mosses have flagellated sperm? __________

c. Do mosses protect the zygote? __________

7. Is the sporophyte generation dependent on the gametophyte generation in the moss? a. __________ Do either of these generations have vascular tissue? b. __________ The capsule contains the sporangium where the cellular process of c. __________ occurs during the production of d. __________. The latter disperse the species. When they germinate, a(n) e. __________ forms to begin the f. __________ generation.

8. The [a.]___________________ anchor the moss gametophyte plant in the soil while absorbing minerals and water. There must be an external source of [b.]___________________ for the sperm to move from the [c.]___________________ to the eggs found in the [d.]___________________.

9. Label this diagram of part of the moss life cycle with the following terms:
   capsule
   gametophyte generation
   rhizoids
   sporophyte generation

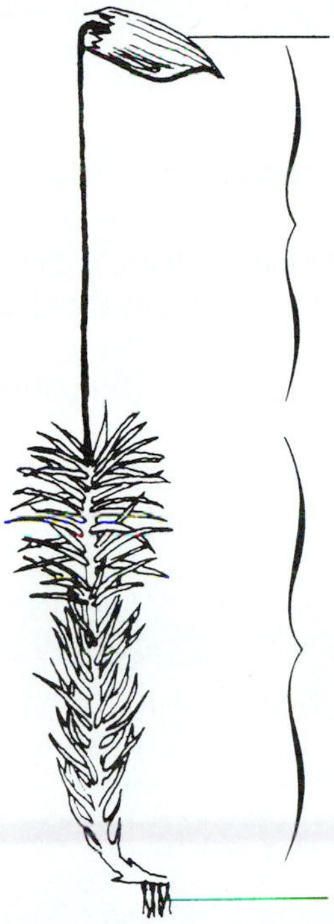

a. 

b. ______________

c. ______________

d. 

## 24.3 Vascular Plants (p. 420)

- In vascular plants, the sporophyte (diploid generation) is dominant and has transport (vascular) tissue.
- In seedless vascular plants, windblown spores disperse the species. In seed plants, seeds, transported by various means, disperse the species.

10. In all vascular plants, the [a.]___________________ generation is the dominant generation, and it is [b.]___________________ (haploid/diploid). [c.]___________________ vascular tissue conducts water and minerals up from the soil. [d.]___________________ vascular tissue transports organic nutrients from one part of the body to another.

11. Indicate whether these statements are true (T) or false (F). Rewrite any false statements to make them true.
    a. _____ All vascular plants produce pollen grains and seeds.
    ______________________________________________________________
    b. _____ Spores disperse seedless vascular plants, while seeds disperse seed plants.
    ______________________________________________________________
    c. _____ In all vascular plants, the gametophyte generation is dependent on the sporophyte.
    ______________________________________________________________

12. Indicate whether these descriptions of rhyniophytes are true (T) or false (F).
    a. _____ had true stem, roots, and leaves
    b. _____ spores were produced without benefit of sporangia
    c. _____ had a forked stem, no roots or leaves

## 24.4 Seedless Vascular Plants (pp. 421–23)

- The seedless vascular plants were much larger and extremely abundant during the Carboniferous period.
- The seedless vascular plants include ferns and other species that do not produce seeds.
- The fern sporophyte is the familiar plant with large fronds that produce windblown spores; the independent gametophyte is a heart-shaped structure that produces flagellated sperm.

13. Match the descriptions to these plants:
    1. whisk ferns
    2. club mosses
    3. horsetails
    4. ferns

    a. _____ Scalelike leaves cover stems and branches; there are terminal strobili.
    b. _____ Whorls of slender branches form at the nodes of the stem where the leaves are; they are sometimes called scouring rushes.
    c. _____ These plants have aerial stems with tiny scales for photosynthesis.
    d. _____ Large fronds subdivide into leaflets.

14. Label this diagram of part of the fern life cycle, using these terms:
    antheridium archegonium rhizoid

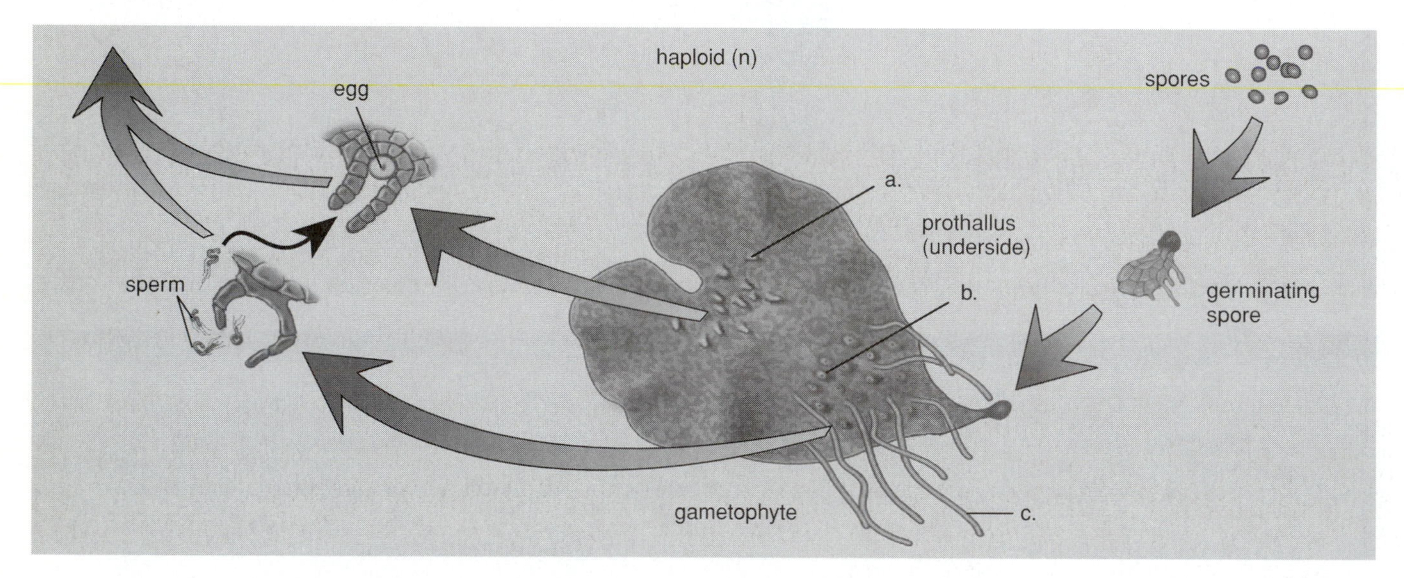


15. In the diagram in question 14, the large, central structure is called the a.__________________. This structure has a(n) b.__________________ shape and represents the c.__________________ generation. It d.__________________ (does/does not) have vascular tissue and e.__________________ (does/does not) have flagellated sperm. The sporophyte f.__________________ (is/is not) the dominant generation. g.__________________ disperse offspring.

16. Label this diagram of part of the life cycle of a fern, using these terms:
fiddlehead    frond    rhizome    sorus/sori

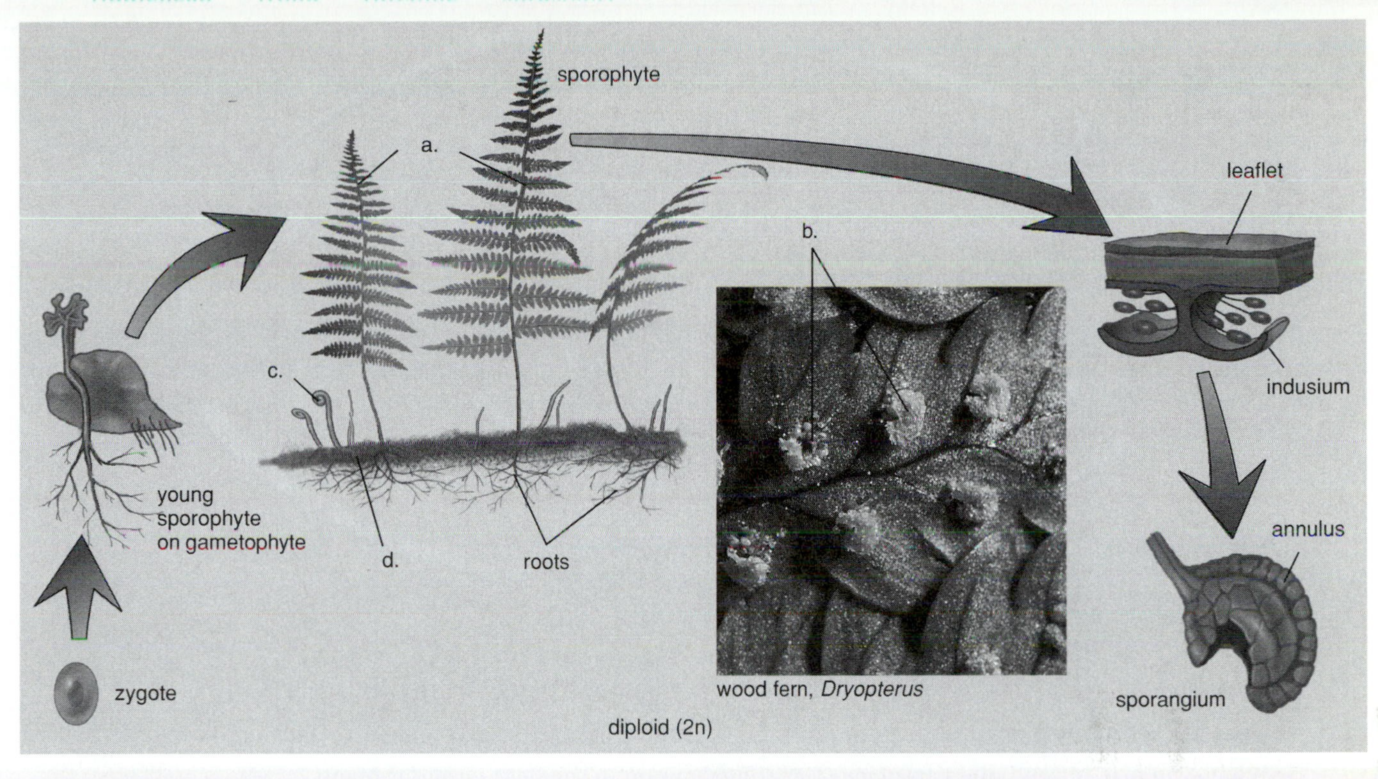


17. The frond is part of the [a] __________________ generation in the fern, and it [b] __________________ (does/does not) have vascular tissue. This generation is [c] __________________ (diploid/haploid).

18. Label the following as a moss (M) or fern (F):
   a. _____ This plant has microphylls.
   b. _____ This plant is used as an ornamental.
   c. _____ Native Americans used this plant during childbirth.

## 24.5 Seed Plants (p. 424)

- Gymnosperms and angiosperms are the vascular plants that produce seeds.
- Seeds contain an embryo and stored food.
- Seed plants are heterosporous.

19. Indicate whether these descriptions of seed plants are true (T) or false (F).
   a. _____ Sperm need water to move through the pollen tube.
   b. _____ Gymnosperms bear cones.
   c. _____ Angiosperms have "naked" seeds.
   d. _____ Seed plants are homosporous.

20. Label the following diagram, showing the alternation of generations in seed plants with the notation n or 2n:

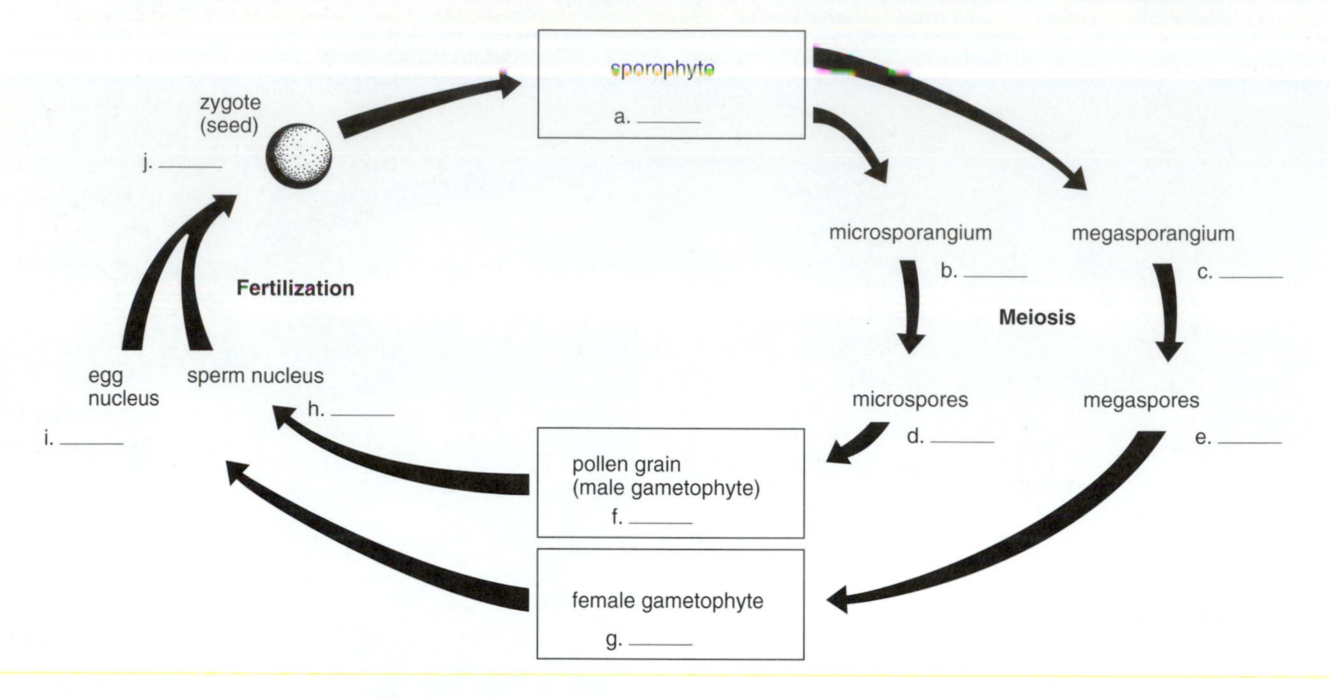


21. a. Is the gametophyte generation dependent on the sporophyte generation in seed plants? ____________

b. Does the sporophyte generation have vascular tissue? ____________

c. What structure in seed plants replaces flagellated sperm in seedless vascular plants? ____________

d. What structure disperses offspring in seed plants? ____________

## 24.6 Gymnosperms (pp. 424–27)

- There are four phyla of gymnosperms: the familiar conifers, and the little-known cycads, ginkgoes, and gnetophytes.
- The conifer sporophyte is usually a cone-bearing tree (e.g., pine tree). The pollen cones produce pollen (male gametophyte), windblown to the seed cone where windblown seeds are produced.

22. Match the descriptions with these types of gymnosperms:

1. conifer
2. cycad
3. ginkgo
4. gnetophytes

a. _____ includes *Ephedra,* the gymnosperm most closely related to angiosperms
b. _____ stout, unbranched stem with large, compound leaves
c. _____ fan-shaped leaves shed in autumn; planted in parks
d. _____ pine trees, hemlocks, spruces

23. Study this diagram of the life cycle of a pine tree and then describe the numbered events.

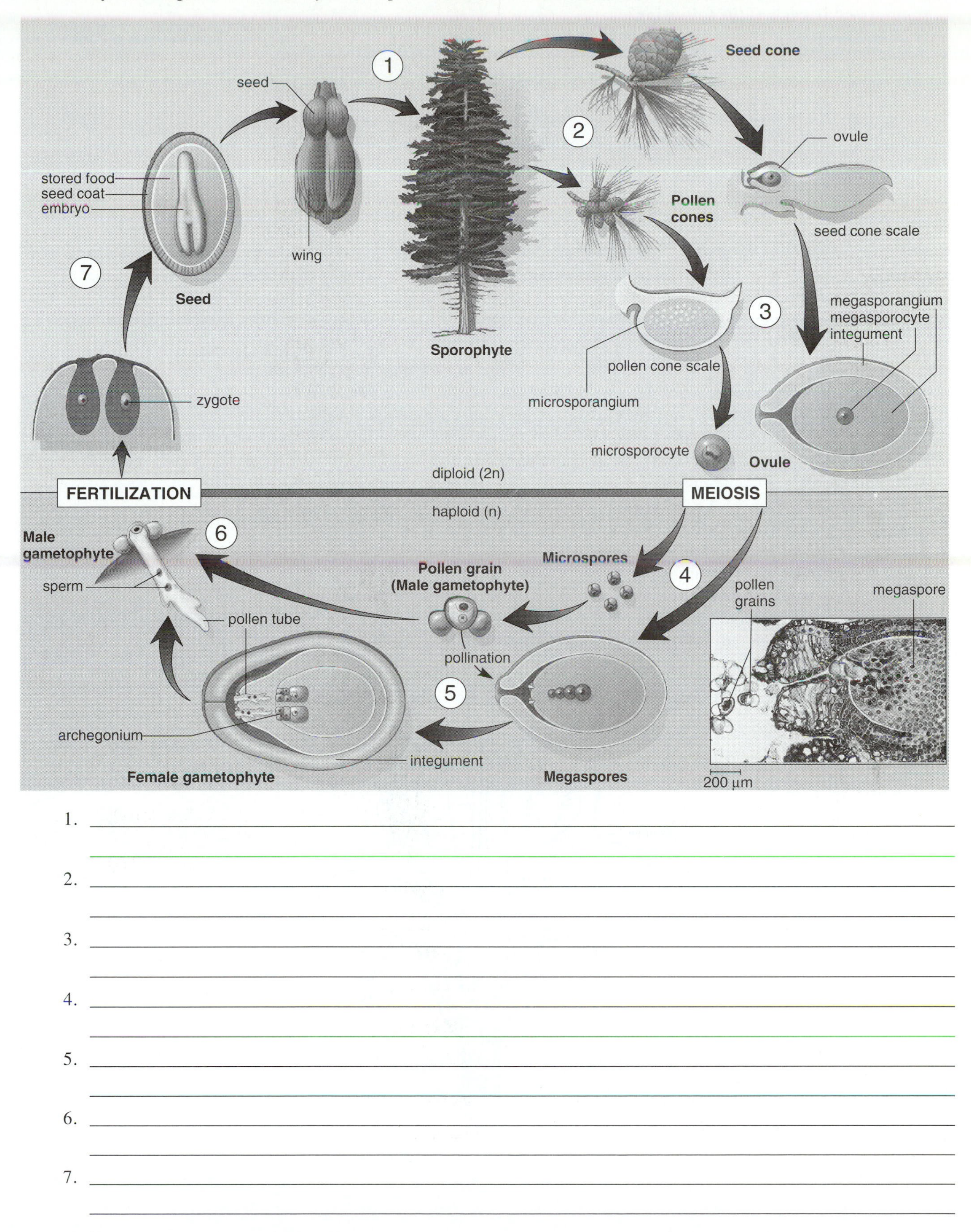


1. ______________________________________________

______________________________________________

2. ______________________________________________

______________________________________________

3. ______________________________________________

______________________________________________

4. ______________________________________________

______________________________________________

5. ______________________________________________

______________________________________________

6. ______________________________________________

______________________________________________

7. ______________________________________________

______________________________________________

24. Gymnosperms have an alternation of generations life cycle. Which structure in the diagram in question 23 is described in each of the following?

a. ____________________ is the sporophyte.

b. ____________________ produces the microspore.

c. ____________________ produces the megaspore.

d. ____________________ is the male gametophyte.

e. ____________________ contains the female gametophyte.

f. ____________________ contains the sperm.

g. ____________________ contains the egg.

h. ____________________ contains the embryonic sporophyte.

## 24.7 Angiosperms (pp. 428–31)

- Angiosperms are the very diversified flowering plants, which may depend on an animal pollinator to carry pollen to another flower.
- The seeds are enclosed within a fruit, which often assists dispersal of seeds.

25. Place the appropriate letter next to each statement.

M—monocot  E—eudicot

a. _____ almost always herbaceous
b. _____ either woody or herbaceous
c. _____ flower parts in fours or fives
d. _____ flower parts in threes
e. _____ net-veined leaves
f. _____ parallel-veined leaves
g. _____ vascular bundles arranged in a circle in the stem
h. _____ vascular bundles scattered in the stem

26. Label this diagram of the structures of a flower using the alphabetized list of terms.

anther
ovary
ovule

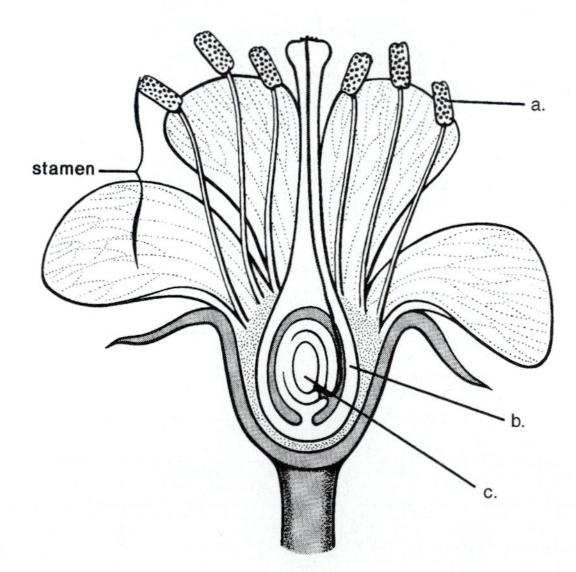

27. Study this diagram of the life cycle of a flowering plant and then describe the numbered events.

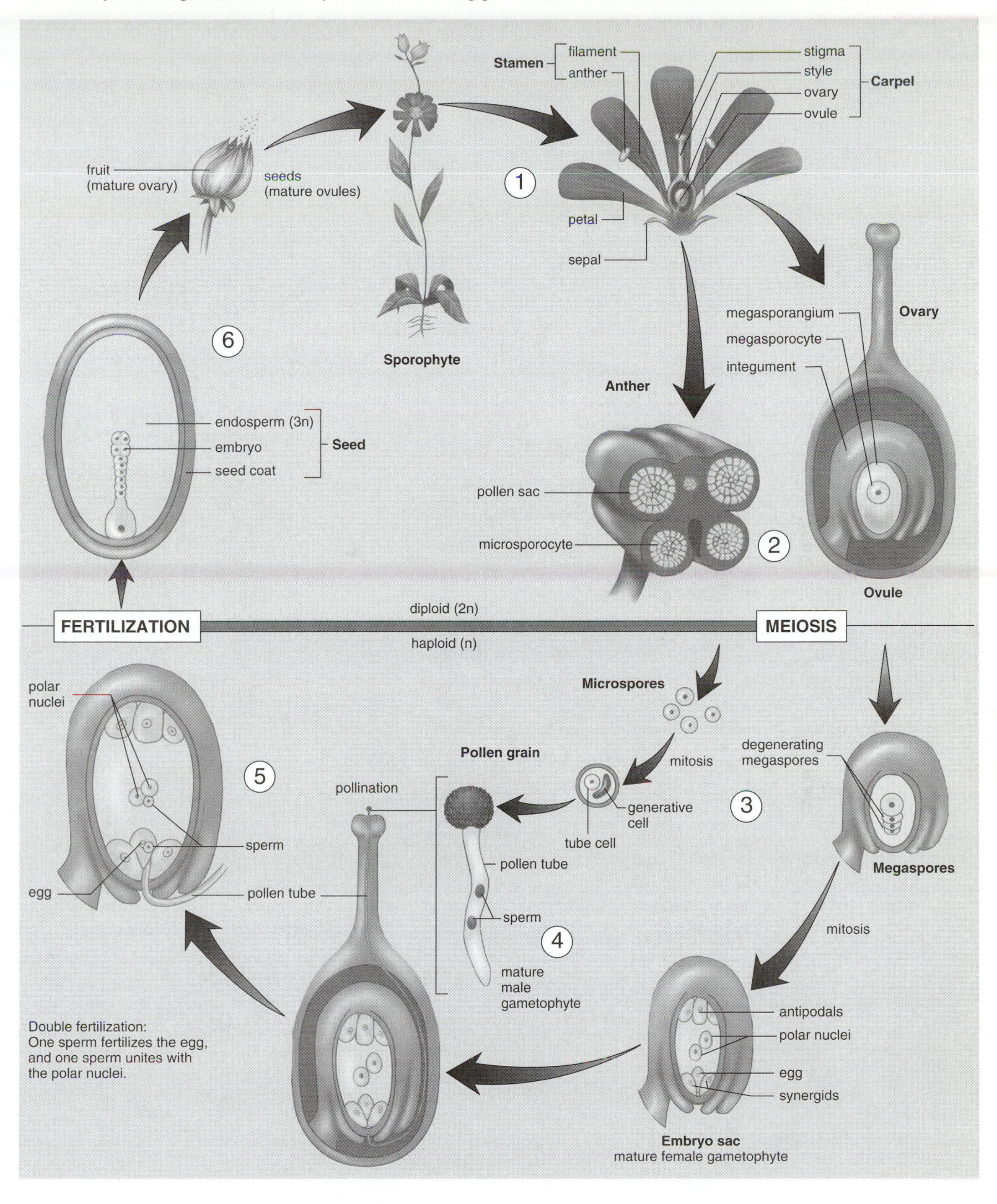

1. ____________________

____________________

2. ____________________

____________________

3. ____________________

____________________

4. ____________________

____________________

5. ____________________

____________________

6. ____________________

____________________

28. Flowering plants have an alternation of generations life cycle. Which structure in the diagram in question 27 is described in each of the following?

a. ____________ is the sporophyte.

b. ____________ produces the microspore.

c. ____________ produces the megaspore.

d. ____________ is the male gametophyte.

e. ____________ contains the female gametophyte.

f. ____________ contains the sperm.

g. ____________ contains the egg.

h. ____________ contains the embryonic sporophyte.

i. ____________ matures to become a fruit.

29. The color and arrangement of flower parts are designed to a.____________.
A flower, necessary to the reproduction of flowering plants, contains b.____________

____________.

A flower produces seeds enclosed by c.____________, which helps d.____________.

30. Complete the following table to compare the different plants:

| Plant | Vascular Tissue (yes or no) | Dominant Generation | Spores or Seeds Disperse Species | Fruit (yes or no) |
|---|---|---|---|---|
| Mosses | | | | |
| Ferns | | | | |
| Gymnosperms | | | | |
| Angiosperms | | | | |

## KeyWord CrossWord

Review key terms by completing this crossword puzzle using the following alphabetized list of terms:

angiosperm
antheridium
archegonium
conifer
eudicotyledon
fruit
gametophyte
gymnosperm
herbaceous
monocotyledon
phloem
pollen
rhizoid
sporophyte
woody
xylem

*Across*

2 flowering plant group; members show one embryonic leaf, parallel-veined leaves, scattered vascular bundles, and other characteristics
6 reproductive organ found in mosses and some vascular plants that produce eggs
7 flowering plant structure, consisting of one or more ripened ovaries, that usually contain seeds
10 vascular tissue that transports water and mineral solutes upward through the plant body
11 plant that contains wood; types include gymnosperms and angiosperms
12 vascular plant producing naked seeds, as in conifers
14 plant that lacks persistant woody tissue
15 male gametophyte in seed plants

*Down*

1 flowering plant; the seeds are borne within a fruit
3 one of the four groups of gymnosperm plants; cone-bearing trees that include pine, cedar, and spruce
4 flowering plant group; members show two embryonic leaves, net-veined leaves, cylindrical arrangement of vascular bundles, and other characteristics
5 diploid generation of the alternation of generations life cycle of a plant; meiosis produces haploid spores that develop into the haploid generation
6 reproductive organ, found in mosses and some vascular plants, that produces flagellated sperm
8 rootlike hair that anchors a plant and absorbs minerals and water from the soil
9 haploid generation of the alternation of generations life cycle of a plant; it produces gametes that unite to form a diploid zygote
13 vascular tissue that conducts organic solutes in plants; it contains sieve-tube members and companion cells

# Chapter Test

## Objective Questions

Do not refer to the text when taking this test.

____ 1. Select the incorrect association.
a. gametophyte—diploid generation
b. gametophyte—produces sex cells
c. sporophyte—diploid generation
d. sporophyte—produces haploid spores

____ 2. Select the nonvascular plant.
a. moss
b. cycad
c. fern
d. rosebush

____ 3. The antheridium is part of the
a. female gametophyte.
b. megasporophyte.
c. male gametophyte.
d. microsporophyte.

____ 4. Rhyniophytes are significant because they
a. are currently the most successful conifers.
b. are currently the most successful flowering plants.
c. were the first to evolve flowers.
d. were the first to evolve vascular tissue.

____ 5. Xylem and phloem are
a. the covering tissues on roots, stems, and leaves.
b. the male and female parts of a flower.
c. two kinds of flowering plants.
d. two types of vascular tissue.

____ 6. Ferns are plants that are
a. nonvascular with seeds.
b. nonvascular without seeds.
c. vascular with seeds.
d. vascular without seeds.

____ 7. Which structure develops into a pollen grain?
a. antheridium
b. archegonium
c. megaspore
d. microspore

____ 8. Select the characteristic NOT descriptive of conifers.
a. can withstand cold winters
b. can withstand hot summers
c. needlelike leaves
d. reproduce through flowers

____ 9. Select the incorrect statement about angiosperms.
a. contain only tracheids in their vascular tissue
b. did not diversify until the Cenozoic era
c. include tiny plants living on pond surfaces
d. the most successful group of plants

____ 10. Select the incorrect association.
a. eudicot—woody or herbaceous
b. eudicot—vascular bundles arranged in a circle within the stem
c. monocot—almost always herbaceous
d. monocot—net-veined leaf

____ 11. In comparing alternation of generations for seedless vascular plants and seed plants, it is apparent that seed plants have
a. flowers only.
b. heterospores.
c. flagellated sperm.
d. female gametophytes and male gametophytes.
e. Both *b* and *d* are correct.

____ 12. The fern is a seedless vascular plant and
a. is a moss.
b. has flagellated sperm.
c. lacks vascular tissue.
d. All of these are correct.

____ 13. The dominant generation in seed plants is the
a. sporophyte.
b. gametophyte.
c. green leafy shoot.
d. flower only.

____ 14. Ferns are restricted to moist places because
a. of the sporophyte generation called the frond.
b. of a sensitive type of chlorophyll.
c. of the water-dependent gametophyte generation.
d. they never grow very tall.

____ 15. In the life cycle of seed plants, meiosis
a. produces the gametes.
b. produces microspores and megaspores.
c. does not occur.
d. produces spores.

____ 16. Double fertilization refers to the fact that in angiosperms
a. two egg cells are fertilized within an ovule.
b. a sperm nucleus fuses with an egg cell and with polar nuclei.
c. two sperm are required for fertilization of one egg cell.
d. a flower can engage in both self-pollination and cross-pollination.

____17. Identify the correct order of life cycle stages in a moss.
  a. gametophyte—sporophyte—spores—zygote—protonema
  b. sporophyte—spores—protonema—gametes—zygote
  c. gametophyte—spores—sporophyte—gametes—zygote
  d. zygote—sporophyte—gametophyte—spores—protonema

____18. The sporangia of a fern is generally
  a. at the tips of the rhizomes.
  b. on the bottom surface of the prothallus.
  c. inside the fiddleheads.
  d. at the point where leaves join the stem.
  e. on the undersides of fronds.

____19. In which of these groups is the gametophyte nutritionally dependent upon the sporophyte?
  a. ferns
  b. angiosperms
  c. gymnosperms
  d. Both *b* and *c* are correct.
  e. All of these are correct.

____20. In pine trees, ______ develop in separate types of cones.
  a. male gametophytes and female gametophytes
  b. pollen and ovules
  c. microspores and megaspores
  d. Both *a* and *b* are correct.
  e. All of these are correct.

## Critical Thinking Questions

Answer in complete sentences.

21. What do you think will be the impact on human life if there is a major extinction of plants during the next century?

22. Should algae be considered plants? Offer reasons why they should and should not be classified this way.

**Test Results:** ______ number correct ÷ 22 = ______ × 100 = ______ %

## Exploring the Internet

The Online Learning Center at *www.mhhe.com/maderbiology8* has additional study material and practice quizzes that can help you master the content of this chapter. You can also find links to websites exploring additional topics in biology. Access to the Online Learning Center is free for those who have purchased a new textbook.

## Answer Key

### Study Exercises

**1. a.** 3, 4 **b.** 1, 2, 3, 4 **c.** 4 **d.** 2, 3, 4 **2. a.** S **b.** IE **c.** VT **d.** F **3. a.** T **b.** F **c.** T **d.** T **4. a.** 2 **b.** 3 **c.** 1 **d.** 4 **e.** 2 **f.** 3 **5. a.** antheridium **b.** egg **c.** archegonium **d.** sperm **6. a.** gametophyte **b.** yes **c.** yes **7. a.** yes **b.** no **c.** meiosis **d.** spores **e.** protonema **f.** gametophyte **8. a.** rhizoids **b.** water **c.** antheridia **d.** archegonia **9. a.** capsule **b.** sporophyte generation **c.** gametophyte generation **d.** rhizoids **10. a.** sporophyte **b.** diploid **c.** Xylem **d.** Phloem **11. a.** F. Only gymnosperms and angiosperms . . . **b.** T **c.** F. In most seed plants . . . **12. a.** F **b.** F **c.** T **13. a.** 2 **b.** 3 **c.** 1 **d.** 4 **14. a.** archegonium **b.** antheridium **c.** rhizoid **15. a.** prothallus **b.** heart **c.** gametophyte **d.** does not **e.** does **f.** is **g.** Spores **16. a.** frond **b.** sorus/sori **c.** fiddlehead **d.** rhizome **17. a.** sporophyte **b.** does **c.** diploid **18. a.** M **b.** F **c.** F **19. a.** F **b.** T **c.** F **d.** F **20. a.** 2n **b.** 2n **c.** 2n **d.** n **e.** n **f.** n **g.** n **h.** n **i.** n **j.** 2n **21. a.** yes **b.** yes **c.** pollen grain **d.** seed **22. a.** 4 **b.** 2 **c.** 3 **d.** 1 **23.** 1. Sporophyte is dominant. 2. pollen cones and seed cones 3. microsporangia on lower surface of pollen cone; megasporangia on upper surface of seed cone inside ovule 4. Meiosis produces microspores/megaspores. 5. Each microspore develops in a male gametophyte (pollen grain); one megaspore develops into a female gametophyte. 6. Pollen tube delivers sperm to egg. 7. Ovule becomes seed. **24. a.** tree **b.** microsporangium on scale of pollen cone **c.** megasporangium inside ovule on scale of seed cone **d.** pollen grain

e. ovule **f.** mature male gametophyte (pollen grain) **g.** female gametophyte inside ovule **h.** seed **25. a.** M **b.** E **c.** E **d.** M **e.** E **f.** M **g.** E **h.** M **26. a.** anther **b.** ovary **c.** ovule **27.** 1. Flower contains stamens and carpel. 2. Megasporocyte produces four haploid megaspores; microsporocyte produces four haploid microspores. 3. One functional megaspore survives and divides mitotically; male gametophyte contains generative cell nucleus and tube cell nucleus. 4. Female gametophyte results and consists of eight haploid nuclei; pollen grain germinates and produces a pollen tube. 5. Double fertilization occurs; one sperm fertilizes the egg, and the other joins with the polar nuclei. 6. Seed contains endosperm, embryo, and seed coat. **28. a.** flowering plant **b.** pollen sac **c.** megasporangium inside ovule **d.** pollen grain **e.** ovule **f.** pollen tube **g.** female gametophyte (embryo sac) **h.** seed **i.** ovary **29. a.** attract pollinators **b.** micro- and megasporangia where microspores and megaspores are produced **c.** fruit **d.** disperse offspring

**30.**

| Vascular Tissue (yes or no) | Dominant Generation | Spores or Seeds Disperse Species | Fruit (yes or no) |
|---|---|---|---|
| no | gametophyte | spores | no |
| yes | sporophyte | spores | no |
| yes | sporophyte | seeds | no |
| yes | sporophyte | seeds | yes |

## KeyWord CrossWord

| | | | | | | | | | | 4 E | | | | | | | | | | |
|---|---|---|---|---|---|---|---|---|---|---|---|---|---|---|---|---|---|---|---|---|
| | | 1 A | | | | | | | | U | | | | | | | | | | |
| 2 M | O | N | O | 3 C | O | T | Y | L | E | D | O | N | | 5 S | | | | | | |
| | | G | | O | | | | | | I | | | | P | | | | | | |
| | | I | | N | | | | 6 A | R | C | H | E | G | O | N | I | U | M | | |
| | | O | | I | | | | N | | O | | | | R | | | | | | |
| | | S | | 7 F | R | U | I | T | | T | | | | O | | | | | 8 R | |
| | | P | | E | | | | H | | Y | | | | P | | | 9 G | | H | |
| | | E | | R | | | | E | | L | | | | H | | | A | | I | |
| | | R | | | | | | R | | E | | | 10 X | Y | L | E | M | | Z | |
| | | M | | | | | | I | | D | | | | T | | | E | | O | |
| | | | | | | | | D | | O | | | | E | | | T | | I | |
| | | | | | | | | I | | N | | | | | | 11 W | O | O | D | Y |
| | | | | | | | | U | | | | | | | | | P | | | |
| | | | | | | 12 G | Y | M | N | O | S | P | E | R | M | | H | | | |
| | | | | | | | | | | | | | | | | | Y | | | |
| | | | | | | | | | | | 13 P | | | | | | T | | | |
| | | | | | | | | | | | 14 H | E | R | B | A | C | E | O | U | S |
| | | | | | | | | | | | L | | | | | | | | | |
| | | | | | | | | | | 15 P | O | L | L | E | N | | | | | |
| | | | | | | | | | | | E | | | | | | | | | |
| | | | | | | | | | | | M | | | | | | | | | |

## Chapter Test

**1.** a **2.** a **3.** c **4.** d **5.** d **6.** d **7.** d **8.** d **9.** a **10.** d **11.** e **12.** b **13.** a **14.** c **15.** b **16.** b **17.** b **18.** e **19.** d **20.** e **21.** The ecological and economical contributions of plants will be lost. Examples include loss of food production and inability to maintain a balance of gases in the atmosphere. **22.** They are photosynthetic producers, so in this way they should be considered plants. However, they lack the adaptations for terrestrial life, so in this way they should not be considered plants.

# 25

# Structure and Organization of Plants

## Chapter Review

Flowering plants are divided into the **monocots,** which have one cotyledon, and the **eudicots,** which have two cotyledons. A flowering plant has three vegetative organs: roots, stems, and leaves. A **root** anchors a plant, absorbs water and minerals, and stores the products of photosynthesis. **Stems** support leaves, conduct materials to and from roots and leaves, and help store plant products. **Leaves** carry on photosynthesis.

Three types of tissue are found in each organ: **epidermal tissue, ground tissue,** and **vascular tissue. Epidermis** is the outer protective covering of plants. **Parenchyma, collenchyma,** and **sclerenchyma** are the types of ground tissue. **Xylem,** which transports water and minerals, and **phloem,** which transports sucrose and organic compounds, are the two types of vascular tissue.

The layers and organization of vascular tissue differ in the roots of the monocot, the herbaceous eudicot, and the woody eudicot. There are several types of roots: **taproots, fibrous roots,** and **adventitious roots.**

In a stem, the **shoot apical meristem** produces new cells that increase stem length. This is primary growth, found in all plants. **Herbaceous stems** exhibit only primary growth.

The stems of woody plants experience primary growth and pronounced secondary growth that increases the girth of trunks, stems, branches, and roots. The cross section of a leaf shows **mesophyll** between upper and lower layers of epidermis. The venation differs between monocots and eudicots. Leaves can be simple or compound. Leaves are diverse and are adapted by various means to environmental conditions.

## Study Exercises

Study the text section by section as you answer the questions that follow.

### 25.1 Plant Organs (pp. 438–39)

- Plants have a root system that contains the roots and a shoot system that contains the stems and the leaves.

1. The root system [a] __________________ the plant in the soil and absorbs [b] __________________ and [c] __________________ from the soil, especially by means of [d] __________________. Stems support the leaves and [e] __________________ water and minerals from roots to the leaves. The leaves carry on [f] __________________, which requires [g] ______________________________________ (list the three requirements).

### 25.2 Monocot Versus Eudicot Plants (p. 440)

- Flowering plants are classified into two groups: the monocots and the eudicots.

2. Complete the following table to indicate the differences between monocot and eudicot plants:

| Plant Part | Monocot | Eudicot |
|---|---|---|
| Leaf veins | | |
| Vascular bundles in stem | | |
| Cotyledons in seed | | |
| Parts in a flower | | |
| Vascular tissue in root | | |

## 25.3 Plant Tissues (pp. 441–43)

- Plant cells are organized into three types of tissues: epidermal tissue, ground tissue, and vascular tissue.

3. Match the cell types with these tissue types:
   1. epidermal tissue
   2. ground tissue
   3. vascular tissue

a. _____ sclerenchyma cells
b. _____ tracheids
c. _____ root hair cells
d. _____ parenchyma cells
e. _____ sieve-tube members
f. _____ vessel elements

4. Indicate whether each of these statements is true (T) or false (F).

**Epidermal Tissue:**

a. _____ Epidermis covers the entire body of nonwoody plants.
b. _____ Epidermis covers the entire body of young woody plants.
c. _____ Epidermis replaces cork in the stems of older plants.
d. _____ Guard cells regulate the entrance of water into the roots.

**Ground Tissue:**

e. _____ Collenchyma cells are a major site of photosynthesis.
f. _____ Collenchyma cells have thinner primary walls than parenchyma.
g. _____ Parenchyma cells are the least specialized cell type.
h. _____ Sclerenchyma cell secondary walls are impregnated with lignin.
i. _____ Sclereids give pears their gritty texture.

**Vascular Tissue:**

j. _____ Phloem transports organic nutrients from leaves to roots.
k. _____ Sieve-tube members are found in xylem.
l. _____ Tracheids are a type of cell in phloem.
m. _____ Xylem transports water from roots to leaves.

## 25.4 Organization of Roots (pp. 444–47)

- In longitudinal section, a eudicot root tip has a zone where new cells are produced, another where they elongate, and another where they differentiate and mature.
- In cross section, eudicot and monocot roots differ in the organization of their vascular tissue.
- Some plants have a taproot, some have fibrous roots, and some have adventitious roots.

5. Label this diagram of a eudicot root tip using the alphabetical list of terms.

cortex
endodermis
epidermis
pericycle
phloem
root cap
root hair
vascular cylinder
xylem
zone of cell division
zone of elongation
zone of maturation

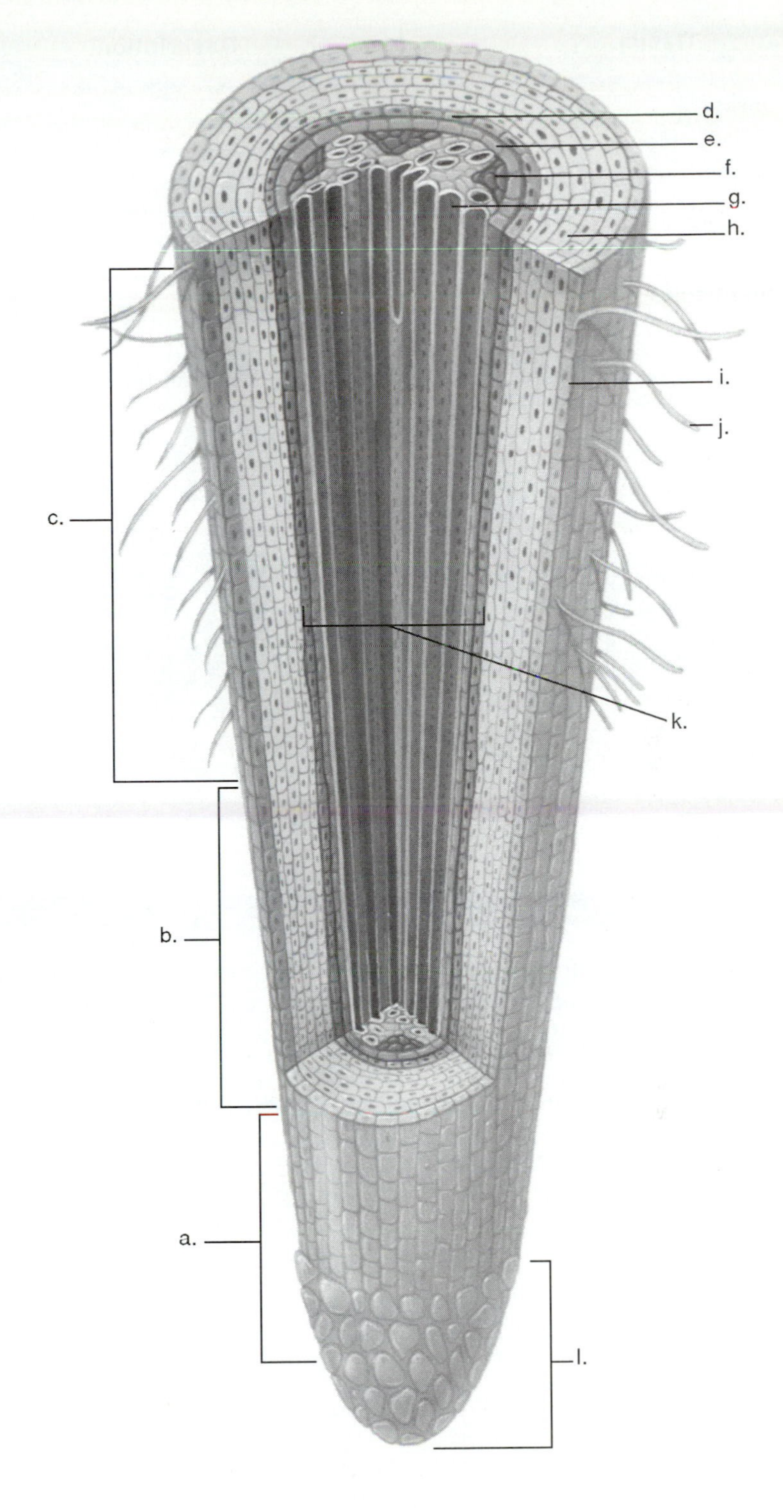


6. Explain what is happening in each of these zones of a root tip.

a. cell division ______________________________

b. elongation ______________________________

c. maturation ______________________________

7. Complete the following table to describe and give a function for the specific tissues within a root:

| Tissue | Description | Function |
|---|---|---|
| Epidermis | | |
| Cortex | | |
| Endodermis | | |
| Vascular tissue | | |

8. Label each type of root shown and describe its special function.

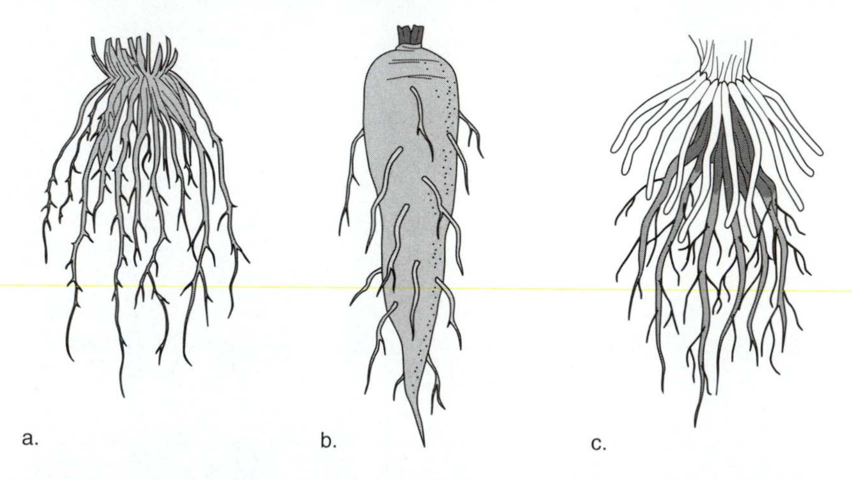


## 25.5 Organization of Stems (pp. 448–52)

- All stems grow in length, but some stems are woody and grow in girth also.
- In cross section, eudicot and monocot herbaceous stems differ in the organization of their vascular tissue.
- Stems are modified in various ways, and some plants have horizontal aboveground or underground stems.

9. Study the diagram and write the term *herbaceous* or *woody* on the lines provided.

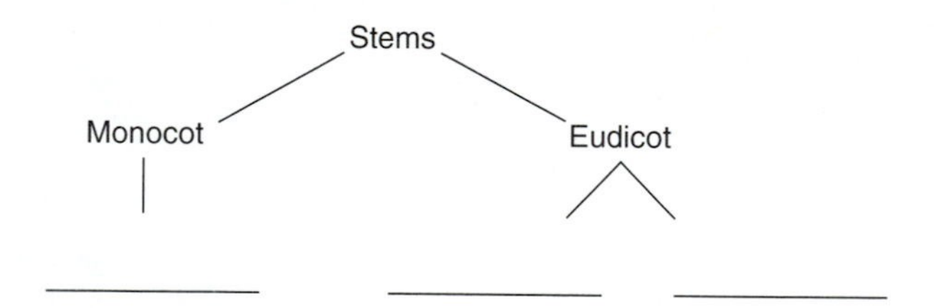

10. Label these diagrams of stems using the alphabetized list of terms (some are used more than once).

bark
cortex
epidermis
phloem
pith
primary xylem
vascular bundle
vascular cambium
wood
xylem

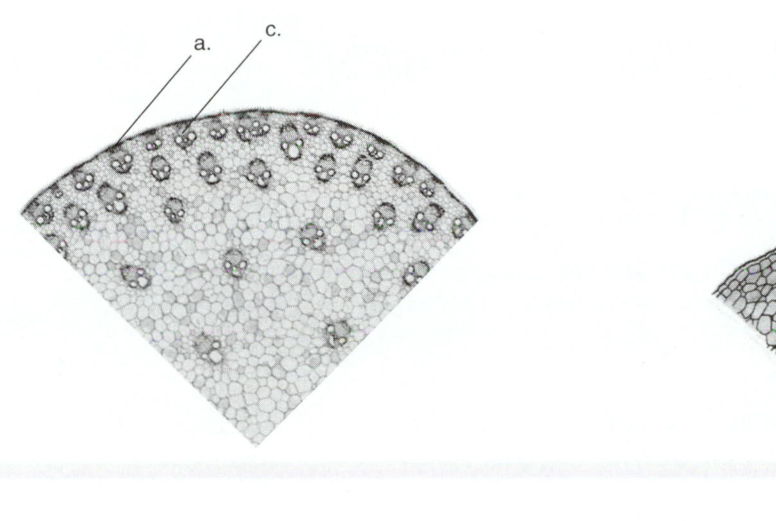


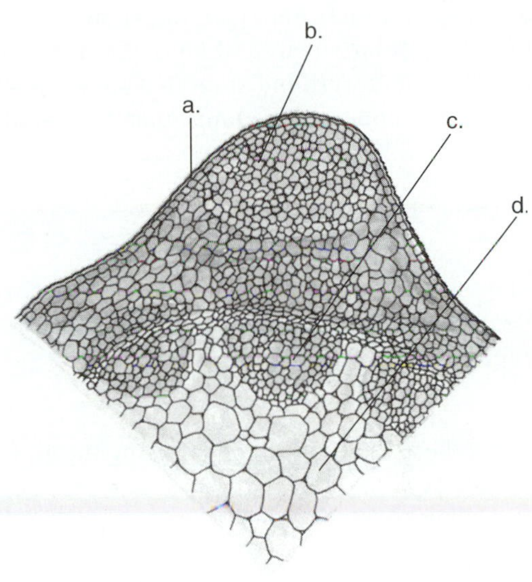


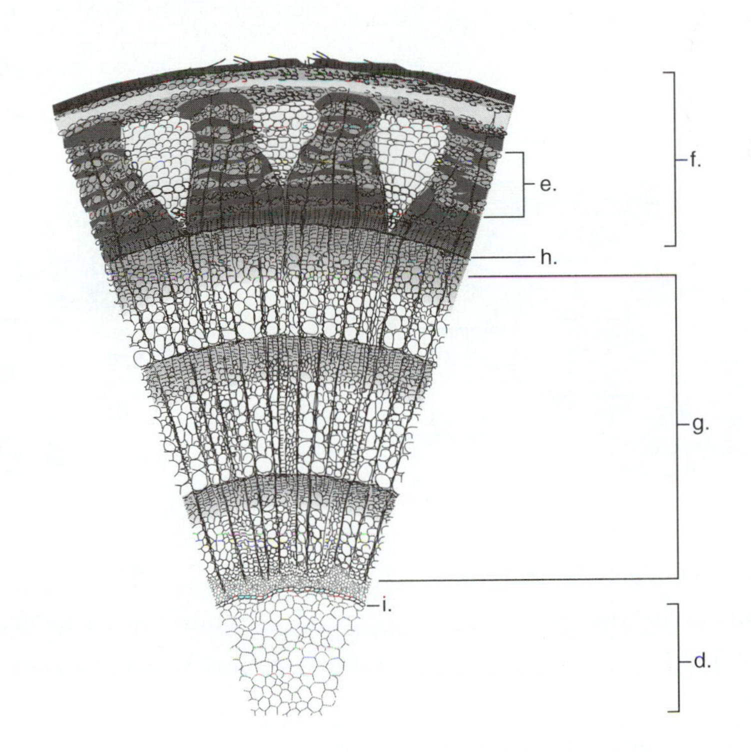

11. a. How would you describe the arrangement of vascular bundles in a herbaceous monocot stem? ____________________

b. In a herbaceous eudicot stem? ____________________

c. Annual rings appear in which type of stem? ____________________

d. What causes annual rings? ____________________

12. Match the descriptions to these types of stems:

1. corm
2. rhizome
3. stolon
4. tuber

a. _____ bulbous underground stem
b. _____ enlarged area of an underground stem that serves as storage area
c. _____ underground stem that survives the winter
d. _____ runner; new plants start from nodes

## 25.6 Organization of Leaves (pp. 454–55 )

- The bulk of a leaf is composed of cells that perform gas exchange and carry on photosynthesis.
- Leaves are modified in various ways; some conserve water, some help a plant climb, and some help a plant capture food.

13. Label these diagrams of a leaf using the alphabetized list of terms (some are used more than once).

cuticle
epidermis
guard cell
leaf vein
palisade mesophyll
spongy mesophyll
stoma
stomata

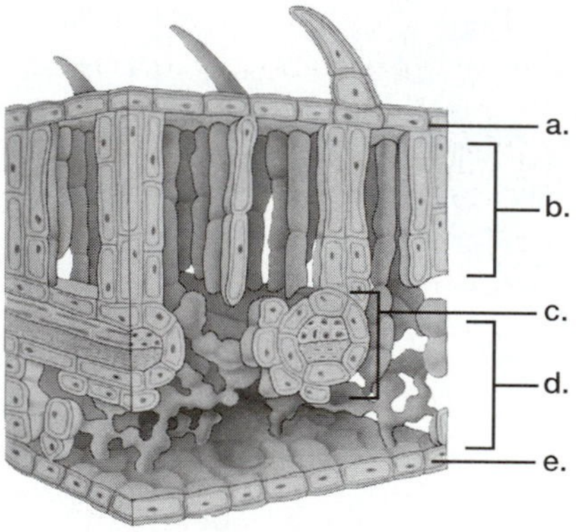

a.
b.
c.
d.
e.

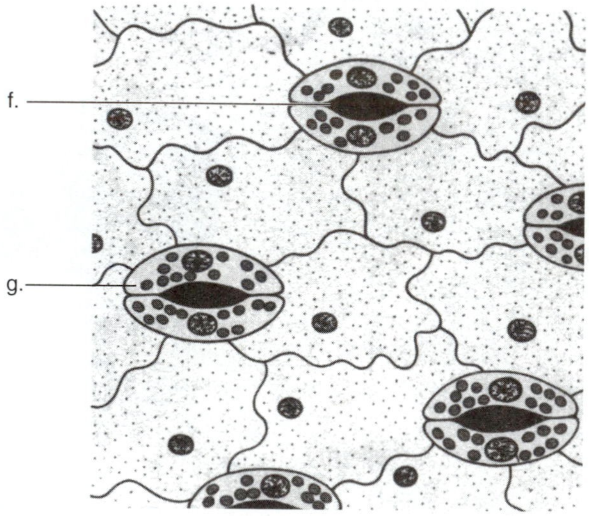

f.
g.

An extension of vascular bundles of stem is the h.____________________. The i.____________________ protects the leaf from drying out. The j.____________________ allow gas exchange. Longitudinal cells that photosynthesize are the k.____________________. Loosely arranged cells that photosynthesize and exchange gases are the l.____________________.

14. Place a check in front of the correct descriptions of leaves.

a. _____ spines of a cactus
b. _____ tendrils of a cucumber
c. _____ trap of a Venus's flytrap
d. _____ stolon of a strawberry plant

## Chapter Test

### Objective Questions

Do not refer to the text when taking this test.

____ 1. Each is an adult tissue in plants EXCEPT
a. epidermal.
b. ground.
c. meristem.
d. vascular.

____ 2. Select the incorrect association.
a. collenchyma—flexible support
b. parenchyma—unspecialized
c. sclerenchyma—tough and hard
d. sieve-tube member—mechanical strength

____ 3. Select the correct association.
a. phloem—minerals
b. phloem—photosynthesis
c. xylem—sugar
d. xylem—water

____ 4. The zone farthest from the root cap is the zone of
a. cell division.
b. elongation.
c. maturation.
d. primary growth.

____ 5. The cortex of the root mainly functions for
a. entrance of substances into the root.
b. photosynthesis.
c. protection.
d. starch storage.

____ 6. The root type in the carrot is the
a. adventitious root.
b. fibrous root.
c. taproot.
d. prop root.

____ 7. In the vascular bundles of a herbaceous stem,
a. phloem develops to the inside.
b. xylem develops to the outside.
c. Both *a* and *b* are correct.
d. Neither *a* nor *b* is correct.

____ 8. A stolon is a(n)
a. aboveground stem.
b. cork covering on a stem.
c. specialized type of leaf.
d. underground stem.

____ 9. The palisade mesophyll
a. stores food in the form of starch for the rest of the plant.
b. absorbs water and minerals.
c. contains elongated cells where photosynthesis takes place.
d. opens to allow gases to move in and out.

____ 10. The wood portion of a woody stem is composed of
a. pith.
b. cambium.
c. bark.
d. secondary xylem.
e. secondary phloem.

____ 11. Trace the path of water from the roots to the leaves.
a. root hairs→cortex→vascular cylinder→vascular bundles→leaf veins
b. root hairs→pith→xylem→phloem→wood→leaf veins
c. cortex→endodermis→xylem→tracheids→parenchyma→leaf veins
d. root hairs→sclerenchyma→vascular cylinder→cortex→leaf veins

____ 12. Stomata are
a. a type of transport tissue.
b. openings in leaf epidermis.
c. found in woody trees only.
d. a universal type of cell.
e. All of these are correct.

____ 13. Which of these comparisons of monocots and eudicots is NOT correct?
**Monocots — Eudicots**
a. net veined—parallel veined
b. one cotyledon—two cotyledons
c. scattered vascular bundles—circular pattern
d. flower parts in threes—flower parts in fours/fives

____ 14. The root cap is produced from the zone of
a. cell division.
b. elongation.
c. maturation.
d. pericycle.
e. vascular bundle.

____ 15. A cross section through the zone of maturation of a eudicot root would show
a. cells of the root cap.
b. the apical meristem.
c. transport tissues.
d. greatly elongated, undifferentiated cells.
e. All of these are correct.

____ 16. Annual rings in woody stems are caused by an increase in rings of the
a. primary phloem.
b. secondary phloem.
c. primary xylem.
d. secondary xylem.

_____ 17. The point of a stem at which leaves or buds are attached is termed the
a. node.
b. internode.
c. lenticel.
d. endodermis.
e. plasmodesmata.

_____ 18. Select the incorrect association.
a. monocot stem—vascular bundles scattered
b. eudicot herbaceous stem—vascular bundles in a ring
c. eudicot woody stem—no vascular tissue
d. All of these are correct.

_____ 19. Water taken up by root hairs has to enter the vascular cylinder through endodermal cells because
a. osmosis draws it in.
b. guard cells regulate the opening of stomata.
c. the Casparian strip is impermeable.
d. Both *a* and *c* are correct.

_____ 20. A leaf is like a root because they both
a. photosynthesize.
b. store the products of photosynthesis in bad times.
c. have vascular tissue.
d. have a double layer of epidermis.

## Critical Thinking Questions

Answer in complete sentences.

21. Do plants have a true organ structure as animals do? What characteristics of their anatomy support your answer?

22. Why is the evolution of xylem such an important adaptation for the success of plants on land?

**Test Results:** _______ number correct ÷ 22 = _______ × 100 = _______ %

## Exploring the Internet

The Online Learning Center at *www.mhhe.com/maderbiology8* has additional study material and practice quizzes that can help you master the content of this chapter. You can also find links to websites exploring additional topics in biology. Access to the Online Learning Center is free for those who have purchased a new textbook.

## Answer Key

### Study Exercises

**1. a.** anchors **b.** water **c.** minerals **d.** root hairs **e.** transport **f.** photosynthesis **g.** carbon dioxide, sunlight, water
**2.**

| Monocot | Eudicot |
|---|---|
| Parallel pattern | Net pattern |
| Scattered | In a ring |
| One | Two |
| In threes | In fours and fives |
| In a ring | In a vascular cylinder |

**3. a.** 2 **b.** 3 **c.** 1 **d.** 2 **e.** 3 **f.** 3 **4. a.** T **b.** T **c.** F **d.** F **e.** F **f.** F **g.** T **h.** T **i.** T **j.** T **k.** F **l.** F **m.** T **5.** See Figure 25.8, page 444, in text. **6. a.** New cells are appearing. **b.** Cells are getting longer. **c.** Cells are mature and specialized.
**7.**

| Description | Function |
|---|---|
| Outer single layer of cells | Root hairs, especially, absorb water and minerals |
| Thin-walled parenchyma cells | Food storage |
| Single layer of rectangular cells | Regulates entrance of minerals into vascular cylinder |
| Xylem and phloem | Transport water, minerals, and organic nutrients |

**8. a.** fibrous root; holds soil **b.** taproot; stores food **c.** prop root; anchors plant **9.** herbaceous, herbaceous, woody **10. a.** epidermis **b.** cortex **c.** vascular bundle **d.** pith **e.** phloem **f.** bark **g.** wood **h.** vascular cambium **i.** primary xylem **11. a.** scattered **b.** ring **c.** woody stem **d.** summer xylem vessels are small, and spring vessels are large **12. a.** 1 **b.** 4 **c.** 2 **d.** 3 **13. a.** epidermis **b.** palisade mesophyll **c.** leaf vein **d.** spongy mesophyll **e.** epidermis **f.** stoma **g.** guard cell **h.** leaf vein **i.** cuticle **j.** stomata **k.** palisade mesophyll **l.** spongy mesophyll **14.** a, b, c

## Chapter Test

**1.** c **2.** d **3.** d **4.** c **5.** d **6.** c **7.** d **8.** a **9.** c **10.** d **11.** a **12.** b **13.** a **14.** a **15.** c **16.** d **17.** a **18.** c **19.** c **20.** c **21.** Several specialized tissues (epidermal, ground, vascular) are integrated structurally in the root, stem, and leaf. All are organs, consisting of two or more tissues working together. **22.** Xylem allows plants to transport water against gravity, from roots to leaves. Without water transport, plants would have to be low lying.

# 26

# Nutrition and Transport in Plants

## Chapter Review

The growth requirements of plants include water, carbon dioxide, oxygen, and some minerals. Plants require eighteen different elements, with carbon, hydrogen, and oxygen making up a substantial amount of a plant's body weight. The other necessary nutrients are taken up by the roots as mineral ions.

Roots are in contact with soil, a mixture of soil particles, **humus** (decayed organic matter), living organisms, air, and water. Humus contributes to the ability of soil to provide mineral ions to plants. **Root hairs** enhance the uptake through active transport. The **root nodules** of leguminous plants fix atmospheric nitrogen.

Water and minerals are transported in **xylem,** and organic nutrients are transported in **phloem.** Water and minerals pass from the soil through the various layers of the root, eventually entering the vascular cylinder. Positive pressure cannot account for water movement in xylem. According to the **cohesion-tension model** of xylem transport, water enters and ascends through the xylem because of tension created by **transpiration** at the leaves. Both cohesion and adhesion prevent water molecules from breaking apart in the water column. Transpiration occurs through the **stomata** of leaves. **Guard cells** regulate the opening and closing of stomata in response to various environmental conditions.

Organic solutes pass through the phloem of the plant from a source to a sink, according to the **pressure-flow model** of phloem transport. The active transport of sucrose at the source causes water to enter the phloem, creating a positive pressure. This causes sap to flow from the area of greater solute concentration to the area of lower solute concentration at the sink.

## Study Exercises

Study the text section by section as you answer the questions that follow.

### 26.1 Plant Nutrition and Soil (pp. 460–63)

- Certain inorganic nutrients (e.g., $NO_3^-$, $K^+$, $Ca^{2+}$) are essential to plants; others that are specific to a type of plant are termed beneficial.
- Soil is built up over time by the weathering of rock and the actions of organisms.
- Soil particles, humus, and living organisms are components of soil that provide oxygen, water, and minerals to plants.
- Soil erosion is a serious threat to agriculture worldwide.

1. Label each of the following as either a macronutrient (MA) or micronutrient (MI). Place the appropriate letters next to each item.

MA—macronutrient MI—micronutrient

a. _____ carbon
b. _____ hydrogen
c. _____ iron
d. _____ manganese
e. _____ potassium
f. _____ boron

2. An investigator burns a plant, examines the ashes, and finds a synthetic element such as plutonium in the ash. What should he conclude? ________________________________________

________________________________________

3. An investigator is studying plant nutrition by means of hydroponics. She finds that the plant apparently is thriving, even though no manganese has been added to the culture water. What should she conclude?________

________________________________________

4. Indicate whether these statements are true (T) or false (F).

a. _____ Clay soil is more permeable than sandy soil.
b. _____ Gravity drains water away from plant roots.
c. _____ Sand is the predominate component in a good agricultural soil.
d. _____ Clay particles retain minerals because they are negatively charged.

5. Indicate whether these statements about topsoil are true (T) or false (F).

a. _____ Topsoil contains humus.
b. _____ Erosion of topsoil is a serious ecological problem.
c. _____ Various organisms are present in topsoil.
d. _____ Minerals are leached from topsoil by rain.

## 26.2 Water and Mineral Uptake (pp. 464–65)

- The tissues of a root are organized so that water and minerals entering between or at the root hairs will eventually enter xylem.
- Mineral ions cross plasma membranes by a chemiosmotic mechanism.
- Plants have various adaptations that assist them in acquiring nutrients. For example, symbiotic relationships are of special interest.

6. The ____________________ of the root regulates the entrance of minerals into the vascular cylinder.

7. Describe the adaptations for each of the following regarding nutrient uptake at the roots:

a. root hairs ________________________________________

b. nodules in leguminous plants ________________________________________

c. mycorrhizas ________________________________________

## 26.3 Transport Mechanisms in Plants (pp. 466–73)

- The vascular system in plants is an adaptation to living on land.
- The vascular tissue xylem transports water and minerals; the vascular tissue phloem transports organic nutrients.

8. Indicate whether these statements are true (T) or false (F). Rewrite any false statements to make them true.
   a. _____ Xylem is composed of sieve-tube members and tracheids. Rewrite:_______________
   b. _____ Xylem is an open pipeline from the roots to the leaves. Rewrite: _______________
   c. _____ Phloem is composed of sieve-tube members connected by strands called fibers. Rewrite:_______________
   d. _____ Sieve-tube members have a nucleus, but companion cells do not. Rewrite: _______________
   e. _____ Water, but not organic nutrients, can move from the leaves to the roots through phloem. Rewrite: ___

## Water Transport (pp. 468–69)

9. What is the proper sequence for these root structures to show the path of water as it enters the vascular cylinder? Indicate by letter. _______________
   a. Casparian strip
   b. cortex
   c. endodermis
   d. epidermis
   e. xylem

10. Water enters the cells of roots by a._______________. b._______________ pressure pushes sap in the xylem in a(n) c._______________ direction. This movement and pressure account for d._______________ along the edges of leaves.

11. Label this diagram, using the alphabetized list of terms.
    adhesion
    cohesion
    root hair
    stoma
    water molecule
    xylem

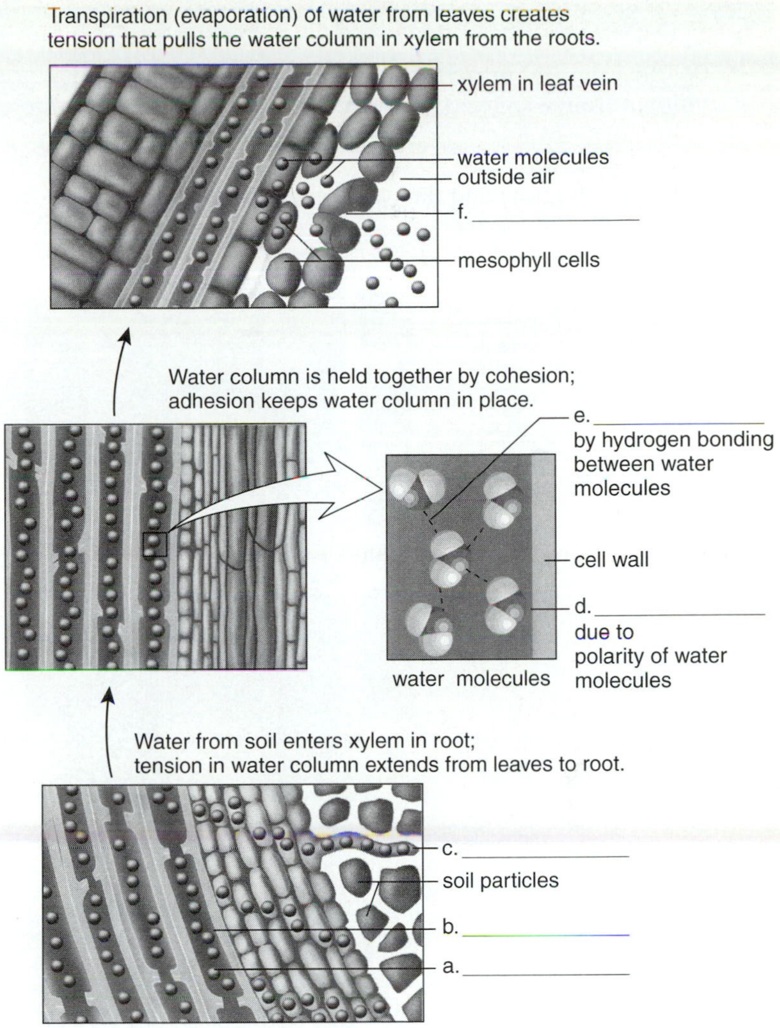

12. Study the diagram, then answer the questions that follow.

    a. Explain why transpiration occurs.
    ______________________________
    ______________________________
    ______________________________
    ______________________________

    b. What does this create, and what effect does it have?
    ______________________________
    ______________________________
    ______________________________
    ______________________________

    c. Why doesn't the water column in xylem break?
    ______________________________
    ______________________________
    ______________________________
    ______________________________

    d. Explain what is meant by the *cohesion-tension model of xylem transport.*
    ______________________________
    ______________________________
    ______________________________
    ______________________________

Transpiration (evaporation) of water from leaves creates tension that pulls the water column in xylem from the roots.

xylem in leaf vein
water molecules
outside air
f. ______________
mesophyll cells

Water column is held together by cohesion; adhesion keeps water column in place.

e. ______________ by hydrogen bonding between water molecules
cell wall
d. ______________ due to polarity of water molecules
water molecules

Water from soil enters xylem in root; tension in water column extends from leaves to root.

c. ______________
soil particles
b. ______________
a. ______________

## Opening and Closing of Stomata (p. 470)

- Stomata must be open for evaporation to occur.

13. Stomata open during photosynthesis when a. ______________ ions are actively transported b. ______________ (*into/out of*) the guard cells. Water now enters the guard cells by c. ______________, causing the cells to buckle out due to their d. ______________ (*thick/thin*) inner walls.

14. What is the proper sequence for steps *a–e* to demonstrate how stomata proceed from closed to open positions? Indicate by letter. ______________
    a. $CO_2$ decreases in leaf.
    b. $K^+$ enters guard cells.
    c. Stomata are closed.
    d. Stomata are open.
    e. Water enters guard cells.

15. What two important events are occurring when stomata are open?
    a. ______________________________
    b. ______________________________

## Organic Nutrient Transport (pp. 472–73)

- Active transport of sucrose into phloem creates a positive pressure that causes organic nutrients to flow in phloem from a source (where sucrose enters) to a sink (where sucrose exits).

16. Study the diagram and then answer the questions that follow.

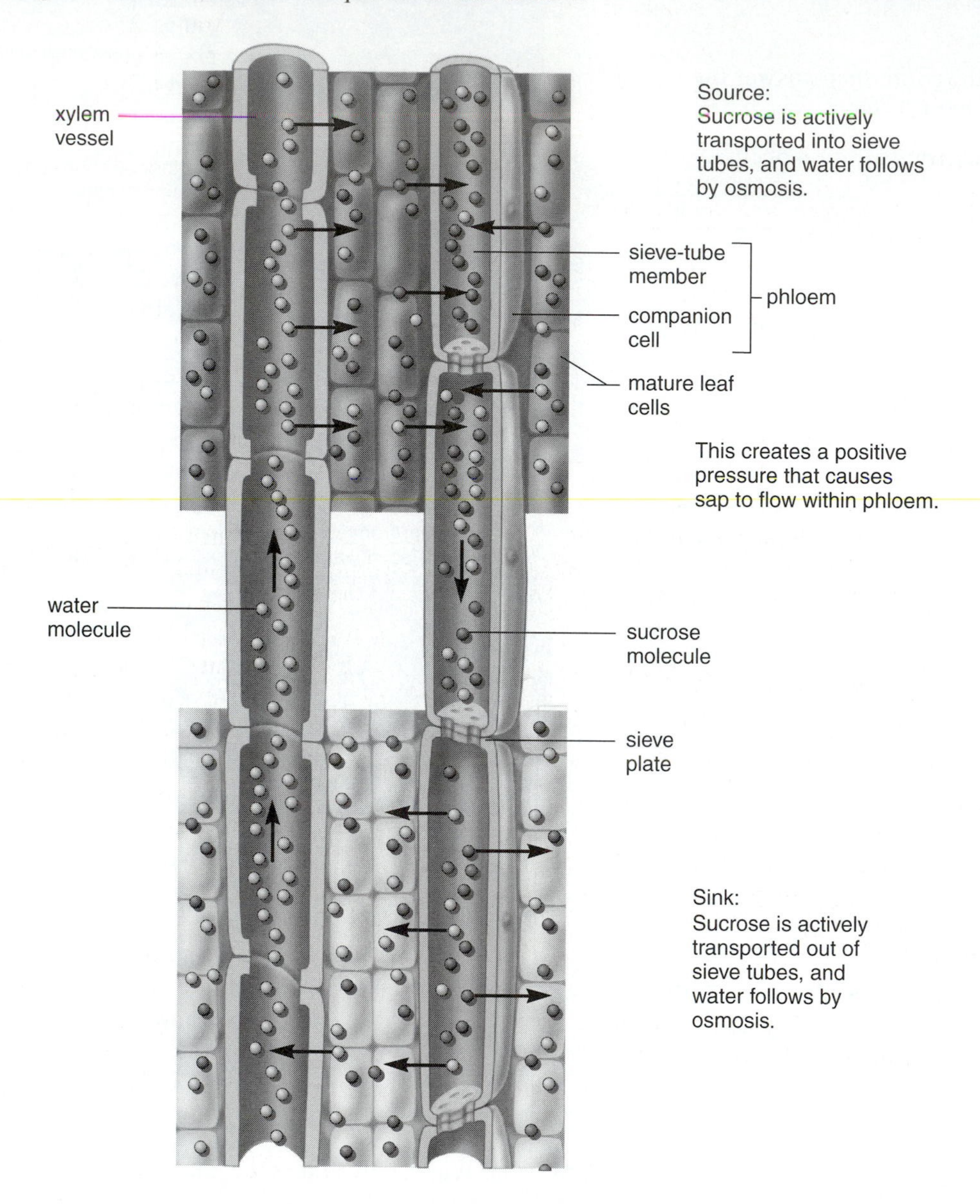


a. What is the "active" part of phloem transport?

______________________________________________

______________________________________________

______________________________________________

b. When water enters phloem at the leaves, what type of pressure is created and what is the effect of this pressure? ______________________________________________

c. Phloem sap moves from a source to a sink. In the summer, what organ is the source of sucrose?

______________________________________________

d. In the spring, before leaves begin to photosynthesize, what organ is the source? ____________________

# Chapter Test

## Objective Questions

Do not refer to the text when taking this test.

In questions 1–10, indicate whether the statement is true (T) or false (F).

____ 1. Water moves through the layers of a root by osmosis.

____ 2. The tracheids of xylem are actively metabolizing cells.

____ 3. Guttation mainly accounts for the upward pull of water through the xylem of a stem.

____ 4. The activity of guard cells regulates the opening and closing of stomata.

____ 5. Calcium is a plant micronutrient.

____ 6. Sulfur is a plant macronutrient.

____ 7. Water leaches elements from the topsoil.

____ 8. The cortex is the second root layer that water encounters as it passes through the root toward the vascular cylinder.

____ 9. By the pressure-flow model of phloem transport, substances travel from the source to the sink.

____ 10. Epiphytes are a specialized type of plant root system.

____ 11. What role does transpiration play in water transport?
- a. no role
- b. pushes the water
- c. pulls the water

____ 12. For transpiration to occur in the leaves,
- a. water must exhibit cohesiveness.
- b. the stomata must be open.
- c. water must evaporate.
- d. All of these are correct.

____ 13. Which statements are true about soil?
- a. Loam contains roughly one-third sand, silt, and clay particles.
- b. Ideally, water clings to soil particles and does not fill the spaces.
- c. Humus is very important to soil quality.
- d. All of these are true.

____ 14. Which is the incorrect association?
- a. xylem—transport of water and minerals
- b. phloem—transport of sucrose
- c. mycorrhizas—atmospheric nitrogen uptake
- d. mineral uptake—ATP required

____ 15. Turgor pressure is important to
- a. stomata opening and closing.
- b. plant cell rigidity.
- c. water flow in xylem.
- d. Both *a* and *b* are correct.
- e. All of these are correct.

____ 16. The Casparian strip causes
- a. water to enter the vascular cylinder.
- b. sucrose to enter phloem.
- c. water to stay within xylem during transport.
- d. All of these are correct.

____ 17. During transpiration, tension is created by
- a. cohesion of water molecules.
- b. active transport of sucrose into root cells.
- c. evaporation of water from the leaves.
- d. All of these are correct.

____ 18. Which is the incorrect association?
- a. xylem transport—transpiration needed
- b. phloem transport—positive pressure created
- c. xylem transport—active transport needed
- d. phloem transport—active transport needed

____ 19. When stomata are open,
- a. carbon dioxide enters leaves.
- b. potassium and water have entered guard cells.
- c. tension pulls water upward.
- d. All of these are correct.

____ 20. During phloem transport, the sink has the
- a. higher solute concentration, accounting for why water flows to it.
- b. lower solute concentration, due to the active transport of sucrose out of it.
- c. higher solute concentration because that is where sucrose is needed.
- d. Both *a* and *c* are correct.

## Critical Thinking Questions

Answer in complete sentences.

21. Physical and chemical laws explain many of the phenomena in biological systems. How is this demonstrated in the ascent of sap through xylem?

22. How are physical and chemical laws demonstrated in the opening and closing of stomata?

**Test Results:** _____ number correct ÷ 22 = _____ × 100 = _____ %

## Exploring the Internet

The Online Learning Center at *www.mhhe.com/maderbiology8* has additional study material and practice quizzes that can help you master the content of this chapter. You can also find links to websites exploring additional topics in biology. Access to the Online Learning Center is free for those who have purchased a new textbook.

## Answer Key

### Study Exercises

**1. a.** MA **b.** MA **c.** MI **d.** MI **e.** MA **f.** MI **2.** The plant probably does not require the plutonium, which it acquired accidentally. **3.** Manganese probably is already in the water. **4. a.** F **b.** T **c.** F **d.** T **5.** All statements are true. **6.** endodermis **7. a.** increase surface area for more uptake **b.** facilitate fixation of nitrogen **c.** increase surface area and facilitate the breakdown of organic matter for nutrient availability **8. a.** F;. . . composed of vessel elements and tracheids. **b.** T **c.** F; . . . strands called plasmodesmata. **d.** F; Companion cells have a nucleus, but sieve-tube members do not. **e.** F; Organic nutrients can move from . . . **9.** d, b, a, c, e **10. a.** osmosis **b.** Root **c.** upward **d.** guttation **11. a.** water molecule **b.** xylem **c.** root hair **d.** adhesion **e.** cohesion **f.** stoma **12. a.** Dry air causes water to evaporate at leaves. **b.** It creates a tension that pulls water from roots to leaves. **c.** Water molecules cling to one another (cohesion) and to sides of vessels (adhesion). **d.** Cohesion refers to water molecules clinging together, and tension refers to the pull created by transpiration. Consequently, water flows upward along the length of the plant. **13. a.** potassium **b.** into **c.** osmosis **d.** thick **14.** c, a, b, e, d **15. a.** Carbon dioxide is being absorbed. **b.** Water is lost in transpiration. **16. a.** active transport of sucrose into phloem **b.** a positive pressure is created, which causes sap to flow in phloem. **c.** leaves **d.** roots

### Chapter Test

**1.** T **2.** F **3.** F **4.** T **5.** F **6.** T **7.** T **8.** T **9.** T **10.** F **11.** c **12.** d **13.** d **14.** c **15.** d **16.** a **17.** c **18.** c **19.** d **20.** b **21.** The evaporation of water and the cohesion between water molecules bring about this process. **22.** Osmosis and the mechanical strength of cell walls are involved in this process.

# 27

# Control of Growth and Responses in Plants

## Chapter Review

Like animals, plants use a reception-transduction-response pathway when they respond to a stimulus. **Tropisms** are growth responses toward or away from unidirectional stimuli. Positive **phototropism** of stems is growth toward light. Negative **gravitropism** of stems is growth away from the direction of gravity. **Thigmotropism** occurs when a plant makes contact with an object. Nastic movements are not directional.

Plants exhibit **circadian rhythms** such as the sleep movements of prayer plants and beans and the closing of stomata. A **biological clock** most likely controls these circadian rhythms.

Both stimulatory and inhibitory **hormones** help control growth patterns. Some hormones stimulate growth **(auxins, gibberellins,** and **cytokinins),** and some inhibit growth **(abscisic acid** and **ethylene).**

**Photoperiodism** is the physiological plant responses to changes in the length of day or night. Flowering in plants is a photoperiodic event, leading to the classification of plants as **short-day, long-day,** or **day-neutral.** The response of the pigment **phytochrome** is believed to be involved in the flowering event.

## Study Exercises

Study the text section by section as you answer the questions that follow.

### 27.1 Plant Responses (pp. 478–81)

- Plants use a reception-transduction-response pathway when they respond to a stimulus.
- Tropisms are growth responses in plants toward or away from unidirectional stimuli such as light or gravity.
- Plants sometimes exhibit circadian rhythms (e.g., closing of stomata) that recur every 24 hours.

1. Match each description with these terms:
   1. reception
   2. transduction
   3. response

   a. _____ The stimulus is changed into a form meaningful to the organism.
   b. _____ The organism reacts to the stimulus.
   c. _____ The organism perceives the stimulus.
   d. _____ Plants have a pigment that detects far-red light.
   e. _____ When far-red light is present, genes produce an enzyme.
   f. _____ When the enzyme is present, chlorophyll is produced, and the plant turns green.

2. Identify the type of response described, and add the word *positive* or *negative* if appropriate.

   a. _______________ A plant tendril encircles an object.
   b. _______________ A root grows down into the soil.
   c. _______________ When sensitive hairs are touched, the leaves of a Venus's flytrap snap shut.
   d. _______________ Leaves track the sun, and the stem bends toward the light.
   e. _______________ Stomata open and close every 24 hours.

## 27.2 Plant Hormones (pp. 482–87)

- Some plant hormones stimulate growth, and other plant hormones inhibit growth.
- The biochemical manner in which auxin and gibberellin function has been studied.
- It now appears that various plant hormones interact to bring about a response to a stimulus.

3. Complete the following table of stimulatory plant hormones:

| Name | Function |
|---|---|
| | |
| | |
| | |

### Auxins (pp. 482–83)

4. Indicate whether these statements concerning oat seedling experiments are true (T) or false (F). Rewrite any false statements to make them true.

   a. _____ Oat seedlings with tips removed still bend toward the light. Rewrite: ____________________

   b. _____ When an agar block containing auxin is placed on one side of a tipless coleoptile, the shoot curves toward that side. Rewrite: ____________________

   c. _____ Auxin normally moves to the shady side, and thereafter the stem bends toward the light. Rewrite: ____________________

5. a. Label the diagrams with these terms. Some are used more than once.
   ATP auxin $H^+$ receptor

   b. Explain what is happening in each diagram.

____________________

____________________

____________________

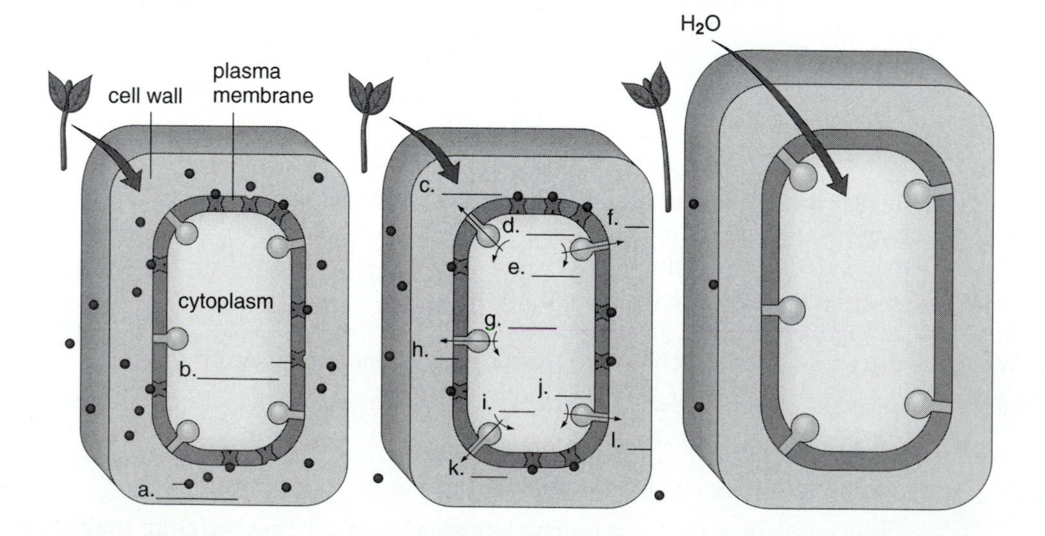

## Gibberellins (p. 484)

6. What effect do gibberellins have on the following?

   a. stems ______________________

   b. dormancy of seeds and buds ______________________

   c. dwarf plants ______________________

7. Study this diagram and then answer the questions that follow concerning how gibberellin stimulates growth of a plant embryo.

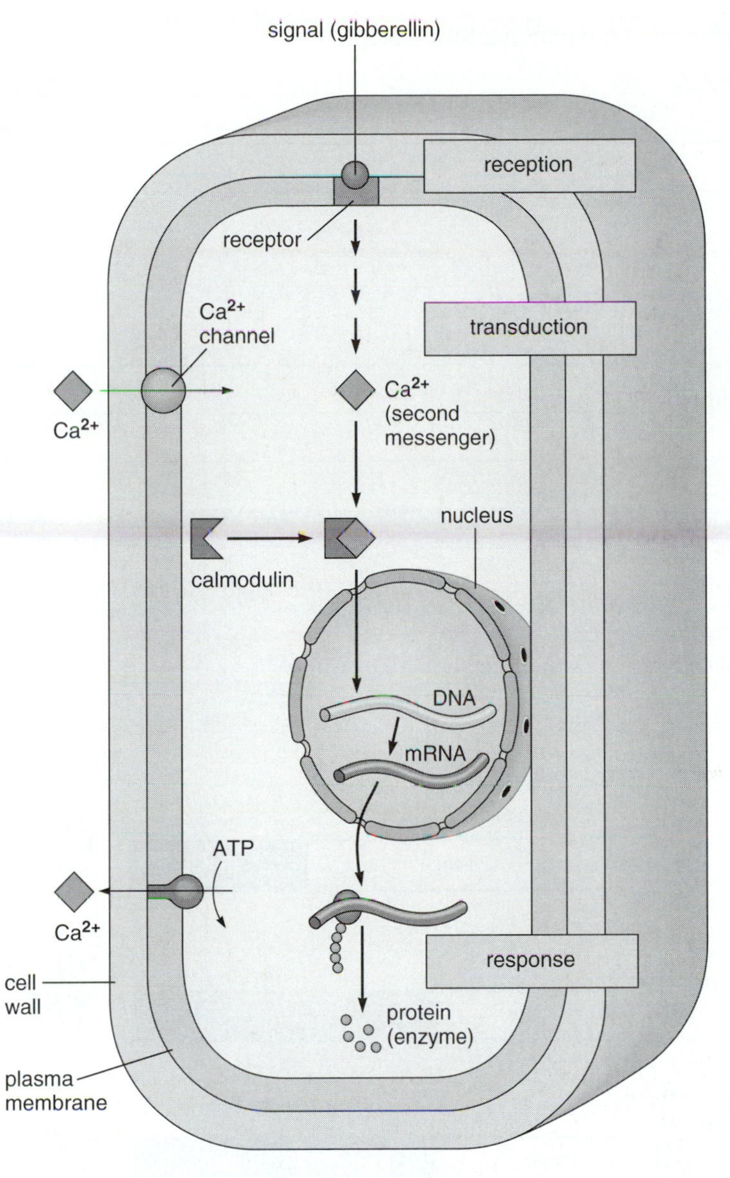


   a. What happens after gibberellin attaches to the plasma membrane? ______________________

   b. How does calcium ($Ca^{2+}$) enter a cell? ______________________

   c. What does the calcium-calmodulin complex do? ______________________

   ______________________

   d. How does this help the embryo grow? ______________________

   ______________________

## Cytokinins (p. 486)

8. Indicate whether these statements about tissue culture experiments that demonstrate plant hormone interaction are true (T) or false (F).
   a. _____ Cellular enzymes affect the activity of plant hormones.
   b. _____ Oligosaccharins function in animals but not in plants.
   c. _____ The acidity of the culture medium affects cell differentiation.
   d. _____ The ratio of auxin to cytokinin affects cell differentiation.

## Abscisic Acid (p. 487)

9. What effect does abscisic acid have on the following?
   a. seed and bud dormancy ______________________________
   b. stomata ______________________________
   c. a bud in the fall ______________________________

## Ethylene (p. 487)

10. What effect does ethylene have on the following?
    a. ripening of fruit ______________________________
    b. growth of house plants ______________________________
    c. abscission ______________________________

# 27.3 Photoperiodism (pp. 488–89)

- Plant responses controlled by the length of day or night (photoperiod) involve the pigment phytochrome.

11. Study the diagram and then answer the questions that follow.

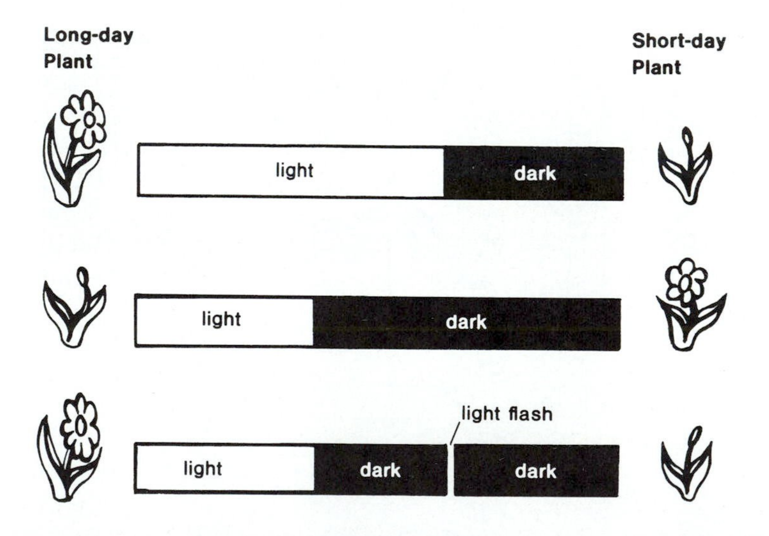


A long-day (short-night) plant flowers when the night is a.__________________ than a critical length.
A short-day (long-night) plant will not flower when the night is b.__________________ than a critical length.
A long-day (short-night) plant will not flower when the night is c.__________________ than a critical length.
A short-day (long-night) plant will flower when the night is d.__________________ than a critical length.

A long-day (short-night) plant will flower if a flash of light interrupts a night that is e.___________________

than a critical length. A short-day (long-night) plant will not flower if a flash of light interrupts a night that is f.___________________ than a critical length.

The conclusion is that it is the length of the g.___________________, and not the h.___________________, that controls flowering.

12. a. Label the three arrows in the diagram with these terms:
metabolic conversion     absorbs red light     absorbs far-red light

$P_r$     $P_{fr}$

b. Place the following terms above and below the arrows as appropriate:
daytime     shade and evening

c. Place the following terms to one side of $P_r$ or $P_{fr}$ as appropriate:
flowering is stimulated     stem elongation is inhibited.

d. The conclusion is that $P_{fr}$ ___________________________________________
___________________________________________.

# Chapter Test

## Objective Questions

Do not refer to the text when taking this test.

____ 1. The main site of auxin production is in the
a. epidermis.
b. ground tissue.
c. shoot apical meristem.
d. vascular tissue.

____ 2. Select the incorrect statement about auxins.
a. They cause breakdown of polysaccharides in the cell wall.
b. Their concentration in leaves and fruits prevents leaves and fruits from falling to the ground.
c. They are produced in the shoot apex of the plant.
d. They are transported to the side of the plant receiving light.

____ 3. The main effect of gibberellins is to
a. hasten the ripening of fruits.
b. inhibit the flowering process.
c. prevent leaf abscission.
d. promote cell division and enlargement.

____ 4. Senescence is the
a. aging of the overall plant.
b. loss of leaf color.
c. propagation of cuttings.
d. process of flowering.

____ 5. Ethylene increases the ripening of fruit by stimulating
a. cell wall production.
b. enzyme activity.
c. pigment production.
d. RNA synthesis.

____ 6. Another name for ABA is the
a. dormancy hormone.
b. flowering inhibitor.
c. growth inhibitor.
d. stress hormone.

____ 7. Select the incorrect association.
a. positive gravitropism—response to gravity
b. phototropism—response to light stimulus
c. negative gravitropism—response to gravity
d. thigmotropism—response to chemical stimulus

____ 8. Which of the following is NOT a response to primarily an internal stimulus?
a. contraction movement
b. nastic movement
c. phototropic movement
d. twining movement

____ 9. Which of the following is NOT classified as a short-day plant?
a. barley
b. cocklebur
c. poinsettia
d. chrysanthemum

____ 10. Interrupting the dark period with a flash of white light prevents flowering in a
a. long-day plant.
b. short-day plant.

____ 11. Which of these would represent transduction following perception of a stimulus?
a. The stem bends toward the light.
b. An enzyme is made.
c. The cell wall weakens.
d. The cell takes in water.

____ 12. Which of these is an example of thigmotropism?
a. A dwarf bean becomes a pole bean.
b. The prayer plant performs a sleep movement.
c. When tendrils are touched, they begin to curve.
d. All of these are correct.

____ 13. Stomata close every 24 hours because
a. auxin moves to the shady side of a stem.
b. cytokinins and auxin are in balance.
c. plants have a biological clock.
d. All of these are correct.

____ 14. Oat seedlings will NOT bend when
a. exposed to artificial plant hormones.
b. coleoptile tips are cut off.
c. $P_r$ has become $P_{fr}$.
d. All of these are correct.

____ 15. The reception of auxin leads to
a. the removal of $H^+$ from the cell.
b. $ATP \rightarrow ADP + P$.
c. the weakening of plant cell walls.
d. All of these are correct.

____ 16. When calcium combines with calmodulin,
a. the amount of amylase in embryonic cells decreases.
b. the amount of amylase in embryonic cells increases.
c. a gene is activated.
d. Both *b* and *c* are correct.

____ 17. Which of these is mismatched?
a. abscisic acid—stomata close
b. ethylene—fruits ripen
c. cytokinin—stems elongate
d. auxin—stems bend toward the light

____ 18. Photoperiodism is a
a. nastic response.
b. circadian rhythm.
c. tropism.
d. light/dark response.

____ 19. Phytochrome
a. stimulates the genes directly.
b. produces an enzyme directly.
c. is involved in perception of stimulus.
d. All of these are correct.

____ 20. In a short-day plant, flowering is believed to depend on
a. proper lighting.
b. the presence of phytochrome.
c. the proper hormonal balance.
d. All of these are correct.

## Critical Thinking Questions

Answer in complete sentences.

21. Based on their response to photoperiod, many plant species are grouped into short-day or long-day categories. Do you think that these are the best titles for these plant categories?

22. Explain what is meant by a reception-transduction-response pathway.

**Test Results:** ______ number correct ÷ 22 = ______ × 100 = ______ %

## Exploring the Internet

The Online Learning Center at *www.mhhe.com/maderbiology8* has additional study material and practice quizzes that can help you master the content of this chapter. You can also find links to websites exploring additional topics in biology. Access to the Online Learning Center is free for those who have purchased a new textbook.

# Answer Key

## Study Exercises

**1. a.** 2 **b.** 3 **c.** 1 **d.** 1 **e.** 2 **f.** 3 **2. a.** thigmotropism **b.** positive gravitropism **c.** nastic movement **d.** positive phototropism **e.** circadian rhythm
**3.**

| Name | Function |
|---|---|
| auxin | phototropism, gravitropism, apical dominance |
| gibberellin | stem elongation, growth of dwarf plants, seed and bud dormancy broken |
| cytokinin | cell division, prevention of leaf senescence, initiation of growth |

**4. a.** F; . . . do not bend toward . . . **b.** F; . . . curves away from that side **c.** T **5. a.** See Figure 27.8, page 483, in text. **b.** Auxin is attaching to receptors. Proton pump is breaking down ATP and pumping $H^+$ out of the cell. $H^+$ causes the cell wall to break down, water enters, and the cell increases in size. **6. a.** makes them grow **b.** breaks dormancy **c.** makes them grow **7. a.** $Ca^{2+}$ enters the cell and attaches to calmodulin. **b.** through a channel in the plasma membrane **c.** activates a gene and amylase is produced **d.** Amylase breaks down starch, providing glucose as a source of energy. **8. a.** F **b.** F **c.** T **d.** T **9. a.** promotes it **b.** closes them **c.** promotes formation of bud scales **10. a.** promotes it **b.** retards it **c.** promotes it **11. a.** shorter **b.** shorter **c.** longer **d.** longer **e.** longer **f.** longer **g.** dark period (night) **h.** light period (day) **12. a.–b.** See Figure 27.16, page 489, in text. **c.** Place both labels beside $P_{fr}$ **d.** is the metabolically active form

## Chapter Test

**1.** c **2.** d **3.** d **4.** b **5.** b **6.** d **7.** d **8.** c **9.** a **10.** b **11.** b **12.** c **13.** c **14.** b **15.** d **16.** d **17.** c **18.** d **19.** c **20.** d **21.** The category titles are somewhat misleading, because the plants are responding mainly to the length of darkness as opposed to the length of daylight. **22.** During reception, a plasma membrane receptor receives the hormone, such as gibberellin; during transduction, a cellular signal such as a second messenger forms; during the response, DNA is activated and a protein, such as amylase, is made so that growth and elongation occur. See also figure on page 213 of Study Guide.

# 28

# Reproduction in Plants

## Chapter Review

All plants have two stages (generations) in their life cycle—the diploid **sporophyte** and the haploid **gametopyte**. In flowering plants, the sporophyte is dominant and is the generation that bears flowers.

The **flower** is unique to angiosperms. It is the reproductive organ of the plant. A complete flower has all flower parts: **sepals, petals, stamens,** and at least one **carpel.** It produces microspores and megaspores. A **microspore** develops into a male gametophyte, which becomes a pollen grain, and a **megaspore** develops into a female gametophyte, an embryo sac located within an ovule. After fertilization, the ovule of the ovary matures into a **seed**. The ovary becomes the **fruit.** Fruits aid dispersal of seeds. There are simple fruits and compound fruits.

**Seed germination** has occurred when the embryo breaks out of the seed coat and becomes a seedling with roots, stems, and leaves.

Asexual means of reproduction in flowering plants include propagation from cuttings and **tissue culture** techniques. Certain agricultural plants have been genetically engineered to be herbicide and/or insect resistant. Transgenic crops of the future are expected to have improved agricultural traits and food qualities, and to have higher yields.

## Study Exercises

Study the text section by section as you answer the questions that follow.

### 28.1 Reproductive Strategies (pp. 494–500)

- Flowering plants have an alternation of generations life cycle. The plant that bears flowers is the sporophyte.
- Flowering plants have two types of gametophytes: the male gametophyte (pollen grain) and the female gametophyte (embryo sac).
- Adaptation to a land environment includes pollen grains that carry sperm to the female gametophyte, which produces an egg.

1. a. What role does the sporophyte play in the plant life cycle? ______________________________

______________________________________________________________

b. What role does the female gametophyte play in the flowering plant life cycle? ______________

______________________________________________________________

c. What role does the male gametophyte play in the flowering plant life cycle? ______________

______________________________________________________________

d. What role does an ovule play in the flowering plant life cycle? ______________________

______________________________________________________________

2. Label this diagram of the life cycle of a flowering plant using the alphabetized list of terms.

   anther
   egg
   embryo in seed
   female gametophyte
   male gametophyte
   megaspore
   microspore
   ovary
   sperm
   sporophyte
   zygote

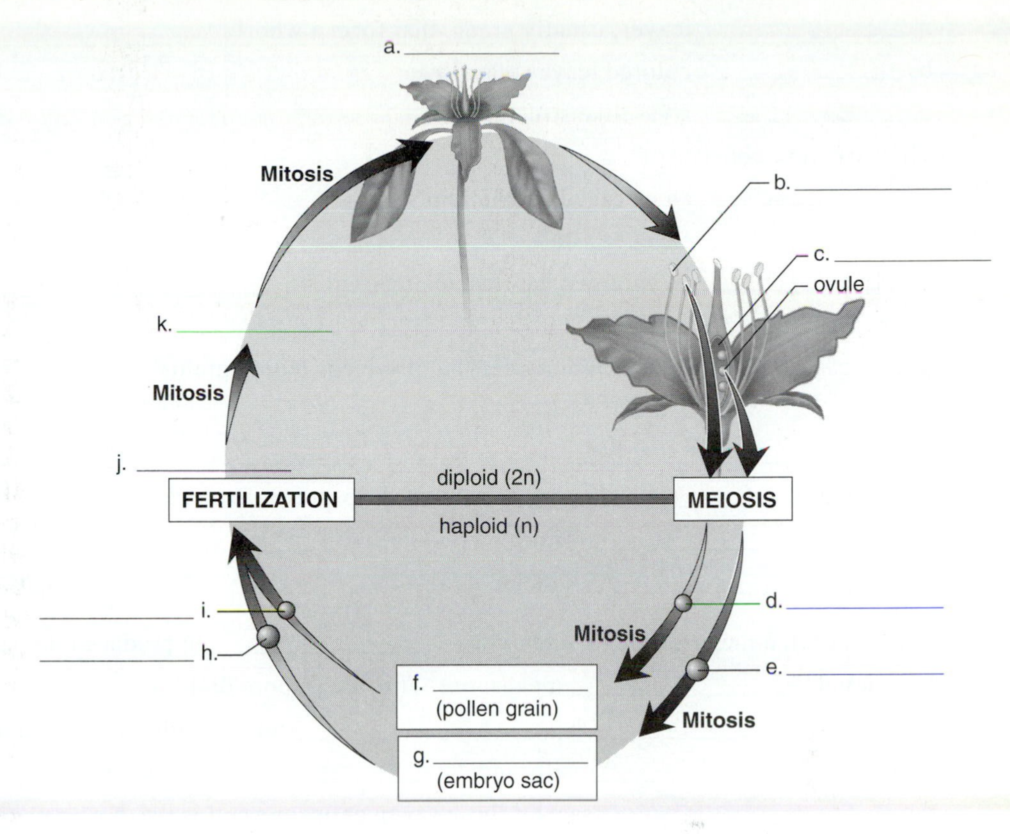


Which generation is dominant? l. ________________ Flowering plants produce two different types of spores: m. ________________ and n. ________________.

3. Label this diagram of a flower using the alphabetized list of terms.

   anther
   carpel
   filament
   ovary
   ovule
   peduncle
   petal
   receptacle
   sepal
   stamen
   stigma
   style

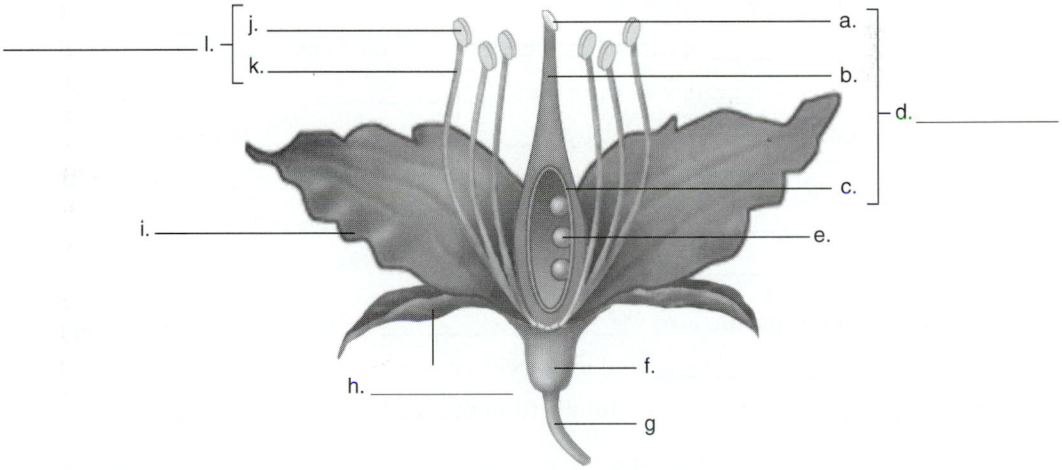

____________ l. j. ____________ k. ____________

a. ____________

b. ____________

c. ____________

d. ____________

e. ____________

f. ____________

g. ____________

h. ____________

i. ____________

4. Name the following parts of a flower:

a. ______________ leaves, usually green, that form a whorl

b. ______________ colored leaves of a flower

c. ______________ a vaselike structure

This structure consists of:

d. ______________ an enlarged, sticky knob

e. ______________ a slender stalk

f. ______________ an enlarged base enclosing ovules

Stamen components include:

g. ______________ a saclike container that produces pollen grains

h. ______________ a slender stalk

5. In the a.____________________, many b.____________________undergo c.____________________, each producing d.____________________ haploid microspores. Each microspore divides once by e.____________________, forming a two-celled f.____________________. This is the g.____________________, which contains the h.____________________cell and the tube cell.

6. In the carpel, a megasporocyte undergoes a.________________to produce one functional b.________________megaspore. This megaspore divides by c.____________________ d.____________________times, resulting in a female gametophyte with e.____________________cells and f.____________________nuclei. This is called the g.____________________sac. Of these seven cells, one cell is the h.____________________; one cell has i.____________________ polar nuclei, two are synergid cells, and three are antipodal cells.

7. Place the appropriate letter next to each statement.

P—pollination F—fertilization

a. _____ can be assisted by the wind

b. _____ sperm nucleus and egg nucleus unite

c. _____ transfer from anther to stigma

d. _____ 3n endosperm cell forms

8. a.____________________occurs when a pollen grain lands on the sticky b.____________________. One cell (tube cell) in a pollen grain will form the c.____________________, which dissolves a path to the ovary. The other cell (generative cell) divides by d.____________________ to form e.____________________ sperm. One will fertilize the egg, forming a(n) f.____________________. The other will unite with two polar nuclei, forming a(n) g.____________________ endosperm cell. This process is unique to flowering plants and is called h.____________________.

9. The zygote develops into a(n) a.____________________ enclosed in a seed coat surrounded by the b.____________________ wall and is called a(n) c.____________________. The endosperm forms d.____________________ food within the seed, which will then germinate to produce a new e.____________________.

## 28.2 Seed Development (p. 501)

- The eudicot embryo goes through a series of stages; once the cotyledons appear, it is possible to distinguish the root and shoot apical meristems.
- The embryo plus its stored food is contained within a seed coat.
- In flowering plants, seeds are enclosed by fruits, which develop from the ovary and accessory organs of a flower.

10. Indicate whether these statements about the development of the eudicot embryo are true (T) or false (F).
    a. _____ The eudicot embryo goes through stages that consist of three different shapes.
    b. _____ Root apical meristem produces underground growth.
    c. _____ Shoot apical meristem produces aboveground growth.
    d. _____ The early embryo is pear shaped.

## 28.3 Fruit Types and Seed Dispersal (pp. 502–5)

- Fruits aid the dispersal of seeds.

11. Label the type of fruit depicted using the alphabetized list of terms.
    achene
    berry
    capsule
    drupe
    follicle
    legume
    nut
    pome

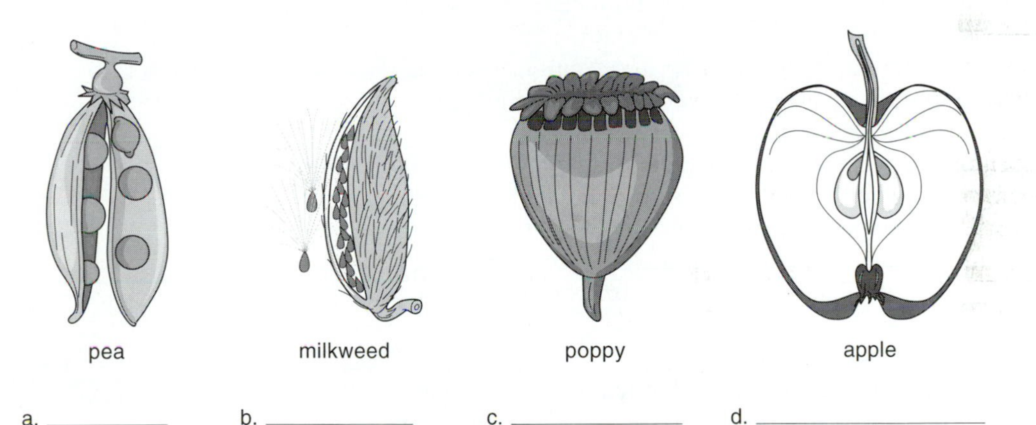


a. __________ b. __________ c. __________ d. __________

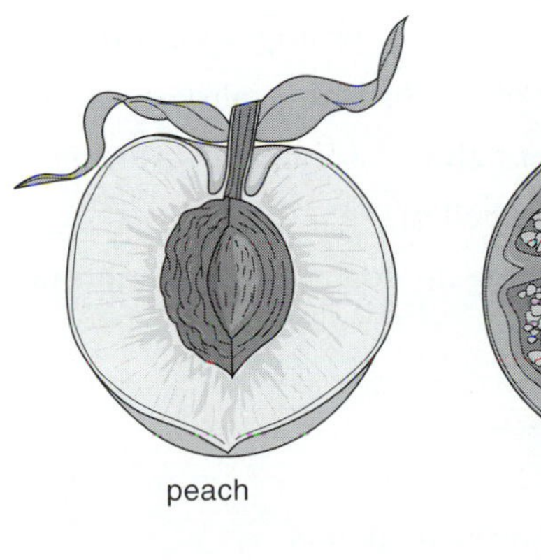


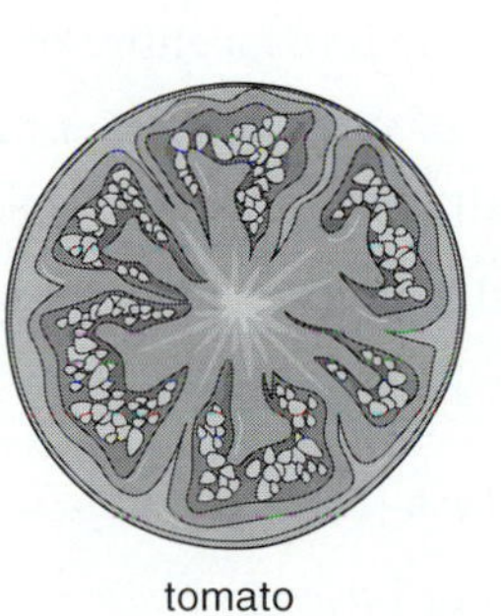

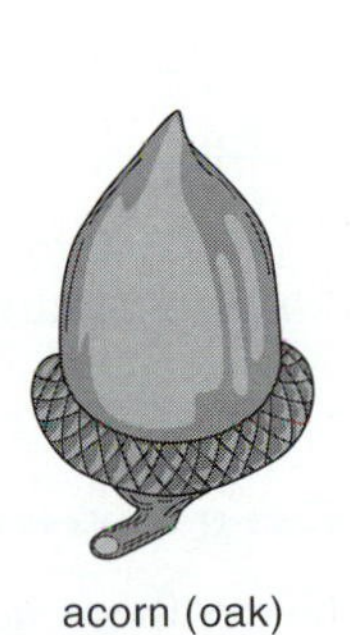


e. __________ f. __________ g. __________ h. __________

12. Explain these terms:

a. dehiscent ____________________

b. indehiscent ____________________

c. simple fruit ____________________

d. dry fruit ____________________

e. compound fruit ____________________

A peach is a simple fruit with a(n) f.____________________ pericarp, while a nut is a simple fruit with a(n) g.____________________ pericarp.

13. List four ways that seeds and fruits are dispersed.

a. ____________________

b. ____________________

c. ____________________

d. ____________________

14. Label each statement as describing the structure and germination of either the bean seed (B) or the corn kernel (C).

a. _____ It is actually a fruit.
b. _____ The plumule is enclosed in a sheath called the coleoptile.
c. _____ Most of the food storage tissue is endosperm.
d. _____ Shoot is hook shaped.

## 28.4 Asexual Reproduction in Plants (pp. 508–11)

- Many flowering plants have an asexual means of propagation (i.e., from stem cuttings, roots, or nodes of rhizomes and stolons).
- Many plants can be regenerated by tissue culture from meristem tissue and from individual cells. This has contributed to the genetic engineering of plants.

15. Place a check in front of the example(s) of vegetative propagation.
a. _____ strawberry plants grown from the nodes of stolons
b. _____ potato plants grown from the eyes of a potato
c. _____ ornamental plants grown from stem cuttings

16. Place a check in front of the characteristics of vegetative propagation.
a. _____ sexual reproduction
b. _____ asexual reproduction
c. _____ new plant genetically identical to original plant
d. _____ new plant genetically dissimilar to original plant

17. Micropropagation is a commercial way to produce thousands of a.____________________ (*identical/dissimilar*) seedlings using b.____________________ culture. A protoplast is a(n) c.____________________ that can go on to develop into an entire plant. Somatic embryos can be grown from flower d.____________________, which is free of viruses. Anther culture, which allows the production of plants that express e.____________________ alleles, uses the tube cell of f.____________________ grains. The process of culturing individual cells—derived from root, stem, or leaf—to produce desirable chemicals is called g.____________________.

### Genetic Engineering of Plants (p. 510)

18. Indicate whether these statements are true (T) or false (F).
a. _____ It is possible to produce transgenic animals but not transgenic plants.
b. _____ Fertile transgenic corn and wheat plants can be grown from protoplasts.
c. _____ A gene gun bombards plant cells with DNA-coated metal particles.
d. _____ Propagation of plants from single cells is a form of asexual reproduction.

# CHAPTER TEST

## OBJECTIVE QUESTIONS

Do not refer to the text when taking this test.

____ 1. Which of these is a true statement?
- a. A microspore develops into an immature male gametophyte.
- b. The dominant generation of the angiosperm is the haploid gametophyte.
- c. Pollination is the union of male and female sex cells.
- d. Most fruits are compound fruits.

____ 2. Which of these is a false statement?
- a. Flower structures are modified leaves.
- b. The hypocotyl is the portion of the stem above the attachment of the cotyledons.
- c. Hooks and spines of clover are a means of dispersal.
- d. A fruit is the mature ovary containing seeds.

____ 3. In the life cycle of flowering plants, there are
- a. two types of spores.
- b. microspores and megaspores.
- c. two types of pollen grains.
- d. Both *a* and *b* are correct.
- e. All of these are correct.

____ 4. The ovule is to the carpel as the
- a. anther is to the stamen.
- b. anther is to the filament.
- c. filament is to the anther.
- d. All of these are correct.

____ 5. The structure immediately preceding the female gametophyte is the
- a. male gametophyte.
- b. ovule.
- c. megaspore.
- d. carpel.

____ 6. Pollination
- a. precedes fertilization.
- b. is the transfer of pollen to the stigma.
- c. requires the presence of water.
- d. Both *a* and *b* are correct.
- e. All of these are correct.

____ 7. The eudicot embryo begins as
- a. two cotyledons.
- b. a single-celled zygote.
- c. a heart-shaped single cell.
- d. All of these are correct.

____ 8. What do a tomato, a peach, and an apple have in common?
- a. They are all fruits.
- b. They are all fleshy fruits.
- c. They are all drupes.
- d. Both *a* and *b* are correct.
- e. All of these are correct.

____ 9. Insects
- a. routinely disperse seeds in angiosperms.
- b. are flower pollinators.
- c. are the only flower pollinators.
- d. All of these are correct.

____ 10. Vegetative propagation
- a. is asexual propagation.
- b. can occur from stems.
- c. requires the use of leaves.
- d. Both *a* and *b* are correct.
- e. All of these are correct.

____ 11. Micropropagation takes its name from the use of
- a. the microscope.
- b. single cells or tissues.
- c. culture media.
- d. All of these are correct.

____ 12. Anther culture
- a. produces homozygous recessive genotypes.
- b. produces genetically similar plants.
- c. must be combined with ovule culture.
- d. Both *a* and *b* are correct.
- e. All of these are correct.

____ 13. Fertile corn and wheat plants cannot be
- a. genetically engineered.
- b. grown from genetically engineered protoplasts.
- c. grown from single cells.
- d. All of these are correct.

____ 14. In the life cycle of flowering plants, the embryo sac
- a. is the equivalent of the pollen grain.
- b. has ten cells.
- c. contains an embryo.
- d. All of these are correct.

____ 15. A(n) ______ produces four microspores.
- a. tube cell
- b. microsporocyte
- c. anther
- d. microsporangium

____ 16. Fertilization in flowering plants
- a. results in a 2n zygote and 3n endosperm.
- b. results in a 3n embryo and 2n endosperm.
- c. involves the polar nuclei but not the egg nucleus.
- d. occurs rarely.

____ 17. Fruits are classified according to whether they are
- a. big or small.
- b. dry or fleshy.
- c. simple or compound.
- d. Both *b* and *c* are correct.
- e. All of these are correct.

____ 18. The pollen grain contains
a. the sperm.
b. the egg.
c. the embryo.
d. It depends on the photoperiod.

____ 19. Fertilization in flowering plants occurs
a. in the flower.
b. prior to seed formation.
c. in the ovary.
d. All of these are correct.

____ 20. A seed typically
a. contains an embryo and stored food.
b. germinates before it starts to grow.
c. has cotyledons.
d. All of these are correct.

## Critical Thinking Questions

Answer in complete sentences.

21. What is one advantage and one disadvantage of asexual plant reproduction compared to sexual plant reproduction?

22. Aside from pollination, how does seed dispersal by birds and mammals represent another example of coevolution?

**Test Results:** ______ number correct ÷ 22 = ______ × 100 = ______ %

## Exploring the Internet

The Online Learning Center at *www.mhhe.com/maderbiology8* has additional study material and practice quizzes that can help you master the content of this chapter. You can also find links to websites exploring additional topics in biology. Access to the Online Learning Center is free for those who have purchased a new textbook.

## Answer Key

### Study Exercises

**1. a.** The sporophyte produce spores; in flowering plants, the sporophyte produces microspores and megaspores. **b.** The female gametophyte, the embryo sac, contains an egg and the polar nuclei. **c.** The male gametophyte, the pollen grain, produces sperm and carries them to the embryo sac. **d.** Ovule first contains the megaspore, then the embryo sac, and then develops into the seed. **2. a.** sporophyte **b.** anther **c.** ovary **d.** microspore **e.** megaspore **f.** male gametophyte **g.** female gametophyte **h.** egg **i.** sperm **j.** zygote **k.** embryo in seed **l.** sporophyte **m.** megaspores **n.** microspores See Figure 28.1, page 494, in text. **3. a.** stigma **b.** style **c.** ovary **d.** carpel **e.** ovule **f.** receptacle **g.** peduncle **h.** sepal **i.** petal **j.** anther **k.** filament **l.** stamen **4. a.** sepals **b.** petals **c.** carpel **d.** stigma **e.** style **f.** ovary **g.** anther **h.** filament **5. a.** anther **b.** microsporocytes **c.** meiosis **d.** four **e.** mitosis **f.** pollen grain **g.** male gametophyte **h.** generative **6. a.** meiosis **b.** haploid **c.** mitosis **d.** four **e.** seven **f.** eight **g.** embryo **h.** egg **i.** two **7. a.** P **b.** F **c.** P **d.** F **8. a.** Pollination **b.** stigma **c.** pollen tube **d.** mitosis **e.** two **f.** zygote **g.** 3n **h.** double fertilization **9. a.** embryo **b.** ovary **c.** fruit **d.** stored **e.** sporophyte **10. a.** T **b.** T **c.** T **d.** F **11. a.** legume **b.** follicle **c.** capsule **d.** pome **e.** drupe **f.** berry **g.** achene **h.** nut **12. a.** splits open **b.** does not split open **c.** developed from a single ovary **d.** pericarp is dry **e.** developed from a group of individual ovaries **f.** fleshy **g.** hard **13. a.** hooks and spines **b.** defecation by birds and mammals **c.** squirrel activity **d.** wind and ocean currents **14. a.** B **b.** C **c.** C **d.** B **15.** a, b, c **16.** b,c **17. a.** identical **b.** tissue **c.** naked cell **d.** meristem **e.** recessive **f.** pollen **g.** cell suspension culture **18. a.** F **b.** F **c.** T **d.** T

## Chapter Test

**1.** a **2.** b **3.** d **4.** a **5.** c **6.** d **7.** b **8.** d **9.** b **10.** d **11.** b **12.** d **13.** b **14.** a **15.** b **16.** a **17.** d **18.** a **19.** d **20.** d **21.** Asexual reproduction offers a quick, effective means of reproducing plants; however, it lacks the mechanism for genetic variability available through sexual reproduction. Through sexual reproduction, more adaptive forms of the organism can be produced and survive, especially if the environment is changing. **22.** The seeds are dispersed over wide distances, promoting their range of survival. The birds and mammals gain a food source while eating the fruit with the seed.

# Part VI Animal Evolution

# 29

# Introduction to Invertebrates

## Chapter Review

Animals are multicellular eukaryotes, locomote by contracting fibers, and ingest their food. They all have the diploid life cycle, but differ in a number of ways by which they are classified. Many animals, such as those discussed in this chapter, are **invertebrates.**

**Sponges** have the cellular level of organization and lack tissues and symmetry. Sponges are **sessile filter feeders** and depend on a flow of water through the body to acquire food, digested in vacuoles within collar cells that line a central cavity.

**Cnidarians** and **comb jellies** have two true tissue layers and are radially symmetrical. Cnidarians have a sac body plan and exist as either polyps (e.g., *Hydra*) or medusae, (e.g., jellyfishes) or they can alternate between the two (e.g., *Obelia*).

**Flatworms** and **ribbon worms** are **bilaterally symmetrical** and have the organ level of organization. They are also acoelomates. Flatworms may be free-living or parasitic. Freshwater planarians have muscles, a ladder-type nervous system, and **cephalization** consistent with a predatory way of life. Parasitic flatworms—namely, tapeworms and flukes—lack cephalization and are otherwise modified for a parasitic lifestyle.

Roundworms and rotifers have a **tube-within-a-tube** body plan and are **pseudocoelomates.** Many roundworms, ranging from pinworms to *Ascaris* and *Trichinella,* are parasitic.

## Study Exercises

Study the text section by section as you answer the questions that follow.

### 29.1 Evolution of Animals (pp. 518–19)

- Animals are multicellular heterotrophs that move about and ingest their food. They have the diploid life cycle.
- Animals are classified according to certain criteria, including level of organization, symmetry, body plan, type of coelom, and the presence or absence of segmentation.

1. The evolution of animals shows that all animals are believed to have descended from a.________________ ancestors. b.______________ may have evolved from a protist c.______________ from the rest.
2. Beside the letters, write the names of the animal phyla depicted in the phylogenetic tree on the following page. Beside the numbers, write the type of symmetry.

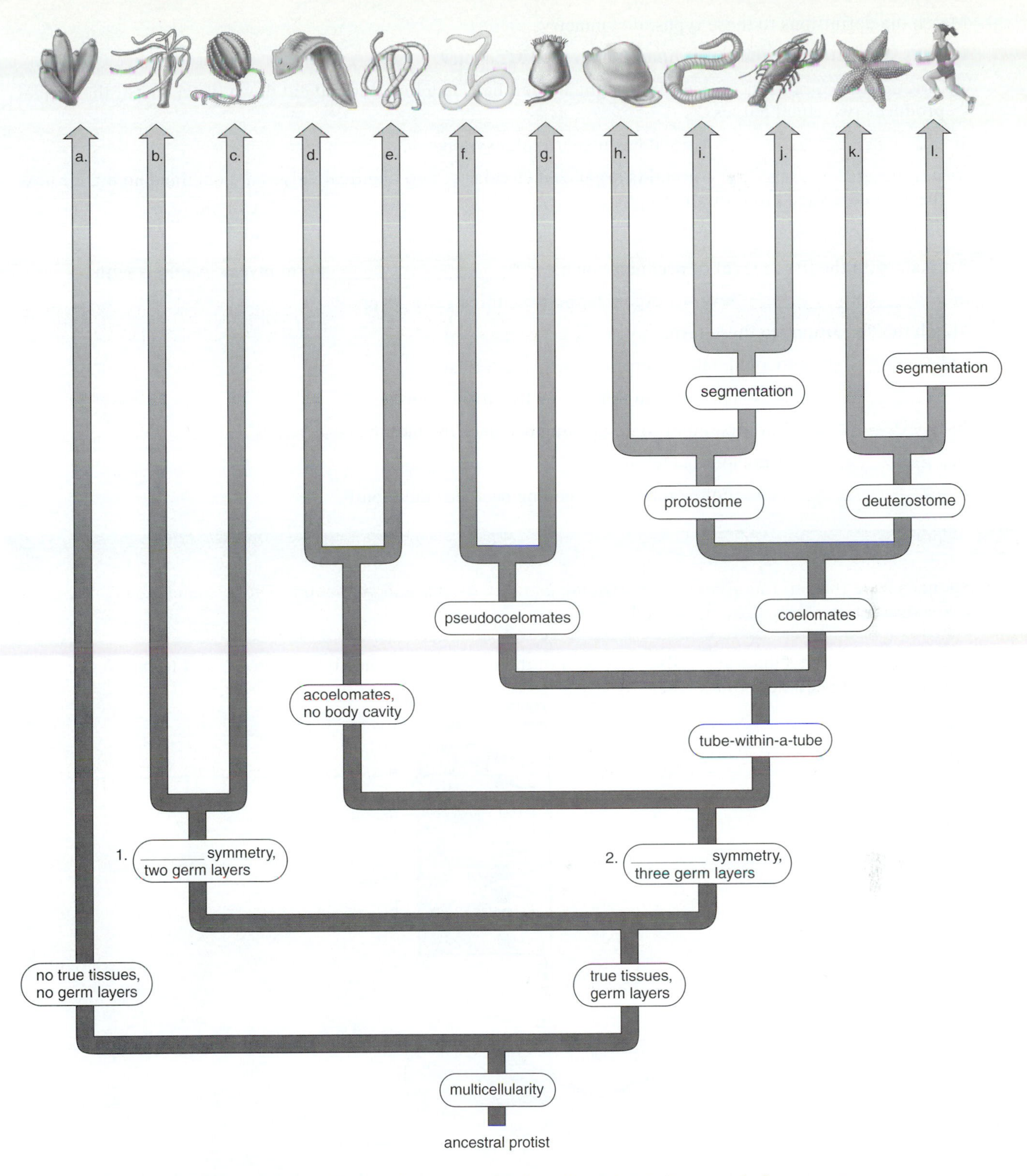


3. Place a check in front of those phrases that correctly describe animal characteristics.

a. _____ autotrophic nutrition
b. _____ haploid life cycle
c. _____ ingest food
d. _____ meiosis produces gametes
e. _____ multicellular organisms

4. Match the definitions to these types of symmetry:

various symmetry    bilateral symmetry    radial symmetry

a. ____________________ A definite right and left half; one longitudinal cut down the center of the animal produces two equal halves.

b. ____________________ Animal has no particular symmetry.

c. ____________________ Animal is organized circularly; two identical halves are obtained no matter how the animal is sliced longitudinally.

5. The three germ layers are called a.________________, b.________________, and c.________________.

6. Animals with the tissue level of organization have a.________________ germ layers. Animals with the b.________________ level of organization have three germ layers.

7. Match the definitions to these terms:

acoelomate    pseudocoelomate    protostome    deuterostome

a. ____________ A cavity is incompletely lined with mesoderm.

b. ____________ The first embryonic opening becomes the mouth.

c. ____________ A coelom is lacking.

d. ____________ The second embryonic opening becomes the mouth.

## 29.2 Multicellularity (pp. 520–21)

- Sponges have the cellular level of organization and lack tissues and symmetry. They depend on a flow of water through the body to acquire food.

8. Label the cells in this diagram of the body wall of the sponge using the alphabetized list of terms. Write the function of each cell next to its label.

amoeboid cell
collar cell
epidermal cell
pore
spicule

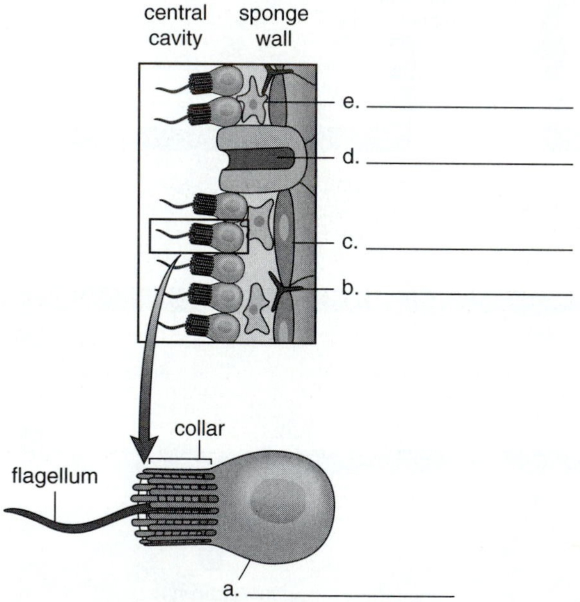

9. Sponges are the only animals to have the a.________________ level of organization. Their bodies are perforated by many b.________________; the phylum name Porifera means c.________________. The beating of the flagella of d.________________ creates water currents, which flow through the pores into the central cavity and out through the e.________________. Food particles brought in by water are digested in food vacuoles and f.________________ cells. The result of sexual reproduction is a ciliated larva that moves (disperses) by g.________________. The two methods of asexual reproduction of sponges are called h.________________ and fragmentation. If a sponge fragments, it can grow whole again by a process called i.________________. The classification of sponges is based on the type of j.________________ they have.

## 29.3 Two Tissue Layers (pp. 522–25)

- Cnidarians and comb jellies have a radially symmetrical, saclike body consisting of two tissue layers from the germ layers ectoderm and endoderm.
- Cnidarians typically are either polyps (e.g., *Hydra*) or medusae (e.g., jellyfishes) or alternate between these two forms, the polyp being an asexual stage and the medusae being a sexual stage.

10. Embryonic comb jellies and cnidarians have a.________________ germ layers, the b.________________ and the c.________________. Because of this, they are said to be d.________________. They exhibit e.________________ symmetry.

11. What is the function of a nematocyst? ________________________________

12. Label this diagram with these terms:

    hydra jellyfish medusa polyp

a. ________________

b. ________________

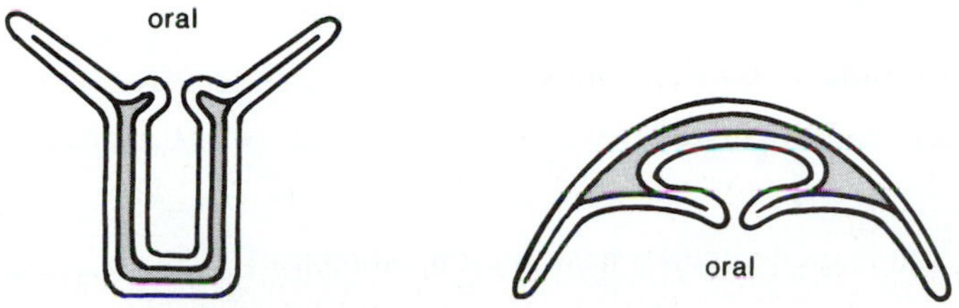

oral

oral

________________ c.

________________ d.

13. In cnidarians, which demonstrate the alternation of generations life cycle, the polyploid stage is sessile and produces a.________________. The medusan stage is motile and produces b.________________ and c.________________.

14. The hydra is a solitary a.________________ that lives in b.________________. Because they contain muscle fibers, the cells of the epidermis of the hydra are called c.________________. Cells capable of becoming other types of cells, allowing the animal to regenerate, are called d.________________ cells. e.________________ secrete digestive juices. The main type of gastrodermal cell is the f.________________ cell. Hydras can reproduce either asexually by g.________________ or sexually.

15. The cnidarians, whose calcium-carbonate skeletons provide a major ocean habitat for thousands of other animals, are the ________________.

## 29.4 Bilateral Symmetry (pp. 526–29)

- Ribbon worms and flatworms have tissues and organs derived also from mesoderm, the third germ layer. They are bilaterally symmetrical and have the organ level of organization. They are also acoelomates.
- Planarians are free-living predators, but flukes and tapeworms are animals adapted to a parasitic way of life.

16. Animals with three germ layers are said to be a.________________, and they exhibit b.________________ symmetry.

17. Flatworms have the a.________________ body plan; ribbon worms have a b.________________ body plan. Of the two types of body plans, which is the most advanced? c.________________ Why? d.________________________________

________________________________

18. Complete the following table to compare a hydra and a planarian:

| | Hydra | Planarian |
|---|---|---|
| Body plan | | |
| Cephalization | | |
| Number of germ layers | | |
| Organs | | |
| Symmetry | | |

## Parasitic Flatworms (p. 528)

19. a. The condition schistosomiasis is caused by a(n) ___________________.

    b. The Chinese liver fluke requires two hosts, the snail and the fish. Humans contract liver flukes when they eat ___________________.

20. The body of the tapeworm is made up of a head region called a(n) a.___________________ and numerous segments called b.___________________. Each segment contains mainly c.___________________. After fertilization, the bags are filled with maturing d.___________________.

21. Label this diagram of the life cycle of a human tapeworm with these terms:

    humans meat pigs/cattle

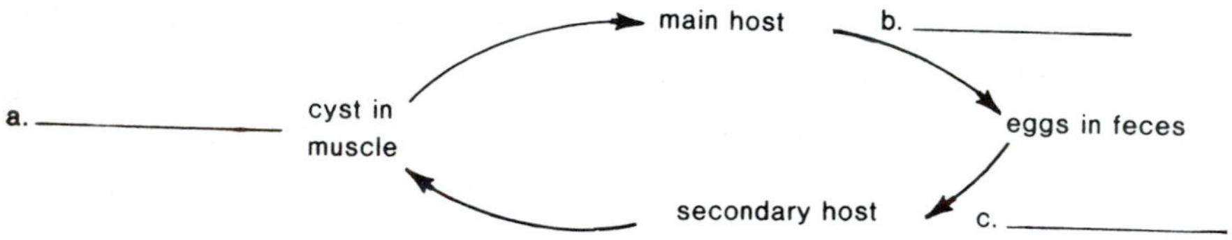

22. Complete the following table to compare the structure of the parasitic tapeworm with that of the planarian:

| Body Structure | Tapeworm | Planarian |
|---|---|---|
| Eyes | | |
| Nervous system | | |
| Digestive system | | |
| Reproductive system | | |

## 29.5 Tube-within-a-Tube (pp. 530–32)

- Roundworms and rotifers have a coelom, a body cavity where organs are found and that can serve as a hydrostatic skeleton. Their coelom is a pseudocoelom because it is not completely lined by mesoderm.
- Roundworms take their name from a lack of segmentation; they are very diverse and include some well-known parasites.

23. Two animal phyla that consist of pseudocoelomates are [a]________________ and [b]________________.

24. Complete the following table to compare the structure of flatworms and roundworms:

| | Flatworms | Roundworms |
|---|---|---|
| Number of germ layers | | |
| Organs | | |
| Sexes separate | | |
| Pseudocoelom | | |
| Body plan | | |

25. Trichinosis is a condition humans get by eating rare [a]________________ containing [b]________________.

26. Complete the following table to compare the four animal phyla discussed in this chapter.

| | Sponges | Cnidarians | Flatworms | Roundworms |
|---|---|---|---|---|
| Level of organization | | | | |
| Number of germ layers | | | | |
| Type of symmetry | | | | |
| Type of body plan | | | | |
| Type of coelom | | | | |

# Chapter Test

## Objective Questions

Do not refer to the text when taking this test.

In questions 1–10, match each description with these animals:

a. sponges
b. cnidarians
c. flatworms
d. roundworms

____ 1. tube-within-a-tube body plan
____ 2. below the tissue level of organization
____ 3. two germ layers present
____ 4. tissue level of organization
____ 5. pseudocoelom present
____ 6. include planarians
____ 7. include *Ascaris*
____ 8. have collar cells and spicules
____ 9. feed by nematocysts
____ 10. some cause elephantiasis

____ 11. Select the incorrect association.
a. amoeboid cell—digestion
b. collar cell—water current
c. epidermal cell—covering
d. spicule cell—reproduction

____ 12. The gastrovascular cavity functions for
a. digestion only.
b. transport only.
c. digestion and transport.
d. neither digestion nor transport.

____ 13. Which of the following is NOT an embryonic germ layer among cnidarians?
a. ectoderm
b. endoderm
c. mesoderm

____ 14. In cnidarians, the epidermis is separated from the cell layers lining the internal cavity by
a. mesoderm.
b. mesoglea.
c. a coelom.
d. a pseudocoelom.
e. endoderm.

____ 15. Which statement is NOT true of *Hydra's* life cycle?
a. It reproduces asexually by means of budding.
b. It alternates between the polyp form and the medusa form.
c. The polyp produces the egg and sperm.
d. Only the polyp form is ever haploid.

____ 16. In contrast to cnidarians, flatworms have
a. a complete digestive tract.
b. sexual reproduction.
c. a mesoderm layer that gives rise to organs.
d. a nervous system.
e. specialized tissues for gas exchange.

____ 17. Cephalization means that
a. a definite head develops.
b. reproduction is sexual.
c. the nervous system is ladderlike.
d. wastes are excreted through flame cells.

____ 18. A fluke is responsible for the condition of
a. pinworms.
b. schistosomiasis.
c. trichinosis.
d. elephantiasis.

____ 19. A parasitic worm living in a vertebrate intestine probably has little need for
a. a high degree of tolerance to pH changes.
b. a digestive tract.
c. reproductive organs.
d. a means of attachment to the host.
e. glycolysis.

____ 20. A true coelom differs from a pseudocoelom in that it
a. has a body cavity for internal organs.
b. is completely lined by mesoderm.
c. is incompletely lined by mesoderm.
d. is found only in segmented worms.

____ 21. The distinction between protostomes and deuterostomes is based on differences in their
a. digestive tracts.
b. nervous systems.
c. embryonic development.
d. circulatory systems.
e. Both *a* and *c* are correct.

____ 22. Humans may become infected with *Ascaris* by
a. soil that contains eggs.
b. consuming muscle tissue that contains a cyst.
c. drinking water contaminated with eggs.
d. Both *a* and *c* are correct.

In questions 23–25, match the descriptions with these organisms (some are used more than once):

a. cnidarians
b. flatworms
c. roundworms

____ 23. sac body plan, radial symmetry
____ 24. alternation of generations
____ 25. some are parasitic

## Critical Thinking Questions

Answer in complete sentences.

26. From sponges to roundworms, evolution has produced a more complex body form. What evidence supports this?

27. A loss of body complexity accompanies parasitism among flatworms. What evidence supports this?

**Test Results:** _____ number correct ÷ 27 = _____ × 100 = _____ %

## Exploring the Internet

The Online Learning Center at *www.mhhe.com/maderbiology8* has additional study material and practice quizzes that can help you master the content of this chapter. You can also find links to websites exploring additional topics in biology. Access to the Online Learning Center is free for those who have purchased a new textbook.

## Answer Key

### Study Questions

**1. a.** protistan **b.** Sponges **c.** separately **2.** See Figure 29.2, page 519, in text. **3.** c, d, e **4. a.** bilateral symmetry **b.** various symmetry **c.** radial symmetry **5. a.** ectoderm **b.** mesoderm **c.** endoderm **6. a.** two **b.** organ **7. a.** pseudocoelomate **b.** protostome **c.** acoelomate **d.** deuterostome **8. a.** collar cell—produces water currents and captures food **b.** spicule—internal skeleton **c.** epidermal cell—protection **d.** pore—entrance of water **e.** amoeboid cell—distributes nutrients and produces gametes **9. a.** cellular **b.** pores **c.** pore bearing **d.** collar cells **e.** osculum **f.** amoeboid **g.** swimming **h.** budding **i.** regeneration **j.** skeleton **10. a.** two **b.** ectoderm **c.** endoderm **d.** diploblasts **e.** radial **11.** to trap or paralyze prey **12. a., b.** polyp, hydra **c., d.** medusa, jellyfish **13. a.** medusae **b.** egg **c.** sperm **14. a.** polyp **b.** freshwater **c.** epitheliomuscular cells **d.** interstitial (embryonic) **e.** Gland cells **f.** nutritive-muscular **g.** budding **15.** corals **16. a.** triploblasts **b.** bilateral **17. a.** sac **b.** tube-within-a-tube **c.** tube-within-a-tube **d.** With a one-way flow of contents, there is the possibility of specialization of parts along the length of the tube.

**18.**

| Hydra | Planarian |
|---|---|
| sac | sac |
| no | yes |
| two | three |
| no | yes |
| radial | bilateral |

**19. a.** blood fluke **b.** raw fish **20. a.** scolex **b.** proglottids **c.** sex organs **d.** eggs **21. a.** meat **b.** humans **c.** pigs/cattle

**22.**

| Tapeworm | Planarian |
|---|---|
| no | yes |
| much reduced | extensive |
| not present | branches |
| well developed | present |

**23. a.** Nematoda (roundworms) **b.** Rotifera (rotifers)

24.

| Flatworms | Roundworms |
| --- | --- |
| three | three |
| yes | yes |
| no | yes |
| no | yes |
| sac | tube-within-a-tube |

25. **a.** pork **b.** encysted *Trichinella* larvae

26.

| Sponges | Cnidarians | Flatworms | Roundworms |
| --- | --- | --- | --- |
| cell | tissue | organ | organ |
| — | two | three | three |
| radial or none | radial | bilateral | bilateral |
| — | sac | sac | tube-within-a-tube |
| — | — | acoelomate | pseudocoelomate |

## Chapter Test

**1.** d **2.** a **3.** b **4.** b **5.** d **6.** c **7.** d **8.** a **9.** b **10.** d **11.** d **12.** c **13.** c **14.** b **15.** b **16.** c **17.** a **18.** b **19.** b **20.** b **21.** c **22.** d **23.** a **24.** a **25.** b, c **26.** Among roundworms, there is a tube-within-a-tube body plan with bilateral symmetry and an organ level of organization. Also, a pseudocoelom develops. **27.** Organ systems may be lost partially or completely in parasites. Only the reproductive system remains well developed. The head region bears hooks and/or suckers instead of sense organs.

# 30

# The Protostomes

## Chapter Review

The **protostomes** (and **deuterostomes**) have a **coelom** completely lined with mesoderm. In protostomes, the mouth appears at or near the blastopore, the first embryonic opening. The body of a **mollusc** is composed of a visceral mass, a **mantle**, and a foot. Molluscs are adapted to various ways of life; for example, clams are sessile filter feeders, squids are active predators in the deep ocean, and snails are terrestrial. **Annelids** are the segmented worms. Polychaetes (e.g., clam worms, tube worms) are marine animals with parapodia for swimming and gas exchange; earthworms live on land; and **leeches** are parasitic and are usually found in fresh water. In the earthworm, the coelom and the nervous, excretory, and circulatory systems all provide evidence of **segmentation.**

**Arthropods** have a ventral solid nerve cord and an external **exoskeleton** made of **chitin,** periodically shed by **molting;** this is different from the internal skeleton of **vertebrates.** Like vertebrates, however, arthropods are segmented and have **jointed appendages.** This combination has led to specialization of parts, and in some arthropods, the body is divided into special regions, each with its own type of appendages. Whereas the crayfish, a **crustacean,** is adapted to a marine existence, the grasshopper, an **insect,** is adapted to a terrestrial existence. Insects have wings and breathe by means of air tubes called **tracheae.** Their circulatory system does not contain a respiratory pigment. The uniramians include insects, centipedes, and millipedes. The chelicerates include horseshoe crabs, scorpions, ticks, mites, and spiders.

## Study Exercises

Study the text section by section as you answer the questions that follow.

### 30.1 Advantages of Coelom in Protostomes and Deuterostomes (pp. 536–37)

- A coelom has many advantages: the digestive system can become more complex, coelomic fluid assists body processes and acts as a hydrostatic skeleton.
- In protostomes, the mouth appears at or near the blastopore, the first embryonic opening, and the coelom develops by a splitting of the mesoderm.

1. What are the advantages of having a coelom in regard to the following?
   a. the digestive tract ______________________
   b. locomotion ______________________
   c. circulation ______________________
   d. reproduction ______________________
2. Place the appropriate letter next to each statement.
   D—deuterostome P—protostome
   a. _____ A schizocoelom develops.
   b. _____ The blastopore becomes the anus.
   c. _____ The blastopore becomes the mouth.
   d. _____ The coelom develops from mesodermal pouches.
   e. _____ The fate of developing cells is fixed and determinate.
   f. _____ The fate of developing cells is indeterminate.
   g. _____ Radial cleavage occurs.
   h. _____ Spiral cleavage occurs.

## 30.2 Molluscs (pp. 538–41)

- The body of a mollusc typically contains a visceral mass, a mantle, and a foot.
- Clams are adapted to a life in sandy or muddy soil, squids to an active life in the sea, and snails are adapted to a life on land.

3. Give an example of each of the following animals:
   a. bivalve ____________________
   b. cephalopod ____________________
   c. gastropod ____________________
4. State the manner in which a clam, a squid, and a snail are adapted to their way of life.
   a. clam ____________________
   ____________________
   b. squid ____________________
   ____________________
   c. snail ____________________
   ____________________
5. Complete the following table to compare the clam and the squid.

| Characteristic | Clam | Squid | Snail |
|---|---|---|---|
| Skeleton | | | |
| Food procurement | | | |
| Locomotion | | | |
| Cephalization | | | |
| Reproduction | | | |

## 30.3 Annelids (pp. 542–44)

- Annelids are the segmented worms with a well-developed coelom, a closed circulatory system, a ventral solid nerve cord, and paired nephridia in each segment.
- Polychaetes include marine predators with a definite head region and filter feeders with terminal tentacles to filter food from the water.
- Oligochaetes include the earthworms that burrow in the soil and use a moist body wall as a respiratory region.

6. Annelids are the a.____________________ worms. The tube-within-a-tube body plan has resulted in a digestive system with b.____________________ parts. Annelids have an extensive c.____________________ circulatory system. The nervous system is a brain and d.____________________ solid nerve cord. Paired e.____________________ in each segment collect and excrete waste.

7. Complete the following table to compare annelids:

| Class | Representative Organism | Setae | Parapodia |
| --- | --- | --- | --- |
| Polychaeta | | | |
| Oligochaeta | | | |
| Hirudinea | | | |

8. Describe the following systems to show that an earthworm is segmented:
   a. nervous system ______________________
   b. excretory system ______________________
   c. circulatory system ______________________
9. Place a check in front of the item(s) that correctly describes earthworm reproduction.
   a. _____ Worms are hermaphroditic.
   b. _____ Worms have separate sexes.
   c. _____ Glands in every segment provide mucus.
   d. _____ The clitellum provides mucus.
   e. _____ Worms exchange sperm and eggs.
   f. _____ Worms exchange sperm.
   g. _____ The embryo develops externally.

## 30.4 Arthropods (pp. 544–51)

- Arthropods are segmented with specialized body regions and an external skeleton that includes jointed appendages.
- Among the many kinds of arthropods, crustaceans are adapted to a life at sea and insects are adapted to a terrestrial existence.

10. Like the annelids, arthropods are a.__________________, but there is specialization of parts. The segments are fused into three regions: b.__________________, c.__________________, and d.__________________. The arthropods have an exoskeleton that contains e.__________________, and there are f.__________________ appendages. Because they have an external skeleton, arthropods have to g.__________________ to grow larger. Arthropods have a(n) h.__________________ solid nerve cord; i.__________________ is apparent, and the head bears sense organs, including j.__________________ eyes in most species.
11. Place the appropriate letters next to the organisms.

    CH—chelicerates CR—crustaceans U—uniramians

    a. _____ copepods and krill
    b. _____ scorpions
    c. _____ centipedes
    d. _____ insects
    e. _____ lobsters and crayfish
    f. _____ spiders
    g. _____ horseshoe crabs
12. Fill in the blanks.
    a. Which subphylum of arthropods lacks antennae, mandibles, and maxillae, and instead has pincerlike appendages and pedipalps for feeding? __________________
    b. Which subphylum of arthropods has an exoskeleton hardened by the addition of calcium and biramous appendages? __________________
    c. Which subphylum of arthropods has only one pair of antennae and appendages with only one branch?

    __________________

13. Label these diagrams of the grasshopper, using the following alphabetized list of terms (some terms are used more than once).

antenna
compound eye
crop
Malpighian tubules
spiracle(s)
tracheae
tympanum

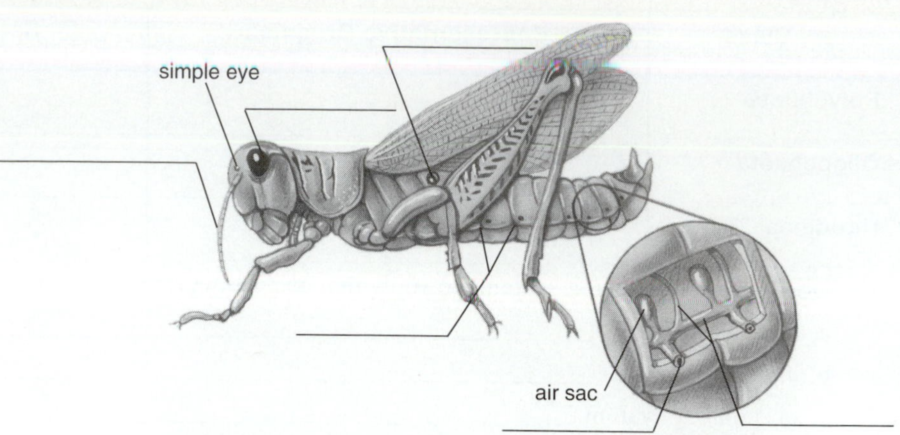


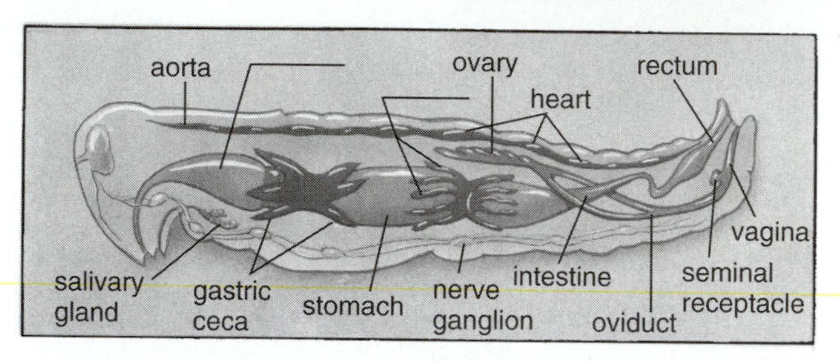


14. Give the function of each of the following structures:

a. crop and gizzard ______________________________

b. Malpighian tubules ______________________________

c. tracheae ______________________________

d. hemocoel ______________________________

e. ovipositor ______________________________

15. Complete the following table to compare the crayfish to the grasshopper, indicating how each is adapted to its way of life.

| System | Crayfish | Grasshopper |
|---|---|---|
| Locomotion | | |
| Excretion | | |
| Digestion | | |
| Reproduction | | |
| Respiration | | |

16. To demonstrate a knowledge of arachnids:

a. circle the arachnids: snails, chiggers, spiders, millipedes
b. circle the appendages of arachnids: walking legs, chelicerae, pedipalps, swimmerets
c. circle the internal organs of arachnids: Malpighian tubules, green glands, salt glands, book lungs
d. circle each habitat where you would expect to find an arachnid: marine waters only, on land, in the frozen tundra, in the tropics

# Chapter Test

## Objective Questions

Do not refer to the text when taking this test.

____ 1. A true coelom differs from a pseudocoelom in that the latter
a. has a body cavity for internal organs.
b. is completely lined by mesoderm.
c. is incompletely lined by mesoderm.
d. is found only in segmented worms.

____ 2. The distinction between protostomes and deuterostomes is based on differences in their
a. digestive tracts.
b. nervous systems.
c. embryonic development.
d. circulatory systems.
e. Both *a* and *c* are correct.

____ 3. The digestive and reproductive systems of a mollusc are located within the
a. mantle cavity.
b. visceral mass.
c. gastrovascular cavity.
d. highly branched coelom.
e. water vascular cavity.

____ 4. A common feature of all classes of molluscs is
a. a muscular foot.
b. a coiled shell, either internal or external.
c. obvious segmentation.
d. their carnivorous nature.
e. All of these are correct.

____ 5. Earthworms are incompletely adapted to life on land because
a. their reproduction takes place in the water.
b. they have to exchange gases with water.
c. they tend to dry out if exposed to dry air.
d. their setae have to be kept wet.

____ 6. What type of skeleton helps an earthworm to move?
a. an exoskeleton
b. a bony endoskeleton
c. a hydrostatic skeleton
d. spiny plates

____ 7. The ______ in the earthworms secretes mucus for deposition of eggs and sperm.
a. typhlosole
b. nephridia
c. clitellum
d. setae
e. gizzard

____ 8. Select the statement NOT descriptive of annelids.
a. excretion by nephridia
b. sac body plan
c. segmented body
d. well-developed coelom

____ 9. Molting by arthropods means that they
a. circulate blood through a closed system.
b. move with jointed appendages.
c. reproduce sexually.
d. shed their exoskeletons.

____ 10. What do a crab, an insect, and a spider have in common?
a. cephalization
b. external skeleton
c. molting
d. All of these are correct.

____ 11. Select the incorrect association.
a. chelicerate—pincerlike appendages
b. uniramian—one branch on appendages
c. crustacean—calcified exoskeleton
d. crustacean—insects
e. chelicerate—spiders

____ 12. Select the incorrect association.
a. insect—scorpion
b. arachnid—spider
c. crustacean—decapods
d. insect—grasshopper

____ 13. Which is NOT generally true of arthropods?
a. exoskeleton contains chitin
b. breathe with tracheae
c. have jointed appendages
d. have compound eyes

____ 14. Select the incorrect association.
a. annelid—earthworm
b. mollusc—spider
c. arthropod—grasshopper
d. annelid—clam worm, *Nereis*

____ 15. Which is NOT a general feature of insects?
a. body divided into head, thorax, and abdomen
b. three pairs of legs
c. respiration typically by book lungs
d. one or two pairs of wings

____ 16. Which of these is NOT a mollusc?
a. squid
b. chiton
c. leech
d. clam

____ 17. Which of these is NOT a characteristic of molluscs?
a. body in three parts—head, thorax, and abdomen
b. a foot modified in various ways
c. a visceral mass that contains the internal organs
d. usually an open circulatory system

____ 18. Select the incorrect association.
a. clam—sessile filter feeder
b. squid—closed circulatory system
c. snail—shell and broad foot
d. clam—well-developed brain

____ 19. What do a scallop, a chambered nautilus, and a chiton have in common?
a. cephalization
b. vertebrate-type eyes
c. a mantle cavity
d. a jointed external skeleton

____ 20. What do an earthworm, a clam worm, and a leech have in common?
a. a clitellum
b. parapodia
c. a closed circulatory system
d. All of these are correct.

## Critical Thinking Questions

Answer in complete sentences.

21. The arthropods are considered the most successful animal phylum inhabiting the earth. What justifies this claim?

22. Compare external features of the earthworm to those of the grasshopper to demonstrate that segmentation leads to specialization of parts.

**Test Results:** ______ number correct ÷ 22 = ______ × 100 = ______ %

## Exploring the Internet

The Online Learning Center at *www.mhhe.com/maderbiology8* has additional study material and practice quizzes that can help you master the content of this chapter. You can also find links to websites exploring additional topics in biology. Access to the Online Learning Center is free for those who have purchased a new textbook.

## Answer Key

### Study Exercises

**1. a.** The digestive tract can move independently of the body wall. The coelom allows specialization in that the tract has room to coil. **b.** With a hydrostatic skeleton, muscular contractions allow the animal to move. **c.** Coelomic fluid can substitute for a circulatory system. **d.** The coelomic cavity can provide room for storage of eggs/sperm. **2. a.** P **b.** D **c.** P **d.** D **e.** P **f.** D **g.** D **h.** P **3. a.** clam, oyster, mussel, scallop **b.** squid, cuttlefish, octopus, nautilus **c.** snail, whelk, conch, periwinkle, sea slug **4. a.** A clam has a hatchet foot for burrowing in the sand and is a sessile filter feeder that lacks cephalization. **b.** A squid has a head-foot—that is, tentacles about the head with vertebrate-type eyes that help the squid actively capture food. **c.** A snail has a broad, flat foot for moving over flat surfaces and a head with a radula for scraping up food from a surface.

**5.**

| Clam | Squid | Snail |
| --- | --- | --- |
| external shell | no external skeleton | external shell |
| filter feeder | active predator | garden snail is herbivorous |
| hatchet foot | jet propulsion and fins | broad, flat foot |
| no | yes | yes |
| separate sexes | separate sexes | hermaphroditic |

**6. a.** segmented **b.** specialized **c.** closed **d.** ventral **e.** nephridia

7.

| Representative Organism | Setae | Parapodia |
|---|---|---|
| clam worm | many | yes |
| earthworm | few | no |
| leech | no | no |

**8. a.** ganglia and lateral nerves in every segment **b.** paired nephridia in every segment **c.** branched blood vessels in every segment **9.** a, d, f, g **10. a.** segmented **b.** head **c.** thorax **d.** abdomen **e.** chitin **f.** jointed **g.** molt **h.** ventral **i.** cephalization **j.** compound **11. a.** CR **b.** CH **c.** U **d.** U **e.** CR **f.** CH **g.** CH **12. a.** chelicerata **b.** crustacea **c.** uniramia **13. a.** antenna **b.** compound eye **c.** tympanum **d.** spiracles **e.** crop **f.** Malpighian tubules **g.** spiracle **h.** tracheae. See Figure 30.14, page 549, in text. **14. a.** crop stores food, gizzard grinds it **b.** excretory tubules that concentrate nitrogenous waste **c.** air tubes that deliver oxygen to muscles **d.** cavity where blood is found **e.** special female appendage for depositing eggs in soil

15.

| Crayfish | Grasshopper |
|---|---|
| swimmerets | hopping legs, wings |
| liquid | solid |
| gastric mill | crop |
| male uses swimmeret to pass sperm | male uses penis to pass sperm |
| gills | trachea |

**16. a.** chiggers, spiders **b.** walking legs, chelicerae, pedipalps **c.** Malpighian tubules, book lungs **d.** on land, in the tropics

## Chapter Test

**1.** c **2.** c **3.** b **4.** a **5.** c **6.** c **7.** c **8.** b **9.** d **10.** d **11.** d **12.** a **13.** b **14.** b **15.** c **16.** c **17.** a **18.** d **19.** c **20.** c **21.** They are the most diversified phylum with a wide variety of species filling numerous ecological niches. Insects far outnumber any other type of animal. **22.** Segmentation is obvious in an earthworm due to the uniform external ring. In an anthropod such as the grasshopper, segmentation has led to specialization of parts. There is a head region, a thorax, and an abdominal region. The segments devoted to the head bear mouthparts; those of the thorax bear legs and wings, and the abdomen has no appendages.

# 31

# The Deuterostomes

## Chapter Review

In **deuterostomes,** the blastopore develops into an anus. Cleavage is radial, and an enterocoelom develops.

The **echinoderms** have evolved radial symmetry, an endoskeleton, **skin gills,** a central nerve ring, and a **water vascular system.** The sea star is a well-known echinoderm showing these specializations.

The four basic **chordate** characteristics are a **notochord; a dorsal tubular nerve cord; pharyngeal pouches;** and a **post-anal tail,** which they exhibit at some point in their life history. Only **lancelets** show all of these characteristics as adults. The **vertebrates** develop a vertebral column in place of a notochord.

Vertebrates are divided into superclasses: the agnathans (jawless fishes) and the gnathostomates (jawed fishes and tetrapods). There are six classes of jawed vertebrates: cartilaginous fishes (e.g., sharks); bony fishes (e.g., trout, cod); amphibians (e.g., frogs, salamanders); reptiles (e.g., snakes, lizards, turtles); birds (e.g., cardinal, eagle); mammals (e.g,. dogs, cats, humans).

The evolution of vertebrates is still being studied. The hagfishes and lampreys are descendants of the original **jawless fishes.** Sharks are modern representatives of the **cartilaginous fishes.** The original **bony fishes** diverged into two groups: the **ray-finned fishes** and the **lobe-finned fishes.**

**Amphibians** were the first **tetrapods.** They evolved from lobe-finned fishes during the Devonian period, and reached their greatest size and diversity in the swamp forests of the Carboniferous period. Most amphibians return to the water to reproduce.

**Reptiles,** believed to have evolved from amphibians, have a shelled amniotic egg. A shelled egg, along with extraembryonic membranes, protects the embryo and makes reproduction on land possible. Stem reptiles gave rise to dinosaurs and birds. The mammals trace their ancestry to another reptilian line of descent.

While amphibians and reptiles are ectothermic—their body temperature is the same as the environment, both birds and mammals are endothermic—they metabolically produce a constant body temperature. Adaptations of **birds** include feathers and hollow bones, both adaptations for flight. **Mammals** have evolved hair and mammary glands. **Monotremes** are egg-laying mammals, and **marsupials** are pouched mammals that give birth to very immature young. While the young of **placental mammals** are more developed at birth, they still require parental care to survive.

## Study Exercises

Study the text section by section as you answer the questions that follow.

### 31.1 Echinoderms (pp. 556–57)

- Echinoderms and chordates are both deuterostomes.
- Echinoderms have radial symmetry and a unique water vascular system for locomotion.

1. Complete the following table to describe the characteristics of echinoderms:

| Characteristic | Description |
|---|---|
| Type of symmetry | |
| Skeletal system | |
| Respiration | |
| Nervous system | |
| Water vascular system | |

2. Complete the following table to describe types of echinoderms:

| Class Name | Examples | Distinctive Features |
|---|---|---|
| Crinoidea | | |
| Holothuroidea | | |
| Echinoidea | | |
| Ophiuroidea | | |
| Asteroidea | | |

3. Label this diagram of a sea star with the following terms (some terms are used more than once).

ampulla
anus
arm
cardiac stomach
central disk
digestive gland
eyespot
gonads
pyloric stomach
sieve plate
skin gill
spine
tube feet

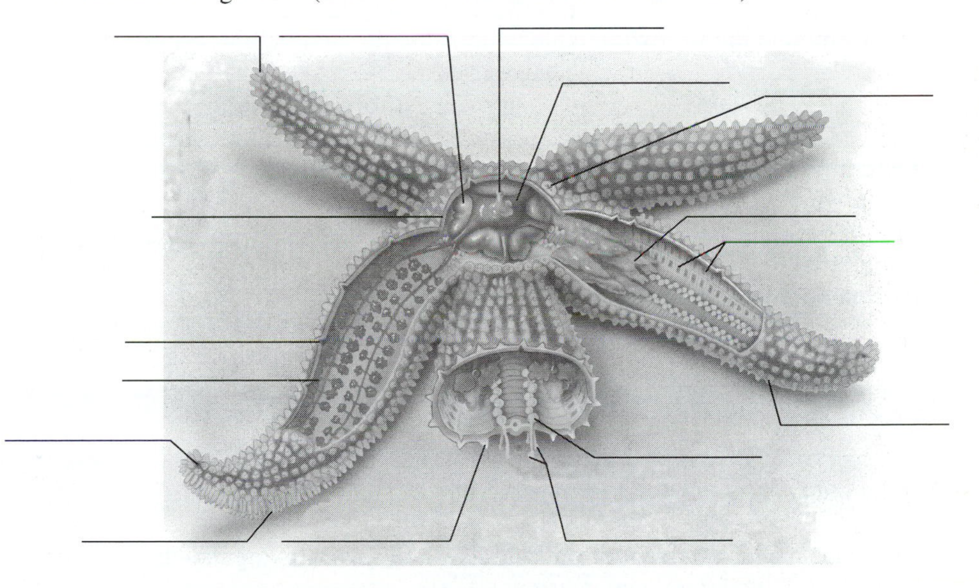

4. Trace the path of water in the water vascular system: sieve plate to a.____________canal to b.____________canal to radial canal to c.____________feet. Each of these feet has a(n) d.____________. The function of the water vascular system is e.____________.

5. A sea star has a two-part stomach. Describe how the sea star feeds on a clam, mentioning both parts of the stomach. ____________________________________________

____________________________________________

## 31.2 Chordates (pp. 557–58)

- Lancelets are invertebrate chordates with the four chordate characteristics as adults: a notochord, a dorsal tubular nerve cord, pharyngeal pouches, and a post-anal tail.

6. Label this diagram with the four primary chordate characteristics.

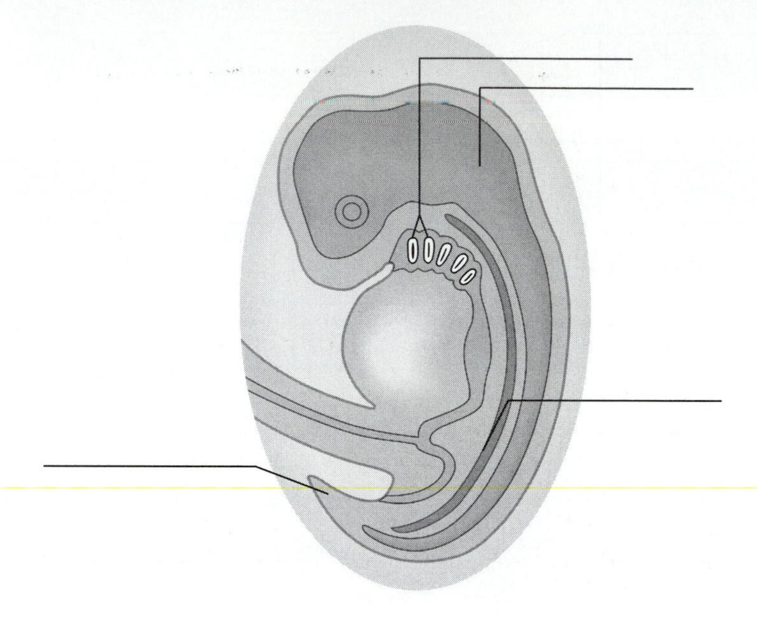

7. What type of evidence suggests that chordates and echinoderms are related?

_______________________________________________

_______________________________________________

8. Complete the following table to describe the invertebrate chordates:

| Name | Chordate Characteristics | Appearance |
|---|---|---|
| Lancelets (subphylum Cephalochordata) | | |
| Tunicates (subphylum Urochordata) | | |

## 31.3 Vertebrates (pp. 560–74)

- In vertebrates, the notochord is replaced by the vertebral column. Most vertebrates also have a head region, endoskeleton, and paired appendages.

9. Place a check in front of the characteristics that distinguish the vertebrates.
   a. _____ bilateral symmetry in all
   b. _____ radial symmetry in some
   c. _____ tube-within-a-tube plus coelom
   d. _____ segmented
   e. _____ vertebral column replaces embryonic notochord
   f. _____ open/closed circulatory system
   g. _____ cephalization with compound eyes
   h. _____ living endoskeleton

10. Label each of the following as being in superclass Agnatha or superclass Gnathostomata and as being a fish or a tetrapod:
   a. lampreys and hagfishes ________________, ________________
   b. class Chondrichthyes ________________, ________________
   c. class Osteichthyes ________________, ________________
   d. class Amphibia ________________, ________________
   e. class Reptilia ________________, ________________
   f. class Aves ________________, ________________
   g. class Mammalia ________________, ________________
11. To what class do each of these belong?
   a. frogs and salamanders ________________
   b. snakes and lizards ________________
   c. sharks and rays ________________
   d. storks and robins ________________
   e. horses and giraffes ________________

## Fishes (pp. 560–63)

- There are three groups of fishes: jawless (e.g., lampreys and hagfishes), cartilaginous (e.g., sharks, rays, and skates), and bony fishes (ray-finned and lobe-finned). Most modern-day fishes are ray-finned (e.g., trout, perch, etc.).

12. Place the appropriate letters next to each statement.
   CF—cartilaginous fishes BF—bony fishes
   a. _____ lateral line system
   b. _____ placoid (toothlike) scales
   c. _____ operculum
   d. _____ scales of bone
   e. _____ swim bladder
13. a. What is the significance of ostracoderms (Cambrian period) in the history of vertebrates?
   ________________________________________________
   b. What is the significance of placoderms (Devonian period) in the history of vertebrates?
   ________________________________________________
   c. What is the significance of lobe-finned fishes (e.g., coelacanths and lungfishes) in the history of vertebrates? ________________________________________________
   ________________________________________________
   d. What is the significance of amphibians in the history of vertebrates? ________________
   ________________________________________________
   e. What is the significance of reptiles in the history of vertebrates? ________________
   ________________________________________________

## Amphibians (pp. 563–65)

- Amphibians (e.g., frogs and salamanders), more numerous during the Carboniferous period, evolved from lobe-finned fishes.

14. Describe each of these features of amphibians.
    a. skin ________________________
    b. lungs ________________________
    c. body temperature ________________________
    d. life cycle ________________________
    e. heart ________________________

## Reptiles (pp. 566–69)

- Modern-day reptiles (e.g., snakes, lizards, turtles) are the remnants of an ancient group that evolved from amphibians.
- The shelled egg of reptiles, which contains extraembryonic membranes, is an adaptation for reproduction on land.

15. The reptile evolved from the [a]______________ at the end of the Paleozoic era. In the Mesozoic era, the reptiles underwent an [b]______________ radiation. There were types of reptiles adapted to life on [c]______________, in the [d]______________, and in the [e]______________. As long as the reptiles, including the dinosaurs, dominated the Earth, the [f]______________ remained insignificant. In addition to mammals, some experts say that birds evolved from the reptiles, and other experts say birds are [g]______________.

16. Describe each of these features of reptiles.
    a. skin ________________________
    b. lungs ________________________
    c. body temperature ________________________
    d. type of egg ________________________
    e. heart ________________________

## Birds (p. 571)

- There is a close evolutionary relationship between birds and reptiles. However, birds are endothermic—that is, they maintain a constant body temperature.
- Birds have feathers and skeletal adaptations that enable them to fly.

17. How do the following characteristics of birds contribute to their ability to fly?
    a. feathers ________________________
    b. horny beak instead of jaws ________________________
    c. keel ________________________
    d. four-chambered heart ________________________
    e. one-way flow of air in lungs ________________________

## Mammals (pp. 572–74)

- Mammals, which evolved from reptiles, were present when the dinosaurs existed. They did not diversify until the dinosaurs became extinct.
- The chief characteristics of mammals are having hair and mammary glands. Mammals are endothermic—hair helps them maintain a constant body temperature.
- Mammals are classified according to their methods of reproduction: monotremes lay eggs, marsupials have a pouch in which newborn mature, and placental mammals retain the offspring in a uterus until birth.

18. Match the types of mammals to these descriptions (some descriptions are used more than once).
    1. All have hair and mammary glands.
    2. All lay eggs.
    3. All have pouches.
    4. All have internal development of their young, to term.

    a. _____ monotremes
    b. _____ marsupials
    c. _____ placental mammals

19. Name a type of mammal adapted to each of the following:

    a. flying in air ______________________________
    b. running on land ______________________________
    c. swimming in the ocean ______________________________
    d. preying on other animals ______________________________
    e. living in trees ______________________________

## *Constructing Office Buildings*

The object of this game is to construct an office building by matching the numbered terms with the organisms in the key (some numbers should be matched to more than one letter). Five correct answers in a row gives you one story. Any wrong answer is a natural disaster that forces you to start from the ground again.

### Office Building One

A fourteen-story office building is possible.

**Key One**

a. Protozoa
b. Porifera
c. Cnidaria
d. Platyhelminthes
e. Nematoda
f. Annelida
g. Mollusca
h. Arthropoda
i. Echinodermata

____ 1. sea star (starfish)
____ 2. clitellum
____ 3. flame cells
____ 4. egg—nymph—adult
____ 5. organ system level of organization
____ 6. ampulla
____ 7. fluke
____ 8. jellyfish
____ 9. leech
____10. octopus
____11. jointed appendages
____12. tube feet
____13. elephantiasis
____14. soft body
____15. five hearts
____16. bilateral symmetry
____17. muscles

____ 18. water vascular system
____ 19. medusa
____ 20. cellular level of organization
____ 21. muscular foot
____ 22. five arms
____ 23. earthworm
____ 24. nerve net
____ 25. clam
____ 26. setae
____ 27. *Trichinella*
____ 28. ladderlike nervous organization
____ 29. horseshoe crab
____ 30. stone canal
____ 31. pseudocoelom
____ 32. coelom
____ 33. mesoglea
____ 34. pore bearers
____ 35. polychaete
____ 36. squid
____ 37. segmentation
____ 38. closed circulatory system
____ 39. open circulatory system
____ 40. mesoderm
____ 41. collar cells
____ 42. trachea
____ 43. trochophore larva
____ 44. hydra
____ 45. *Hirudo*
____ 46. Malpighian tubules
____ 47. mantle
____ 48. *Dirofilaria*—filarial worm
____ 49. nematocysts
____ 50. metamorphosis
____ 51. gills
____ 52. typhlosole
____ 53. *Ascaris*
____ 54. sieve plate (madreporite)
____ 55. hermaphroditic
____ 56. acoelomate
____ 57. pseudocoelomate
____ 58. pyloric stomach
____ 59. worms
____ 60. molting
____ 61. nephridia
____ 62. visceral mass
____ 63. cephalization
____ 64. spicules
____ 65. green gland
____ 66. exoskeleton
____ 67. wings
____ 68. sessile
____ 69. sac body plan
____ 70. tube-within-a-tube body plan

How many stories is your building? _______

## Office Building Two

An eight-story office building is possible.

**Key Two**

a. sponge
b. hydra
c. planarian
d. *Ascaris*
e. clam
f. earthworm
g. lobster
h. sea star
i. grasshopper
j. *Obelia*

____ 1. diploblastic
____ 2. coelomate
____ 3. segmented
____ 4. nematocyst
____ 5. sac body plan
____ 6. tube-within-a-tube body plan
____ 7. bilateral symmetry
____ 8. radial symmetry
____ 9. organs
____ 10. closed circulatory system
____ 11. insect
____ 12. belong to the same phylum
____ 13. flame cells
____ 14. Malpighian tubules
____ 15. green gland
____ 16. nephridia
____ 17. tracheal tubes
____ 18. gills
____ 19. body wall for respiration
____ 20. skin gills
____ 21. spicules
____ 22. shell
____ 23. carapace
____ 24. "spiny skin"
____ 25. planula larva
____ 26. metamorphosis
____ 27. nymph
____ 28. molt
____ 29. ovipositor
____ 30. asexual reproduction
____ 31. alternation of generations
____ 32. clitellum
____ 33. bilateral larva but radial adult
____ 34. nerve ring
____ 35. ganglia in foot and visceral mass
____ 36. dorsal tubular nerve cord
____ 37. ventral solid nerve cord
____ 38. no nervous system
____ 39. nerve net
____ 40. ladder-type nervous organization

How many stories is your building? _______

## Office Building Three

A ten-story office building is possible.

**Key Three**

a. jawless fishes
b. cartilaginous fishes
c. bony fishes
d. amphibians
e. reptiles
f. birds
g. mammals

____ 1. four-chambered heart
____ 2. frogs and salamanders
____ 3. air sacs
____ 4. lampreys and hagfish
____ 5. infant dependency
____ 6. Chondrichthyes
____ 7. hair
____ 8. smooth, nonscaly skin
____ 9. ectothermic
____ 10. two-chambered heart
____ 11. differentiated teeth
____ 12. some are parasitic
____ 13. highly developed brain
____ 14. feathers
____ 15. fish, but no operculum
____ 16. evolved from amphibians
____ 17. endothermic
____ 18. snakes, lizards
____ 19. primates
____ 20. epidermal placoid (toothlike) scales
____ 21. shelled egg
____ 22. whales and dolphins
____ 23. class Aves
____ 24. lateral line system
____ 25. mammary glands
____ 26. some are filter feeders
____ 27. metamorphosis
____ 28. sharks, rays, and skates
____ 29. marsupials
____ 30. Osteichthyes
____ 31. ray-finned fishes
____ 32. monotremes
____ 33. scales of bone
____ 34. dinosaurs
____ 35. operculum
____ 36. evolved from reptiles
____ 37. lobe-finned fishes
____ 38. double circulatory loop
____ 39. paired pelvic and pectoral fins
____ 40. one-way path through lungs
____ 41. three-chambered heart
____ 42. lungs
____ 43. expandable rib cage
____ 44. gills as an adult
____ 45. wings
____ 46. single circulatory loop
____ 47. molt
____ 48. usually four limbs
____ 49. amniotic egg
____ 50. cartilaginous skeleton

How many stories is your building? ______

## Definitions Wordmatch

Review key terms by completing this matching exercise, selecting from the following alphabetized list of terms:

chordate
deuterostome
echinoderm
lancelet
mammal
marsupial
placental mammal
water vascular system

a. ________________ Series of canals that takes water to the tube feet of an echinoderm, allowing them to expand.

b. ________________ Group of coelomate animals in which the first embryonic opening is associated with the anus and the second embryonic opening is associated with the mouth.

c. ________________ Invertebrate chordate that has a body resembling a lancet and retains the four chordate characteristics as an adult.

d. ________________ Endothermic vertebrate characterized especially by the presence of hair and mammary glands.

e. ________________ Member of a group of mammals bearing immature young nursed in a marsupium, or pouch—for example, kangaroo and opossum.

# Chapter Test

## Objective Questions

Do not refer to the text when taking this test.

____ 1. Each is a vertebrate characteristic EXCEPT
  a. bilateral symmetry.
  b. coelom development.
  c. open circulatory system.
  d. segmentation.

____ 2. The earliest vertebrate fossils came from the
  a. amphibians.
  b. bony fishes.
  c. cartilaginous fishes.
  d. jawless fishes.

____ 3. Select the incorrect association.
  a. Cenozoic—humans evolve
  b. Mesozoic—dinosaurs dominate
  c. Paleozoic—amphibians live in swamp forests
  d. Cenozoic—fishes evolve

____ 4. The skin of amphibians functions mainly for
  a. circulation.
  b. excretion.
  c. reproduction.
  d. respiration.

____ 5. The extraembryonic membrane in the reptile egg promotes
  a. additional reinforcement from drying out.
  b. complete independence from the water for reproduction.
  c. enhanced elimination of wastes from the embryo.
  d. increased hardness to prevent breakage.

____ 6. Bird feathers are modified
  a. fish fins.
  b. mammalian hair.
  c. reptilian scales.
  d. vertebrate teeth.

____ 7. The hair of mammals is an adaptation for
  a. camouflage in all species.
  b. control of body temperature.
  c. faster locomotion.
  d. regulation of waste elimination.

____ 8. The most successful mammals are the
  a. marsupials.
  b. monotremes.
  c. lancelets.
  d. placental mammals.

____ 9. Which is true of echinoderms?
  a. contain a dorsal tubular nerve cord
  b. have internal organs in a visceral mass
  c. move by a water vascular system
  d. All of these are true.

____ 10. Which is found only among echinoderms?
  a. deuterostome developmental pattern
  b. radial symmetry
  c. exoskeleton
  d. tube feet
  e. ventral mouth

____ 11. Which feature is NOT found among fishes?
  a. endoskeleton
  b. closed circulatory system
  c. warm blood
  d. dorsal tubular nerve cord

____ 12. The most important reason amphibians are incompletely adapted to life on land is that
  a. they depend on water for external fertilization.
  b. they must reproduce in water.
  c. their skin is more important than the lungs for gas exchange.
  d. their means of locomotion is poorly developed.

____ 13. A four-chambered heart is seen among
  a. fishes.
  b. amphibians.
  c. birds.
  d. mammals.
  e. Both *c* and *d* are correct.

____ 14. Which pair of statements correctly contrasts birds and mammals?
  a. Birds are cold-blooded. Mammals are warm-blooded.
  b. Birds are egg-laying. No mammals are egg-laying.
  c. Birds have air sacs in addition to lungs. Mammals have no such sacs.
  d. Birds lack a septum between the ventricles. Mammals have such a septum.

____ 15. What do echinoderms and chordates have in common?
a. radial symmetry
b. pharyngeal pouches
c. second embryonic opening is the mouth
d. All of these are correct.

____ 16. Extraembryonic membranes
a. are found during the development of all vertebrates.
b. are found during the development of reptiles, birds, and mammals.
c. have exactly the same function in all vertebrates.
d. Both *a* and *b* are correct.
e. Both *a* and *c* are correct.

____ 17. Which is NOT a distinguishing feature of vertebrates?
a. dorsal notochord
b. jointed internal skeleton
c. extreme cephalization
d. open circulatory system
e. efficient respiration

____ 18. The type of mammal that lays eggs while nourishing its young with milk is called
a. a monotreme.
b. a marsupial.
c. placental.
d. hermaphroditic.

____ 19. Which is NOT true of echinoderms?
a. external skeleton
b. tube feet
c. skin gills
d. gonads in arms

____ 20. Which is NOT an echinoderm?
a. sea lily
b. sea urchin
c. sea cucumber
d. sea horse

## Critical Thinking Questions

Answer in complete sentences.

21. Compare the success of chordate evolution to arthropod evolution. What are the similarities and differences?

22. Compare the adaptations of amphibians and reptiles to a land existence.

**Test Results:** ______ number correct ÷ 22 = ______ × 100 = ______ %

## Exploring the Internet

The Online Learning Center at *www.mhhe.com/maderbiology8* has additional study material and practice quizzes that can help you master the content of this chapter. You can also find links to websites exploring additional topics in biology. Access to the Online Learning Center is free for those who have purchased a new textbook.

# Answer Key

## Study Exercises

1.

| Description |
|---|
| radial |
| spine-bearing, calcium-rich plates in an endoskeleton |
| gas exchange across skin gills and tube feet |
| central nerve ring plus radial nerves |
| series of canals that ends at tube feet; a means of locomotion |

2.

| Examples | Distinctive Features |
|---|---|
| sea lilies, feather stars | branched arms for filter feeding |
| sea cucumbers | resemble cucumber, have tentacles for feeding |
| sea urchins, sand dollars | spines for locomotion, defense, burrowing |
| brittle stars | long, flexible arms |
| sea stars | five arms |

**3.** See Figure 31.1, page 556, in text. **4. a.** stone **b.** ring **c.** tube **d.** ampulla **e.** locomotion **5.** The sea star everts its cardiac stomach, puts it in the shell, and secretes enzymes; partly digested food is taken up, and digestion is completed in the pyloric stomach. **6.** See page 557 in text. **7.** They are both deuterostomes and develop similarly. See also Figure 30.1, page 536 in text.

**8.**

| Chordate Characteristics | Appearance |
|---|---|
| all four | lancet-shaped |
| all four (larva); gill slits (adult) | thick-walled, squat sac |

**9.** a, c, d, e, h **10. a.** Agnatha, fish **b.** Gnathostomata, fish **c.** Gnathostomata, fish **d.** Gnathostomata, tetrapod **e.** Gnathostomata, tetrapod **f.** Gnathostomata, tetrapod **g.** Gnathostomata, tetrapod **11. a.** Amphibia **b.** Reptilia **c.** Chondrichthyes **d.** Aves **e.** Mammalia **12. a.** CF, BF **b.** CF, BF **c.** BF **d.** BF **e.** BF **13. a.** first vertebrates **b.** first jawed vertebrates **c.** closest living relatives of modern amphibians **d.** first vertebrates to live on land **e.** first vertebrates to reproduce on land **14. a.** smooth, nonscaly, and used for respiration **b.** small and poorly developed **c.** ectothermic **d.** undergo metamorphosis from tadpole to adult **e.** three chambers **15. a.** amphibians **b.** adaptive **c.** land **d.** air **e.** sea **f.** mammals **g.** dinosaurs **16. a.** thick, dry, scaly **b.** more developed than in amphibians **c.** ectothermic **d.** amniotic **e.** nearly or completely four chambered **17. a.** provide broad, flat surfaces **b.** reduces weight **c.** attaches flight muscles **d.** provides good delivery of $O_2$-rich blood to muscles **e.** provides good oxygenation of blood **18. a.** 1,2 **b.** 1,3 **c.** 1,4 **19. a.** bat **b.** horse **c.** whale **d.** lion **e.** monkey

## Office Building One

**1.** i **2.** f **3.** d **4.** h **5.** d, e, f, g, h **6.** i **7.** d **8.** c **9.** f **10.** g **11.** h **12.** i **13.** e **14.** c, d, f, g **15.** f **16.** d, e, f, g, h **17.** d, e, f, g, h, i **18.** i **19.** c **20.** a, b **21.** g **22.** i **23.** f **24.** c **25.** g **26.** f **27.** e **28.** d **29.** h **30.** i **31.** e **32.** f, g, h, i **33.** c **34.** b **35.** f **36.** g **37.** f, g, h **38.** f **39.** g, h, i **40.** d, e, f, g, h, i **41.** b **42.** h **43.** f, g, h **44.** c **45.** f **46.** h **47.** g **48.** e **49.** c **50.** h **51.** g, h, i **52.** f **53.** e **54.** i **55.** d, f **56.** b, c **57.** e **58.** i **59.** d, e, f **60.** h **61.** f **62.** g **63.** d, h **64.** b **65.** h **66.** g, h **67.** h **68.** b, c **69.** c, d **70.** e, f, g, h, i

## Office Building Two

**1.** b, j **2.** d, e, f, g, h, i **3.** f, g, i **4.** b **5.** b, c **6.** d, e, f, g, h, i **7.** c, d, e, f, g, i **8.** b, h, j **9.** c, d, e, f, g, h, i **10.** f **11.** i **12.** b and j; g and i **13.** c **14.** i **15.** g **16.** f **17.** i **18.** e, g, h **19.** f **20.** h **21.** a **22.** e **23.** g **24.** h **25.** b **26.** i **27.** i **28.** g, i **29.** i **30.** b **31.** j **32.** f **33.** h **34.** h **35.** e **36.** none **37.** f, g **38.** a **39.** b, j **40.** c

## Office Building Three

**1.** e, f, g **2.** d **3.** f **4.** a **5.** g **6.** b **7.** g **8.** a, d **9.** a, b, c, d, e **10.** a, b, c **11.** g **12.** a **13.** g **14.** f **15.** a, b **16.** e **17.** f, g **18.** e **19.** g **20.** b **21.** e, f **22.** g **23.** f **24.** b, c **25.** g **26.** b, g **27.** d **28.** b **29.** g **30.** c **31.** c **32.** g **33.** c **34.** e **35.** c **36.** f, g **37.** c **38.** c, d, e, f, g **39.** c **40.** f **41.** d **42.** d, e, f, g **43.** e, f, g **44.** a, b, c **45.** f **46.** a, b, c **47.** e **48.** d, e, f, g **49.** e, f **50.** a, b

## Definitions Wordmatch

**a.** water vascular system **b.** deuterostome **c.** lancelet **d.** mammal **e.** marsupial

## Chapter Test

**1.** c **2.** d **3.** d **4.** d **5.** b **6.** c **7.** b **8.** d **9.** c **10.** d **11.** c **12.** a **13.** e **14.** c **15.** c **16.** b **17.** d **18.** a **19.** a **20.** d **21.** Each phylum contains numerous diversified species adapted to a variety of environments. Arthropods have more species. **22.** Amphibians reproduce in the water and have a larval stage that develops in the water. The skin must be kept moist because it supplements the lungs for gas exchange. Reptiles reproduce on land because they lay a shelled egg with extraembryonic membranes. The skin can prevent desiccation because it is dry and scaly. The lungs are moderately developed, and a rib cage helps ventilate the lungs.

# 32

# Human Evolution

## Chapter Review

**Primates** are adapted to an **arboreal life.** They have evolved opposable thumbs; their digits bear nails, not claws; they have binocular vision; and the forebrain is enlarged. Primates bear one offspring at a time, and much of their behavior is learned during an extended childhood.

The classification of humans reflects their evolutionary history. Humans belong to the order Primates, along with **prosimians,** monkeys, and apes. The anthropoids includes monkeys, apes, and humans. The superfamily of **hominoids** includes only apes and humans. **Hominids** (family Hominidae) includes the bipedal *Sahelanthropus; Ardipithecus; Australopithecus;* and *Homo,* humans. The genus ***Homo*** includes only extinct and living human species.

Several early hominid fossils dated around the time of the split between apes and humans have now been found. These fossils have a chimp-sized braincase, but are believed to have walked erect.

An **australopithecine** may be a direct ancestor for humans. The australopithecines were diverse and bipedal. Some species lived at the same time as early *Homo.* ***Australopithecus afarensis*** could walk erect, but had only a small brain. Later-appearing australopithecines may have manufactured stone tools, but ***H. habilis*** certainly did. ***H. erectus***, the first fossil to have a brain size of about 1,000 cc, migrated from Africa into Europe and Asia. *H. erectus* controlled fire and may have been big-game hunters.

Two contradicting hypotheses have been suggested about the origin of modern humans. The **multiregional continuity hypothesis** says that modern humans originated separately in Asia, Europe, and Africa. The **out-of-Africa hypothesis** says that modern humans originated in Africa and, after migrating into Europe and Asia, replaced the archaic *Homo* species found there. The Cro-Magnon were modern humans who entered Europe and Asia from Africa.

## Study Exercises

Study the text section by section as you answer the questions that follow.

### 32.1 Evolution of Primates (pp. 579–82)

- Primates include prosimians, monkeys, apes, and humans.
- Primate characteristics include an enlarged brain, opposable thumbs, binocular vision, and an emphasis on learned behavior.
- The hominoid common ancestor, *Proconsul,* was prevalent in Africa during the Miocene epoch.

1. List an adaptation for arboreal life in relation to the following:
   a. vision ____________________
   b. digits ____________________
   c. brain size ____________________
   d. birth number ____________________

#### Phylogenetic Tree (pp. 580–81)

2. Beside each taxonomic category, place the following terms as appropriate: monkeys, apes, humans
   a. anthropoids ______________, ______________, ______________
   b. hominoids ______________, ______________

3. In the following chart, list characteristics that distinguish New World monkeys from Old World monkeys:

| New World Monkeys | Old World Monkeys |
| --- | --- |
| | |
| | |

4. What is the name given to the hominoid believed to be ancestral to today's hominids. ________________

5. In the following chart, compare the anatomy of a human to that of an ape:

| | Human | Ape |
| --- | --- | --- |
| Spine shape | | |
| Pelvis shape | | |
| Femur angle | | |
| Knee: weight-bearing capacity | | |
| Foot shape | | |

## 32.2 Evolution of Hominids (pp. 583–87)

- The split between the ape and human lineage occurred about 6 MYA.
- To be classified a hominid, a fossil must have the anatomy suitable for bipedalism.

6. Australopithecines are well known [a]______________. When applied to australopithecines, the term *gracile* means [b]______________ and the term *robust* means [c]______________. Australopithecines walked [d]______________, but their brain was [e]______________. Their remains show that they evolved the first humanlike feature, [f]________________. Australopithecines exhibit mosaic evolution, meaning that their [g]________________________________________________________________.

7. *Homo* [a]______________ have larger brains than the australopithecines. They show evidence of making and using primitive [b]______________.

8. *Homo* [a]______________, which evolved from *Homo* [b]______________ made more advanced tools and could control [c]______________.This species first appeared in [d]______________, then migrated into Asia and Europe.

## 32.3 Evolution of Modern Humans (pp. 588–90)

- The multiregional continuity hypothesis proposes that modern humans evolved separately in Asia, Africa, and Europe.
- The out-of-Africa hypothesis proposes that modern humans evolved in Africa, and then migrated into Asia and Europe, where they replaced the archaic *Homo* species, including the Neanderthals.
- Cro-Magnon people were modern humans who made sophisticated tools and definitely had culture.

9. Which hypothesis—the out-of-Africa hypothesis or the multiregional continuity hypothesis—states that *H. erectus* and then, later, humans left Africa? a.______________ Which hypothesis states that *H. erectus* left Africa and that modern humans then simultaneously arose in Europe, Asia, and Africa? b.______________ With which hypothesis would you expect more similarity between fossils dated 100,000 BP? c.______________ The fossil record shows several varieties of humans in Asia and Europe dated prior to 100,000 BP. These are called the d. "______________ *Homo*." One example of an "archaic *Homo*" is a e.______________.

10. Match the *Homo* species with the following phrases that describe their way of life. (Numbers can be used more than once.)
    1. brain size less than 1,000 cc
    2. brain size 1,000 cc or larger
    3. more likely scavenged meat
    4. more likely hunted animals
    5. certainly had speech and culture
    6. most likely had speech and culture
    7. perhaps had speech and culture
    8. made tools
    9. had upright posture

    a. *Homo habilis* ______________

    b. *Homo erectus* ______________

    c. Neanderthals ______________

    d. Cro-Magnons ______________

11. a.______________ in humans may be the result of adaptations to local b.______________. Mitochondrial c.______________ shows that all ethnic groups have a recent common d.______________.

# Chapter Test

## Objective Questions

Do not refer to the text when taking this test.

____ 1. Which of these are NOT primates?
a. lemurs
b. monkeys
c. gorillas
d. humans
e. All of these are primates.

____ 2. Which is NOT true of primates?
a. They have diversified and live in the air, on land, and in the water.
b. They are adapted to life in trees.
c. They have opposable thumbs.
d. They have jointed appendages.

____ 3. Which is NOT an ape?
a. gibbon
b. gorilla
c. orangutan
d. chimpanzee
e. All of these are apes.

____ 4. Modern humans are more closely related to
a. monkeys than apes.
b. African apes than Asian apes.
c. Neanderthals than Cro-Magnons.
d. whales than prosimians.

____ 5. Which of these is NOT a correct contrast between human and ape anatomy?

| | Human | Ape |
|---|---|---|
| a. | short pelvis | long pelvis |
| b. | long legs | short legs |
| c. | bipedal | knuckle walking |
| d. | sloping face | straight face |

____ 6. Australopithecines
a. were remarkable for their lack of body hair.
b. may have lived at the same time as *H. habilis*.
c. were all robust and most closely related to the Neanderthals.
d. had dentition similar to that of modern humans.

____ 7. Australopithecines were
a. apelike below the waist and humanlike above the waist.
b. humanlike below the waist and apelike above the waist.
c. generally apelike.
d. generally humanlike.

____ 8. *Homo erectus*
a. evolved from *Homo habilis*.
b. fossils have been found in Africa, Asia, and Europe.
c. has a larger brain size than early *Homo,* but not as large as a Neanderthal.
d. All of these are correct.

____ 9. Dentition tells us much about
a. the brain size of hominoids.
b. how long ago modern humans left Africa.
c. the foods hominids ate.
d. Both *a* and *c* are correct.

In questions 10–14, match the descriptions with the following hominids:
a. australopithecines
b. *H. habilis*
c. *H. erectus*
d. Cro-Magnon

____10. species most likely NOT to have tools
____11. species most likely to have traveled out of Africa
____12. species most likely to have been able to control fire first
____13. species most likely to have made tools first
____14. species known to have painted and sculpted

____15. The Neanderthals
a. preceded the Cro-Magnon.
b. lived in Europe and Asia during the last Ice Age.
c. buried their dead with flowers and tools.
d. had massive brow ridges, a protruding nose, and jaws with no chin.
e. All of these are correct.

____16. A likely hypothesis is that *H. erectus*
a. never migrated out of Africa; Lucy's son did.
b. still climbed trees and ate only fruit.
c. hunted animals and had a home base.
d. All of these are likely.

____17. The multiregional continuity hypothesis says that
a. modern humans arose in several different places.
b. no interbreeding took place between different types of humans.
c. Neanderthals, and not *H. erectus,* migrated out of Africa.
d. the Neanderthals were more modern in appearance than once thought.

____18. The fact that humans adapted to various climates shows that
a. not all humans can reproduce with one another.
b. humans are adaptable.
c. we need to rethink some of the tenets of human evolution.
d. All of these are correct.

## Critical Thinking Questions

Answer in complete sentences.

19. In what way do australopithecines illustrate mosaic evolution?

20. What types of evidence would convince you that a particular fossil should be classified in the genus *Homo?*

**Test Results:** ______ number correct ÷ 20 = ______ × 100 = ______ %

## Exploring the Internet

The Online Learning Center at *www.mhhe.com/maderbiology8* has additional study material and practice quizzes that can help you master the content of this chapter. You can also find links to websites exploring additional topics in biology. Access to the Online Learning Center is free for those who have purchased a new textbook.

# ANSWER KEY

## STUDY EXERCISES

**1. a.** eyes forward with stereoscopic vision **b.** nails, not claws, and opposable thumbs **c.** large, well developed **d.** a single offspring at a time **2. a.** monkeys, apes, humans **b.** apes, humans

**3.**

| New World Monkeys | Old World Monkeys |
|---|---|
| long, prehensile tail | no prehensile tail |
| flat nose | protruding nose |

**4.** *Proconsul*

**5.**

| Human | Ape |
|---|---|
| s-shaped | only a slight curve |
| bowl-shaped | long, narrow |
| angles inward to knees | angles out |
| more | less |
| arched | no arch |

**6. a.** hominids **b.** slender **c.** powerful **d.** erect **e.** small **f.** bipedalism **g.** body parts evolved at different rates **7. a.** *habilis* **b.** stone tools **8. a.** *erectus* **b.** *habilis* **c.** fire **d.** Africa **9. a.** out-of-Africa **b.** multiregional continuity **c.** multiregional continuity **d.** archaic **e.** Neanderthal **10. a.** 1, 3, 7, 8, 9 **b.** 2, 4, 6, 8, 9 **c.** 2, 4, 6, 8, 9 **d.** 2, 4, 5, 8, 9 **11. a.** Variations **b.** environments **c.** DNA **d.** ancestry

## CHAPTER TEST

**1.** e **2.** a **3.** e **4.** b **5.** d **6.** b **7.** b **8.** d **9.** c **10.** a **11.** c **12.** c **13.** b **14.** d **15.** e **16.** c **17.** a **18.** b **19.** Mosaic evolution occurs when various parts evolve at different rates. Australopithecines walked erect but had a small brain; therefore, the lower half of their body was more like that of modern humans than the upper half. **20.** A hominid should be classified in the genus *Homo* if it has erect posture, dentition like that of modern humans, a brain of 1,000 cc or more, a high forehead, and a projecting chin.

# Part VII Comparative Animal Biology

# 33

# Animal Organization and Homeostasis

## Chapter Review

Human tissues are categorized into four groups. **Epithelial tissue** covers the body and lines its cavities. The different types of epithelial tissue (**squamous, cuboidal,** and **columnar**) can be simple or stratified and have cilia or microvilli. Also, columnar cells can be pseudostratified. Epithelial cells sometimes form glands that secrete either into ducts or into blood.

**Connective tissues,** in which cells are separated by a **matrix,** often bind body parts together. **Loose fibrous connective tissue** has both white and yellow fibers. Adipose tissue has fat cells. **Dense fibrous connective tissue,** such as in **tendons** and **ligaments,** contains closely packed white fibers. Both **cartilage** and **bone** have cells within lacunae, but the matrix for cartilage is more flexible than that for bone, which contains calcium salts. In bone, the lacunae lie in concentric circles within an osteon, about a central canal. Blood is a connective tissue in which the matrix is a liquid called **plasma.**

**Muscular tissue** is of three types. Both **skeletal muscle** and **cardiac muscle** are **striated;** both cardiac and **smooth muscle** are involuntary. Skeletal muscle is found in muscles attached to bones, and smooth muscle is found in internal organs. Cardiac muscle makes up the heart.

**Nervous tissue** has one main type of cell, the **neuron.** Each neuron has dendrites, a cell body, and an axon. The brain and spinal cord contain complete neurons, while the nerves contain only fibers. Neurons and their fibers are specialized to conduct nerve impulses.

Tissues are joined together to form organs, each one having a specific function. **Skin** is a two-layered organ that waterproofs and protects the body. The **epidermis** contains a germinal layer that produces new epithelial cells that become keratinized as they move toward the surface. The **dermis,** a largely fibrous connective tissue, contains epidermally-derived glands and hair follicles, nerve endings, and blood vessels. Receptors for touch, pressure, temperature, and pain are present. A **subcutaneous layer,** made up of loose connective tissue containing **adipose** cells, lies beneath the skin.

Organs are grouped into organ systems. In the dorsal cavity, the brain is in the cranial cavity, and the spinal cord is in the vertebral canal. Other internal organs are located in the ventral cavity, where the thoracic cavity is separated from the abdominal cavity by the diaphragm.

**Homeostasis** is the relative constancy of the internal environment. All organ systems contribute to the constancy of tissue fluid and blood. Special contributions are made by the liver, which keeps blood glucose constant, and the kidneys, which regulate the pH. The nervous and endocrine systems regulate the other systems. Both of these are controlled by a feedback mechanism, which results in fluctuation above and below the desired levels illustrated by body temperature.

## Study Exercises

Study the text section by section as you answer the questions that follow.

### 33.1 Types of Tissues (pp. 596–601)

- Animals have the following levels of organization: molecules—cells—tissues—organs—organ systems—organism.
- Animal tissues can be organized into four major types: epithelial, connective, muscular, and nervous tissue.

1. What four categories of tissues does the text recognize? ______________, ______________, ______________, ______________

## Epithelial Tissues (pp. 596–97)

- Epithelial tissues, which line body cavities and cover surfaces, are specialized in structure and function.

2. a. Draw a diagram of squamous epithelium.

   b. Name one place in the human body where squamous epithelial tissue can be found. ______________________

   c. What is the function of this tissue? ______________________

3. a. Draw a diagram of cuboidal epithelial tissue.

   b. Name one place in the human body where cuboidal epithelial tissue can be found. ______________________

   c. What is the function of this tissue? ______________________

4. a. Draw a diagram of simple columnar epithelial tissue.

   b. Name one place in the human body where simple columnar epithelial tissue can be found. ________________

   c. What is the function of this tissue? ______________________

5. The windpipe is lined by pseudostratified ciliated columnar epithelium. Describe this tissue.

## Connective Tissues (pp. 598–99)

- Connective tissues, which protect, support, and bind other tissues, include cartilage and bone and also blood, the only liquid tissue.

6. a. Draw a diagram of loose fibrous connective tissue and include fibroblasts and fibers.

b. Where in the body do you find this type of tissue? ______________________________

7. Which type of fiber predominates in dense fibrous connective tissue? a. ____________________

Tendons are a type of dense fibrous connective tissue that join b. ____________________ to c. ______________________________. Ligaments join d. __________________ to e. __________________.

8. a. Draw a diagram of cartilage, and label lacunae and the matrix.

b. Name one place in the human body where cartilage can be found. ______________________________

9. Draw a diagram of compact bone, and label lacunae, osteons, and central canals.

10. a. Draw a diagram of blood, and include red blood cells, white blood cells, and platelets.

One function of red blood cells is to b. ____________________; one function of white blood cells is to c. ____________________; and one function of platelets is to d. ____________________.

## Muscular Tissue (p. 600)

- Muscular tissues, which contract, make body parts move.

11. Complete the following table to compare types of muscular tissue:

| | Fiber Appearance | Location | Control |
|---|---|---|---|
| Skeletal | | | |
| Smooth | | | |
| Cardiac | | | |

## Nervous Tissue (p. 601)

- Nervous tissues coordinate the activities of the other tissues and body parts.

12. The brain and spinal cord are made up of cells called a.________________________. Outside the central nervous system, connective tissue binds the long fibers of these cells to form b.________________________. The function of a neuron is to c.________________________. The other type of cells in nervous tissue are d.________________________. This type of cell provides e.________________________ and ________________________ to neurons and keeps tissue free of debris. Neuroglia outnumber neurons f.________________________ to one and take up more than g.________________________ the volume of the brain.

## 33.2 Organs and Organ Systems (pp. 602–5)

- Organs usually contain several types of tissues. For example, although skin is composed primarily of epithelial tissue and connective tissue, it also contains muscle and nerve fibers.
- Organs are grouped into organ systems, each of which has specialized functions.
- The coelom, which arises during development, is later divided into various cavities where specific organs are located.

13. Label this diagram of skin using the alphabetized list of terms.
    adipose tissue
    arrector pili muscle
    artery
    connective tissue
    dermis
    epidermis
    hair root
    hair shaft
    nerve
    oil gland
    sense organs
    subcutaneous layer
    sweat gland
    vein

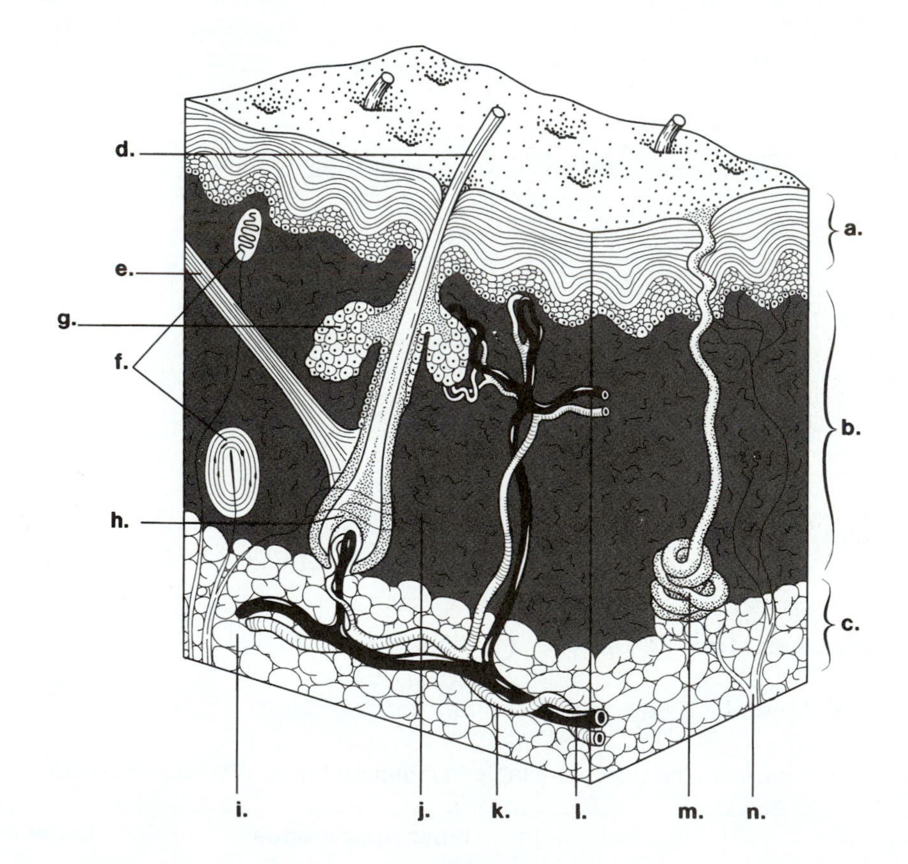


14. Name the system that performs each of the following functions:
    a. transports materials by the blood ________________________
    b. breaks down substances in the diet ________________________
    c. distributes and exchanges gases ________________________
    d. pulls on bones to produce movements ________________________
    e. coordinates body activities through hormones ________________________

15. Label the following as located in the thoracic cavity (T) or the abdominal cavity (A):
   a. _____ small intestine
   b. _____ ovaries
   c. _____ bladder
   d. _____ heart
   e. _____ lungs
   f. _____ stomach
   g. _____ liver
   h. _____ kidneys

## 33.3 Homeostasis (pp. 606–7)

- Homeostasis is the dynamic equilibrium of the internal environment. All organ systems contribute to homeostasis in animals.

16. a. What is homeostasis? ____________________
    b. What is the internal environment? ____________________
    c. Give an example of homeostasis. ____________________

17. How does each of the following systems contribute to homeostasis?
    a. digestive system ____________________
    b. respiratory system ____________________
    c. urinary system (i.e., the kidneys) ____________________

18. The two systems of the body that control homeostasis are [a] ____________________ and [b] ____________________.

19. Put the following terms in the proper sequence to describe a negative feedback cycle:
    effector    sensory receptor    regulatory center    response

    ____________________

20. a. The body maintains a relatively steady body temperature. If the body cools, what events raise body temperature?
    ____________________
    ____________________
    b. If the body heats up, what events lower body temperature? ____________________
    ____________________

# Chapter Test

## Objective Questions

Do not refer to the text when taking this test.

____ 1. Cells working together form
   a. organs.
   b. tissues.
   c. systems.
   d. organisms.

____ 2. Which of these is true of humans?
   a. They have a vertebral column.
   b. The ventral cavity is divided into two cavities.
   c. The endocrine and nervous systems help coordinate the internal organs.
   d. Muscle and bones make up most of the body weight.
   e. All of these are true.

____ 3. Select the incorrect association.
   a. fat—subcutaneous layer
   b. sense organs—dermis
   c. keratinization—epidermis
   d. nerves and blood vessels—epidermis

____ 4. Which type of epithelial tissue is composed of flat cells?
   a. squamous
   b. cuboidal
   c. columnar
   d. striated

____ 5. What type of epithelial cells are found in the epidermis?

a. squamous
b. columnar
c. cuboidal
d. Both *b* and *c* are correct.

____ 6. Which of these tissues lines body cavities?

a. muscular
b. epithelial
c. connective
d. tendons

____ 7. Both cartilage and bone

a. have a flexible matrix.
b. have cells within lacunae.
c. contain calcium salts.
d. are a type of epithelial tissue.

____ 8. Which of these is a connective tissue?

a. bone
b. cartilage
c. blood
d. All of these are correct.

____ 9. Which of these is a loose connective tissue?

a. tendons
b. ligaments
c. adipose tissue
d. All of these are correct.

____ 10. Which of these contains blood vessels?

a. lacuna
b. matrix
c. central canal
d. Both *a* and *b* are correct.

____ 11. Which of the following are cell fragments?

a. platelets
b. red blood cells
c. white blood cells
d. Both *a* and *b* are correct.

____ 12. What type of tissue does the dermis consist of?

a. epithelial
b. connective
c. muscle
d. All of these are correct.

____ 13. A muscle tissue that is both striated and involuntary is

a. smooth.
b. skeletal.
c. cardiac.
d. All of these are correct.

____ 14. Which of these types of muscles helps maintain posture?

a. skeletal
b. smooth
c. cardiac
d. involuntary

____ 15. Which of these is a nerve cell?

a. neuroglia
b. dendrite
c. cell body
d. neuron

____ 16. Which of these is the uppermost region of skin?

a. fat
b. subcutaneous
c. dermis
d. epidermis

____ 17. Adipose tissue is found in the

a. dermis.
b. epidermis.
c. subcutaneous.
d. All of these are correct.

____ 18. Which of these is located in the abdominal cavity?

a. heart
b. digestive system
c. much of the reproductive system
d. Both *b* and *c* are correct.

____ 19. Which of these is an example of homeostasis?

a. Muscle tissue is specialized to contract.
b. Normal body temperature is always about 37° C.
c. There are more red blood cells than white blood cells.
d. All of these are correct.

____ 20. In a negative feedback control system,

a. homeostasis is impossible.
b. there is a constancy of the internal environment.
c. there is a fluctuation about a mean.
d. Both *a* and *c* are correct.

____ 21. When body temperature rises, sweat glands become ______ and blood vessels ______.

a. active; constrict
b. inactive; dilate
c. active; dilate
d. inactive; constrict

## Critical Thinking Questions

Answer in complete sentences.

22. How is each level of organization of the human body more than merely the sum of its parts?

23. Why is it important to body health to maintain a relatively constant environment?

**Test Results:** _____ number correct ÷ 23 = _____ × 100 = _____ %

## Exploring the Internet

The Online Learning Center at *www.mhhe.com/maderbiology8* has additional study material and practice quizzes that can help you master the content of this chapter. You can also find links to websites exploring additional topics in biology. Access to the Online Learning Center is free for those who have purchased a new textbook.

## Answer Key

### Study Exercises

**1.** epithelial, connective, muscular, nervous **2. a.** See Figure 33.2, page 597, in text. **b.** blood vessels, esophagus **c.** protection **3. a.** See Figure 33.2, page 597, in text. **b.** kidney tubules **c.** absorption **4. a.** See Figure 33.2, page 597, in text. **b.** lining of intestine **c.** protection and absorption **5.** appears to be many layers but is actually only one; cells have little hairs **6. a.** See Figure 33.3, page 598, in text. **b.** beneath skin and most epithelial layers **7. a.** collagen fibers **b.** muscles **c.** bones **d.** bones **e.** bones **8. a.** See Figure 33.3, page 598, in text. **b.** nose, ears, ends of bones **9.** See Figure 33.3, page 598, in text. **10. a.** See Figure 33.4, page 599, in text. **b.** carry oxygen **c.** fight infection **d.** aid in blood clotting

**11.**

| Fiber Appearance | Location | Control |
|---|---|---|
| striated | skeleton | voluntary |
| spindle-shaped | internal organs | involuntary |
| striated | heart | involuntary |

**12. a.** neurons **b.** nerves **c.** transmit nerve impulses **d.** neuroglia **e.** support and protection **f.** nine **g.** half **13. a.** epidermis **b.** dermis **c.** subcutaneous layer **d.** hair shaft **e.** arrector pili muscle **f.** sense organs **g.** oil gland **h.** hair root **i.** adipose tissue **j.** connective tissue **k.** vein **l.** artery **m.** sweat gland **n.** nerve **14. a.** cardiovascular **b.** digestive **c.** respiratory **d.** musculoskeletal **e.** endocrine, nervous **15. a.** A **b.** A **c.** A **d.** T **e.** T **f.** A **g.** A **h.** A **16. a.** the relative constancy of the internal environment **b.** tissue fluid **c.** body temperature remaining around 37° C **17. a.** provides nutrient molecules **b.** removes carbon dioxide and adds oxygen to the blood **c.** eliminates wastes and salts **18. a.** nervous **b.** endocrine **19.** sensory receptor, regulatory center, effector, response **20. a.** Blood vessels constrict and sweat glands become inactive. **b.** Blood vessels dilate and sweat glands become active.

### Chapter Test

**1.** b **2.** e **3.** d **4.** a **5.** a **6.** b **7.** b **8.** d **9.** c **10.** c **11.** a **12.** b **13.** c **14.** a **15.** d **16.** d **17.** c **18.** d **19.** b **20.** c **21.** c **22.** The parts are integrated, interacting to form a more complex structure. For example, cardiac tissue functions in a way that the individual cells cannot function. **23.** A constant environment refers to the optimal conditions at which the body functions best, such as body temperature or glucose concentration in the blood.

# 34

# Circulation

## Chapter Review

Among invertebrates, cnidarians and flatworms supply cells with oxygen through diffusion in a gastrovascular cavity. An internal transport system is lacking. Other invertebrates have such a circulatory system, which may be either open or closed.

Among vertebrates, fish have a single circulatory loop that works with the gills to form the **cardiovascular system.** The other vertebrates have both **pulmonary** and **systemic** circulation. In either type of circuit, blood flows through the following series of vessels: **arteries, arterioles, capillaries, venules,** and **veins.** The oxygenation of blood is most efficient in the four-chambered **heart** of birds and mammals.

In humans, the movement of blood in the cardiovascular system is dependent on the beat of the heart. During the **cardiac cycle,** the SA node **(cardiac pacemaker)** initiates the beat and causes the atria to contract.

The AV node conveys the stimulus and initiates contraction of the ventricles. The heart supplies the **blood pressure**, which keeps blood moving. Skeletal muscle contraction pushes blood in the veins toward the heart. Veins have valves that prevent the backward flow of blood.

The blood consists of the **plasma** and cells. Plasma contains mostly water and proteins, but it also contains nutrients and wastes. **Red blood cells** transport oxygen. Several types of **white blood cells** fight infection and establish immunity. One type, **neutrophils,** have phagocytic ability. **Platelets** promote the clotting of blood. Their activity leads to the formation of fibrin, which traps red blood cells, forming the clot.

**Hypertension** and **atherosclerosis** are two cardiovascular disorders that lead to stroke and heart attack.

## Study Exercises

Study the text section by section as you answer the questions that follow.

### 34.1 Transport in Invertebrates (pp. 612–13)

- Some invertebrates do not have a circulatory system, and others have an open, as opposed to a closed, system.

1. Match the animals to these means of circulation:
   1. gastrovascular cavity
   2. open circulatory system
   3. closed circulatory system

   a. _____ planarian
   b. _____ earthworm
   c. _____ insect

2. Match the animals to these types of blood:
   1. colorless
   2. pigmented
   3. none

   a. _____ earthworm
   b. _____ insect
   c. _____ planarian

## 34.2 Transport in Vertebrates (pp. 614–15)

- Vertebrates have a closed circulatory system: arteries take blood away from the heart to the capillaries, where exchange occurs, and veins take blood to the heart.
- Fishes have a single-loop circulatory pathway, whereas the other vertebrates have a double-loop circulatory pathway—to and from the lungs and also to and from the tissues.

3. Label this diagram of the blood vessels using the following alphabetized list of terms.
   arterioles
   artery
   capillaries
   heart
   vein
   venules

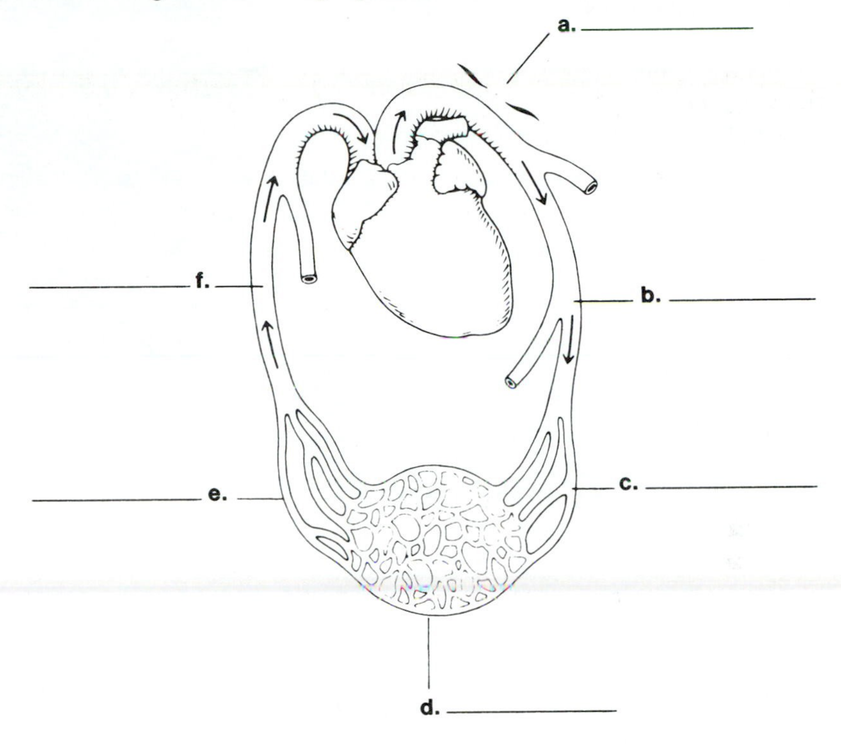


4. Match the animals with these types of vertebrate hearts:
   1. one atrium and one ventricle
   2. two atria and one ventricle
   3. two atria and two ventricles

   a. _____ frog
   b. _____ fish
   c. _____ mammal

## 34.3 Transport in Humans (pp. 616–21)

- The right side of the heart pumps blood to the lungs, and the left side pumps blood to the tissues.
- Blood pressure causes blood to flow in the arteries and arterioles. Skeletal muscle contraction causes blood to flow in the venules and veins, and valves prevent the backward flow of blood.

5. Label this diagram of the heart with the following terms:

aorta
atrioventricular (mitral) valve
atrioventricular (tricuspid) valve
chordae tendineae
inferior vena cava
left atrium
left pulmonary artery
left pulmonary veins
left ventricle
pulmonary trunk
right atrium
right pulmonary arteries
right pulmonary veins
right ventricle
semilunar valves
septum
superior vena cava

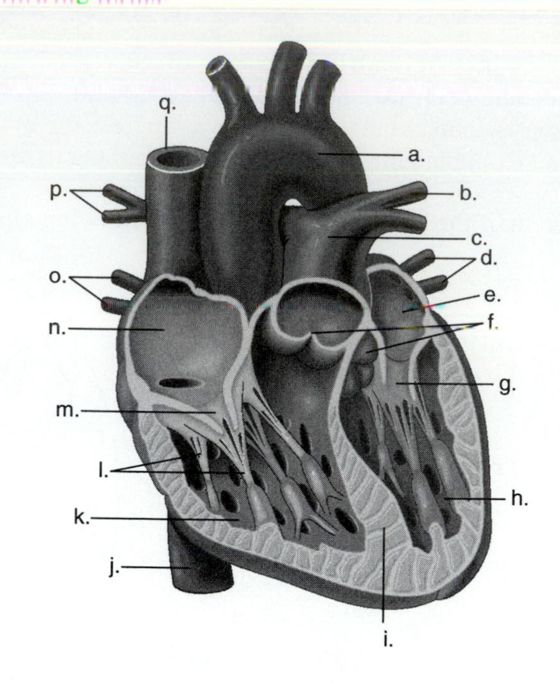


6. Trace the path of blood through the heart

a. from the vena cava to the lungs. ____________________

____________________

b. from the lungs to the aorta.

____________________

7. Use the numbers 1–10 to trace the sequence of blood flow throughout the body.

a. _____ left atrium
b. _____ left ventricle
c. _____ pulmonary artery
d. _____ pulmonary vein
e. _____ lungs
f. _____ right atrium
g. _____ right ventricle
h. _____ systemic artery
i. _____ systemic capillary
j. _____ systemic vein

## The Heartbeat (pp. 618–19)

8. The heart beats about 70 times a minute. The a. ____________________ node initiates the contraction of the b. ____________________ (chambers). This stimulus is picked up by the c. ____________________ node, which initiates the contraction of the d. ____________________ (chambers). When the chambers are not contracting, they are relaxing. Contraction is scientifically termed e. ____________________, and the resting is termed f. ____________________. Contraction of the atria forces the blood through the g. ____________________ valves into the h. ____________________. The closing of these valves is the *lub* sound. Next, the ventricles contract and force the blood into the arteries. Now the i. ____________________ valves close; this is the *dub* sound.

9. Complete the following table with the words *systole* and *diastole* to show what occurs during the 0.85 second of one heartbeat:

| Time | Atria | Ventricles |
|---|---|---|
| 0.15 sec | | |
| 0.30 sec | | |
| 0.40 sec | | |

10. How does the hepatic portal system differ from the pulmonary and systemic circuits?

_______________

Questions 11–14 are based on the adjacent diagram:

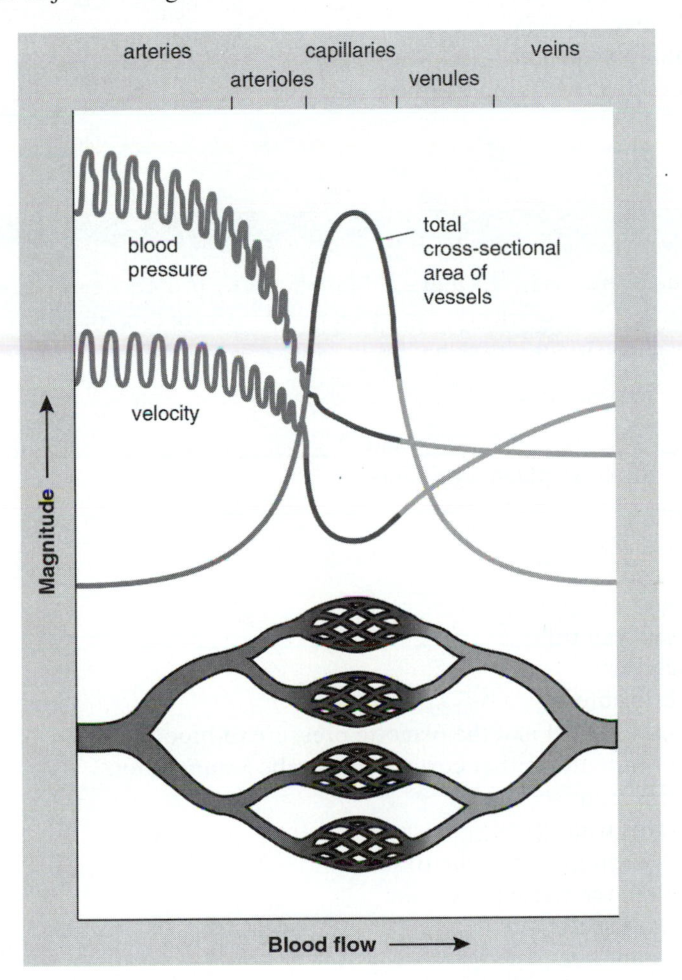


11. What force accounts for blood flow in arteries?

_______________

12. Why does this force fluctuate?

_______________

13. What causes the blood pressure and velocity to drop off?

_______________

14. What force accounts for blood flow in the veins?

_______________

## 34.4 Cardiovascular Disorders (pp. 622–24)

- Although the cardiovascular system is very efficient, it is still subject to degenerative disorders.

15. Which of these would help prevent cardiovascular disease?
    a. _____ Don't smoke.
    b. _____ Don't exercise.
    c. _____ Don't gain excessive weight.
    d. _____ Don't have a diet rich in cholesterol.
16. Match the cardiovascular disorders to these terms:
    1. myocardial infarction
    2. hypertension
    3. varicose veins
    4. hemorrhoids
    5. angina pectoris
    6. familial hypercholesterolemia
    7. embolus in cranial artery

    a. _____ heart attack
    b. _____ varicose veins in rectum
    c. _____ weakened valves in legs
    d. _____ elevated blood pressure
    e. _____ inherited condition leading to atherosclerosis
    f. _____ stroke
    g. _____ chest pain

## 34.5 Blood, a Transport Medium (pp. 625–27)

- In humans, blood is composed of cells and a fluid containing proteins and various other molecules and ions.
- Blood clotting is a series of reactions that produce a clot—fibrin threads in which red blood cells are trapped.
- Exchange of substances between blood and tissue fluid across capillary walls supplies cells with nutrients and removes wastes.

17. Plasma is mostly a.__________________ and b.__________________.
18. Match the functions with these plasma proteins:
    1. fibrinogen
    2. albumin
    3. globulins
    4. all plasma proteins

    a. _____ transports cholesterol
    b. _____ helps blood clot
    c. _____ transports bilirubin
    d. _____ helps maintain the pH and the osmotic pressure of blood
19. Place a check in front of the items that correctly describe hemoglobin.
    a. _____ heme contains iron
    b. _____ globin contains iron
    c. _____ becomes oxyhemoglobin in the tissues
    d. _____ loses oxygen in the tissues
    e. _____ makes red blood cells red
    f. _____ makes eosinophils red
20. The red blood cells, scientifically called a.__________________, are made in the b.__________________. Upon maturation, they are small, biconcave disks that lack a c.__________________ and contain d.__________________. After about 120 days, red blood cells are destroyed in the e.__________________ and f.__________________. An insufficient number of red blood cells or not enough hemoglobin characterizes the conditon of g.__________________.
21. What is erythropoietin?

    ______________________________________________________________

22. White blood cells, scientifically called a.__________________, are made in the b.__________________.

23. Three differences between red blood cells and white blood cells are that white blood cells are a.__________________ in size than red blood cells, have a(n) b.__________________, and do not contain c.__________________.

24. Match the descriptions with these types of white blood cells:
   1. basophil
   2. lymphocyte
   3. monocyte
   4. neutrophil

   a. _____ an agranular leukocyte with a large, round nucleus; the B cells produce antibodies and the T cells destroy cells that contain viruses
   b. _____ an abundant granular leukocyte with a multilobed nucleus that phagocytizes foreign material
   c. _____ a large agranular leukocyte that transforms into a macrophage
   d. _____ a granular leukocyte with dark blue staining granules

25. The following shows the reactions that occur as blood clots:

   platelets → prothrombin activator

   prothrombin → thrombin

   fibrinogen → fibrin threads

   a. Does the left side or the right side list substances always present in the blood? __________________
   b. Which substances are enzymes? __________________
   c. Which substance is the actual clot? __________________

26. Label this diagram of capillary exchange using the alphabetized list of terms.

| | |
|---|---|
| amino acid | osmotic pressure (two times) |
| arterial end | oxygen |
| blood pressure (two times) | tissue fluid |
| carbon dioxide | venous end |
| glucose | wastes |
| net pressure in | water (two times) |
| net pressure out | |

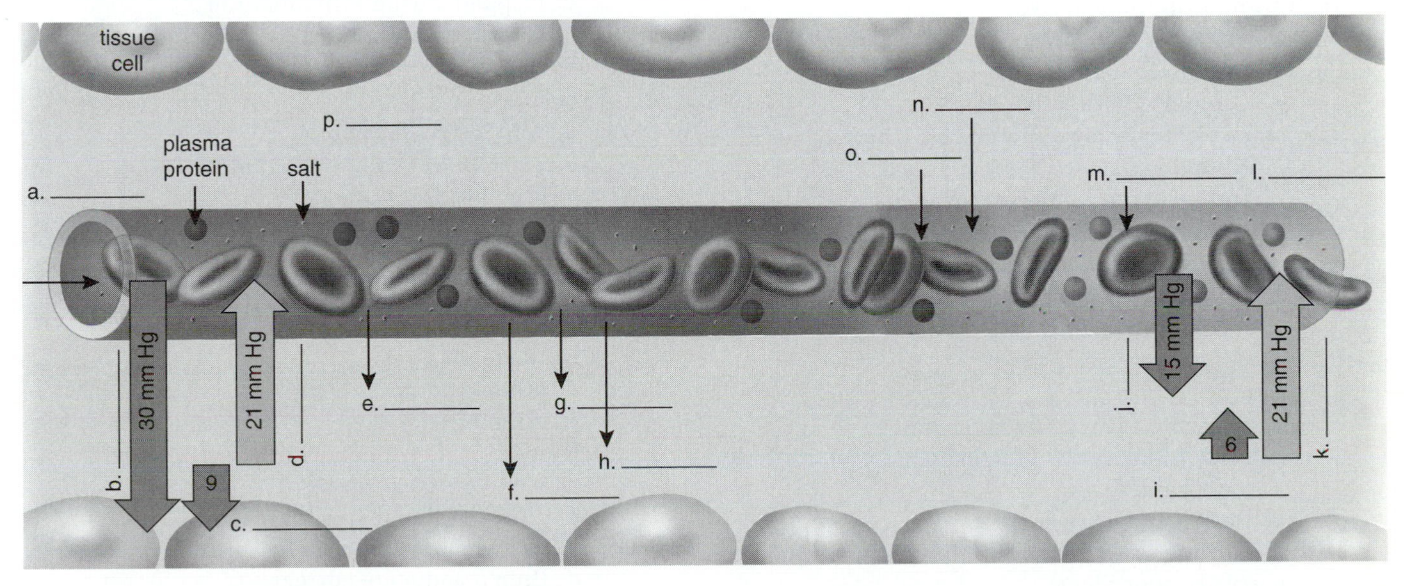

# Chapter Test

## Objective Questions

Do not refer to the text when taking this test.

____ 1. Which organism transports by a gastrovascular cavity?
 a. cnidarian
 b. earthworm
 c. insect
 d. mollusc

____ 2. Indicate the correct pathway of blood flow.
 a. arteries, capillaries, veins
 b. arteries, veins, capillaries
 c. veins, arteries, capillaries
 d. veins, capillaries, arteries

____ 3. The function of the heart valves is to
 a. prevent the backward flow of blood.
 b. pump the blood.
 c. separate the two sides of the heart.
 d. signal the chambers to contract.

____ 4. When the atria contract, the ventricles are in
 a. diastole.
 b. systole.

____ 5. Which chamber has the thickest walls?
 a. right atrium
 b. right ventricle
 c. left atrium
 d. left ventricle

____ 6. The heart sounds are due to
 a. blood flowing.
 b. the closing of the valves.
 c. the heart muscle contracting.
 d. Both *a* and *c* are correct.

____ 7. The chamber of the heart that receives blood from the pulmonary veins
 a. is the right atrium.
 b. is the left atrium.
 c. contains $O_2$-rich blood.
 d. contains $O_2$-poor blood.
 e. Both *a* and *d* are correct.
 f. Both *b* and *c* are correct.

____ 8. The SA node
 a. works only when it receives a nerve impulse.
 b. is located in the left atrium.
 c. initiates the heartbeat.
 d. All of these are correct.

____ 9. Arteries
 a. carry blood away from the heart.
 b. carry blood toward the heart.
 c. have valves.
 d. Both *b* and *c* are correct.

____10. All arteries carry $O_2$-rich blood, and all veins carry $O_2$-poor blood.
 a. true
 b. false

____11. Which of these vessels have the weakest walls?
 a. arteries
 b. veins
 c. Both are the same.

____12. The venae cavae
 a. carry blood to the right atrium.
 b. carry blood away from the right atrium.
 c. join with the aorta.
 d. have a high blood pressure.

____13. The coronary arteries carry blood
 a. from the aorta to the heart tissues.
 b. from the heart to the brain.
 c. directly to the heart from the pulmonary circuit.
 d. from the lungs directly to the left atrium.

____14. At the capillary, fluid is forced out of the vessel by the
 a. blood pressure.
 b. osmotic pressure.

____15. Gas exchange occurs in
 a. pulmonary capillaries.
 b. renal capillaries.
 c. coronary capillaries.
 d. all capillaries.

____16. For a blood pressure reading of 130/90, 130 is the
 a. diastolic pressure.
 b. systolic pressure.

____17. Select the incorrect statement about red blood cells.
 a. contain hemoglobin
 b. contain iron
 c. respond during inflammation
 d. transport oxygen

____18. Select the incorrect statement about white blood cells.
 a. activate prothrombin
 b. exist in agranular and granular forms
 c. lymphocyte is one type
 d. neutrophil is one type

____19. For the coagulation of blood, fibrinogen is converted to
 a. calcium.
 b. fibrin.
 c. prothrombin.
 d. thrombin.

____20. Blood flow is slow in capillaries because
 a. blood pressure drops off.
 b. cross-sectional area increases.
 c. valves prevent a backward flow of blood.
 d. Both *a* and *b* are correct.

## Critical Thinking Questions

Answer in complete sentences.

21. How do you think lower osmotic pressure would affect capillary exchange?

22. What do you think would happen to the heartbeat if the SA node did not stimulate the AV node?

**Test Results:** ______ number correct ÷ 22 = ______ × 100 = ______ %

## Exploring the Internet

The Online Learning Center at *www.mhhe.com/maderbiology8* has additional study material and practice quizzes that can help you master the content of this chapter. You can also find links to websites exploring additional topics in biology. Access to the Online Learning Center is free for those who have purchased a new textbook.

## Answer Key

### Study Exercises

**1. a.** 1 **b.** 3 **c.** 2 **2. a.** 2 **b.** 1 **c.** 3 **3. a.** heart **b.** artery **c.** arterioles **d.** capillaries **e.** venules **f.** vein **4. a.** 2 **b.** 1 **c.** 3 **5. a.** aorta **b.** left pulmonary artery **c.** pulmonary trunk **d.** left pulmonary veins **e.** left atrium **f.** semilunar valves **g.** atrioventricular (mitral) valve **h.** left ventricle **i.** septum **j.** inferior vena cava **k.** left ventricle **l.** chordae tendineae **m.** atrioventricular (tricuspid) valve **n.** right atrium **o.** right pulmonary veins **p.** right pulmonary arteries **q.** superior vena cava; see also Figure 34.6, page 617, in text **6. a.** vena cava, right atrium, atrioventricular valve, right ventricle, pulmonary semilunar valve, pulmonary artery, lungs **b.** lungs, pulmonary veins(s), left atrium, atrioventricular valve, left ventricle, aortic semilunar valve, aorta **7. a.** 10 **b.** 1 **c.** 7 **d.** 9 **e.** 8 **f.** 5 **g.** 6 **h.** 2 **i.** 3 **j.** 4 **8. a.** SA **b.** atria **c.** AV **d.** ventricles **e.** systole **f.** diastole **g.** atrioventricular **h.** ventricles **i.** semilunar

**9.**

| Atria | Ventricles |
|---|---|
| systole | diastole |
| diastole | systole |
| diastole | diastole |

**10.** The hepatic portal system begins and ends in the capillaries. A pulmonary circuit carries blood to the lungs, and from the lungs back to the heart. The systemic circuit takes blood throughout the body. **11.** blood pressure due to systole of left ventricle **12.** systole is followed by diastole of left ventricle **13.** distance from heart and increase in cross-sectional area of blood vessels **14.** skeletal muscle contraction **15.** a, c, d **16. a.** 1 **b.** 4 **c.** 3 **d.** 2 **e.** 6 **f.** 7 **g.** 5 **17. a.** water **b.** plasma proteins **18. a.** 3 **b.** 1 **c.** 2 **d.** 4 **19.** a, d, e **20. a.** erythrocytes **b.** red bone marrow **c.** nucleus **d.** hemoglobin **e.** liver **f.** spleen **g.** anemia **21.** a molecule that stimulates production of red blood cells **22. a.** leukocytes **b.** bone marrow **23. a.** larger **b.** nucleus **c.** hemoglobin **24. a.** 2 **b.** 4 **c.** 3 **d.** 1 **25. a.** left side **b.** prothrombin activator and thrombin **c.** fibrin threads **26. a.** arterial end **b.** blood pressure **c.** net pressure out **d.** osmotic pressure **e.** water **f.** oxygen **g.** amino acid **h.** glucose **i.** net pressure in **j.** blood pressure **k.** osmotic pressure **l.** venous end **m.** water **n.** wastes **o.** carbon dioxide **p.** tissue fluid (See also Figure 34.13, page 627, in text.)

### Chapter Test

**1.** a **2.** a **3.** a **4.** a **5.** d **6.** b **7.** f **8.** c **9.** a **10.** b **11.** b **12.** a **13.** a **14.** a **15.** d **16.** b **17.** c **18.** a **19.** b **20.** d **21.** Tissue fluid would not be returned as efficiently to the venous end of the capillary. It would remain in the tissue spaces, and edema would result. **22.** The ventricles would not be signaled to contract. Therefore, ventricular systole would not take place, and $O_2$-rich blood would not effectively flow to the body's tissues.

# 35

# Lymph Transport and Immunity

## Chapter Review

The **lymphatic system** consists of lymphatic vessels and lymphoid organs. The lymphatic vessels absorb fat molecules at intestinal villi and excess tissue fluid at the cardiovascular capillaries. Eventually, two main ducts empty into the subclavian veins.

Lymphocytes are produced and accumulate in the lymphoid organs (**lymph nodes, spleen, thymus gland, tonsils,** and **red bone marrow**).

**Immunity** involves nonspecific and specific defenses. Nonspecific defenses include barriers to entry, the **inflammatory reaction,** natural killer cells, and protective proteins.

Specific defenses require lymphocytes, produced in the bone marrow. **B lymphocytes** mature in the bone marrow and undergo **clonal selection** in the lymph nodes and the spleen. **T lymphocytes** mature in the thymus.

B cells are responsible for **antibody-mediated immunity.** An **antibody** is a Y-shaped molecule that has two binding sites. Each antibody is specific for a particular **antigen.** Activated B cells become antibody-secreting **plasma cells** and **memory B cells.** Memory B cells respond if the same antigen enters the body at a later date.

The two main types of T cells are cytotoxic T cells and helper T cells. **Cytotoxic T cells** kill cells on contact; **helper T cells** stimulate other immune cells and produce lymphokines.

Immunity can be induced in various ways. **Vaccines** are available to promote long-lived, **active immunity,** and antibodies sometimes are available to provide an individual with short-lived, **passive immunity.**

**Cytokines,** notably **interferon** and **interleukins,** are used in an attempt to promote the body's ability to recover from cancer and to treat AIDS.

**Allergies** result when an overactive immune system forms antibodies to substances not normally recognized as foreign. Cytotoxic T cells attack transplanted organs as nonself; therefore, immunosuppressive drugs must be administered.

Blood transfusions require compatible blood types. The antigens (A and B) are on the red blood cells and the antibodies (anti-A and anti-B) are in the plasma. The Rh antigen is also particularly important because an Rh-negative mother may produce anti-Rh antibodies that will attack the red blood cells of an Rh-positive fetus.

**Autoimmune diseases** occur when antibodies and T cells attack the body's own tissues.

## Study Exercises

Study the text section by section as you answer the questions that follow:

### 35.1 The Lymphatic System (pp. 632–34)

- The lymphatic vessels form a one-way system, which transports lymph from the tissues and fat from the lacteals to certain cardiovascular veins.
- The lymphoid organs (lymph nodes, tonsils, spleen, thymus gland, and red bone marrow) play critical roles in defense mechanisms.

1. Give three functions of the lymphatic system.
   a. ______________________________
   b. ______________________________
   c. ______________________________

2. Indicate whether these statements about the structure/function of lymphatic vessels and lymphoid organs are true (T) or false (F).
   a. _____ Bone marrow lacks lymphoid tissue.
   b. _____ Lymph lobules are subdivided into sinus-containing nodes.
   c. _____ The contraction of skeletal muscles blocks the return of lymph to the bloodstream.
   d. _____ The sinuses of the spleen are filled with lymph.
   e. _____ Lymphatic vessels are similar to cardiovascular veins.
   f. _____ Lymphatic vessels contain valves.

3. Indicate whether the following statements are true (T) or false (F). Rewrite any false statements to make them true.
   a. _____ Vessels of the lymphatic system begin with cardiovascular capillaries. Rewrite: _______________
   _______________________________________________
   b. _____ Lymph most closely resembles arterial blood. Rewrite: _______________________________
   c. _____ The right thoracic duct serves the lower extremities, abdomen, one arm, and one side of the head and neck. Rewrite: _______________________________________
   d. _____ Lymphatic capillaries merge directly to form a particular lymphatic duct. Rewrite: _____________
   _______________________________________________

4. Place the appropriate letter(s) next to each statement.

   TG— thymus gland  S—spleen  RBM—red bone marrow

   a. _____ contains red pulp and white pulp
   b. _____ contains stem cells
   c. _____ is located along the trachea
   d. _____ is located in the upper left abdominal cavity
   e. _____ produces hormones believed to stimulate the immune system
   f. _____ cleanses blood
   g. _____ site of origin for all types of blood cells

## 35.2 The Immune System (pp. 634–44)

- Immunity consists of nonspecific and specific defenses to protect the body against disease.
- Nonspecific diseases consist of barriers to entry, the inflammatory reaction, and protective proteins.
- Specific defenses require two types of lymphocytes: B lymphocytes and T lymphocytes.

5. Match the descriptions to these defense mechanisms:

   1. barrier to entry  2. inflammatory reaction  3. complement system  4. natural killer cells

   a. _____ accompanied by swelling and redness
   b. _____ cilia action in the respiratory tract
   c. _____ produces holes in bacterial cell walls
   d. _____ stomach secretions
   e. _____ histamine increases capillary permeability
   f. _____ injured cells release kinins
   g. _____ vagina is inhabited by nonpathogenic bacteria
   h. _____ stimulates inflammatory reaction, binds to some bacteria and punches holes in others
   i. _____ neutrophils and macrophages carry out phagocytosis
   j. _____ secretions of the oil, or sebaceous, glands
   k. _____ kills cells infected with a virus and tumor cells

6. Fill in this table with *yes* or *no*.

| | Cytotoxic T Cell | B Cell |
|---|---|---|
| Ultimately derived from stem cells in bone marrow | a. | b. |
| Pass through thymus | c. | d. |
| Carry antigen receptors on membrane | e. | f. |
| Cell-mediated immunity | g. | h. |
| Antibody-mediated immunity | i. | j. |

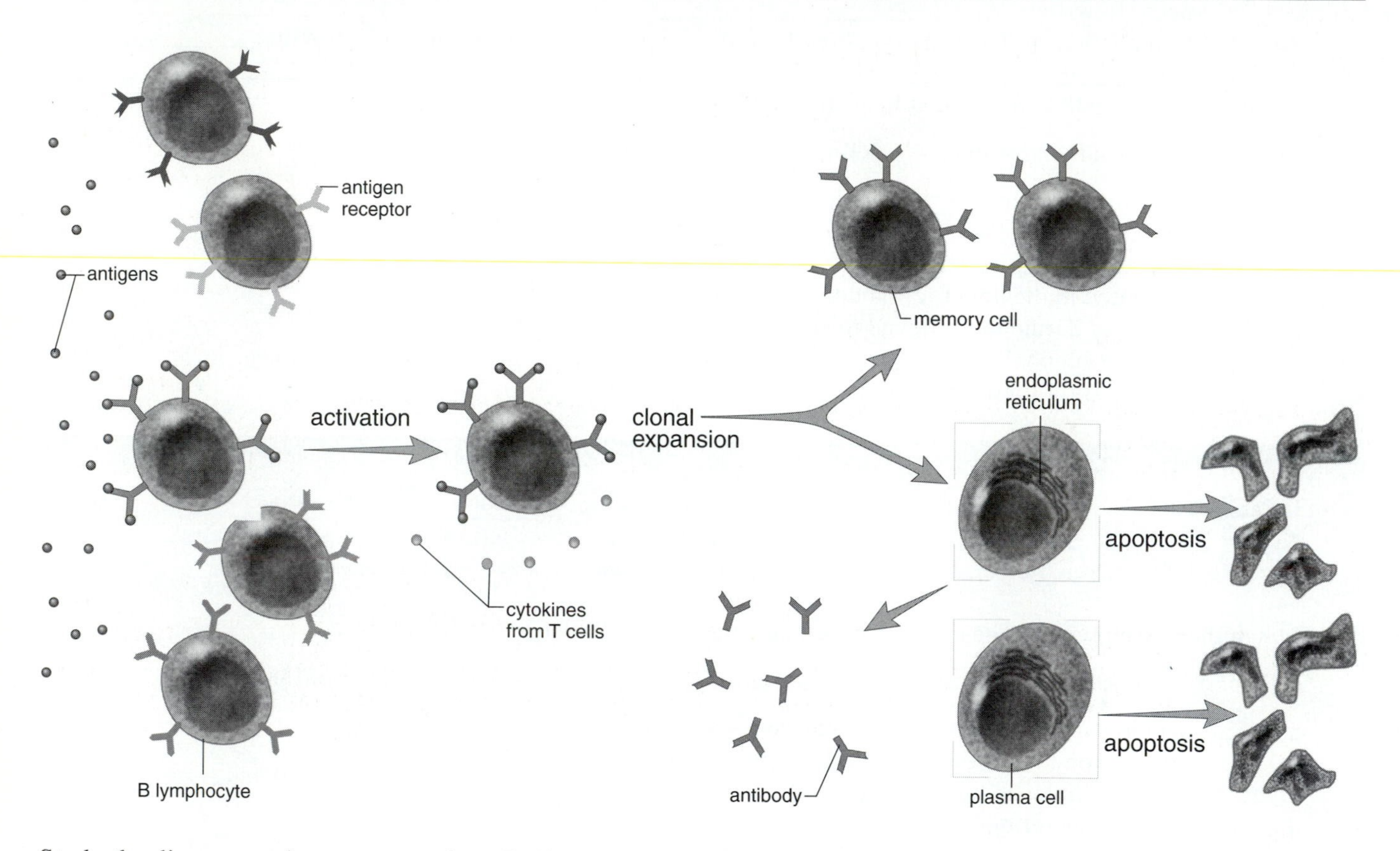


Study the diagram and answer questions 7–10.

7. Which of these makes a B cell undergo clonal expansion?
   a. when an antigen binds to its antigen receptor
   b. when fever is present

8. Explain the expression *clonal selection theory.* ______________________________

______________________________________________________________

9. Which of these are cells that result from the clonal expansion of B cells?
   a. plasma cells
   b. memory cells
   c. both types of cells

10. What happens to these cells once the infection is under control?
    a. memory cells ______________________________
    b. plasma cells ______________________________

11. Label this diagram of an antibody molecule using these terms:
    antigen-binding site
    constant region
    heavy chain
    light chain
    variable region

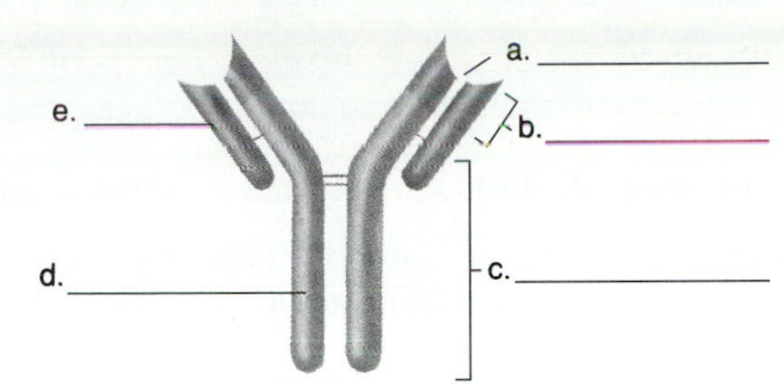


f. What is the function of antibodies? ____________________

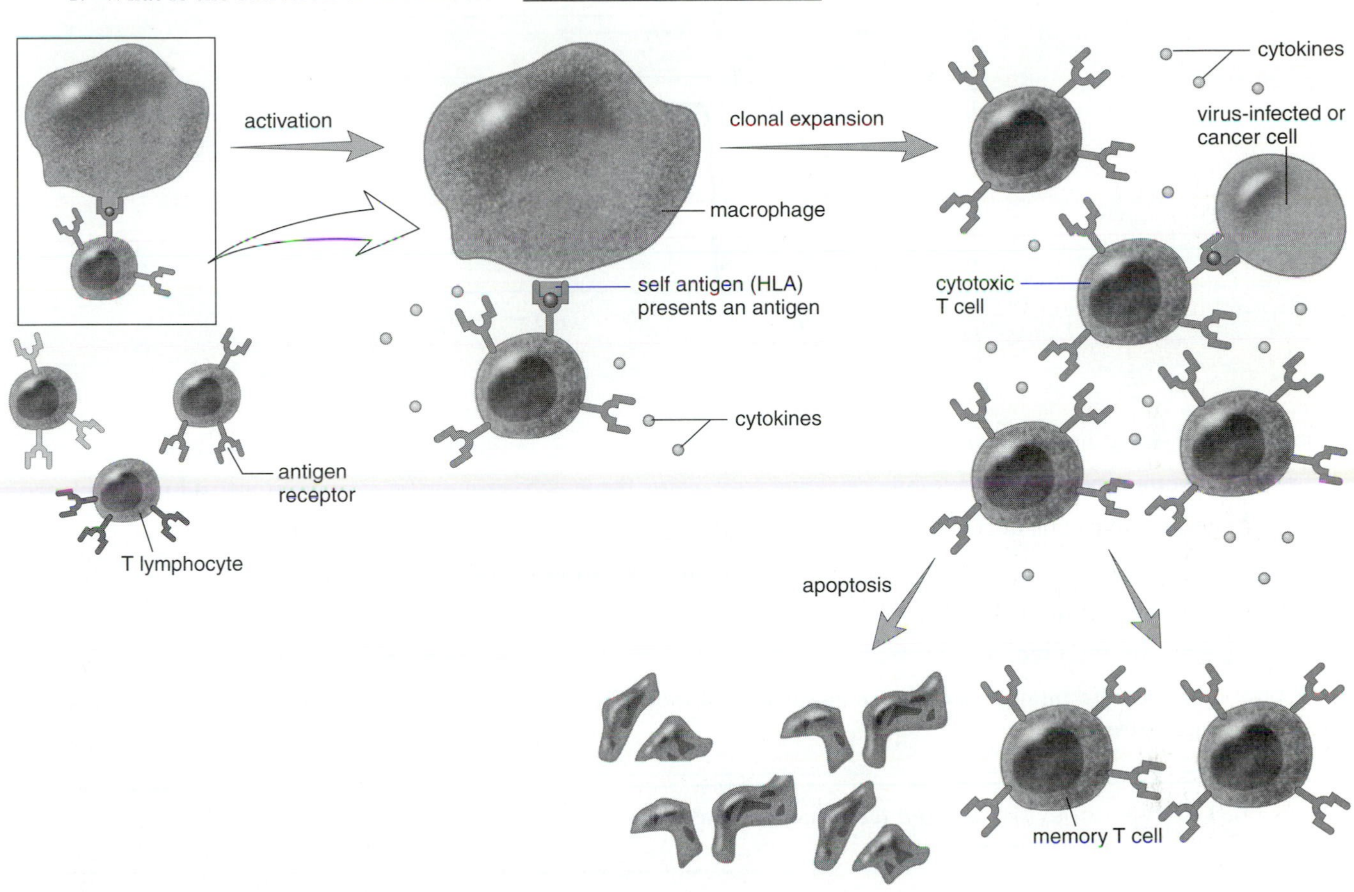


Study the diagram and answer questions 12–14.

12. Which of these makes a particular T cell undergo clonal expansion?
    a. when the antigen is presented to the T cells by an APC
    b. when a T cell encounters an antigen

13. a. What is the significance of HLA antigens? ____________________

    ____________________

    b. Why are they called antigens? ____________________

    ____________________

14. a. What happens after a helper T cell recognizes an antigen? ____________________

    ____________________

    b. What happens after a cytotoxic T cell recognizes an antigen? ____________________

    ____________________

    c. What happens to T cells (except for memory T cells) after the infection is past? ____________________

    ____________________

## 35.3 Induced Immunity (pp. 644–46)

- Induced immunity for medical purposes involves the use of vaccines to achieve long-lasting immunity and the use of antibiotics to provide temporary immunity.

15. Label this diagram, using the following alphabetized list of terms.

first exposure to vaccine plasma antibody concentration primary response
secondary response second exposure to vaccine

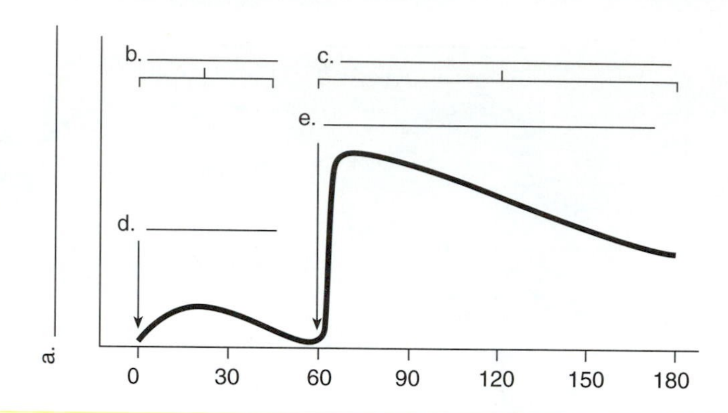


In question 15*f–i,* fill in the blanks.

The secondary exposure f. ______________________ the g. ____________________ titer to a high level.

Active immunity is dependent on the number of h. ____________________ cells and

i. ____________________ cells capable of responding to lower doses of an antigen.

In questions 16–19, fill in the blanks.

16. When an individual receives antibodies from another, as when a baby breast-feeds, it is called ____________.

17. Why is it better to immunize children against childhood diseases in advance, rather than wait until the disease has been contracted?

_______________________________________________________________

18. Name two types of cytokines, and tell what role they play in immunotherapy.

a. _______________________________________________________________

b. _______________________________________________________________

19. What are monoclonal antibodies? ______________________________________

_______________________________________________________________

## 35.4 Immunity Side Effects (pp. 646–49)

- While immunity preserves our existence, it also is responsible for certain undesirable effects, such as tissue rejection, allergies, and autoimmune disease.

20. Relate the immune response to each of these:

    a. allergy ______________________________

    b. tissue rejection ______________________________

    c. autoimmune disease ______________________________

21. The following table indicates the blood types. Fill in the fourth and fifth columns by using this formula: The donor's antigen(s) must not be of the same type as the recipient's antibody (antibodies).

| Blood Type | Antigen | Antibody | Can Receive From | Can Donate To |
|---|---|---|---|---|
| A | A | Anti-B | a. | b. |
| B | B | Anti-A | c. | d. |
| AB | A, B | ______ | e. | f. |
| O | None | Anti-A and B | g. | h. |

22. This diagram shows the results of typing someone's blood. What is the blood type? ______________

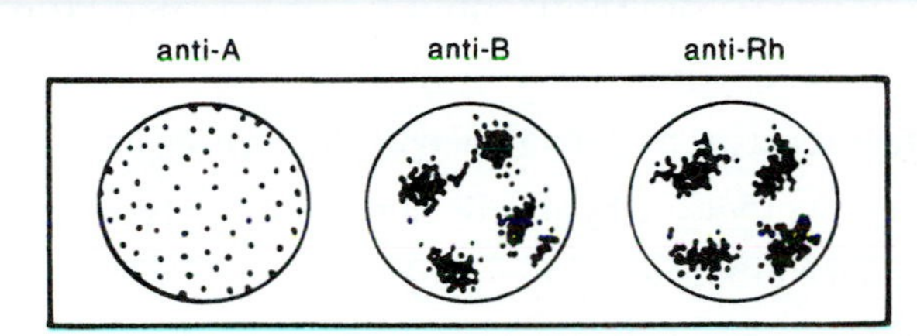


23. Draw a similar diagram showing the results if someone has AB-negative blood.

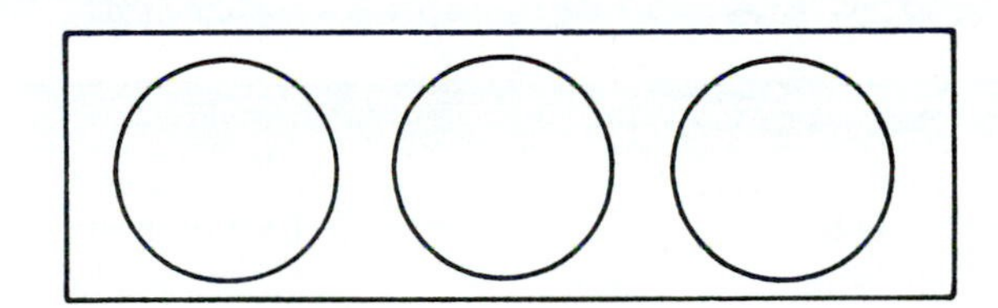

24. Consider these possible combinations of mates:

    Rh$^+$ mother and Rh$^-$ father Rh$^-$ mother and Rh$^-$ father
    Rh$^+$ mother and Rh$^+$ father Rh$^-$ mother and Rh$^+$ father

    Which of these combinations can cause pregnancy difficulties and why?

    ______________________________

    ______________________________

## Definitions Wordsearch

Review key terms by using the following alphabetized list of terms to fill in the blanks. Then complete the wordsearch.

```
L E N I M A T S I H E R T M
I G S D C O M P L E M E N T
E A N T I B O D Y D L P A O
N H U I P R K J N E G H U D
I P H L L A R N O M J U T V
K O C E A D H E R A L O O B
O R M C H Y K D E H E L I Y
H C E E H K U F F P M E M M
P A P P L I O R R S A V M B
M M M U D N G H E E M Y U D
Y Y D O G I E T T S H I N C
L E A S E N T Y N P P L E C
M U D S L S D E I Y H N T D
```

antibody
autoimmune
complement
edema
histamine
interferon
kinins
lymph
lymphokine
macrophage

a. ______________ Swelling due to the accumulation of tissue fluid.
b. ______________ A system of plasma proteins that are a nonspecific defense.
c. ______________ A disease caused by the immune system attacking the person's own body.
d. ______________ Substances found in damaged tissues that initiate nerve impulses, triggering pain.
e. ______________ A molecule released by T cells that enhances the abilities of other immune system cells.
f. ______________ Large phagocytic cell.
g. ______________ Protein released by cells infected with viruses.
h. ______________ A protein produced by B cells in response to foreign antigens.
i. ______________ Substance produced by basophils or mast cells in an allergic reaction.
j. ______________ Tissue fluid inside lymphatic vessels.

## Chapter Test

### Objective Questions

Do not refer to the text when taking this test.

____ 1. The two collecting ducts of the lymphatic system empty into
  a. cardiovascular arteries.
  b. cardiovascular veins.
  c. pulmonary arteries.
  d. pulmonary veins.

____ 2. The structure of a lymphatic vessel is most similar to that of a
  a. cardiovascular artery.
  b. cardiovascular arteriole.
  c. cardiovascular vein.
  d. skeletal muscle fiber.

____ 3. Lymph nodes house
  a. neutrophils and monocytes.
  b. lymphocytes and macrophages.
  c. granular leukocytes.
  d. red blood cells.

____ 4. Lymph is ______________ in lymphatic vessels.
  a. blood
  b. serum
  c. tissue fluid
  d. plasma

____ 5. All of the following are functions of macrophages EXCEPT
a. liberate a colony-stimulating factor to increase leukocytes.
b. phagocytize bacteria.
c. scavenge dead and decaying tissue.
d. transport oxygen in the blood.

In questions 6–9, match the descriptions to these terms:
a. thymus gland b. spleen
c. lymph node d. red bone marrow

____ 6. Causes maturation of T cells
____ 7. Purifies lymph
____ 8. Purifies blood
____ 9. Formation of agranular and granular leukocytes
____ 10. The spleen
a. contains stem cells from the bone marrow.
b. is located along the trachea.
c. produces a hormone believed to stimulate the immune system.
d. contains red pulp and white pulp.
____ 11. The thymus
a. contains all types of stem cells from the bone marrow.
b. is located along the trachea.
c. produces a hormone believed to stimulate the immune system.
d. Both *b* and *c* are correct.
____ 12. Activity of the complement system is an example of nonspecific defense by
a. barriers to entry.
b. phagocytic cells.
c. protective proteins.
d. Both *a* and *c* are correct.
____ 13. Secretions of the oil glands are an example of nonspecific defense by a(n)
a. barrier to entry.
b. protective protein.
c. phagocytic cell.
d. acidic pH.
____ 14. Interferon is produced by cells in response to the presence of
a. chemical irritants.
b. viruses.
c. bacterial infection.
d. malarial parasite in blood.
____ 15. The most active white blood cell phagocytes are
a. neutrophils and macrophages.
b. neutrophils and eosinophils.
c. lymphocytes and macrophages.
d. lymphocytes and neutrophils.
____ 16. The white blood cells primarily responsible for specific immunity are
a. neutrophils.
b. eosinophils.
c. macrophages.
d. lymphocytes.
____ 17. Which of these is NOT a valid contrast between T cells and B cells?
**T cells** **B cells**
a. mature in the thymus/mature in bone marrow
b. antibody-mediated immunity/cell-mediated immunity
c. antigen must be presented by APC/direct recognition
d. cytokines/do not produce cytokines
____ 18. A particular antibody can
a. attack any type of antigen.
b. attack only a specific type of antigen.
c. be produced by any B lymphocyte.
d. be produced by any T lymphocyte.
____ 19. The clonal selection theory refers to the
a. presence of four different types of T lymphocytes in the blood.
b. response of only one type of B lymphocyte to a specific antigen.
c. occurrence of many types of plasma cells, each producing many types of antigens.
____ 20. The portions of an antibody molecule that pair up with the foreign antigens are the
a. heavy chains.
b. light chains.
c. variable regions.
d. constant regions.
____ 21. Which of these pairs is incorrect?
a. helper T cells—orchestrate the immune response
b. cytotoxic T cells—stimulate B cells to produce antibodies
c. memory T cells—long-lasting active immunity
d. suppressor T cells—shut down the immune response
____ 22. A person receiving an injection of gamma globulin as a protection against hepatitis is an example of
a. naturally acquired active immunity.
b. naturally acquired passive immunity.
c. artificially acquired passive immunity.
d. artificially acquired active immunity.
____ 23. A person vaccinated to produce immunity to the flu is an example of
a. naturally acquired active immunity.
b. naturally acquired passive immunity.
c. artificially acquired passive immunity.
d. artificially acquired active immunity.

____ 24. Allergies are caused by
a. strong toxins in the environment.
b. autoimmune diseases.
c. the overproduction of IgE.
d. the receipt of IgA in breast milk.

____ 25. Which is NOT true of an autoimmune response?
a. responsible for such diseases as multiple sclerosis and myasthenia gravis and perhaps type 1 diabetes
b. occurs when self-antibodies attack self-tissues
c. interferes with the transplantation of organs between one person and another
d. All of these are true.

## Critical Thinking Questions

Answer in complete sentences.

26. How do we know that one aspect of specific immunity is the ability of the body to recognize self as opposed to nonself?

27. Describe how the lymphatic system aids the activities of the integumentary system, and vice versa.

**Test Results:** ______ number correct ÷ 27 = ________ × 100 = ______%

## Exploring the Internet

The Online Learning Center at *www.mhhe.com/maderbiology8* has additional study material and practice quizzes that can help you master the content of this chapter. You can also find links to websites exploring additional topics in biology. Access to the Online Learning Center is free for those who have purchased a new textbook.

# Answer Key

## Study Exercises

**1. a.** return of excess tissue fluid to bloodstream **b.** receive lipoproteins at intestinal villi **c.** defense against disease **2. a.** F **b.** F **c.** F **d.** F **e.** T **f.** T **3. a.** F, . . . begin with lymph capillaries **b.** F, . . . resembles tissue fluid that has entered the lymph vessels **c.** F, The right lymphatic duct serves the right arm, the right side of the head and neck, and the right thoracic area. **d.** F, . . . capillaries form lymphatic vessels first, and these merge before entering a particular lymphatic duct **4. a.** S **b.** RBM **c.** TG **d.** S **e.** TG **f.** S **g.** RBM **5. a.** 2 **b.** 1 **c.** 3 **d.** 1 **e.** 2 **f.** 2 **g.** 1 **h.** 3 **i.** 2 **j.** 1 **k.** 4 **6.**

| Cytotoxic T Cell | | B Cell | |
|---|---|---|---|
| a. | yes | b. | yes |
| c. | yes | d. | no |
| e. | yes | f. | yes |
| g. | yes | h. | no |
| i. | no | j. | yes |

**7.** a **8.** antigen selects the B cell that will clone **9.** c **10. a.** remain in body ready to produce more antibodies when needed **b.** undergo apoptosis and die off **11. a.** antigen-binding site **b.** variable region **c.** constant region **d.** heavy chain **e.** light chain **f.** combine with antigens and mark them for destruction **12.** a **13. a.** identify cell as belonging to an individual **b.** They are antigenic in someone else's body. **14. a.** secretes cytokines that stimulate other immune cells **b.** destroys cells infected with a virus **c.** undergo apoptosis and die **15. a.–e.** See Figure 35.9*b*, p. 644, in text. **f.** boosts **g.** antibody **h.** memory B **i.** memory T **16.** passive immunity **17.** It is better to prevent a disease than to try to treat it with antibiotics. Resistant strains of bacteria and allergies to antibiotics are two side effects, and viruses do not respond to antibiotic therapy. **18. a.** Interleukins activate and maintain killer activity of T cells. **b.** Interferon causes other cells to resist a viral infection. **19.** Same type of antibodies produced by the same lymphocyte. **20. a.** Antigen attaches to IgE antibodies on mast cells, and histamine release causes allergic response. **b.** Antibodies and cytotoxic T cells attack foreign antigens. **c.** Viral infection tricks immune cells into attacking tissues of self. **21. a.** A, O **b.** A, AB **c.** B, O **d.** B, AB **e.** A, B, AB, O **f.** AB **g.** O **h.** A, B, AB, O **22.** $B^+$ **23.** Clumping should occur for anti-A and anti-B. **24.** $Rh^-$ mother and $Rh^+$ father because the mother might form antibodies to destroy red blood cells of this or a future baby who is $Rh^+$.

## Definitions Wordsearch

```
  E N I M A T S I H
  G     C O M P L E M E N T
E A N T I B O D Y D     A
N H             N E     U
I P             O M     T
K O             R A     O
O R             E       I
H C     H K     F       M
P A   P   I     R       M
M M M     N     E       U
Y Y       I     T       N
L         N     N       E
          S     I
```

**a.** edema **b.** complement **c.** autoimmune **d.** kinins **e.** lymphokine **f.** macrophage **g.** interferon **h.** antibody **i.** histamine **j.** lymph

## Chapter Test

**1.** b **2.** c **3.** b **4.** c **5.** d **6.** a **7.** c **8.** b **9.** d **10.** d **11.** d **12.** c **13.** a **14.** b **15.** a **16.** d **17.** b **18.** b **19.** b **20.** c **21.** b **22.** c **23.** d **24.** c **25.** c **26.** Ordinarily, antibodies and T cells attack only foreign antigens. If and when they attack the body's own cells, illness results. **27.** The lymphatic system drains excess tissue fluid from the skin and protects against infections. The skin serves as a barrier to entry of pathogens.

# 36

# Digestion and Nutrition

## Chapter Review

As feeders, animals can be continuous or discontinuous. Throughout the animal kingdom, the digestive tract is either incomplete (e.g., planarians) or complete (e.g., earthworms). The teeth are an important digestive structure in all mammals. The several different shapes of teeth are each specialized for a different function. Specializations develop throughout the digestive system, such as the varying length of the intestine, shorter in carnivores and longer in herbivores.

In humans, food passes through a series of chambers: mouth, pharynx, **esophagus,** stomach, **small intestine,** and **large intestine.** In the mouth, chewing breaks down food, and **salivary amylase** initiates the chemical digestion of carbohydrates. The chemical digestion of proteins begins in the stomach through the action of **pepsin.** Food is also stored there. The chemical and physical breakdown of food produces chyme. The action of other enzymes and **bile** in the small intestine finishes the chemical breakdown of all food molecules. The products of chemical digestion (fatty acids, amino acids, monosaccharides) are absorbed in this region. Water and minerals are reabsorbed in the large intestine. Wastes are eliminated from the body after passing through the large intestine, or **colon.**

Several structures near the small intestine produce substances secreted into the digestive tract and influence its activity: the **liver** (bile production), the **gallbladder** (bile storage), and the **pancreas** (enzyme production). The pancreas also produces the hormones insulin and glucagon to regulate carbohydrate metabolism.

Several hormones—gastrin, secretin, and CCK—regulate digestive processes.

The nutrients released by the digestive process should provide us with an adequate amount of energy and all the necessary **vitamins** and **minerals.**

The bulk of the diet should be carbohydrates (like bread, pasta, and rice) and fruits and vegetables. The vitamins A, E, and C, found in fruits and vegetables, are antioxidants that protect cell contents from damage due to free radicals.

## Study Exercises

Study the text section by section as you answer the questions that follow.

### 36.1 Digestive Tracts (pp. 654–56)

- An incomplete digestive tract with only one opening has little specialization of parts; a complete digestive tract with two openings does have specialization of parts.
- Discontinuous feeders, rather than continuous feeders, need a storage area for food.
- The dentition of herbivores, carnivores, and omnivores is adapted to the type of food they eat.

1. Label each of the following as describing a planarian (P) or an earthworm (E):
   a. _____ incomplete tract
   b. _____ pharynx, intestine
   c. _____ complete tract
   d. _____ pharynx, crop, gizzard, intestine
2. A complete gut has both a(n) a.__________________ and a(n) b.__________________.
3. Label each of the following as describing a clam (C) or a squid (S):
   a. _____ siphons
   b. _____ no jaws
   c. _____ stomach only
   d. _____ tentacles
   e. _____ jaws
   f. _____ stomach and cecum
4. a. Which type of feeder tends to take in small particles or, in the case of parasites, food that needs no digestion? __________________
   b. Which type of feeder is more likely to require a storage area for food? __________________
5. Label each of the following as describing an herbivore (H) or a carnivore (C):
   a. _____ large, flat molars
   b. _____ chew thoroughly
   c. _____ sharp canines
   d. _____ bolt food
6. a. On what type of food do herbivores feed? __________________
   b. Carnivores? __________________
   c. Which type of food is harder to digest? __________________
   d. Which type of feeder needs to chew food thoroughly and requires a more complicated digestive tract with accessory parts? __________________

## 36.2 Human Digestive Tract (pp. 657–63)

- The human digestive tract has many specialized parts and several accessory organs of digestion, which contribute in their own way to the digestion of food.
- The digestive enzymes are specific and have an optimum temperature and pH at which they function best.
- The products of digestion are small molecules, such as amino acids and glucose, that can cross plasma membranes.

In questions 7–13, fill in the blanks using this diagram.

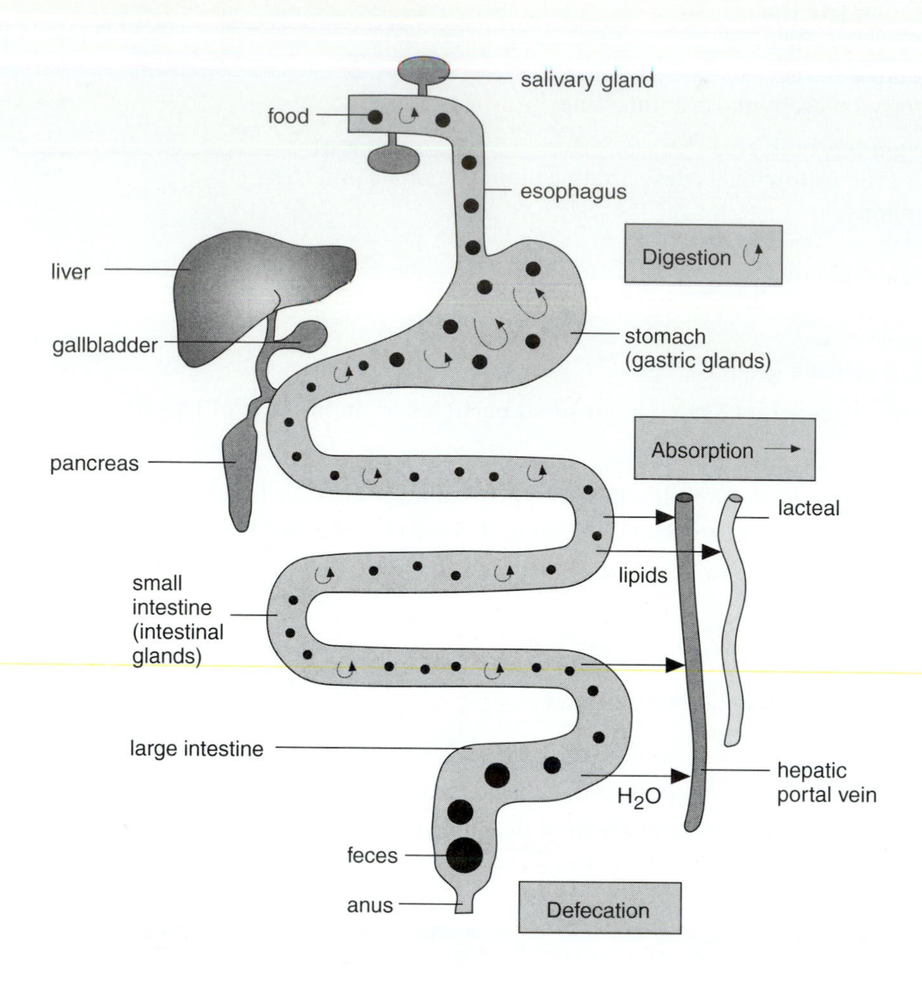


7. Food is received by the [a]_______________, which possesses [b]_______________ to grind and break up particles. [c]_______________ glands secrete [d]_______________ that moistens food and binds it together for swallowing. The taste of food is detected by [e]_______________ located on the [f]_______________.
8. At the back of the mouth, the muscular [a]_______________ is involved in swallowing food. The epiglottis covers the [b]_______________ during the swallowing.
9. The esophagus propels food toward the [a]_______________, using muscular contractions called [b]_______________.
10. The stomach secretes a protective [a]_______________. Otherwise, the lining would be injured by the strong [b]_______________ acid secreted by [c]_______________ glands. Gastric juice also contains [d]_______________ that breaks down proteins.
11. Food mixing with gastric juice in the stomach becomes a._______________, which enters the [b]_______________, the first part of the small intestine.
12. The surface of the small intestine is folded into [a]_______________, which, in turn, have minute projections from individual cells called [b]_______________.
13. [a]_______________ and [b]_______________ enter the blood capillaries of a villus. Glycerol and [c]_______________ enter the epithelial cells of the villi. They join and are packaged as lipoproteins, which enter a [d]_______________.

14. State the purpose of microvilli. ______________________________

______________________________

In questions 15–19, fill in the blanks.

15. A number of hormones control the secretions of digestive juices. a. ______________ is secreted in response to protein in foods and enhances gastric gland output, while b. ______________ inhibits gastric secretion. c. ______________ is secreted in response to acidic chyme. When fats are present in chyme, d. ______________ triggers the release of bile from the gallbladder.

16. Pancreatic juice contains a mix of a. ______________ solution to neutralize stomach acid and digestive b. ______________ to further break down food.

17. The liver produces a greenish substance called a. ______________, stored in the b. ______________. The liver detoxifies substances entering the c. ______________. Other functions of the liver include (list three):
d. ______________________________
e. ______________________________
f. ______________________________

18. a. ______________ is an inflammatory disease of the liver caused by a viral infection, while b. ______________ is damage caused by chronic alcohol abuse.

19. The large intestine stores a. ______________, eliminated at the anus. Unusual outgrowths of the lining of the colon, called b. ______________, can be either benign or cancerous. Colon cancer incidence may increase for people who do not have enough c. ______________ in their diets.

20. Complete this table.

**Major Digestive Enzymes**

| Enzyme | Produced by | Site of Action | Optimum pH | Digestion |
|---|---|---|---|---|
| a. ______________ | salivary glands | b. ______________ | neutral | starch + $H_2O \longrightarrow$ maltose |
| pancreatic amylase | c. ______________ | small intestine | d. ______________ | |
| e. ______________ | small intestine | f. ______________ | basic | maltose + $H_2O \longrightarrow$ glucose + glucose |
| pepsin | g. ______________ | stomach | h. ______________ | protein + $H_2O \longrightarrow$ peptides |
| i. ______________ | pancreas | j. ______________ | basic | |
| peptidases | k. ______________ | small intestine | l. ______________ | peptide + $H_2O \longrightarrow$ amino aacids |
| m. ______________ | pancreas | n. ______________ | basic | RNA and DNA + $H_2O \longrightarrow$ nucleotides |
| nucleosidases | o. ______________ | small intestine | p. ______________ | nucleotide + $H_2O \longrightarrow$ base + sugar + phosphate |
| q. ______________ | pancreas | r. ______________ | basic | fat droplet + $H_2O \longrightarrow$ glycerol + fatty acids |

21. Match the enzyme to the food or breakdown product.

salivary amylase/pancreatic amylase pepsin/trypsin lipase nuclease maltase peptidase nucleosidases

a. ______________ protein
b. ______________ maltose
c. ______________ RNA and DNA
d. ______________ starch
e. ______________ fat
f. ______________ nucleotides
g. ______________ peptides

## 36.3 Nutrition (pp. 664–66)

- Describe how the different classes of nutrients enter into general circulation within the body.
- Discuss proper nutrition and how carbohydrates, protein, and fat should be proportioned in the diet.
- Discuss the vitamin and mineral requirements in the diet.

22. Fill in the blanks of the food guide pyramid. Include the following food groups, and indicate the recommended number of daily servings from each group:

    bread, rice, pasta
    dairy
    fruit
    meat, poultry, fish, and beans
    sweets, fats, and oils
    vegetables

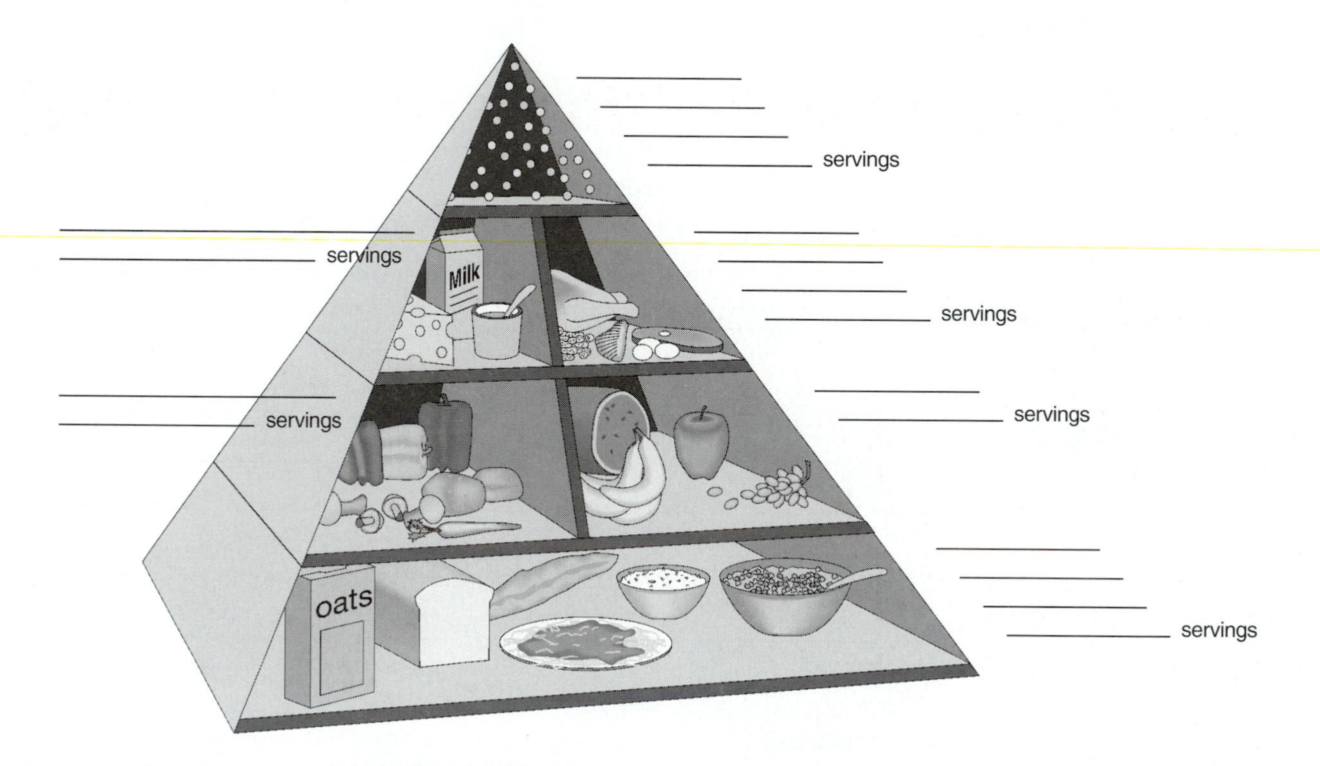


23. Place a check beside the most nutritious food.

    a. _____ fat

    b. _____ beef

    c. _____ vegetables

24. Place a check beside the type of food that should be the bulk of any diet.

    a. _____ vegetables

    b. _____ fruits

    c. _____ dairy

    d. _____ complex carbohydrates

25. Indicate whether these statements are true (T) or false (F).

a. _____ Complex carbohydrates provide energy and fiber.

b. _____ Meat provides proteins, but it also contains saturated fats.

c. _____ The body can make all the essential amino acids it needs.

d. _____ Saturated fats, whether in butter or margarine, can lead to more LDL, and this can cause plaque buildup in the arteries.

In questions 26–27, fill in the blanks.

26. Many vitamins function as a. _______________ in various metabolic pathways. Vitamin b. _______________ is important as a visual pigment, and vitamin c. _______________ becomes a compound that enhances calcium absorption. Vitamins d. _______________ are antioxidants.

27. Too much of the mineral a. _______________ can lead to hypertension. The mineral b. _______________ is a major component of bones and teeth.

## KeyWord CrossWord

Review key terms by completing this crossword puzzle using the following alphabetized list of terms.

amylase
anus
esophagus
fiber
gallbladder
gastric
lacteal
lipase
liver
pharynx

*Across*

1 Fat-digesting enzyme secreted by the pancreas.
3 Indigestible plant material that may lower risk of colon cancer.
4 Muscular passageway at the back of the mouth where swallowing occurs.
5 Starch-digesting enzyme.
8 Final canal of the large intestine.
9 Type of gland found in the stomach.

*Down*

1 Large, multifunctioned organ in the abdominal cavity; secretes bile.
2 Muscular tube leading from the pharynx to the stomach.
6 Lymphatic vessel inside a villus of the small intestine.
7 Muscular sac that stores bile.

# Chapter Test

## Objective Questions

Do not refer to the text when taking this test.

____ 1. Select the correct terms to describe digestion in planarians.
 a. complete, saclike
 b. complete, tubelike
 c. incomplete, saclike
 d. incomplete, tubelike

____ 2. Select the incorrect association.
 a. canine—tearing
 b. incisor—biting
 c. molar—crushing
 d. premolar—swallowing

____ 3. Salivary amylase speeds up the conversion of
 a. protein to amino acids.
 b. protein to peptides.
 c. starch to glucose.
 d. starch to maltose.

____ 4. Select the correct sequence for the passage of food.
 a. pharynx, esophagus, stomach
 b. large intestine, small intestine, stomach
 c. pharynx, stomach, esophagus
 d. stomach, large intestine, small intestine

For questions 5–7, select from the following options:
 a. mouth
 b. esophagus
 c. stomach
 d. small intestine
 e. large intestine

____ 5. Protein digestion begins here.

____ 6. Nutrients are absorbed here.

____ 7. Starch is digested here.

____ 8. Villi serve to
 a. increase surface area for absorption.
 b. increase the synthesis of enzymes.
 c. speed up elimination of wastes.
 d. speed up loss of water from the body.

____ 9. Select the function NOT performed by the pancreas.
 a. detoxifies poisons
 b. makes enzymes
 c. makes insulin
 d. secretes sodium bicarbonate

____ 10. Select the function NOT performed by the liver.
 a. makes blood proteins
 b. makes new red blood cells
 c. produces bile
 d. stores glucose as glycogen

____ 11. As a result of the digestive process, ______ are absorbed into the body.
 a. proteins
 b. fats
 c. starches
 d. proteins and fats
 e. amino acids, glucose, and fatty acids

____ 12. The two enzymes involved in the digestion of proteins are
 a. salivary amylase and lipase.
 b. trypsin and hydrochloric acid.
 c. pancreatic amylase and bile.
 d. pepsin and trypsin.

____ 13. Bile
 a. is an important enzyme for the digestion of fats.
 b. is made by the gallbladder.
 c. contains products from hemoglobin breakdown.
 d. emulsifies fat.
 e. Both *c* and *d* are correct.

____ 14. HCl
 a. is an enzyme.
 b. creates an acidic environment necessary for pepsin to work.
 c. is found throughout the intestinal tract.
 d. digests fats.

____ 15. Pancreatic juice is directly regulated by
 a. the presence of food in the intestine.
 b. the sight of food.
 c. the thought of food.
 d. the smell of food.
 e. secretin.

____ 16. The large intestine
 a. digests all types of food.
 b. is the longest part of the intestinal tract.
 c. absorbs water.
 d. is connected to the stomach.

____ 17. Which of the following organs does NOT produce digestive enzymes?
 a. salivary glands
 b. stomach
 c. pancreas
 d. small intestine
 e. large intestine

____ 18. Which type of food group supplies quick energy and should comprise the bulk of the diet?
 a. milk and cheese
 b. lipids
 c. carbohydrates
 d. proteins

____19. Which of the following are considered antioxidants?
a. calcium and sodium
b. B vitamins and selenium
c. vitamin E and iron
d. vitamins C, E, and A

____20. Select the mineral that can help develop strong bones in young persons.
a. calcium
b. magnesium
c. potassium
d. sodium

## Critical Thinking Questions

Answer in complete sentences.

21. The well-preserved skull of a mammal shows well-developed molars and poorly developed canine teeth. What can you conclude about the probable diet of this organism?

22. Why is it advantageous for the human body to have a variety of hormones to control digestion?

**Test Results:** ______ number correct ÷ 22 = ________ × 100 = ______%

## Exploring the Internet

The Online Learning Center at *www.mhhe.com/maderbiology8* has additional study material and practice quizzes that can help you master the content of this chapter. You can also find links to websites exploring additional topics in biology. Access to the Online Learning Center is free for those who have purchased a new textbook.

## Answer Key

### Study Exercises

**1. a.** P **b.** P **c.** E **d.** E **2. a.** mouth **b.** anus **3. a.** C **b.** C **c.** C **d.** S **e.** S **f.** S **4. a.** continuous **b.** discontinuous **5. a.** H **b.** H **c.** C **d.** C **6. a.** vegetation **b.** meat **c.** vegetation **d.** herbivore **7. a.** mouth **b.** teeth **c.** Salivary **d.** saliva **e.** taste buds **f.** tongue **8. a.** pharynx **b.** glottis **9. a.** stomach **b.** peristalsis **10. a.** mucus **b.** hydrochloric **c.** gastric **d.** pepsin **11. a.** chyme **b.** duodenum **12. a.** villi **b.** microvilli **13. a.** Sugars **b.** amino acids **c.** fatty acids **d.** lacteal **14.** To increase the surface area available for absorption in the small intestine. **15. a.** Gastrin **b.** gastric inhibitory peptide **c.** Secretin **d.** cholecystokinin **16. a.** bicarbonate **b.** enzymes **17. a.** bile **b.** gallbladder **c.** blood **d., e., f.** (any three answers) produces plasma proteins, destroys old red blood cells, stores glucose as glycogen, converts hemoglobin to bilirubin and biliverdin, produces urea **18. a.** Hepatitis **b.** cirrhosis **19. a.** digestive wastes (feces) **b.** polyps **c.** fiber **20. a.** salivary amylase **b.** mouth **c.** pancreas **d.** basic **e.** maltase **f.** small intestine **g.** gastric glands **h.** acidic **i.** trypsin **j.** small intestine **k.** small intestine **l.** basic **m.** nuclease **n.** small intestine **o.** small intestine **p.** basic **q.** lipase **r.** small intestine (See Table 36.2, page 660, in text.) **21. a.** pepsin/trypsin **b.** maltase **c.** nuclease **d.** salivary amylase/pancreatic amylase **e.** lipase **f.** nucleosidases **g.** peptidase **22.** See Figure 36.11. page 664, in text. **23.** c **24.** d **25. a.** T **b.** T **c.** F **d.** T **26. a.** coenzymes **b.** A **c.** D **d.** C, E, and A **27 . a.** sodium **b.** calcium

## KeyWord CrossWord

**Across**

**1.** lipase **3.** fiber **4.** pharynx **5.** amylase **8.** anus **9.** gastric

**Down**

**1.** liver **2.** esophagus **6.** lacteal **7.** gallbladder

## Chapter Test

**1.** c **2.** d **3.** d **4.** a **5.** c **6.** d **7.** a **8.** a **9.** a **10.** b **11.** e **12.** d **13.** e **14.** b **15.** e **16.** c **17.** e **18.** c **19.** d **20.** a **21.** Molars are adapted for grinding, the kind of action needed to break down plant food. The animal was an herbivore. Canine teeth, seen in carnivores, are needed more for the tearing of flesh. **22.** Digestion does not need to occur continuously. Certain digestive responses are needed when food intake warrants it (e.g., pepsin secretion in the stomach with the arrival of protein there, signaled by gastrin secretion).

# 37

# Respiration

## Chapter Review

For some aquatic organisms (i.e., hydra), the entire body surface functions for gas exchange. More advanced animals have evolved specialized respiratory organs. **Gills** in the water and **tracheal systems** on land are two examples. **Lungs** are another gas exchange structure among some species of land animals. Muscular contractions usually produce pressure changes that either draw air into the lungs or force air out of the lungs. As another solution for gas exchange, birds have evolved a series of air sacs for the one-way flow of air.

The breathing cycle of humans begins with the contraction of the **diaphragm** and rib-elevating muscles. Their contraction creates a negative pressure, drawing air into the lungs **(inspiration).** A reverse of this process drives air out of the lungs **(expiration).** The rate of the breathing cycle depends on the level of carbon dioxide in the blood, as detected by chemoreceptors.

Once in the **alveoli** of the lungs, carbon dioxide and oxygen are exchanged with the blood by diffusion. Diffusion is also the process whereby body cells receive oxygen and give up carbon dioxide. Oxygen is transported to the tissues in combination with **hemoglobin** as **oxyhemoglobin** ($HbO_2$). Carbon dioxide is mainly transported to the lungs as part of the **bicarbonate ion.**

The respiratory tract can become infected with viruses or bacteria. Pneumonia and tuberculosis are infections that can be controlled with antibiotics. Cigarette smoking is the primary cause of the lung diseases **emphysema** and **lung cancer.**

## Study Exercises

Study the text section by section as you answer the questions that follow.

### 37.1 Gas Exchange Surfaces (pp. 670–73)

- Respiration comprises breathing, external and internal respiration, and cellular respiration.
- Most animals have a special, localized gas exchange area, such as the gills in aquatic animals and lungs in terrestrial animals.
- Amphibians use positive pressure, but other vertebrates use negative pressure, to ventilate the lungs.
- Birds have a complete ventilation system in that there is a one-way flow of air over the gas exchange area.

1. Match the processes with these descriptions.
   1. inspiration and expiration
   2. exchange at respiratory surface
   3. exchange in tissues

   a. _____ breathing
   b. _____ external respiration
   c. _____ internal respiration
2. Label each of the following as characteristic of air (A) or water (W):
   a. _____ Mammals use 1–2% of their energy to breathe here.
   b. _____ It is more difficult for animals to obtain oxygen here.
   c. _____ More energy for breathing is required here.
   d. _____ The hydra carries out respiration in this environment.
   e. _____ This is the less dense medium.

3. Which of the classes of vertebrates use(s) positive pressure to fill their lungs with air? ______________________

4. Which of the classes use(s) negative pressure to fill their lungs with air? ______________________

5. Contrast ventilation by negative pressure with ventilation by positive pressure. ______________________

______________________________________________________________________

6. How is the means of ventilation more complete in birds, compared to the incomplete means in amphibians, reptiles, and mammals? ______________________________________________

______________________________________________________________________

7. Complete the following table to summarize the means of ventilation in the adult animals listed:

| Class | Organ(s) of Exchange | Special Features |
|---|---|---|
| Hydra | | |
| Planarian | | |
| Fish | | |
| Amphibian | | |
| Reptile | | |
| Bird | | |
| Mammal | | |

## 37.2 Human Respiratory System (pp. 674–77)

- There is a continuous pathway for air from the nose (or mouth) to the lungs in humans.
- The breathing rate is regulated and increases when there is a greater demand of exchange of gases in the lungs.

8. Label this diagram using the alphabetized list of terms.
   epiglottis
   glottis
   larynx
   nasal cavity
   pharynx
   trachea

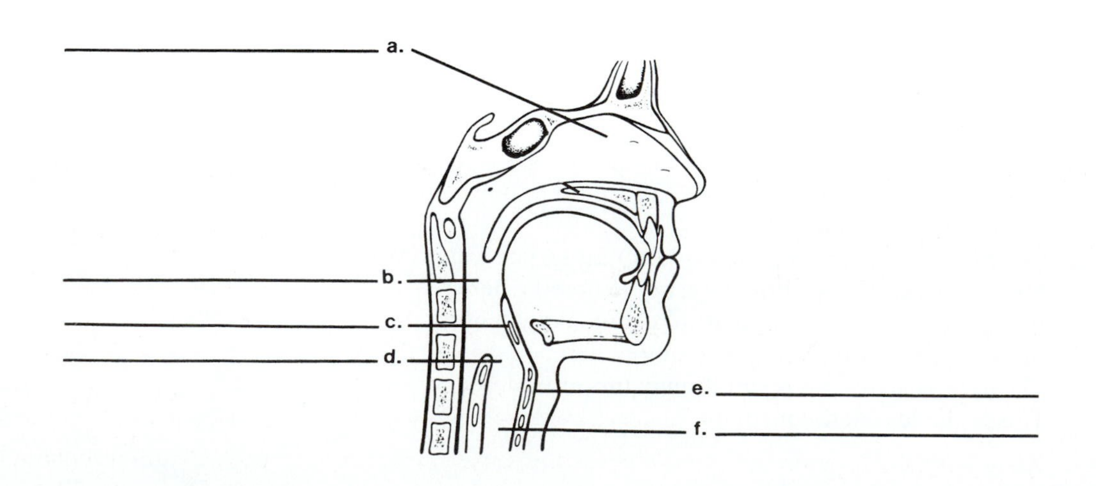

9. Name the structure that functions for each of the following:
   a. gas exchange with blood flowing to and from the heart ______________________
   b. sound production ______________________
   c. filtering, moistening, and warming the air ______________________
   d. conduction of air into the thoracic cavity ______________________
   e. passage of air into the lung ______________________

10. Place the appropriate letter next to each phrase

    I—inspiration   E—expiration

    a. _____ lungs expanded
    b. _____ muscles (diaphragm and ribs) relaxed
    c. _____ diaphragm dome-shaped
    d. _____ chest enlarged
    e. _____ less air pressure in lungs than in outside environment

11. Indicate whether these statements about the control of breathing rate are true (T) or false (F).
    a. _____ Chemoreceptors are sensitive to changes in carbon dioxide in the blood.
    b. _____ Chemoreceptors are sensitive to changes in hydrogen ions in the blood.
    c. _____ The concentration of oxygen in arterial blood is a major stimulus.
    d. _____ The respiratory center is located in the hypothalamus of the brain.

## Gas Exchange and Transport (p. 677)

- The respiratory pigment hemoglobin transports oxygen from the lungs to the tissues and aids the transport of carbon dioxide from the tissues to the lungs.

12. a. Write the equation that describes how oxygen is transported in the blood. Label one arrow *lungs* and the reverse arrow *tissues*.

    b. Write the equation that describes how most of the carbon dioxide is transported in the blood. Label one arrow *lungs* and the reverse arrow *tissues*.

    c. What is the name of the enzyme that speeds up this reaction? ______________________

    d. Carbon dioxide transport produces hydrogen ions. Why does the blood not become acidic?

    ______________________________________________

13. a. How is hemoglobin remarkably suited to the transport of oxygen in animals? ______________________

    ______________________________________________

    b. How does hemoglobin transport carbon dioxide? ______________________

    ______________________________________________

## 37.3 Respiration and Health (pp. 678–79)

- The respiratory tract is especially subject to infections. Cigarette smoking contributes to two major lung disorders—emphysema and cancer.

14. Indicate whether these statements are true (T) or false (F).
    a. _____ A sneeze can transfer bacteria or viruses.
    b. _____ Emphysema develops from a lung infection.
    c. _____ Antibiotics cannot normally control pneumonia.
    d. _____ The moist mucous membrane of the respiratory tract is normally exposed to environmental air.

# Chapter Test

## Objective Questions

Do not refer to the text when taking this test.

____ 1. Oxygen is more difficult to obtain by respiratory organs in the
   a. air.
   b. water.

____ 2. A major adaptation for gas exchange in planarians is the
   a. countercurrent exchange in the lungs.
   b. lack of a gastrovascular cavity.
   c. presence of a thin, flattened body.
   d. thin surface in parapodia.

____ 3. The major respiratory organ of the earthworm is the
   a. entire body surface.
   b. nephridia.
   c. parapodia.
   d. tracheae with spiracles.

____ 4. The frog inhales air by a ______ pressure.
   a. negative
   b. positive

____ 5. During inspiration, contractions
   a. raise the diaphragm and lower the ribs.
   b. raise the diaphragm and raise the ribs.
   c. lower the diaphragm and lower the ribs.
   d. lower the diaphragm and raise the ribs.

____ 6. Air is exhaled from the body by passing through which order of structures?
   a. alveolus, bronchiole, bronchus, trachea
   b. bronchus, bronchiole, trachea, pharynx
   c. pharynx, larynx, trachea, bronchus
   d. trachea, alveolus, bronchus, bronchiole

____ 7. Ventilation in birds is complete because inhaled air
   a. does not meet exhaled air.
   b. does not pass through all structures of the tract.
   c. passes through all structures of the tract.
   d. passes through the lungs by a positive pressure.

____ 8. The structure(s) that receive(s) air after the bronchi is (are) the
   a. pharynx.
   b. trachea.
   c. bronchioles.
   d. villi.

____ 9. Select the incorrect association.
   a. alveolus—gas exchange
   b. bronchus—air enters the lungs
   c. larynx—sound production
   d. nasal cavity—air dried and cooled

____ 10. External respiration is defined as
   a. an exchange of gases in the lungs.
   b. breathing.
   c. an exchange of gases in the tissues.
   d. cellular respiration.

____ 11. When the lungs recoil,
   a. inspiration occurs.
   b. external respiration occurs.
   c. internal respiration occurs.
   d. expiration occurs.
   e. All of these are correct

____ 12. The chest is
   a. expanded during inspiration.
   b. closed off from the abdominal cavity by the diaphragm.
   c. divided into an area for the lungs and an area for the heart.
   d. All of these are correct.

____ 13. The crossing of the digestive and respiratory tracts in the pharynx creates a need for
   a. swallowing.
   b. external nares.
   c. an epiglottis.
   d. a diaphragm.
   e. olfactory epithelium.

____ 14. Chemoreceptors in the body are sensitive to
   a. carbon dioxide.
   b. hydrogen ions.
   c. Both *a* and *b* are correct.
   d. Neither *a* nor *b* is correct.

____ 15. Most carbon dioxide in the blood is transported in the
   a. bicarbonate ion.
   b. carbon dioxide molecule.
   c. hemoglobin molecule.
   d. oxygen molecule.

____ 16. Which of the following statements is NOT true?
   a. The respiratory center is located in the medulla oblongata.
   b. Breathing increases with exercise because of the reduced amount of oxygen in the blood.
   c. The chemoreceptors communicate with the respiratory center.
   d. When appropriate, the respiratory center increases the breathing rate.

____ 17. Hemoglobin carries
   a. $O_2$.
   b. $CO_2$.
   c. hydrogen ions.
   d. All of these are correct.

____ 18. Hemoglobin combines with ______ in the _____.
a. oxygen; lungs
b. carbon dioxide; tissues
c. oxygen; tissues
d. carbon dioxide; lungs
e. Both *a* and *b* are correct.

____ 19. ______ is a condition in which the lungs are inflated due to trapped air caused by bronchiole destruction.
a. Pneumonia
b. Tuberculosis
c. Emphysema
d. Lung cancer
e. Pulmonary fibrosis

____ 20. Smoking cigarettes
a. causes tuberculosis.
b. leads to emphysema and cancer.
c. increases the vital capacity of the lungs.
d. All of these are correct.

## Critical Thinking Questions

Answer in complete sentences.

21. How are the alveoli examples of a structure admirably suited to function?

22. The lungs of mammals do not contain any muscle tissue. How do they passively respond to the pressure changes around them?

**Test Results:** ______ number correct ÷ 22 = ______ × 100 = ______ %

## Exploring the Internet

The Online Learning Center at *www.mhhe.com/maderbiology8* has additional study material and practice quizzes that can help you master the content of this chapter. You can also find links to websites exploring additional topics in biology. Access to the Online Learning Center is free for those who have purchased a new textbook.

## Answer Key

### Study Exercises

**1. a.** 2 **b.** 1 **c.** 3 **2. a.** A **b.** W **c.** W **d.** W **e.** A **3.** amphibians **4.** reptiles, birds, mammals **5.** With positive pressure, the air is being forced into the lungs; with negative pressure, the air is being drawn in. **6.** Fresh, oxygen-rich air passes through the lungs in a one-way direction in birds. This incoming air does not mix with outgoing air, as it does in amphibians, reptiles, and mammals.

**7.**

| Organ(s) of Exchange | Special Features |
|---|---|
| diffusion of gases across body layers | gastrovascular cavity |
| diffusion of gases across body layers | flattened, thin body |
| gills | countercurrent exchange |
| lungs and skin | use of positive pressure to fill lungs |
| lungs | jointed ribs, use of negative pressure to fill lungs |
| lungs | air sacs allow a one-way flow of air |
| lungs | diaphragm, use of negative pressure to fill lungs |

**8.** **a.** nasal cavity **b.** pharynx **c.** epiglottis **d.** glottis **e.** larynx **f.** trachea **9.** **a.** alveolus **b.** larynx/vocal cords **c.** nose **d.** trachea **e.** bronchus **10.** **a.** I **b.** E **c.** E **d.** I **e.** I **11.** **a.** T **b.** T **c.** F **d.** F

**12.** **a.** $Hb + O_2 \underset{\text{tissues}}{\overset{\text{lungs}}{\rightleftharpoons}} HbO_2$

**b.** $CO_2 + H_2O \underset{\text{lungs}}{\overset{\text{tissues}}{\rightleftharpoons}} H_2CO_3 \underset{\text{lungs}}{\overset{\text{tissues}}{\rightleftharpoons}} H^+ + HCO_3^-$

**c.** carbonic anhydrase **d.** Hemoglobin combines with the excess hydrogen ions. **13.** **a.** It easily combines with oxygen in the lungs, and it easily gives up oxygen in the tissues. **b.** It combines with carbon dioxide somewhat, forming carbaminohemoglobin; it also picks up hydrogen ions. **14.** **a.** T **b.** F **c.** F **d.** T

## Chapter Test

**1.** b **2.** c **3.** a **4.** b **5.** d **6.** a **7.** a **8.** c **9.** d **10.** a **11.** d **12.** d **13.** c **14.** c **15.** a **16.** b **17.** d **18.** e **19.** c **20.** b **21.** They are numerous, providing a greater total surface area for gas exchange. Their thin, moist surfaces maximize the diffusion for gas exchange. **22.** The rib cage rises and the pressure inside the lungs decreases when the diaphragm lowers. Then air rushes into lungs. The rib cage lowers and the pressure inside of the lungs increases when the diaphragm rises. Then air is forced out of lungs.

# 38

# Body Fluid Regulation and Excretion

## Chapter Review

Most animals have a means to maintain normal solute and water concentration in body fluids. Marine bony fishes constantly drink water, excrete salt at the gills, and pass an isotonic urine. Freshwater bony fishes never drink water; they take in salt at the gills and pass a hypotonic urine. The form in which animals excrete nitrogen depends on the type of environment inhabited. In a water environment, the form is usually **ammonia.** On land, it is either **urea** or **uric acid.**

Different groups of animals have evolved different excretory organs (i.e., **Malpighian tubules** in insects). Humans have a urinary system that includes the **kidneys** as the excretory organs. Microscopically, a kidney is made up of microscopic units called **nephrons.** A nephron makes **urine** through the processes of glomerular filtration, tubular reabsorption, and tubular secretion. Humans excrete a hypertonic urine through the action of the **loop of the nephron.**

Three hormones are involved in maintaining the blood volume and blood pressure. **Antidiuretic hormone (ADH)** regulates the reabsorption of water by regulating the permeability of the collecting duct. **Aldosterone** causes the reabsorption of sodium ($Na^+$), and therefore water. **Atrial natriuretic hormone (ANH)** prevents the secretion of renin and aldosterone.

The kidneys control the pH of the blood by regulating the excretion of hydrogen ions ($H^+$) and ammonia ($NH_3$), together with the reabsorption of sodium ($Na^+$) and bicarbonate ions ($HCO_3^-$).

## Study Exercises

Study the text section by section as you answer the questions that follow.

### 38.1 Body Fluid Regulation (pp. 684–85)

- The mechanism for maintaining water balance differs according to the environment of the organism.

1. Indicate whether these statements are true (T) or false (F).
   - a. _____ Birds and reptiles living near the sea excrete a concentrated salt solution through salt glands.
   - b. _____ Cartilaginous fishes live in an environment that is nearly isotonic to their internal body fluids.
   - c. _____ Freshwater bony fishes are prone to gaining body water.
   - d. _____ Freshwater bony fishes live in an environment that is hypertonic to their internal body fluids.
   - e. _____ Marine bony fishes actively transport salt into the seawater through their gills.
   - f. _____ Marine bony fishes are prone to losing water.
   - g. _____ Marine bony fishes live in an environment that is hypotonic to their internal body fluids.
   - h. _____ The kangaroo rat forms a very dilute urine.

## 38.2 Nitrogenous Waste Products (p. 686)

- Nitrogenous waste products differ as to the amount of water and energy required to excrete them.

2. a. Beneath the nitrogenous wastes listed, draw an arrow pointed in the direction of the waste that requires the most water to excrete.

   ammonia      urea      uric acid

   b. Beneath these wastes, draw an arrow pointed in the direction of the waste that requires the most energy for an animal to produce.

   ammonia      urea      uric acid

## 38.3 Organs of Excretion (p. 687)

- Complex animals have organs of excretion that maintain the water balance of the body and rid the body of waste molecules.

3. Explain the action of flame cells in planarians. ______________________________

______________________________

4. Explain the action of nephridia in earthworms. ______________________________

______________________________

5. Explain the action of Malpighian tubules in insects. ______________________________

______________________________

6. Associate the excretory organs mentioned in questions 3–5, with these phrases:
   a. _____ usually gets rid of excess water
   b. _____ usually conserves water
   c. _____ excretes uric acid
   d. _____ present in each segment

## 38.4 Urinary System in Humans (pp. 688–93)

- The urinary system of humans consists of organs that produce, store, and rid the body of urine.
- The work of an organ is dependent on its microscopic anatomy; nephrons within the human kidney produce urine.

7. Rearrange letters *a–d* to indicate the direction that urine is transported through the human urinary system.

   ______________

   a. bladder
   b. kidney
   c. ureter
   d. urethra

8. Label this diagram of the kidney using the alphabetized list of terms.
   - collecting duct
   - nephrons
   - renal cortex
   - renal medulla
   - renal pelvis
   - renal pyramid
   - ureter

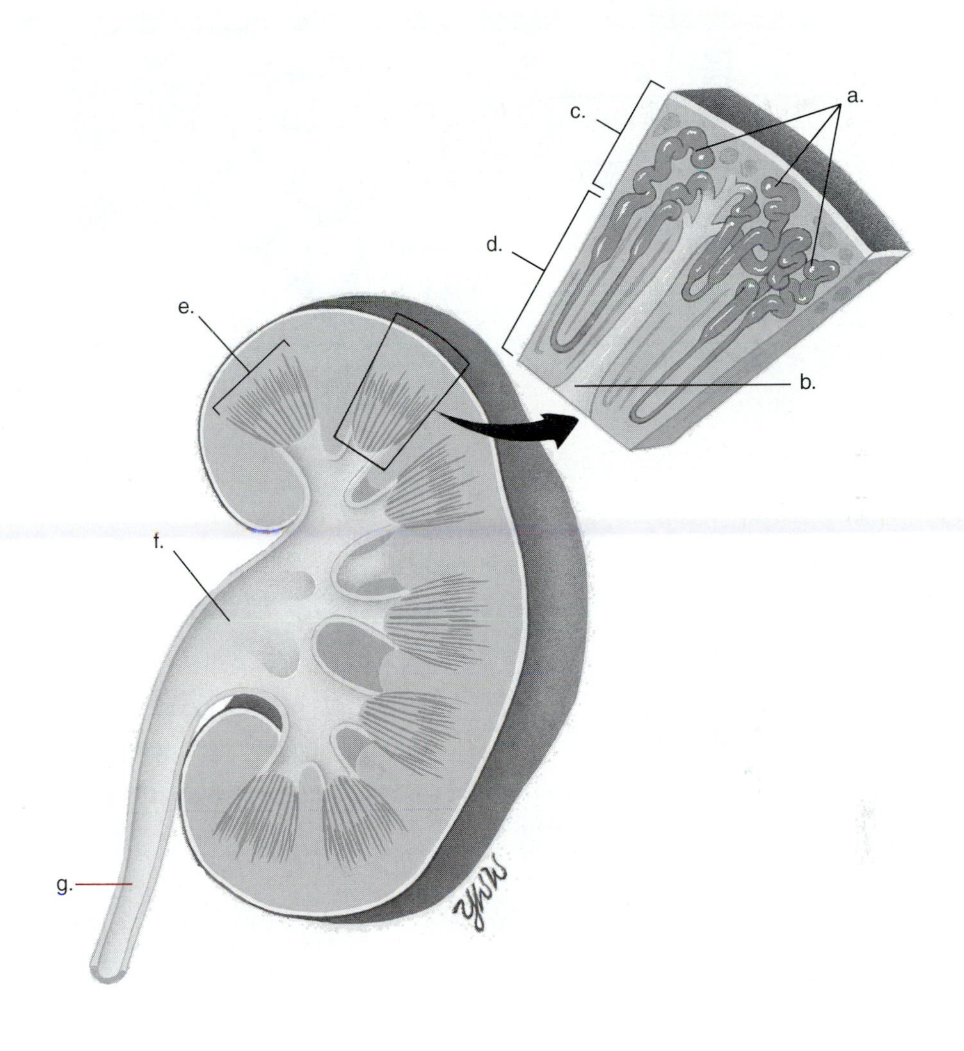

9. Label this diagram of the parts of the nephron using the following alphabetized list of terms.

afferent arteriole
collecting duct
distal convoluted tubule
efferent arteriole
glomerular capsule
glomerulus
loop of the nephron
peritubular capillaries
proximal convoluted tubule

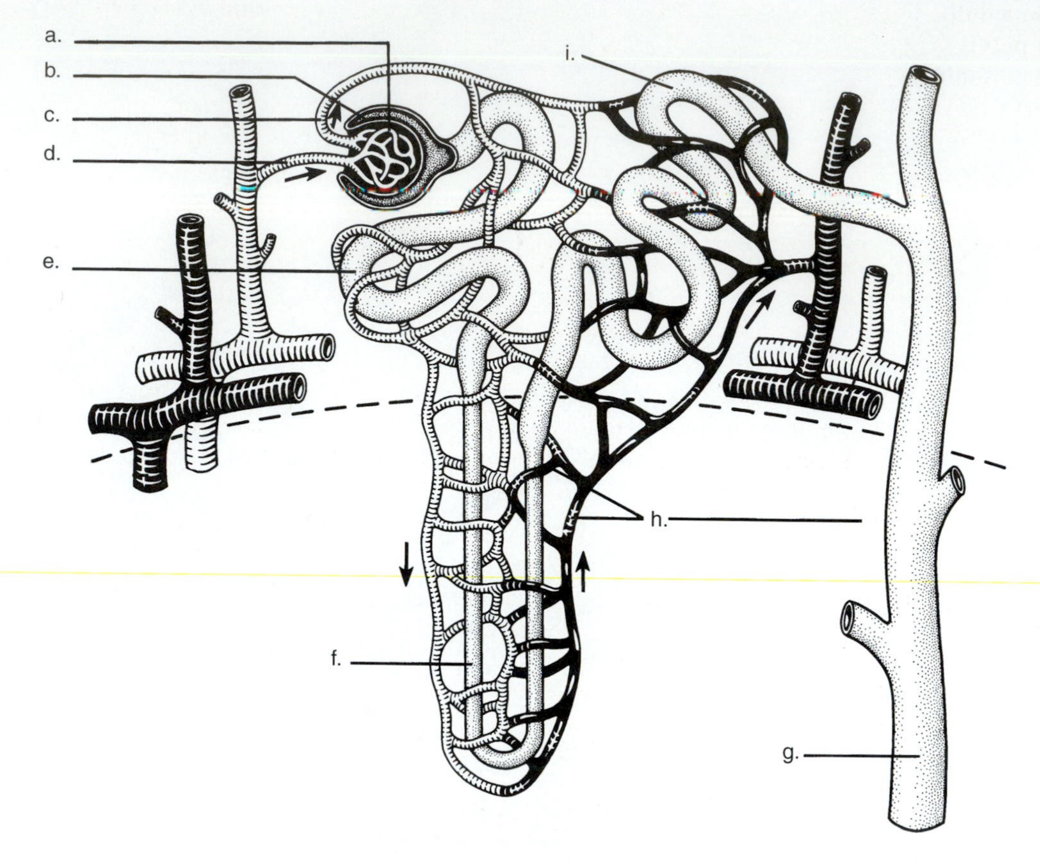


## Urine Formation (pp. 690–93)

- Like many physiological processes, urine formation in humans is a multistep process.
- In addition to ridding the body of waste molecules and maintaining water-salt balance, the human kidneys maintain the normal pH of the blood.

10. In this diagram, add the steps in urine formation.

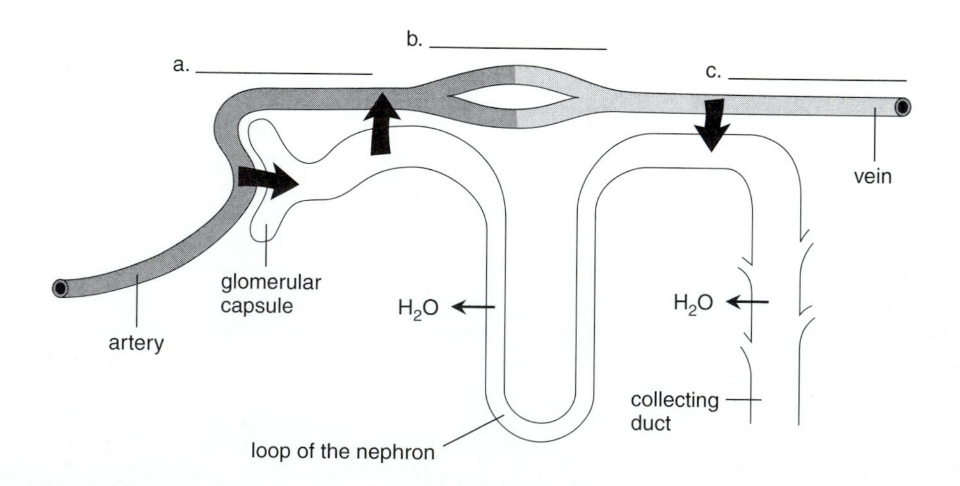

11. Blood approaches the glomerulus in the afferent arteriole. It contains the following:

*small molecules*—water, nutrients (glucose, amino acids), salts, wastes (urea)

*other materials*—blood cells, proteins, hydrogen ions

a. Will the *small materials* or the *other materials* enter the glomerular capsule to become the filtrate? ______________________

b. Which of the molecules in the filtrate will probably be absorbed? ______________________

c. Which of the *other materials* will enter the tubule during the process called tubular secretion? ______________________

12. a. According to the diagram in question 10, where is water primarily reabsorbed? ______________________

______________________________________________

b. What effect does this high degree of reabsorption have on the composition of urine? ______________________

______________________________________________

13. In the diagram in question 10, darken the portion of the loop of the nephron that is impermeable to water. Add to the diagram one set of colored arrows to indicate the active transport of sodium out of the ascending limb. Color the arrows that indicate the movement of water out of the loop and the collecting duct.

## Urine Formation and Homeostasis (pp. 692–93)

14. Examine the following diagram and answer the questions that follow:

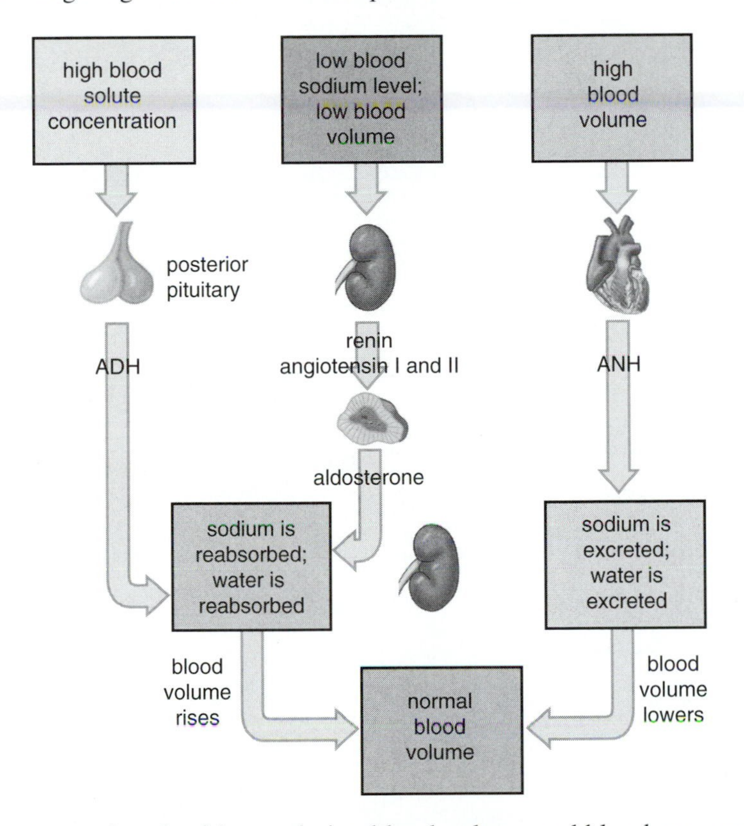


a. Name three hormones involved in regulating blood volume and blood pressure? ______________________

b. Which two hormones act to increase blood volume and blood pressure? ______________________

c. How does ADH function? ______________________

d. After ADH is secreted, the volume of urine produced (increases or decreases)? ______________________

e. How does aldosterone function? ______________________

f. Which hormone opposes the action of ADH and aldosterone? ______________________

15. If the blood is acidic, a.__________________ ions are excreted in combination with b.__________________, while c.__________________ and bicarbonate ions are reabsorbed. If the blood is basic, fewer d.__________________ ions are excreted and fewer e.__________________ and bicarbonate ions are reabsorbed.

## *Excretion Elimination*

In the table, place an X beside the component of blood if the following descriptions pertain to it:

a. in the afferent arteriole
b. in the filtrate
c. in the efferent arteriole
d. reabsorbed into the peritubular capillary
e. secreted from the peritubular capillary
f. present in urine
g. absent from urine
h. in venous blood

| | a | b | c | d | e | f | g | h |
|---|---|---|---|---|---|---|---|---|
| 1. Plasma proteins | | | | | | | | |
| 2. Blood cells | | | | | | | | |
| 3. Glucose | | | | | | | | |
| 4. Amino acids | | | | | | | | |
| 5. Sodium chloride | | | | | | | | |
| 6. Water | | | | | | | | |
| 7. Urea | | | | | | | | |
| 8. Uric acid | | | | | | | | |

There are 36 correct answers. There are 64 possible errors of omission or commission. Any 10 errors and you're ELIMINATED!

# Chapter Test

## Objective Questions

Do not refer to the text when taking this test.

____ 1. Uric acid is ______ soluble in water and ______ toxic.
a. highly; highly
b. highly; slightly
c. slightly; highly
d. slightly; slightly

____ 2. Ammonia is the main means of excreting nitrogen in each of the following groups EXCEPT
a. amphibian adults.
b. amphibian larvae.
c. aquatic invertebrates.
d. bony fishes.

____ 3. Marine bony fishes face the problem of water loss from the body.
a. true
b. false

____ 4. Select the incorrect association.
a. flame cell—hydra
b. kidney—human
c. Malpighian tubule—insect
d. nephridia—earthworm

____ 5. The innermost hollow chamber of the kidney is the
a. renal cortex.
b. glomerulus.
c. renal medulla.
d. renal pelvis.

____ 6. The collecting ducts are primarily found in the
a. renal cortex.
b. renal medulla.
c. renal pelvis.
d. afferent arteriole.

____ 7. Kidneys are organs of homeostasis because they
a. regulate the blood volume.
b. regulate the pH of the blood.
c. help maintain the correct concentration of ions in the blood.
d. excrete nitrogenous wastes.
e. All of these are correct.

____ 8. Nitrogenous wastes are
a. metabolic wastes.
b. toxic.
c. formed from the breakdown of amino acids.
d. All of these are correct.

____ 9. Glomerular filtration occurs at the ______, and reabsorption occurs at the ______.
a. glomerular capsule; proximal convoluted tubule
b. distal convoluted tubule; glomerular capsule

____ 10. By tubular reabsorption, substances pass from the
a. blood into the tubule.
b. tubule into the blood.

____ 11. The hormone aldosterone promotes the
a. excretion of potassium only.
b. reabsorption of sodium only.
c. excretion of potassium and reabsorption of sodium.
d. elimination of water.

____ 12. The kidneys oppose an acidic blood condition by ______ hydrogen ions and ______ sodium and bicarbonate ions.
a. excreting; reabsorbing
b. reabsorbing; excreting

____ 13. Tubular secretion occurs at the
a. glomerular capsule.
b. proximal convoluted tubule.
c. loop of the nephron.
d. distal convoluted tubule.

____ 14. Glucose
a. is in the filtrate and urine.
b. is in the filtrate and not in the urine.
c. undergoes tubular secretion and is in the urine.
d. undergoes tubular secretion and is not in urine.

____ 15. A person who lacks ADH
a. has low blood volume and pressure.
b. will have sugar in the urine.
c. has diabetes mellitus.
d. All of these are correct.

____ 16. Renin is an enzyme that converts
a. angiotensinogen to angiotensin I.
b. angiotensin I to angiotensin II.
c. angiotensin II to converting enzyme.
d. All of these are correct.

____ 17. Sodium is removed from the loop of the nephron by
a. passive reabsorption.
b. active transport.
c. an attraction to $Cl^-$.
d. tubular secretion.

____ 18. The region between the base of the loop of the nephron and the collecting duct has a(n)
a. very low solute concentration.
b. intermediate solute concentration.
c. very high solute concentration.
d. very high water concentration.

## Critical Thinking Questions

Answer in complete sentences.

19. Glomerular filtration is unselective; tubular reabsorption is very selective. Support this statement.

20. How are mitochondria and microvilli major cellular adaptations of the cells of the proximal convoluted tubule of the nephron?

**Test Results:** ______ number correct ÷ 20 = ______ × 100 = ______ %

## Exploring the Internet

The Online Learning Center at *www.mhhe.com/maderbiology8* has additional study material and practice quizzes that can help you master the content of this chapter. You can also find links to websites exploring additional topics in biology. Access to the Online Learning Center is free for those who have purchased a new textbook.

## Answer Key

### Study Exercises

**1. a.** T **b.** T **c.** T **d.** F **e.** T. **f.** T **g.** F **h.** F **2. a.** Arrow should be pointed left. **b.** Arrow should be pointed right. **3.** Planarians live in fresh water. The movement of flame-cell cilia propels hypotonic fluid through excretory canals and out of body. **4.** Fluid from the coelom is propelled through the nephridia by beating cilia; certain substances are reabsorbed by a network of capillaries. **5.** Malpighian tubules take up water and uric acid from the hemolymph and pass them to the gut where water is reabsorbed. **6. a.** flame cells **b.** Malpighian tubules **c.** Malpighian tubules **d.** nephridia **7.** b, c, a, d **8.** See Figure 38.6, p. 688, in text. **9.** See Figure 38.7, page 689, in text. **10. a.** glomerular filtration **b.** tubular reabsorption **c.** tubular secretion **11. a.** small molecules **b.** nutrients, salts, water **c.** hydrogen ions **12. a.** descending limb of the loop of the nephron and collecting duct. **b.** The urine is more concentrated. **13.** See Figure 38.10, p. 692, in text. **14. a.** antidiuretic hormone (ADH), aldosterone, atrial natriuretic hormone (ANH) **b.** antidiuretic hormone (ADH) and aldosterone **c.** Increases permeability of collecting duct so that water is reabsorbed. **d.** decreases **e.** Causes sodium to be reabsorbed and water follows **f.** ANH **15. a.** hydrogen **b.** ammonia **c.** sodium **d.** hydrogen **e.** sodium

## EXCRETION ELIMINATION

| | a | b | c | d | e | f | g | h |
|---|---|---|---|---|---|---|---|---|
| 1. Plasma proteins | x | | x | | | | x | x |
| 2. Blood cells | x | | x | | | | x | x |
| 3. Glucose | x | x | | x | | | x | x |
| 4. Amino acids | x | x | | x | | | x | x |
| 5. Sodium chloride | x | x | | x | | x | | x |
| 6. Water | x | x | x | x | | x | | x |
| 7. Urea | x | x | | | | x | | |
| 8. Uric Acid | x | x | | | x | x | | |

## CHAPTER TEST

**1.** d **2.** a **3.** a **4.** a **5.** d **6.** b **7.** e **8.** d **9.** a **10.** b **11.** c **12.** a **13.** d **14.** b **15.** a **16.** a **17.** b **18.** c **19.** With the exception of the blood cells and plasma proteins, all smaller components of the blood are forced out of the glomerulus in large quantities. Reabsorption returns them to the blood in varying amounts, however, depending on the needs of the body. **20.** The mitochondria increase the energy capacity for the active transport mechanisms carried out by the tubule cells. The microvilli increase the surface area of the cells to make contact with and return materials to the blood by reabsorption.

# 39

# Neurons and Nervous Systems

## Chapter Review

A comparative study of the invertebrates shows a gradual increase in the complexity of the nervous system. In vertebrates, the central nervous system consists of a brain and a spinal cord. The human nervous system is divided into the central nervous system and the peripheral nervous system.

There are three types of **neurons: motor, sensory,** and **interneurons.** As in the motor neuron, each neuron has a **cell body, axon,** and **dendrites.** When an axon is not conducting a nerve impulse, the inside of the neuron is negative compared to the outside. This is the **resting potential,** when there is a concentration of $Na^+$ outside and a concentration of $K^+$ inside the axon, due to the work of the sodium-potassium pump. When a nerve impulse occurs, an **action potential** travels along an axon. Depolarization (inside becomes positive) is due to the movement of $Na^+$ to the inside, and then repolarization (inside becomes negative again) is due to the movement of $K^+$ to the outside.

Transmission of the nerve impulse from one neuron to another takes place across a **synapse.** Synaptic vesicles release a neurotransmitter that binds to receptors in the postsynaptic membrane, causing either stimulation or inhibition.

The **central nervous system** consists of the spinal cord and brain. Humans have the most well-developed cerebral cortex of all the vertebrates. Each lobe contains association areas for more complex levels of consciousness. The lobes of the cortex can be mapped for sensory and motor functions.

In mammals, such as humans, the **peripheral nervous system** contains the somatic system and the autonomic system and consists of cranial and spinal nerves. The somatic system controls the skeletal muscles, and the autonomic system controls the smooth muscles, cardiac muscles, and glands. In spinal nerves, reflex actions occur when nerve impulses begin at a receptor and then move from a sensory neuron to an interneuron to a motor neuron, which stimulates an effector to react. In the autonomic system, the sympathetic division is associated with reactions to stress, and the parasympathetic division is associated with internal responses during relaxation.

## Study Exercises

### 39.1 Evolution of the Nervous System (pp. 698–700)

- A survey of invertebrates shows a gradual increase in the complexity of the nervous system.
- All vertebrates have a well-developed brain, but the forebrain is the largest in mammals, particularly humans.

1. Match the phrases with these animals:
   1. hydra
   2. planarian
   3. earthworm
   4. squid
   5. cat

   a. _____ ladderlike nervous system
   b. _____ ventral solid nerve cord
   c. _____ nerve net
   d. _____ dorsal tubular nerve cord
   e. _____ giant nerve fibers
2. Why is it said that planarians have cephalization? _______________
_______________
3. In what way does the nervous organization in the animals listed in question 1 suggest a central and peripheral nervous system? _______________
_______________

## 39.2 Nervous Tissue (pp. 701–5)

- Nervous tissue is made up of cells called neurons, specialized to carry nerve impulses, and neuroglia, which support and nourish neurons.

4. Every neuron has the three parts listed here. What is the function of each?
   a. dendrite ______
   b. cell body ______
   c. axon ______
5. Label the parts of the sensory neuron, the interneuron, and the motor neuron, using the following alphabetized list of terms (some terms may be used more than once).

   axon　cell body　dendrite　effector　myelin sheath

   neurofibril node (node of Ranvier)　sensory receptor

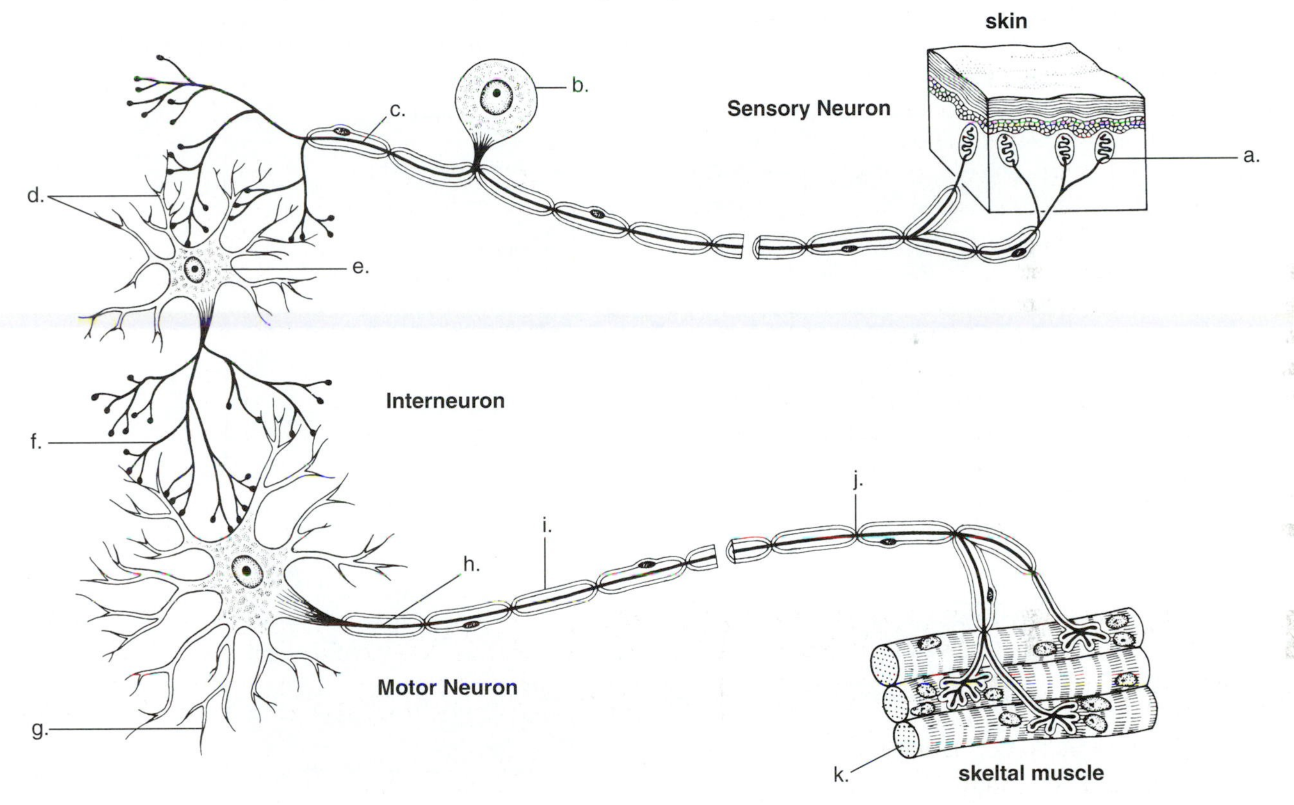


6. State the function of the complete sensory neuron. ______
7. State the function of the complete interneuron. ______
8. State the function of the complete motor neuron. ______

## Transmission of the Nerve Impulses (pp. 702–3)

- A nerve impulse is an electrochemical change that travels across an axomembrane.

9. Examine this diagram depicting the resting potential, and then answer the questions that follow:
   a. What is an oscilloscope? ______________________________

   b. The oscilloscope is reading –65 mV. Why is the inside of an axon negative compared to the outside?

   c. At the time of the resting potential, are there more or less $Na^+$ ions outside the axon than inside the axon?

   d. At the time of the resting potential, are there more or less $K^+$ ions outside the axon than inside the axon?

   e. What mechanism accounts for the unequal distribution of ions across the axomembrane?

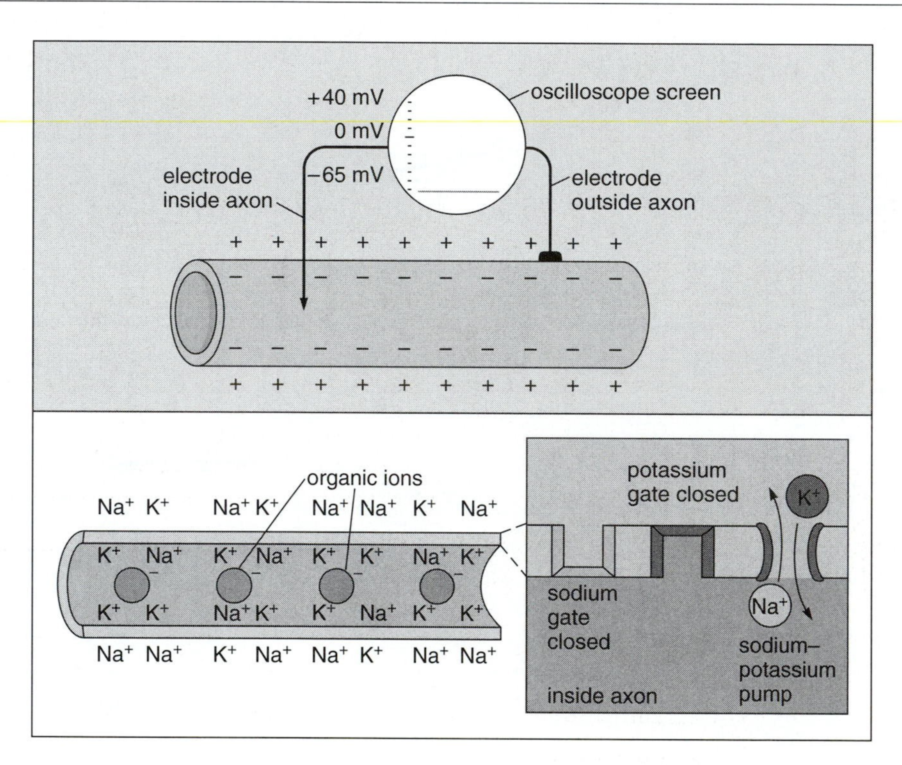


10. Label this diagram depicting the trace on the oscilloscope screen as an action potential occurs using the alphabetized list of terms.
    action potential
    depolarization
    repolarization
    resting potential
    threshold

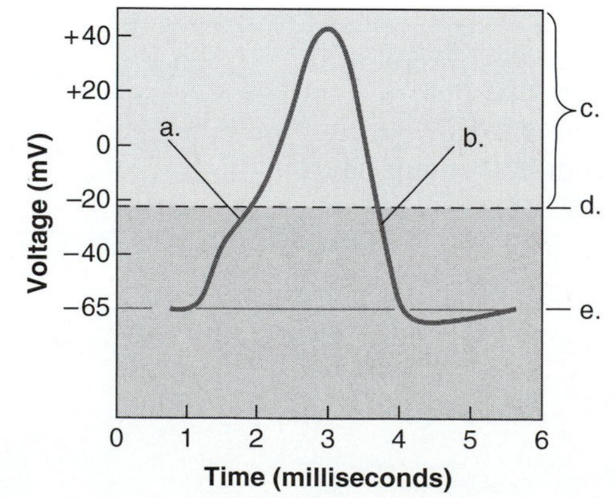

11. Examine this diagram depicting an action potential and then answer the questions *a* through *g* that follow.

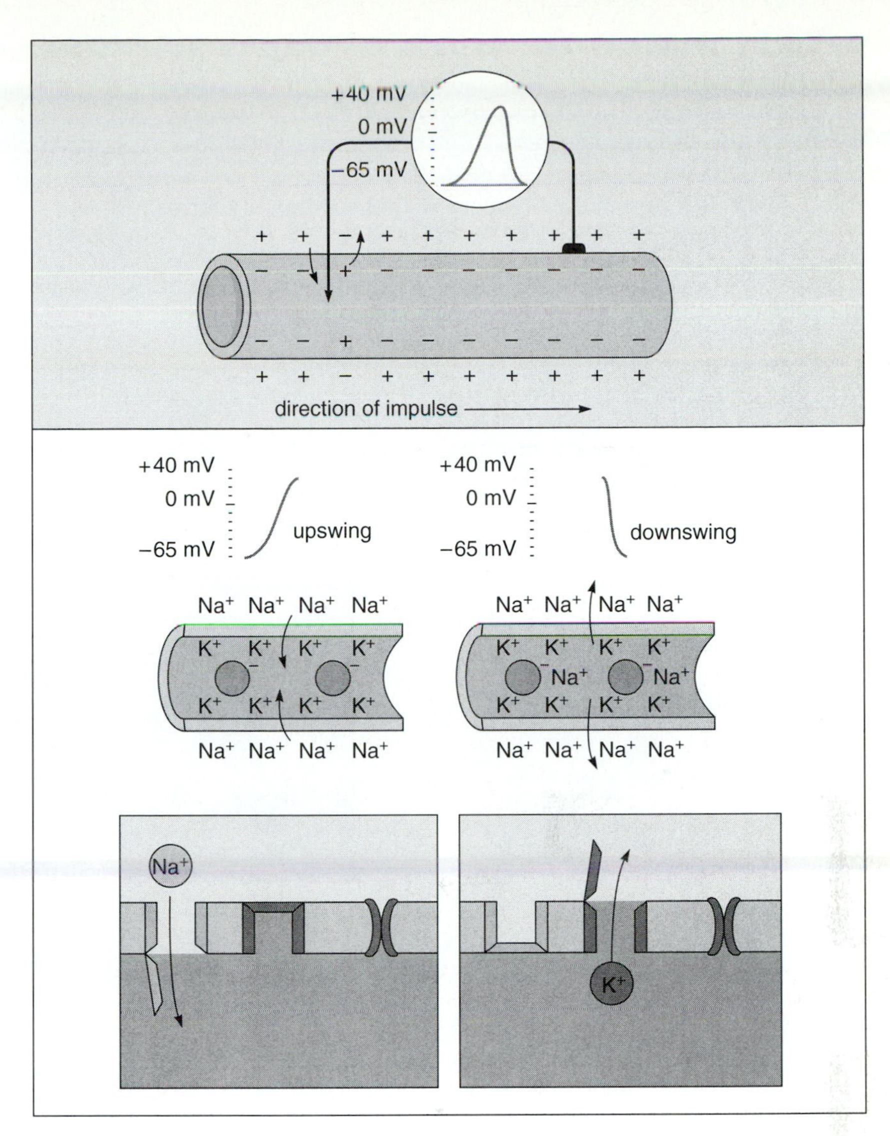


a. What does depolarization refer to? ______________________________

______________________________

b. What does repolarization refer to? ______________________________

______________________________

c. In the top portion of the diagram, there is an arrow leading from a plus sign (+) to the interior of the axon and an arrow leading from a minus sign (–) to the exterior of the axon. Explain with relation to the trace on the oscilloscope screen. ______________________________

______________________________

d. In the middle portion of the diagram to the left, there is an arrow leading from $Na^+$ to the interior of the axon. Relate this arrow to the "upswing" of the action potential. ______________________________

______________________________

e. By what means does $Na^+$ cross the axomembrane? ______________________________

______________________________

f. In the middle portion of the diagram to the right, there is an arrow leading from $K^+$ to the exterior of the axon. Relate this arrow to the "downswing" of the action potential. ______________________________

______________________________

g. By what means does $K^+$ cross the axomembrane? ______________________________

______________________________

## Transmission Across a Synapse (p. 705)

- Transmission of impulses between neurons is usually accomplished by means of chemicals called neurotransmitters.

12. Label this diagram of a synapse with the following terms:
    axon bulbs
    dendrite
    neurotransmitter
    synaptic cleft
    synaptic vesicles

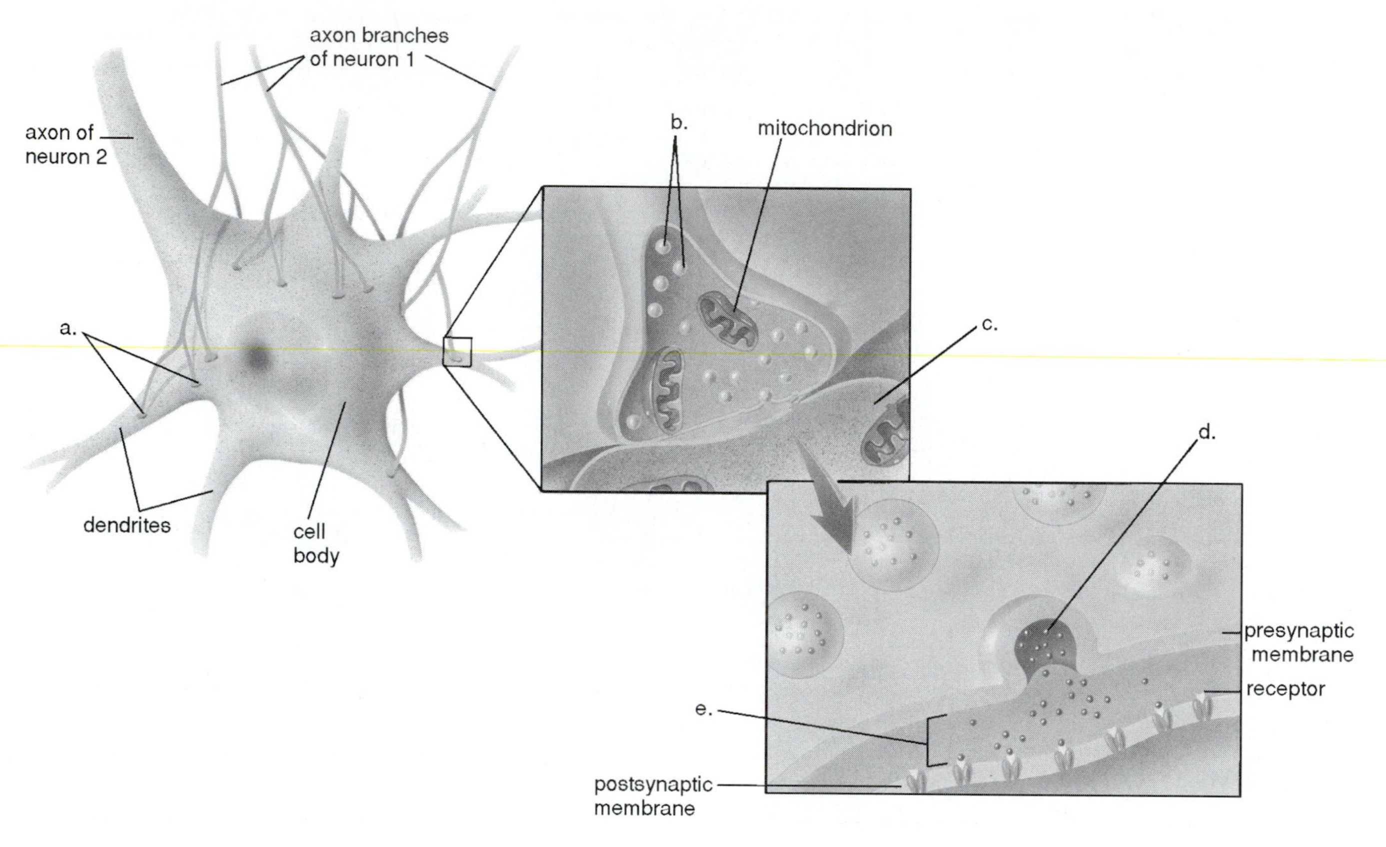


13. a. What causes transmission of the nerve impulse across a synapse?

    ______________________________________________

    ______________________________________________

    b. What is integration?

    ______________________________________________

    ______________________________________________

## 39.3 Central Nervous System: Brain and Spinal Cord (pp. 706–9)

- The central nervous system controls the other systems of the body and coordinates body functions.

14. The central nervous system consists of the a.__________________ and b.__________________.

15. Label the parts of the brain, using the following alphabetized list of terms.

cerebellum    cerebrum    corpus callosum    medulla oblongata    pituitary gland    pons    thalamus

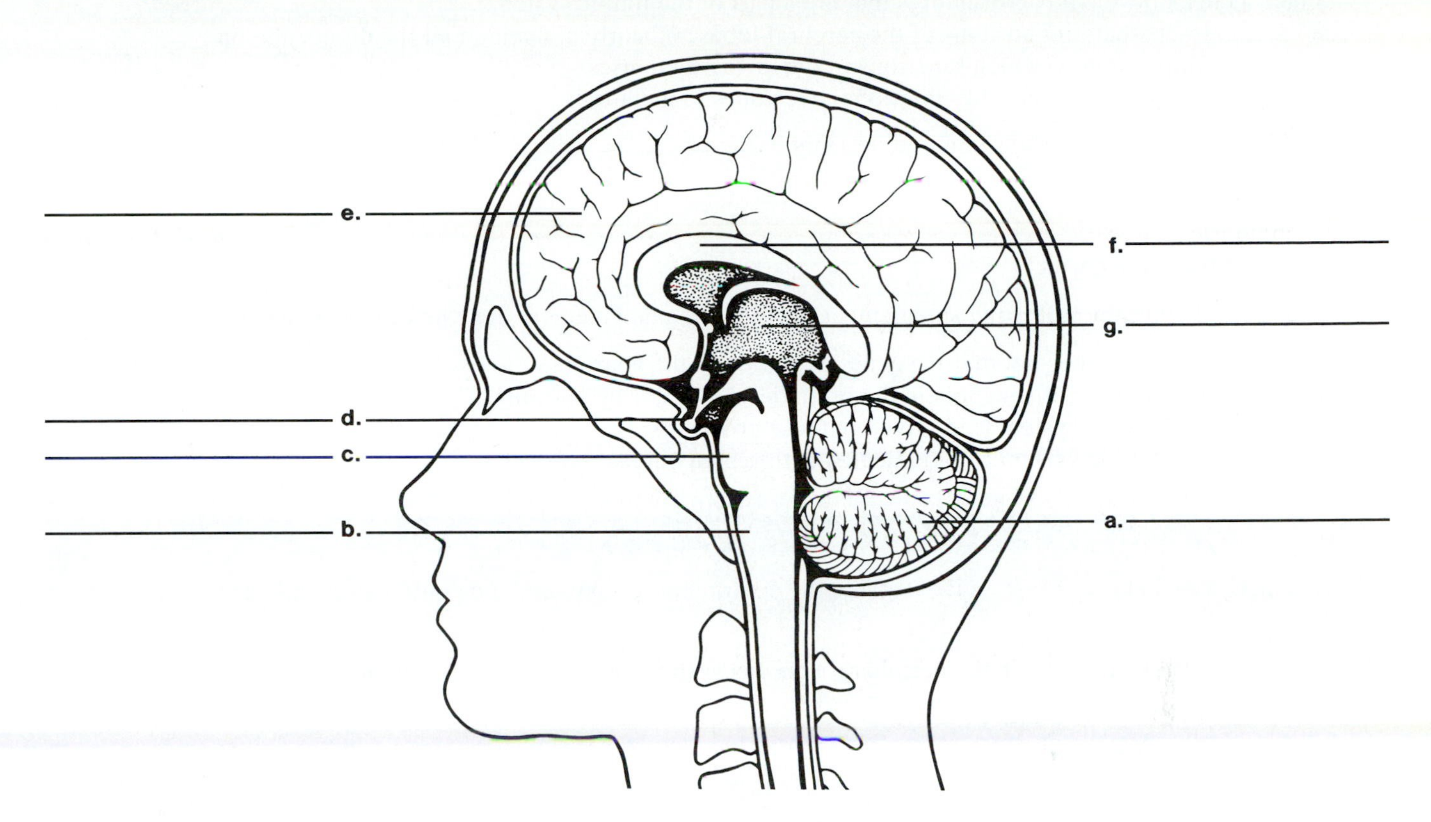


16 . Fill in the following table to indicate the functions of the parts of the brain.

| Brain Part | Function |
|---|---|
| Cerebrum | a. |
| Thalamus | b. |
| Hypothalamus | c. |
| Cerebellum | d. |
| Medulla oblongata | e. |

17. Match each description to the correct name of the lobe using these terms (some terms may be used more than once).

all lobes    frontal lobe    occipital lobe    parietal lobe    temporal lobe

a. ________________ Contains primary motor area, which controls voluntary motions.

b. ________________ Contains primary somatosensory area, which receives sensory information from the skin and skeletal muscles.

c. ________________ Contains a primary visual area.

d. ________________ Contains a primary auditory area.

e. ________________ Contains a primary association area.

f. ________________ Carries on higher mental functions such as reasoning and critical thinking.

18. The outer part of the cerebrum is called the a ________________, and the two halves of the cerebrum are called the b ________________.

19. Place a check beside those structures that are a part of the limbic system.
   a. _____ tracts that join portions of the cerebral lobes, subcortical nuclei, and the diencephalon
   b. _____ hippocampus, which functions in retrieving memories
   c. _____ amygdala, which adds emotional overtones to memories

20. Which of these best describes the limbic system? ______________
   a. reasoning
   b. emotions
   c. memories
   d. All of these are correct.

21. Indicate whether the following statements about learning and memory are true (T) or false (F):
   a. _____ Learning does not involve gene regulation in the neuron.
   b. _____ The number of synapses in the brain decreases during learning.
   c. _____ The limbic system is involved in these processes.
   d. _____ The limbic system communicates with sensory areas.

## 39.4 Peripheral Nervous System (pp. 710–13)

- The peripheral nervous system lies outside the central nervous system and contains cranial nerves.

22. Name two types of nerves in the peripheral nervous system: a. ____________ and b. ____________ nerves. What is a nerve? c. ______________________________. Why is a spinal nerve called a mixed nerve? d. ______________________________

### Somatic System (p. 711)

23. Label this diagram of the reflex arc using the following alphabetized list of terms.
   central canal
   dorsal root ganglion
   effector
   gray matter
   interneuron
   motor neuron
   receptor
   sensory neuron
   ventral horn of gray matter
   white matter

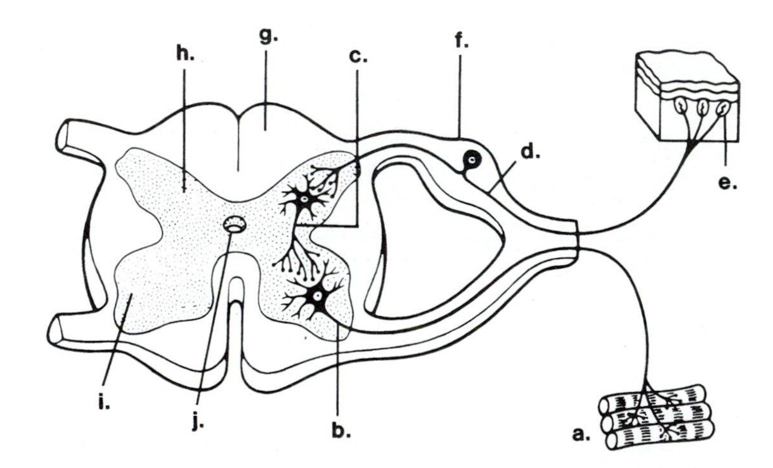

24. During a spinal reflex, a stimulus is received by a(n) a.______________, which initiates an impulse in the b.______________ neuron. This neuron takes the message to the spinal cord and transmits it to the c.______________. This neuron passes the impulse to the d.______________ neuron, which takes the message from the spinal cord and innervates a muscle, causing a reaction to the stimulus.

## Autonomic System (pp. 712–13)

25. Indicate three ways in which the sympathetic and parasympathetic divisions are similar.
   a. ______________________________
   b. ______________________________
   c. ______________________________
26. Complete the following table to show how the two divisions differ:

| | Sympathetic | Parasympathetic |
|---|---|---|
| Type of situation | | |
| Neurotransmitter | | |
| Ganglia near spinal cord, or ganglia near organ? | | |
| Spinal nerves only, or spinal nerves plus vagus? | | |

27. Label each of the following as somatic system (S) or autonomic system (A):
   a. _____ activation of skeletal muscles
   b. _____ parasympathetic and sympathetic divisions
   c. _____ control of internal organs

## Simon Says About Nervous Conduction 

For each correct answer, Simon says, "You may move one step forward." Total possible number of steps forward is 10 steps.

1. Which of these would NOT be used when studying nerve conduction?
   a. voltmeter
   b. oscilloscope
   c. electron microscope
   d. electrodes
   e. electric current
2. Which one is NOT *directly* needed for nerve conduction?
   a. dendrites
   b. axons
   c. plasma membrane
   d. nucleus
   e. cytoplasm of the axon
   f. ions
3. Which one does NOT move during nerve conduction?
   a. sodium
   b. potassium
   c. plus charges
   d. minus charges
4. Which one does NOT accurately describe a resting neuron?
   a. positive on the outside of the membrane and negative on the inside
   b. $Na^+$ on the outside of the membrane and $K^+$ on the inside
   c. –65 mV inside
   d. negative on both sides of the membrane
5. Which one is NOT involved with an action potential?
   a. resting potential
   b. permeability
   c. sodium-potassium pump
   d. plasma membrane
   e. acetylcholine
   f. ions
   g. glycogen
6. Which one does NOT conduct a nerve impulse?
   a. sensory neurons
   b. osteocytes
   c. motor neurons
   d. sensory nerves
   e. motor nerves
7. Which one is improperly matched?

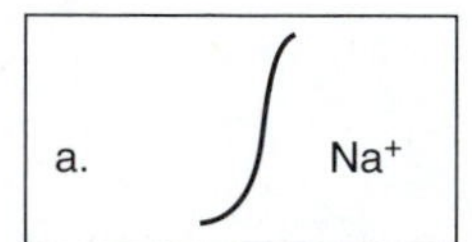


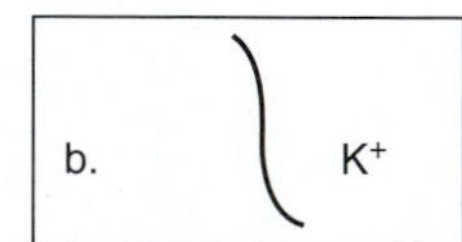


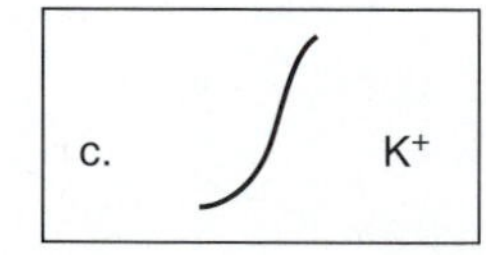


d. $\frac{Na^+}{K^+} \quad \frac{Na^+}{K^+}$

8. Which number could NOT be associated with an action potential?
   a. –65 millivolts
   b. 0 millivolts
   c. +40 millivolts
   d. –40 watts
9. Which one is improperly matched?
   a. ($e^-$) (nerve impulse)
   b. (-sodium-potassium pump) (resting potential)
   c. (+) ($Na^+$)
   d. (–) ($K^+$)
   e. (plasma membrane) (semipermeable)
10. Which one is NOT true?

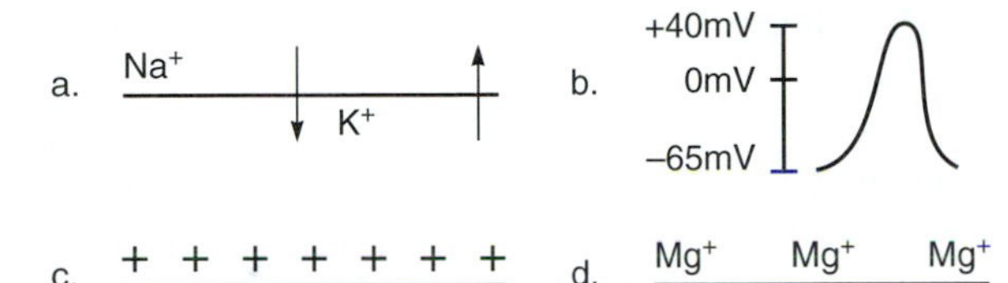


How many steps were you allowed by Simon? ________________

# Chapter Test

## Objective Questions

Do not refer to the text when taking this test.

____ 1. Which of these contains the nucleus?
   a. axon
   b. dendrite
   c. cell body
   d. Both *a* and *b* are correct.

____ 2. The neuron found wholly and completely within the CNS is the
   a. motor neuron.
   b. sensory neuron.
   c. interneuron.
   d. All of these are correct.

In questions 3–6, match the functions to these structures (some letters are not used):
   a. dendrites
   b. axons
   c. synaptic vesicles
   d. neurofibril node
   e. ganglia
   f. motor neurons

____ 3. contain neurotransmitters
____ 4. are nerve cell bodies outside the CNS
____ 5. terminate at muscles
____ 6. are gaps in the myelin sheath

____ 7. Synaptic vesicles are
   a. at the ends of dendrites and axons.
   b. at the ends of axons only.
   c. along the length of all long fibers.
   d. All of these are correct.

____ 8. The sodium-potassium pump maintains the resting potential.
   a. true
   b. false

____ 9. What is involved in a nerve impulse?
   a. ions
   b. electrons
   c. atoms
   d. molecules

____ 10. The downswing of the nerve impulse is caused by the movement of
   a. sodium ions to the inside of a neuron.
   b. sodium ions to the outside of a neuron.
   c. potassium ions to the inside of a neuron.
   d. potassium ions to the outside of a neuron.

____ 11. Rapid conduction of a nerve impulse in vertebrates is due to
   a. the large diameters of the axons.
   b. gaps in the myelin sheath.
   c. an abundance of synapses.
   d. the high permeability of neuronal membranes to ions.
   e. All of these are correct.

____ 12. Acetylcholine
   a. is a neurotransmitter.
   b. crosses the synaptic cleft.
   c. is broken down by acetylcholinesterase.
   d. All of these are correct.

____ 13. From an evolutionary perspective, which group of animals is the first to exhibit a central nerve cord system?
a. cnidarian
b. planarian
c. annelids and arthropods
d. vertebrates

____ 14. A spinal nerve is a __________ nerve.
a. motor
b. sensory
c. mixed
d. All of these are correct.

____ 15. The autonomic system has two divisions called the ______ divisions.
a. central nervous and peripheral nervous
b. somatic and skeletal
c. efferent and afferent
d. sympathetic and parasympathetic

____ 16. Which of the following motor neurons would be found in the autonomic division of the peripheral nervous system?
a. motor neurons ending at skeletal muscle
b. motor neurons leading to the esophagus
c. motor neurons at the surface of the skin
d. All of these are correct.

____ 17. Sensory neurons
a. take impulses to the CNS.
b. have a long dendrite and a short axon.
c. take impulses away from the CNS.
d. Both *b* and *c* are correct.

____ 18. Which system is involved during stress?
a. parasympathetic
b. sympathetic
c. somatic
d. Both *a* and *b* are correct.

____ 19. The neurotransmitter of the parasympathetic division is
a. norepinephrine.
b. acetylcholine.
c. cholinesterase.
d. Both *a* and *b* are correct.

____ 20. Automatic responses to specific external stimuli require
a. rapid impulse transmission along the spinal cord.
b. the involvement of the brain.
c. simplified pathways called reflex arcs.
d. the involvement of the autonomic system.

____ 21. What portion of the nervous system is required for a reflex arc?
a. mixed spinal nerve
b. gray matter of spinal cord
c. cerebrum
d. Both *a* and *b* are correct.
e. All of these are correct.

____ 22. Which is the largest part of the human brain?
a. cerebrum
b. cerebellum
c. medulla oblongata
d. thalamus

____ 23. The function of the cerebellum is
a. consciousness.
b. motor coordination.
c. homeostasis.
d. sense reception.

____ 24. Which portion of the brain is involved in judgment?
a. cerebellum
b. frontal lobe of cerebrum
c. medulla oblongata
d. parietal lobe of cerebrum

____ 25. Drugs of abuse primarily affect the
a. cerebellum.
b. medulla oblongata.
c. limbic system.
d. thalamus.

## Critical Thinking Questions

Answer in complete sentences.

26. How is the structure of the neuron well suited for its functions?

27. How does motor control of internal organs differ from motor control of skeletal muscles?

**Test Results:** ______ number correct ÷ 27 = ______ × 100 = ______ %

## Exploring the Internet

The Online Learning Center at *www.mhhe.com/maderbiology8* has additional study material and practice quizzes that can help you master the content of this chapter. You can also find links to websites exploring additional topics in biology. Access to the Online Learning Center is free for those who have purchased a new textbook.

## Answer Key

### Study Exercises

**1. a.** 2 **b.** 3 **c.** 1 **d.** 5 **e.** 4 **2.** Cephalization is a definite head region that bears sense organs. **3.** The ganglia or a brain and a longitudinal nerve cord suggest a central nervous system; transverse nerves suggest a peripheral nervous system. **4. a.** sends signal to cell body **b.** control center **c.** takes impulse away from cell body **5. a.** sensory receptor **b.** cell body **c.** axon **d.** dendrite **e.** cell body **f.** axon **g.** dendrite **h.** axon **i.** myelin sheath **j.** neurofibril node (node of Ranvier) **k.** effector **6.** to take nerve impulses to CNS **7.** to take nerve impulses from one part of CNS to another **8.** to take nerve impulses away from CNS **9. a.** an instrument that measures potential differences in millivolts **b.** Large organic ions are inside the membrane (also there is a slow leakage of $K^+$ to outside). **c.** more $Na^+$ outside than inside **d.** less $K^+$ outside than inside **e.** sodium-potassium pump **10. a.** depolarization **b.** repolarization **c.** action potential **d.** threshold **e.** resting potential **11. a.** change from a negative potential to a positive potential inside the axon **b.** change from a positive potential inside the axon back to negative again **c.** refers to the reversal of potential that occurs during the action potential **d.** The movement of $Na^+$ to the inside of the axon causes the inside to become positive compared to outside. **e.** gate of $Na^+$ channel opens **f.** The movement of $K^+$ to the outside causes the inside to become negative again. **g.** gate of $K^+$ channel opens **12. a.** axon bulbs **b.** synaptic vesicles **c.** dendrite **d.** neurotransmitter **e.** synaptic cleft **13. a.** neurotransmitter is received by receptor of the next neuron **b.** a summing up of the effects of various neurotransmitters—some are stimulatory and some are inhibitory **14. a.** brain **b.** spinal cord **15. a.** cerebellum **b.** medulla oblongata **c.** pons **d.** pituitary gland **e.** cerebrum **f.** corpus callosum **g.** thalamus **16. a.** motor control, higher levels of thought **b.** integrates and sends sensory information to cerebrum **c.** homeostasis **d.** motor coordination **e.** control of internal organs **17. a.** frontal lobe **b.** parietal lobe **c.** occipital lobe **d.** temporal lobe **e.** all lobes **f.** frontal lobe **18. a.** cortex **b.** cerebral hemisphere **19.** a, b, c **20.** d **21. a.** F **b.** F **c.** T **d.** T **22. a.** cranial **b.** spinal **c.** bundle of fibers (axons) **d.** it contains both sensory and motor fibers **23.** See Figure 39.11, page 711, in text. **24. a.** receptor **b.** sensory **c.** interneuron **d.** motor **25. a.** function automatically and involuntarily **b.** innervate internal organs **c.** use two motor neurons and one ganglion for each impulse
**26.**

| Sympathetic | Parasympathetic |
|---|---|
| emergency activity | normal activity |
| Norepinephrine (NE) | Acetylcholine (ACH) |
| near spinal cord | near organ |
| spinal nerves only | spinal and vagus nerves |

**27. a.** S **b.** A **c.** A

### Simon Says About Nervous Conduction

**1.** c **2.** d **3.** d **4.** d **5.** g **6.** b **7.** c **8.** d **9.** a **10.** d

### Chapter Test

**1.** c **2.** c **3.** c **4.** e **5.** f **6.** d **7.** b **8.** a **9.** a **10.** d **11.** b **12.** d **13.** c **14.** c **15.** d **16.** b **17.** a **18.** b **19.** b **20.** c **21.** d **22.** a **23.** b **24.** b **25.** c **26.** It is a long, thin cell that reaches over distances in the body and is capable of delivering signals rapidly to a body region. **27.** Internal organs are controlled by two branches of the autonomic system that have opposite effects; for example, the sympathetic division causes the heart to speed up and the parasympathetic division causes the heart to slow down. In the somatic system, motor impulses always cause skeletal muscles to contract.

# 40

# Sense Organs

## Chapter Review

Sense organs detect changes in the environment and generate nerve impulses that travel along sensory neural pathways to reach the brain where they are interpreted. This chapter discusses chemoreceptors, photoreceptors, and mechanoreceptors.

Chemoreception is almost universally found in animals. Human **taste buds** and **olfactory cells** are **chemoreceptors** sensitive to chemicals in liquids, food, and air.

Eyes contain **photoreceptors.** The compound eye of arthropods is made up of many independent visual units, whereas the human eye is a **camera-type eye** with a single lens. The photoreceptors in humans are the **rod cells** and the **cone cells** located in the **retina** of the eye. The rods work in minimum light and detect motion, but they do not detect color. The cones require bright light and do detect color.

**Rhodopsin,** the pigment found in rod cells, is a molecule composed of opsin and retinal. When rhodopsin absorbs light energy, opsin is released. A cascade of reactions then causes ion channels to close. Thereafter, nerve impulses go to the visual areas of the cerebral cortex. There are three types of cone cells, containing blue, green, or red pigments. Each pigment is composed of retinal and a particular opsin.

**Mechanoreceptors** in the ear are varied, but many are hair cells with cilia, such as those found in the lateral line of fishes as well as the inner ear of humans. The inner ear contains mechanoreceptors for balance as well as hearing. The stereocilia of hair cells located within the **cochlea** are embedded in the tectorial membrane. When we hear, pressure waves cause these cilia to bend and nerve impulses are carried by the auditory nerve to the brain.

## Study Exercises

Study the text section by section as you answer the questions that follow.

### 40.1 Chemical Senses (pp. 720–21)

- Chemoreceptors for sensing chemical substances in food, liquids, and air are almost universally found in animals.
- Human taste buds and olfactory cells are chemoreceptors that respond to chemicals in food and in the air, respectively.

1. Match the descriptions to these terms:

   taste receptors    smell receptors    both taste and smell receptors

   a. ______________________ receptor proteins combine with chemical
   b. ______________________ brain senses impulses as a weighted average
   c. ______________________ taste buds with microvilli house receptor proteins
   d. ______________________ salty receptor proteins on tip of tongue
   e. ______________________ olfactory cells
   f. ______________________ are not effective when you have a cold
   g. ______________________ easily adapt to outside stimuli
   h. ______________________ involved in enjoyment of food
   i. ______________________ a characteristic combination of receptor proteins are activated

2. Match the descriptions to these sensory receptors.
   1. taste buds
   2. olfactory cells
   3. both

   a. _____ Found along walls of papillae
   b. _____ Ends in a tuft of cilia
   c. _____ Information goes directly to the cerebrum and also to the brain stem
   d. _____ Plasma membrane receptors
   e. _____ Processing begins outside the brain

## 40.2 Sense of Vision (pp. 722–27)

- The eye of arthropods is a compound eye made up of many individual units; the human eye is a camera-type eye with a single lens.
- Sensory receptors for sight contain visual pigments that respond to light rays.
- In the human eye, the rod cells work in minimal light and detect motion; the cone cells require bright light and detect color.
- A great deal of integration occurs in the retina of the human eye before nerve impulses are sent to the brain.

3. Place the appropriate letter next to each statement.

   A—arthropod eye    H—compound eye

   a. _____ consists of three layers
   b. _____ contains one lens for image formation
   c. _____ has many independent visual units
   d. _____ is a compound eye

4. If image-forming eyes are found only among cnidarians, annelids, molluscs, and arthropods, what do the "eyespots" of planarians do? ______________________________________________

   ______________________________________________

5. Why is the eye of vertebrates and certain molluscs (squid and octopus) called a camera-type of eye? _________

   ______________________________________________

6. Label the parts of the eye using the alphabetized list of terms. On the answer blanks provided on the next page, state the name and function of each part of the eye indicated in the illustration.

   choroid
   ciliary body
   cornea
   fovea centralis
   iris
   lens
   optic nerve
   retina
   sclera

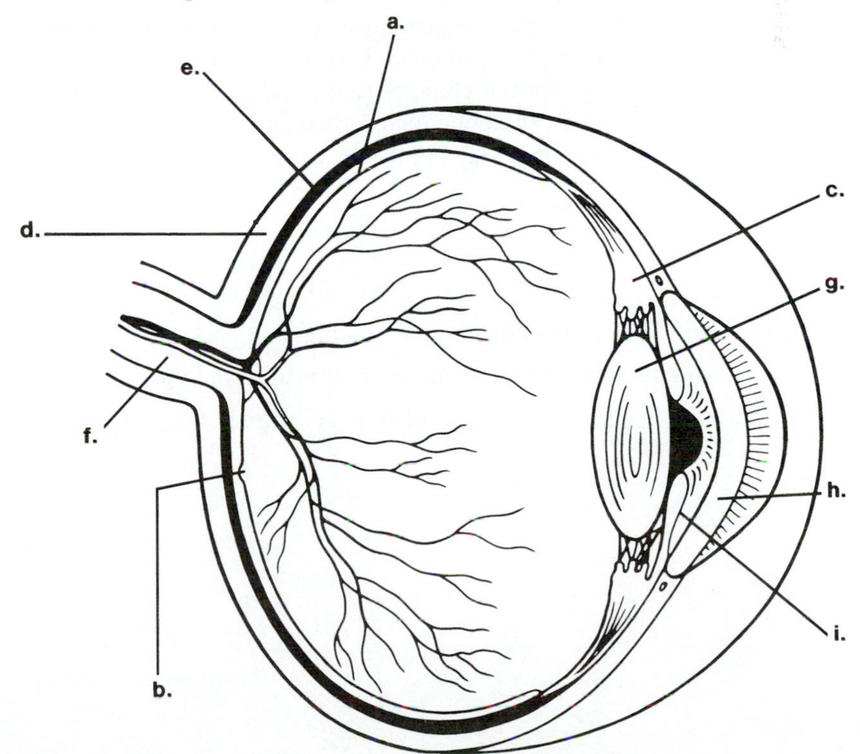

| | Structure | Function |
|---|---|---|
| a. | ____________________ | ____________________ |
| b. | ____________________ | ____________________ |
| c. | ____________________ | ____________________ |
| d. | ____________________ | ____________________ |
| e. | ____________________ | ____________________ |
| f. | ____________________ | ____________________ |
| g. | ____________________ | ____________________ |
| h. | ____________________ | ____________________ |
| i. | ____________________ | ____________________ |

7. In the retina of the eye, the rod cells and cone cells are closest to the choroid. They pass nerve impulses to [a.]__________________ cells, which pass them to [b.]__________________ cells. There are many more rod cells and cone cells than [c.]__________________ cells. Therefore, some [d.]__________________ occurs in the retina before nerve impulses reach the optic nerve. Where the optic nerve passes through the retina, there is a [e.]__________________ spot.

8. The lens is [a.]__________________ for distant objects and [b.]__________________ for near objects. This is called [c.]__________________.

9. Fill in the blanks in this table:

| Name | Description | Where Is Image Focused? | Correction |
|---|---|---|---|
| Nearsightedness | See nearby objects | a. | Concave lens |
| Farsightedness | b. | c. | d. |
| Astigmatism | Cannot focus | Image not focused | e. |

10. Put these in the proper order to describe generation of nerve impulses resulting in sight.
    a. _____ A cascade of reactions leads to closure of ion channels.
    b. _____ When light strikes retinal, it changes shape and rhodopsin is activated.
    c. _____ Rhodopsin contains a pigment called retinal and opsin.
    d. _____ Ganglionic cells send messages to the brain.

11. Only dim light is required to stimulate [a.]__________________, and therefore they are responsible for [b.]__________________ vision. The [c.]__________________, located primarily in the [d.]__________________ of the retina, detect the detail and [e.]__________________ of an object. There are three different kinds of cone cells that contain pigments called the [f.]__________, [g.]__________, and [h.]__________ pigments. Each pigment is made up of retinal and opsin, but the opsin structure in each is slightly different. [i.]__________________ of cone cells are believed to be stimulated by in-between shades of color.

## 40.3 Senses of Hearing and Balance (pp. 728–31)

- The inner ear of humans contains mechanoreceptors for the senses of balance and hearing.
- The mechanoreceptors for hearing are hair cells in the cochlea of the inner ear, which respond to pressure waves.
- The mechanoreceptors for balance are hair cells in the vestibule and semicircular canals of the inner ear, which respond to the tilt of the head and the movement of the body, respectively.

12. Label the parts of the ear using the alphabetized list of terms. On the answer blanks provided, state the name and function of each part of the ear indicated in the illustration.

    auditory canal
    auditory tube
    cochlea
    cochlear nerve
    malleus (hammer)
    pinna
    semicircular canal
    stapes (stirrup)
    tympanic membrane
    vestibule

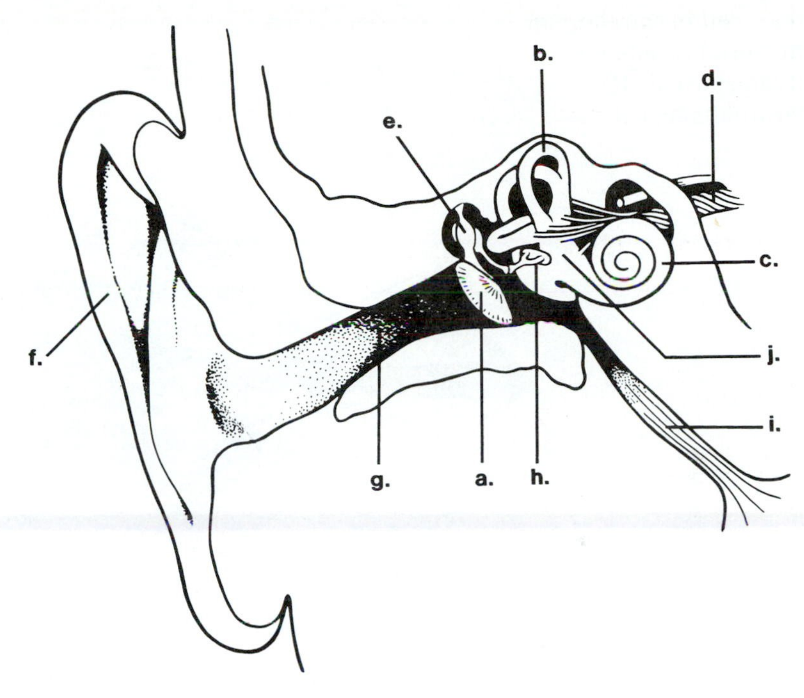


| | Structure | Function |
|---|---|---|
| a. | ______________________ | ______________________ |
| b. | ______________________ | ______________________ |
| c. | ______________________ | ______________________ |
| d. | ______________________ | ______________________ |
| e. | ______________________ | ______________________ |
| f. | ______________________ | ______________________ |
| g. | ______________________ | ______________________ |
| h. | ______________________ | ______________________ |
| i. | ______________________ | ______________________ |
| j. | ______________________ | ______________________ |

13. Match the descriptions with these structures:
   1. semicircular canals
   2. utricle and saccule

   a. _____ contains otoliths
   b. _____ rotational equilibrium
   c. _____ gravitational equilibrium

14. Label this diagram of the cochlea (as it might appear if unwound) using the alphabetical list of terms.
   basilar membrane
   cochlear canal
   cochlear nerve
   hair cell in spiral organ
   tectorial membrane
   tympanic canal
   vestibular canal

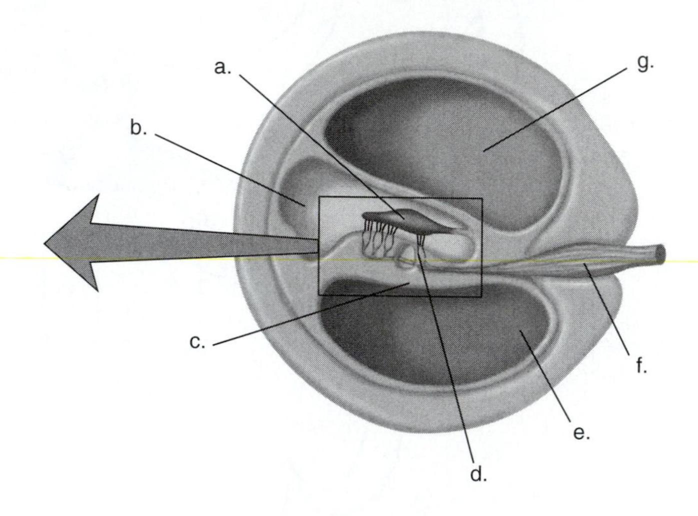


15. What happens when the basilar membrane moves up and down? ______________________________________________

______________________________________________________________________

16. Rearrange the letters of items *a–f* to describe how we hear. ____________________
   a. nerve impulses
   b. tympanic membrane oscillates
   c. fluid pressure waves in cochlear canal
   d. stapes hits oval window
   e. temporal lobe of brain
   f. stimulation of spiral organ

17. Which of these pairs is true about the lateral line of fishes?
   Structure:
   a. _____ Like a Pacinian corpuscle, a lateral line receptor is concentric layers of connective tissue wrapped around the end of a sensory neuron.
   b. _____ Collection of hair cells with cilia embedded in a gelatinous cupula.

   Function:
   c. _____ Detects currents and pressure waves from nearby objects.
   d. _____ Is more sensitive to temperature or chemicals than mechanical stimuli.

# Chapter Test

## Objective Questions

Do not refer to the text when taking this test.

____ 1. Taste cells and olfactory cells are both
 a. somatic senses.
 b. mechanoreceptors.
 c. pseudociliated epithelium.
 d. chemoreceptors.

____ 2. Select the incorrect association.
 a. compound eye—one lens, insect
 b. camera-type eye—one lens, squid
 c. compound eye—many lenses, arthropods
 d. photoreceptor—eyespot, planarian

____ 3. The blind spot is
 a. a nontransparent area on the lens.
 b. a nontransparent area on the cornea.
 c. the area on the retina where there are no rods and cones.
 d. called the fovea centralis.

In questions 4–7, match the questions to these structures:
 a. retina
 b. optic nerve
 c. lens
 d. cerebrum
 e. All of these are correct.

____ 4. Which is (are) necessary to proper vision?
____ 5. Which contain(s) the receptors for sight?
____ 6. Which focus(es) light?
____ 7. In which is the sensation of sight realized?

____ 8. The current theory of color vision proposes that
 a. three primary colors are associated with color vision.
 b. cone cells respond selectively to different wavelengths of light.
 c. rod cells are responsible for nighttime color vision.
 d. Both *a* and *b* are correct.
 e. All of these are correct.

____ 9. Acute vision is made possible by the
 a. choroid.
 b. fovea centralis.
 c. ciliary muscle.
 d. ciliary body.

____10. If you are nearsighted, the image is focused
 a. in front of the retina.
 b. behind the retina.
 c. on the retina.
 d. at the blind spot.

____11. The disorders of nearsightedness and farsightedness are due to
 a. an eyeball of incorrect length.
 b. a cloudy lens.
 c. pressure increase.
 d. a torn retina.

____12. The lateral line of fishes
 a. has receptors with cilia embedded in a cupula.
 b. detects water currents and pressure waves.
 c. is a mechanoreceptor.
 d. All of these are correct.

____13. Select the incorrect association.
 a. outer ear—auditory canal
 b. middle ear—otoliths
 c. inner ear—spiral organ
 d. inner ear—semicircular canals

____14. The cochlear nerve is associated with the
 a. spiral organ.
 b. ossicles.
 c. tympanic membrane.
 d. auditory tube.

In questions 15–17, match the questions to these structures (some are used more than once):
 a. ossicles
 b. otoliths
 c. cochlea
 d. auditory canal

____15. Which has (have) nothing to do with hearing?
____16. In which do you find the receptors for hearing?
____17. Which of these is (are) concerned with balance?

____18. In the utricle and saccule,
 a. stereocilia of hair cells are embedded in an otolithic membrane.
 b. bending of stereocilia tells the direction of head movement.
 c. stereocilia of hair cells are embedded in the tectorial membrane.
 d. Both *a* and *b* are correct.

____19. Which part of the ear has receptors for hearing?
 a. outer
 b. middle
 c. inner
 d. All of these are correct.

____20. When we hear,
 a. the basilar membrane vibrates.
 b. stereocilia of hair cells embedded in the tectorial membrane bend.
 c. nerve impulses travel in the cochlear nerve to the brain.
 d. All of these are correct.

## Critical Thinking Questions

Answer in complete sentences.

21. Damage to the retina, optic nerve, and brain prevents vision in a person. Only one is a sense organ. Why are all three necessary?

22. The cochlea, vestibule, and semicircular canals are three distinct structures of the inner ear. How are they remarkably similar at the receptor cell and stimulus level?

**Test Results:** _________ number correct ÷ 22 = _____ × 100 = _____ %

## Exploring the Internet

The Online Learning Center at *www.mhhe.com/maderbiology8* has additional study material and practice quizzes that can help you master the content of this chapter. You can also find links to websites exploring additional topics in biology. Access to the Online Learning Center is free for those who have purchased a new textbook.

## Answer Key

### Study Exercises

**1. a.** both **b.** taste receptors **c.** taste receptors **d.** taste receptors **e.** smell receptors **f.** smell receptors **g.** smell receptors **h.** both **i.** smell receptors **2. a.** 1 **b.** 2 **c.** 1 **d.** 3 **e.** 2 **3. a.** H **b.** H **c.** A **d.** A **4.** seek light **5.** They use a single, adjustable lens to bring images into focus. **6. a.** retina; photoreceptors for sight **b.** fovea centralis; makes acute vision possible **c.** ciliary body; holds lens in place; accommodation **d.** sclera; protects and supports eyeball **e.** choroid; absorbs stray light rays **f.** optic nerve; transmits nerve impulse to brain **g.** lens; refracts and focuses light **h.** cornea; refracts light rays **i.** iris; regulates entrance of light **7. a.** bipolar **b.** ganglionic **c.** ganglionic **d.** integration **e.** blind **8. a.** flat **b.** rounded **c.** accommodation **9. a.** in front of retina **b.** see distant objects **c.** behind retina **d.** convex lens **e.** irregular lens **10.** c, b, a, d **11. a.** rod cells **b.** night **c.** cone cells **d.** fovea **e.** color **f.** blue **g.** green **h.** red **i.** Combinations **12. a.** tympanic membrane; starts vibration of ossicles **b.** semicircular canal; rotational equilibrium **c.** cochlea; contains mechanoreceptors for hearing **d.** cochlear nerve; transmission of nerve impulse **e.** malleus (hammer); transmits vibrations **f.** pinna; reception of sound waves **g.** auditory canal; collection of sound waves **h.** stapes (stirrup); transmits vibrations to oval window **i.** auditory tube; connects middle ear to pharynx **j.** vestibule; gravitational equilibrium **13. a.** 2 **b.** 1 **c.** 2 **14. a–g.** See Figure 40.10, page 729, in text. **15.** Stereocilia of hair cells embedded in tectorial membrane bend, and nerve impulses begin. **16.** b, d, c, f, a, e **17.** b, c

### Chapter Test

**1.** d **2.** a **3.** c **4.** e **5.** a **6.** c **7.** d **8.** d **9.** b **10.** a **11.** a. **12.** d **13.** d **14.** a **15.** b **16.** c **17.** b **18.** d **19.** c **20.** d **21.** The photoreceptors of the retina generate nerve impulses. A sensory pathway along the optic nerve must deliver a signal to the brain. The brain (third component) must interpret this input for the sense of vision to occur. **22.** All three structures have sensory hair cells that must be disturbed to create the necessary stimulus for hearing (cochlea), rotational equilibrium (semicircular canals), and gravitational equilibrium (vestibule).

# 41

# Support Systems and Locomotion

## Chapter Review

Three types of skeletons have evolved among the different species of the animal kingdom: the **hydrostatic skeleton,** the **exoskeleton,** and the **endoskeleton.** Skeletons generally provide support, protection, and assistance in movement. Additional functions in the human skeleton are mineral storage and blood cell production.

Bone is constantly being renewed: **osteoblasts** build new bone, and **osteoclasts** break down bone. **Osteocytes** are found in the lacunae of osteons.

The human skeleton is divided into two parts: (1) the **axial skeleton,** made up of the skull; the ribs; the sternum; and the vertebrae; and (2) the **appendicular skeleton,** composed of the **pectoral girdle,** the **pelvic girdle,** and their appendages. **Joints** are regions where bones are connected. Joints are classified as fibrous, cartilaginous, or synovial.

Whole skeletal muscles get shorter when they contract; therefore, muscles work in antagonistic pairs. Muscle fibers are cells that contain **myofibrils** in addition to the usual cellular components. Longitudinally, myofibrils are divided into **sarcomeres,** where it is possible to note the arrangements of **actin** and **myosin** filaments. The **sliding filament theory** of muscle contraction says that **myosin** filaments have cross-bridges, which attach to and detach from **actin** filaments, causing them to slide and the **sarcomere** to shorten. The H zone disappears as actin filaments approach one another.

Innervation of a muscle fiber begins at a **neuromuscular junction.** Here, synaptic vesicles release ACh into the synaptic cleft. When the **sarcolemma** receives ACh, impulses move down the T tubules to the sarcoplasmic reticulum. When calcium ions are released from the sarcoplasmic reticulum, contraction occurs. When nerve impulses to the muscle cease, calcium ions are actively transported back into the calcium storage sites, and muscle relaxation occurs.

## Study Exercises

Study the text section by section as you answer the questions that follow.

### 41.1 Diversity of Skeletons (pp. 736–37)

- Animals have one of three types of skeletons: a hydrostatic skeleton, an exoskeleton, or an endoskeleton.
- The strong but flexible skeleton of arthropods and vertebrates is adaptive for living on land.

1. Cnidarians, flatworms, roundworms, and annelids have a.______________________ skeletons. With this type of skeleton, muscles contract against b.______________________ compartments. A limitation of this type of skeleton is that it does not provide c.______________________.
   Molluscs and arthropods have another type of skeleton called a(n) d.______________________. Although this offers protection similar to a suit of e.______________________, it must be f.______________________ periodically, allowing the animal to grow. Echinoderms and vertebrates have (a/an) g.______________________, primarily made of h.______________________ and i.______________________. This type of skeleton protects j.__________________, can k.__________________ with the animal, and also allows l.__________________ movements.

## 41.2 The Human Skeletal System (pp. 738–43)

- The human skeleton has many functions that contribute to homeostasis.
- Bone is formed during development, and bone renewal continues in adults.
- There are two types of bone tissue, compact bone and spongy bone, that differ in structure and function.
- In the human skeleton, the axial skeleton consists of the skull, the vertebral column, the sternum, and the ribs; the appendicular skeleton contains bones within the pectoral and pelvic girdles and the attached limbs.

2. List five functions of the human skeletal system.
   a. ____________________
   b. ____________________
   c. ____________________
   d. ____________________
   e. ____________________
3. Label this diagram of a long bone using the alphabetized list of terms.
   cartilage
   compact bone
   medullary cavity
   spongy bone

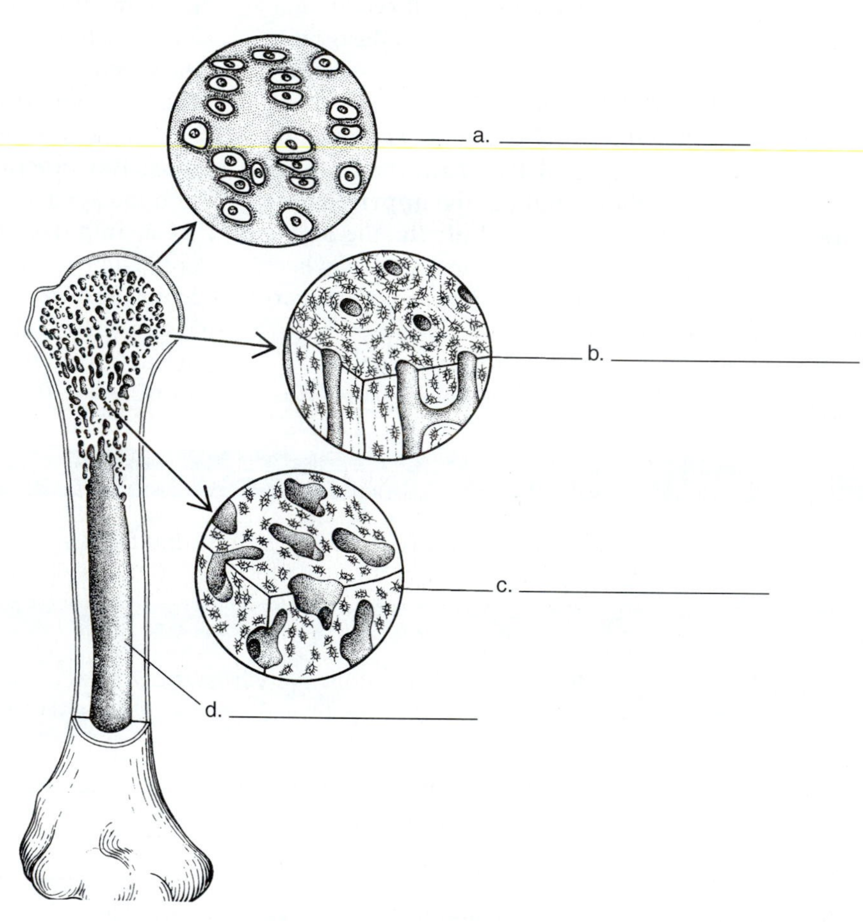


In questions 4–6, fill in the blanks.

4. Which part of the long bone shown in the previous illustration is associated with red bone marrow? a. ____________ Which part of bone is the hardest? b. ____________ Which part of bone is the most flexible? c. ____________ Which part of a long bone is associated with yellow bone marrow? d. ____________

5. a. What is the relationship between osteoblasts and osteocytes? ____________

   b. What role is played by the cartilaginous disks of long bones? ____________

   c. What is the function of an osteoclast? ____________

   d. What is the process of bone formation called? ____________

6. During the continual process of bone remodeling, osteoclasts remove worn-out bone cells. At the same time, a.____________ is released into the bloodstream. The two factors that affect bone thickness are b.____________ and c.____________.

7. Axial versus appendicular skeleton. Write *ax* in front of all bones belonging to the axial skeleton; write *ap* in front of all bones belonging to the appendicular skeleton. Write *pec* in front of all bones belonging to the pectoral girdle; write *pel* in front of all bones belonging to the pelvic girdle. Some items have more than one answer.

   a.________ coxal bone
   b.________ sternum
   c.________ humerus
   d.________ scapula
   e.________ skull
   f.________ femur
   g.________ ribs
   h.________ radius
   i.________ clavicle
   j.________ tibia
   k.________ fibula
   l.________ ulna

8. The radius and ulna are to the lower arm as the a.____________ and b.____________ are to the lower leg. The femur is to the upper leg as the c.____________ is to the upper arm. The metacarpals are to the palm as the d.____________ are to the foot.

9. Give the scientific terms for the common names.

   a. Shinbone ____________
   b. Collarbone ____________
   c. Hipbone ____________
   d. Thighbone ____________

## Classification of Joints (p. 743)

- Bones are connected at the joints; the joints differ in movability.

10. Match the descriptions to these types of joints:

    fibrous cartilaginous synovial

    a. ____________ lined with synovial membrane
    b. ____________ immovable, as in a suture
    c. ____________ slightly movable
    d. ____________ freely movable
    e. ____________ connected by hyaline cartilage

11. Fill in the table to describe the shoulder and elbow joints:

| Joint | Anatomical Type | Degree of Movement |
|---|---|---|
| Shoulder joint | | |
| Elbow joint | | |

## 41.3 The Human Muscular System (pp. 745–49)

- Macroscopically, human skeletal muscles work in antagonistic pairs; muscles at rest exhibit tone.
- Microscopically, muscle fiber contraction is dependent on actin and myosin filaments, and a ready supply of calcium ions and ATP.
- Motor nerve fibers release ACh at a neuromuscular junction, and thereafter a muscle fiber contracts.

12. Label this diagram of muscles and bones in the arm using the alphabetical list of terms.
    - biceps brachii
    - humerus
    - radius
    - scapula
    - triceps brachii
    - ulna

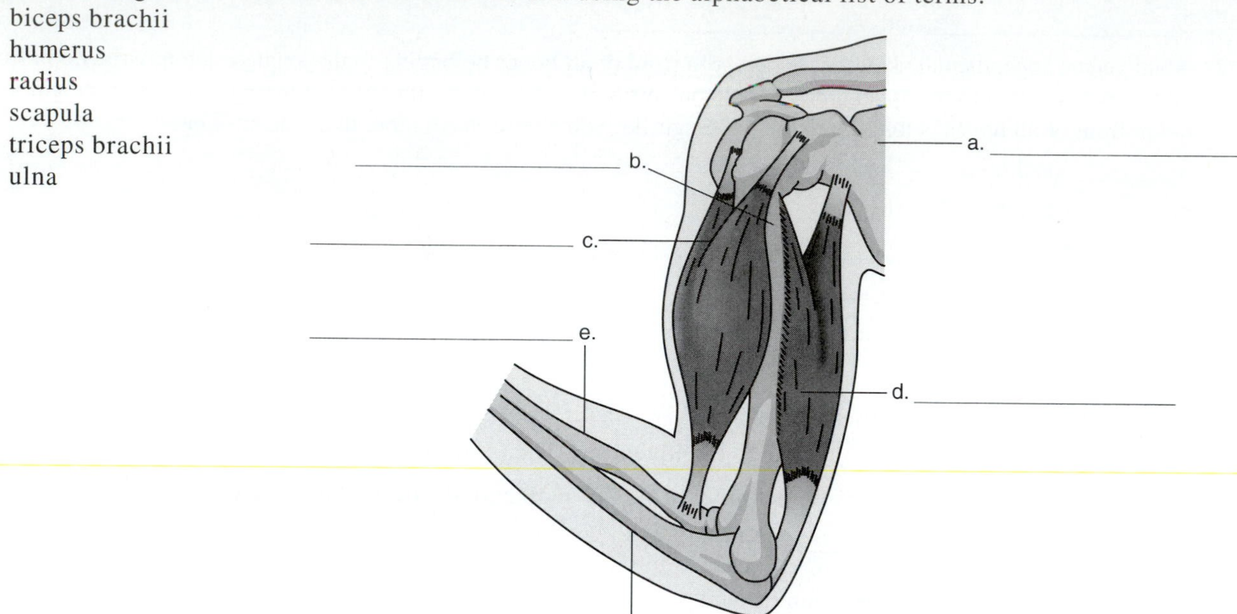


13. The biceps brachii and triceps brachii are antagonistic pairs. The biceps brachii raises the lower arm and the triceps brachii lowers the lower arm. Why do muscles work in antagonistic pairs?

______________________________

14. Label this diagram of a portion of a muscle fiber using the following alphabetized list of terms.
    - muscle fiber
    - myofibril
    - sarcolemma
    - sarcomere
    - sarcoplasmic reticulum
    - T tubule (used twice)
    - Z line

a. ______ b. ______ c. ______ d. ______ e. ______ f. ______ g. ______ h. ______

15. Label this diagram of a sarcomere using the following alphabetized list of terms.
    actin filament
    H zone
    myosin filament
    Z line

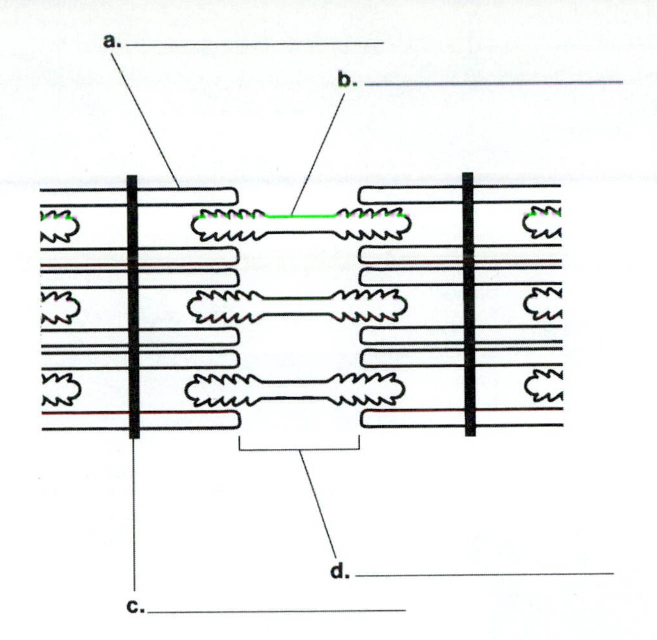


In question 16, fill in the blanks.

16. a. Which of your labels in question 15 is a thin filament? ____________________
    b. Which of your labels is a thick filament? ____________________
    c. Which of your labels is reduced in size when a sarcomere contracts? ____________________
    d. Which component has cross-bridges? ____________________
    e. Which of your labels is the filament that moves when the sarcomere contracts? ____________________
    f. What molecule immediately supplies energy for muscle contraction? ____________________
    g. What molecule is a storage form of high-energy phosphate in muscles? ____________________

## Muscle Innervation (p. 748)

17. What is the proper sequence for these phrases to describe what occurs at the neuromuscular junction to trigger muscle contraction? Indicate by letter. ____________________
    a. receptor sites on sarcolemma
    b. nerve impulse
    c. release of calcium from sarcoplasmic reticulum
    d. the neurotransmitter acetylcholine is released
    e. sarcomeres shorten
    f. synaptic cleft
    g. spread of impulses over sarcolemma to T tubules

18. Study this diagram and fill in the blanks that follow.

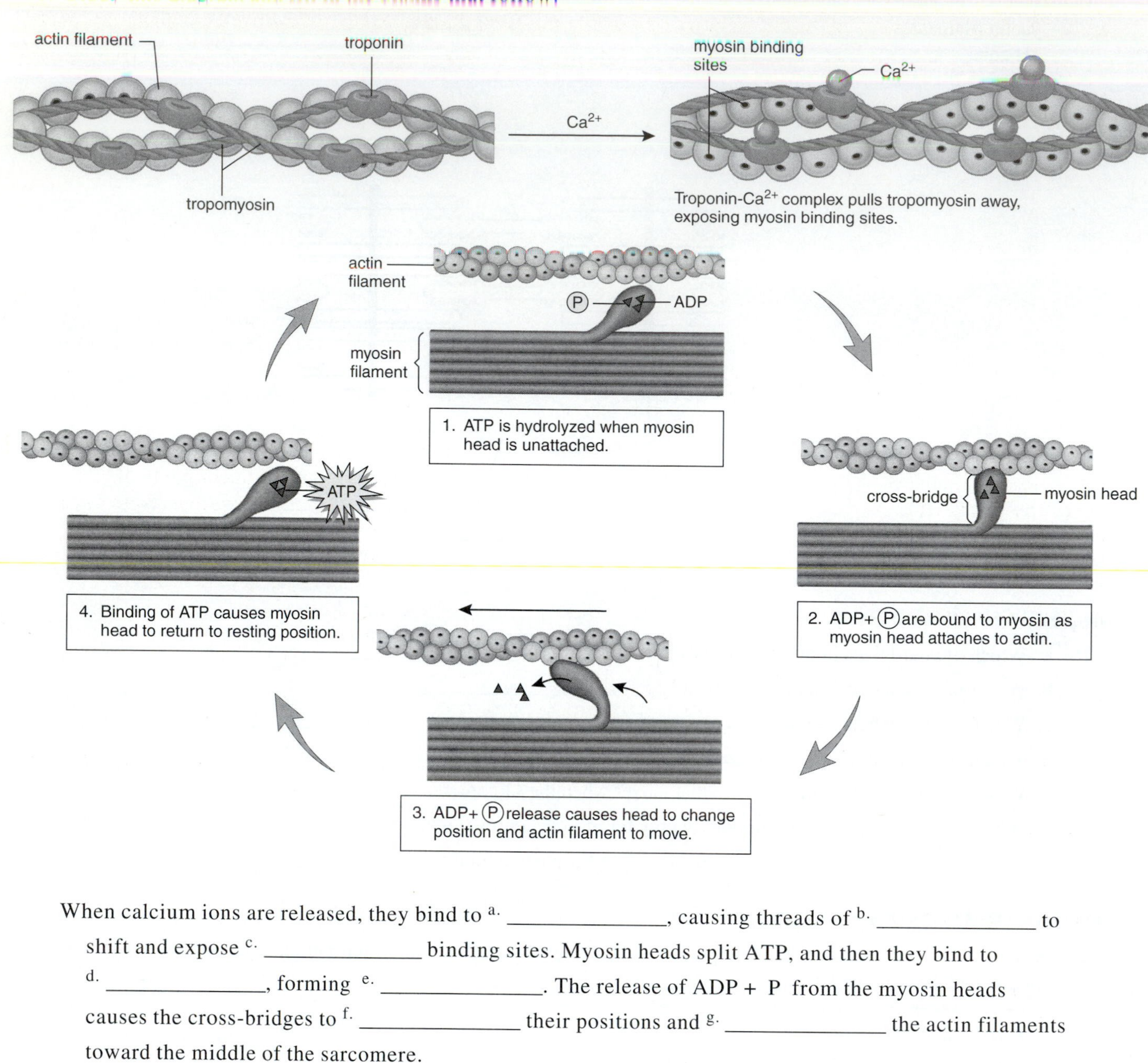


When calcium ions are released, they bind to [a] _______________, causing threads of [b] _______________ to shift and expose [c] _______________ binding sites. Myosin heads split ATP, and then they bind to [d] _______________, forming [e] _______________. The release of ADP + P from the myosin heads causes the cross-bridges to [f] _______________ their positions and [g] _______________ the actin filaments toward the middle of the sarcomere.

# Chapter Test

## Objective Questions

Do not refer to the text when taking this test.

____ 1. Which of these contains osteons?
   a. compact bone
   b. spongy bone
   c. red bone marrow
   d. yellow bone marrow

____ 2. Which of these is a function of the skeleton?
   a. support
   b. protection
   c. production of red blood cells
   d. All of these are correct.

____ 3. Which of these is NOT in the appendicular skeleton?
   a. clavicle
   b. coxal bone
   c. metatarsals
   d. vertebrae

____ 4. Facial bones include the
   a. frontal bone.
   b. occipital bone.
   c. mandible.
   d. All of these are correct.

____ 5. Ligaments join
   a. bone to bone.
   b. muscle to muscle.
   c. muscle to bone.
   d. All of these are correct.

____ 6. Vertebrae have
   a. immovable joints.
   b. freely movable joints.
   c. slightly movable joints.
   d. joints that vary according to the person.

____ 7. Which type of cell is found in lacunae of osteons?
   a. osteoblasts
   b. osteoclasts
   c. osteocytes
   d. Any of these are correct.

____ 8. When muscles are antagonistic,
   a. one pushes and one pulls on a bone.
   b. they have opposite actions on a bone.
   c. one is in the axial skeleton and the other is in the appendicular skeleton.
   d. Both *a* and *b* are correct.

____ 9. When sarcomeres contract, they get shorter, and this requires that muscles work in antagonistic pairs.
   a. true
   b. false

____ 10. Myosin is
   a. the thick filament of a sarcomere.
   b. a protein.
   c. an ATPase enzyme.
   d. All of these are correct.

____ 11. According to the sliding filament model,
   a. actin moves past myosin.
   b. myosin moves past actin.
   c. actin and myosin move past each other.
   d. actin both moves and breaks down ATP.
   e. None of these are correct.

____ 12. The release of calcium from the sarcoplasmic reticulum
   a. causes sarcomeres to relax.
   b. causes sarcomeres to contract.
   c. is the result of a transmission across a synapse.
   d. Both *b* and *c* are correct.
   e. All of these are correct.

____ 13. Which of these is the smallest unit?
   a. muscle fiber
   b. myofibril
   c. sarcomere
   d. actin

____ 14. The junction between nerve and muscle is
   a. called the neuromuscular junction.
   b. where a neurotransmitter travels from the nerve fiber to a muscle fiber.
   c. the region where nerves innervate muscles.
   d. All of these are correct.

____ 15. Creatine phosphate is
   a. used by sarcomeres.
   b. used to change ADP to ATP.
   c. a molecule found in DNA.
   d. All of these are correct.

____ 16. Muscle fatigue
   a. follows summation and tetanus.
   b. involves the buildup of lactate.
   c. occurs only in the laboratory.
   d. Both *a* and *b* are correct.
   e. All of these are correct.

In questions 17–20, match the descriptions to these structures:

   a. ligament
   b. tendon
   c. knee
   d. Z line
   e. sacroplasmic reticulum

____ 17. attaches skeletal muscle to the skeleton
____ 18. stabilizes bones at joints
____ 19. synovial joint
____ 20. an organelle of a muscle fiber

## Critical Thinking Questions

Answer in complete sentences.

21. Support the following statement: Contraction of a skeletal muscle is a good example of a whole structure working through the collection of its parts.

22. Support the following statement: A motion such as bending the forearm is not as simple as it appears to be.

**Test Results:** ______ number correct ÷ 22 = ______ × 100 = ______ %

## Exploring the Internet

The Online Learning Center at *www.mhhe.com/maderbiology8* has additional study material and practice quizzes that can help you master the content of this chapter. You can also find links to websites exploring additional topics in biology. Access to the Online Learning Center is free for those who have purchased a new textbook.

## Answer Key

### Study Exercises

**1. a.** hydrostatic **b.** fluid-filled **c.** protection **d.** exoskeleton **e.** armor **f.** shed **g.** endoskeleton **h.** cartilage **i.** bone **j.** internal organs **k.** grow **l.** flexible **2. a.** support **b.** protection of internal organs **c.** flexible movement **d.** blood cell production **e.** mineral storage **3. a.** cartilage **b.** compact bone **c.** spongy bone **d.** medullary cavity **4. a.** spongy bone **b.** compact bone **c.** cartilage **d.** medullary cavity **5. a.** Bone-forming osteoblasts eventually become mature osteocytes. **b.** They increase in length and allow bone to grow longer. **c.** to break down bone **d.** ossification **6. a.** calcium **b.** exercise **c.** hormones **7. a.** ap, pel **b.** ax **c.** ap **d.** ap, pec **e.** ax **f.** ap **g.** ax **h.** ap **i.** ap, pec **j.** ap **k.** ap **l.** ap **8. a.** tibia **b.** fibula **c.** humerus **d.** metatarsals **9. a.** tibia **b.** clavicle **c.** coxal bone **d.** femur **10. a.** synovial **b.** fibrous **c.** cartilaginous **d.** synovial **e.** cartilaginous

**11.**

| Anatomical Type | Degree of Movement |
|---|---|
| ball and socket | freely movable |
| hinge | freely movable |

**12. a.** scapula **b.** humerus **c.** biceps brachii **d.** triceps brachii **e.** radius **f.** ulna **13.** Muscles can only pull and cannot push. **14. a.** muscle fiber **b.** T tubule **c.** sarcoplasmic reticulum **d.** T tubule **e.** myofibril **f.** sarcolemma **g.** sarcomere **h.** Z line **15. a.** actin filament **b.** myosin filament **c.** Z line **d.** H zone **16. a.** actin filament **b.** myosin filament **c.** H zone **d.** myosin **e.** actin filament **f.** ATP **g.** creatine phosphate **17.** b, d, f, a, g, c, e **18. a.** troponin **b.** tropomyosin **c.** myosin **d.** actin **e.** cross-bridges **f.** change **g.** pull

### Chapter Test

**1.** a **2.** d **3.** d **4.** c **5.** a **6.** c **7.** c **8.** b **9.** a **10.** d **11.** a **12.** d **13.** d **14.** d **15.** b **16.** d **17.** b **18.** a **19.** c **20.** e **21.** Several levels of organization are involved. A muscle shortens through the shortening of its cells. At the subcellular level, thousands of sarcomeres containing actin/myosin filaments shorten, to shorten the cell and produce the contraction of the entire muscle. **22.** It is complex because a number of muscles interact to bring this about. As one muscle contracts, an opposing muscle must relax. Other muscles work to support the major contracting muscle.

# 42

# Hormones and the Endocrine System

## Chapter Review

**Hormones** are chemical signals that affect the activity of other glands or tissues. Endocrine glands secrete hormones into the bloodstream, and from there they are distributed to target organs or tissues. Some other tissues also produce hormones.

Chemical signals are active at various levels of communication between body parts, between cells, and even between individuals. Some chemical signals, such as traditional endocrine hormones and secretions of neurosecretory cells, act between body parts. Others, such as prostaglandins, growth factors, and neurotransmitters, act locally. Pheromones, well known in other animals, are chemical signals that act at a distance between individuals. Research suggests that humans may also produce pheromones.

**Steroid hormones** enter the nucleus and combine with a receptor molecule, and the complex attaches to and activates DNA. Transcription and translation lead to protein synthesis. The **peptide hormones** are usually received by a receptor located in the plasma membrane. Most often their reception leads to the activation of an enzyme that changes ATP to **cyclic AMP** (cAMP). cAMP then activates another enzyme, which activates another, and so forth.

The endocrine system works with the nervous system to maintain homeostasis. The secretion of hormones involved in maintaining homeostasis is controlled in two ways: by negative feedback and/or by antagonistic hormonal actions.

Neurosecretory cells in the hypothalamus produce **antidiuretic hormone** and **oxytocin,** stored in axon endings in the **posterior pituitary** until they are released.

The hypothalamus produces **hypothalamic-releasing** and **hypothalamic-inhibiting hormones,** which pass to the anterior pituitary by way of a portal system. The **anterior pituitary** produces at least six types of hormones, and some of these stimulate other hormonal glands to secrete hormones. Therefore, the anterior pituitary is sometimes called the master gland.

The **thyroid gland** produces thyroxine and triiodothyronine, hormones that play a role in the growth and development of immature forms; in mature individuals, they increase the metabolic rate. The thyroid gland also produces **calcitonin,** which helps lower the blood calcium level. The **parathyroid glands** raise the blood calcium and decrease the blood phosphate level.

The adrenal glands respond to stress. Immediately, the adrenal medulla secretes **epinephrine** and **norepinephrine,** which bring about responses we associate with emergency situations. On a long-term basis, the adrenal cortex primarily produces the **glucocorticoids (cortisol)** and the **mineralocorticoids (aldosterone)**. Cortisol stimulates the hydrolysis of proteins to amino acids, converted to glucose; in this way, it raises the blood glucose level. Aldosterone causes the kidneys to reabsorb sodium ions ($Na^+$) and to excrete potassium ions ($K^+$).

The **pancreatic islets** secrete insulin, which lowers the blood glucose level, and glucagon, which has the opposite effect. The most common illness due to hormonal imbalance is diabetes mellitus, caused by the failure of the pancreas to produce insulin or the failure of the cells to take it up.

The gonads produce the sex hormones; the **thymus gland** secretes thymosins, which stimulate T lymphocyte production and maturation; the **pineal gland** produces **melatonin,** whose function in humans is uncertain—it may be involved in **circadian rhythms** and the development of the reproductive organs.

## Study Exercises

Study the text section by section as you answer the questions that follow.

### 42.1 Chemical Signals (pp. 754–55)

- In general, there are three types of chemical signals, and hormones are one of these types.
- Hormones influence the metabolism of their target cells.

1. Write either *peptide hormone* or *steroid hormone* on the lines above each diagram. Using the following alphabetized list of terms, place an appropriate word or phrase on the lines within each diagram. Terms may be used more than once.

   active   cyclic AMP   hormone-receptor complex   protein synthesis

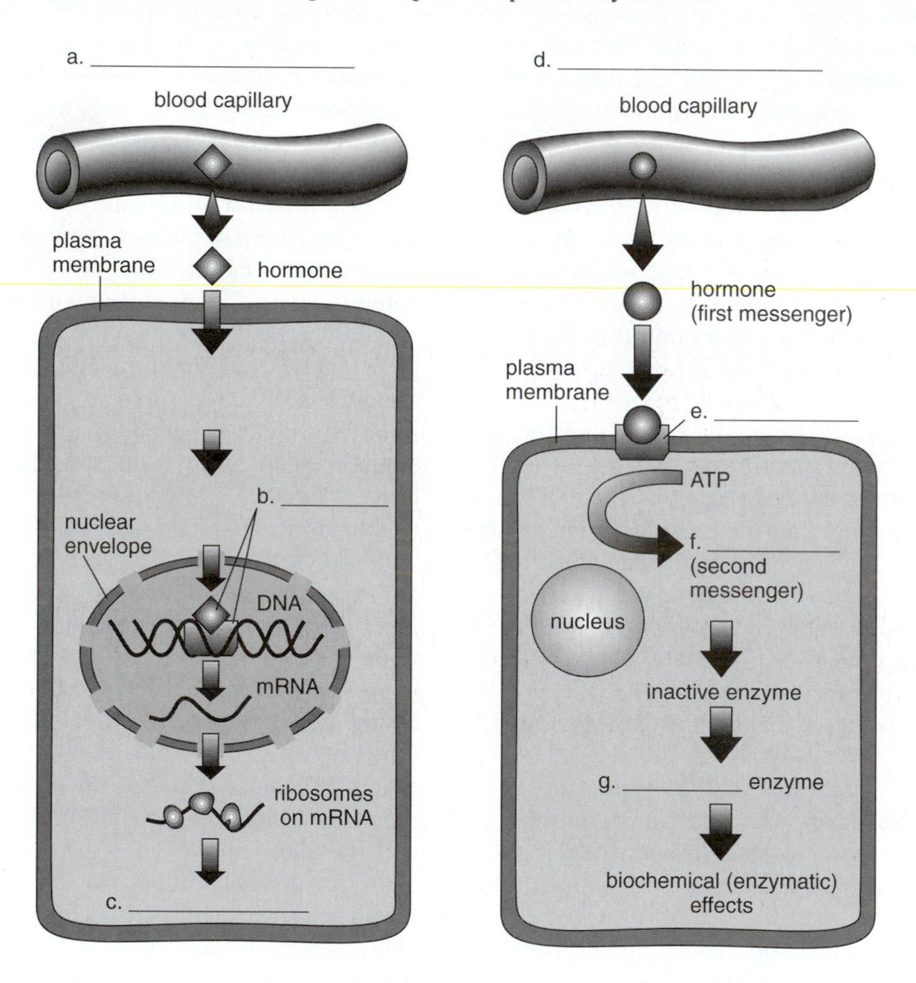


2. Place the appropriate letters next to each description.

   ES—endocrine system   NS—nervous system

   a. _____ uses chemical signals that bind to receptor proteins
   b. _____ sometimes acts from a distance and sometimes acts locally
   c. _____ uses releasing hormones
   d. _____ uses neurotransmitters
   e. _____ uses hormones distributed in the blood

## 42.2 Human Endocrine System (pp. 756–69)

- The endocrine system works with the other systems of the body to maintain homeostasis.

3. Label the endocrine glands in this diagram, and name at least one hormone produced by each.

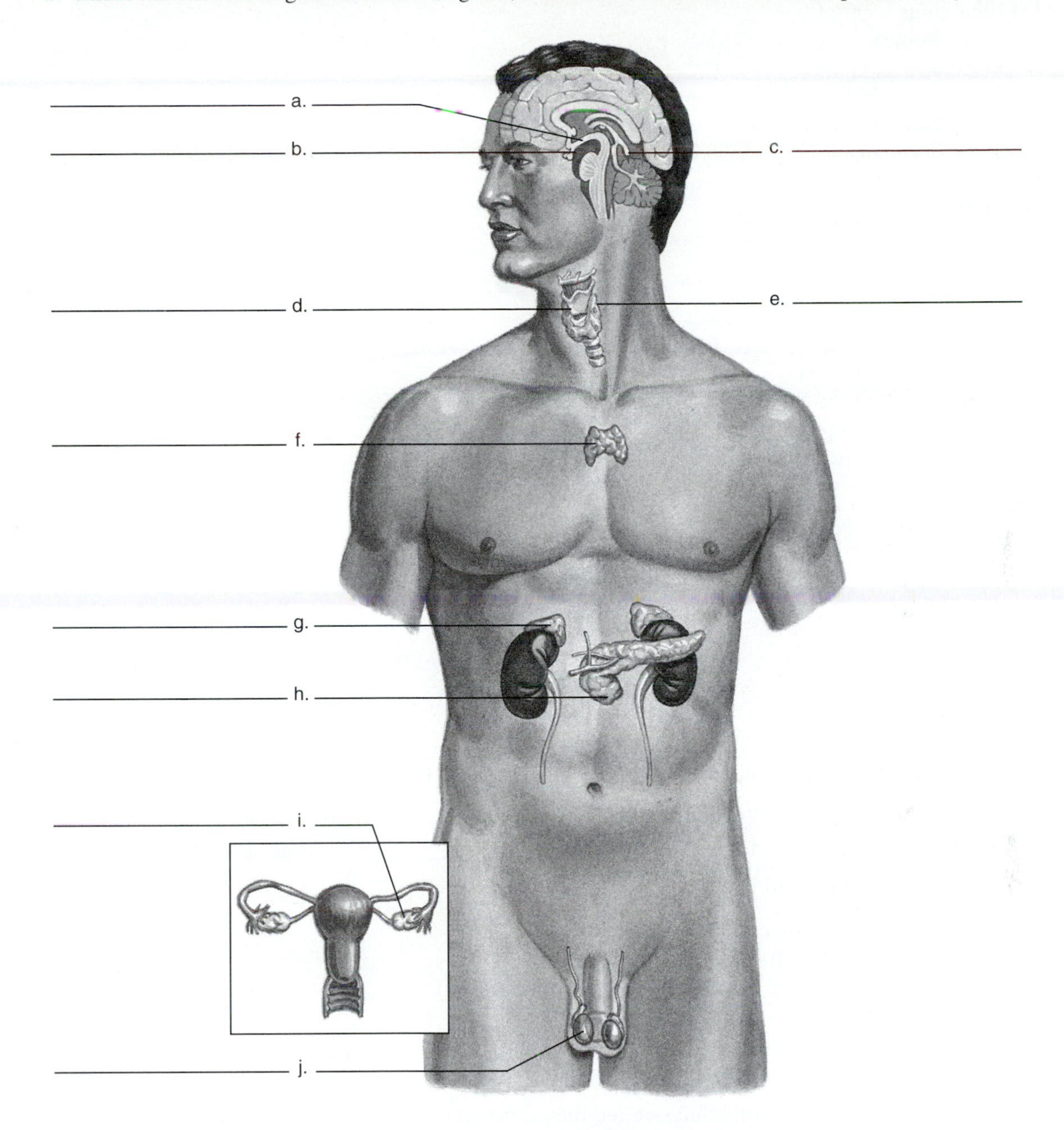


In question 4, fill in the blanks.

4. Control of hormone release. Often the release of a hormone is dependent upon the blood level of the substance it is controlling. When the level of a substance increases, generally this causes the hormone secretion to a. ______________________. This is an example of b. ______________________ feedback. In other instances, c. ______________________ hormones oppose each other's actions, thus regulating the target substance in the body.

## HYPOTHALAMUS AND PITUITARY GLAND (PP. 758–60)

- The hypothalamus controls the function of the pituitary gland, which, in turn, controls several other glands.

5. Match the descriptions to the numbers in the diagram.
   a. _____ hypothalamus
   b. _____ anterior pituitary
   c. _____ target gland
   d. _____ feedback that inhibits hypothalamus
   e. _____ feedback that inhibits anterior pituitary
   f. _____ releasing hormone
   g. _____ stimulating hormone
   h. _____ target gland hormone

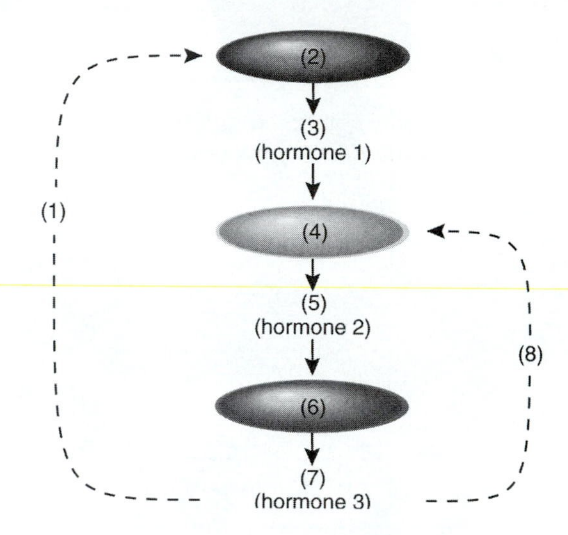


In question 6, fill in the blanks.

6. If the "gland" in the diagram in question 3 is the thyroid, then the first hormone is a.__________________, the second is b.________________, and the third is c.________________. If the "gland" in the diagram in question 5 is the adrenal cortex, then the first hormone is d.__________________, the second is e.__________________, and the third is f.__________________.

7. Place the appropriate letters next to each statement.
   AP—anterior pituitary PP—posterior pituitary
   a. _____ connected to hypothalamus by nerve fibers
   b. _____ connected to hypothalamus by blood vessels
   c. _____ secretes hormones produced by hypothalamus
   d. _____ controlled by releasing hormones produced by hypothalamus

8. To show why the anterior pituitary is sometimes called the master gland, complete this table:

| Anterior Pituitary Produces | Gland Controlled | Hormone Produced by Gland |
|---|---|---|
| TSH | a. | b. |
| ACTH | c. | d. |
| Gonadotropic hormones | | |
| Female | e. | f. |
| Male | g. | h. |

In question 9, fill in the blanks.

9. The anterior pituitary produces three other hormones. The hormone a. ______________ causes the mammary glands to develop and produce milk. b. ______________ hormone causes skin color changes in lower vertebrates. Growth hormone (GH) promotes cell division, protein synthesis, and c. ______________ growth. If too little GH is produced during childhood, the individual becomes a pituitary d. ______________. If too much is produced, the individual is a pituitary e. ______________. If there is an overproduction of GH in the adult, f. ______________ results, and the face, hands, and feet g. ______________.

## Thyroid and Parathyroid Glands (pp. 761–62)

- Parathyroid glands are embedded in the thyroid but have an entirely separate structure and function.
- The thyroid produces two hormones that speed metabolism and another that lowers the blood calcium level.
- The parathyroid glands produce a hormone that raises the blood calcium level.

10. Match the phrases to these conditions. Some conditions are used more than once.
    cretinism    exophthalmic goiter    simple goiter    myxedema

    a. ______________ hypothyroidism (choose more than one)
    b. ______________ hyperthyroidism
    c. ______________ hypothyroidism since birth
    d. ______________ hypothyroidism in the adult
    e. ______________ lack of iodine

11. Match the descriptions to these phrases. Some phrases are used more than once.
    1. low blood $Ca^{2+}$
    2. high blood $Ca^{2+}$
    3. $Ca^{2+}$ is deposited in bones
    4. $Ca^{2+}$ is deposited in blood

    a. _____ Calcitonin is present.
    b. _____ Parathyroids are mistakenly removed during an operation.
    c. _____ Calcitonin will be released.
    d. _____ PTH will be released.
    e. _____ PTH is present.
    f. _____ Calcitonin will not be released.
    g. _____ PTH will not be released.

## Adrenal Glands (pp. 763–65)

- The adrenal medulla and the adrenal cortex are separate parts of the adrenal glands that have functions in relation to stress.

12. Place the appropriate letters next to each description.
    AM—adrenal medulla    AC—adrenal cortex

    a. _____ inner portion of adrenal gland
    b. _____ outer portion of adrenal gland
    c. _____ hypothalamus sends nervous impulses
    d. _____ hypothalamus sends releasing hormone to anterior pituitary, and anterior pituitary sends ACH to target gland
    e. _____ releases glucocorticoids and mineralocorticoids
    f. _____ releases epinephrine and norepinephrine
    g. _____ short-term reaction to stress
    h. _____ long-term reaction to stress

13. Distinguish between cortisol and aldosterone by writing *yes* or *no* on each line.

| | | Cortisol | Aldosterone |
|---|---|---|---|
| Controlled by ACTH | a. | ________ | ________ |
| Glucocorticoid | b. | ________ | ________ |
| Mineralocorticoid | c. | ________ | ________ |
| $Na^+/K^+$ balance | d. | ________ | ________ |
| Amino acids → glucose | e. | ________ | ________ |
| Controlled by angiotensin II | f. | ________ | ________ |

In question 14, fill in the blanks.

14. When there is low blood $Na^+$, the kidneys secrete [a] ______________, an enzyme that converts angiotensinogen to [b] ______________, which later becomes [c] ______________ in lung capillaries. The latter causes the adrenal cortex to release [d] ______________. Blood pressure now [e] ______________. The heart releases a hormone called [f] ______________, antagonistic to aldosterone.

15. Place the appropriate letters next to these symptoms:
    AD—Addison disease CS—Cushing syndrome
    a. _____ cannot handle bodily stress
    b. _____ cannot maintain blood glucose level
    c. _____ tendency toward diabetes mellitus
    d. _____ low blood pressure because of a low blood sodium level
    e. _____ high blood pressure because of a high blood sodium level
    f. _____ edema because of too much sodium in system
    g. _____ bronzing of skin
    h. _____ thin arms and legs; enlarged trunk

## Pancreas (pp. 766–67)

- The pancreas secretes hormones that help control the blood glucose level.
- Diabetes mellitus occurs when cells are unable to take up glucose and it spills over into the urine.

16. Write the word *insulin* or *glucagon* on the appropriate arrow.
    a. glycogen ←―――――― glucose molecules
    b. storage in liver ――――――→ in the blood

17. Complete each of the following statements with the term *increases* or *decreases*.
    Glucagon [a] ________________ blood sugar concentration. In type 1 diabetes, insulin production from the pancreas [b] ________________. In type 2 diabetes, the response of body cells to the influence of insulin [c] ________________.

## Other Endocrine Glands (pp. 768–69)

- The gonads produce the sex hormones that control secondary sexual characteristics.

18. Place the appropriate letters next to each description.

    M—male F—female

    a. _____ testosterone
    b. _____ ovaries
    c. _____ estrogen
    d. _____ androgens
    e. _____ facial hair
    f. _____ breast development

19. Match these numbered items to the glands. There is more than one match for each gland, and answers may be used more than once.

    1. T lymphocytes
    2. melatonin
    3. testosterone
    4. circadian rhythm
    5. males
    6. females
    7. estrogen and progesterone
    8. thymosins
    9. secondary sex characteristics

    a. _____ testes
    b. _____ ovaries
    c. _____ thymus
    d. _____ pineal gland

20. Indicate whether these statements are true (T) or false (F).

    a. _____ The pineal gland secretes MSH.
    b. _____ The pineal gland produces melatonin.
    c. _____ The thymus gland contributes to cell-mediated immunity.
    d. _____ The thymus gland increases in size with maturity.

# Hormone Hockey

For every five correct answers in sequence, you have scored one goal.

First goal: Match the hormone to the glands (*a–i*).

Glands:

a. anterior pituitary
b. thyroid
c. parathyroids
d. adrenal cortex
e. adrenal medulla
f. pancreas
g. gonads
h. pineal gland
i. posterior pituitary

Hormones:

1. ___ insulin
2. ___ oxytocin
3. ___ melatonin
4. ___ cortisol
5. ___ thyroxine

Second goal: Match the condition to the glands (*a–i*).

Conditions:

6. ___ diabetes mellitus
7. ___ cretinism
8. ___ Addison disease
9. ___ hypertension
10. ___ giantism

Third goal: Match the function to the hormones (*a–l*).

a. melatonin
b. estrogen
c. androgen
d. insulin
e. glucagon
f. epinephrine
g. aldosterone
h. cortisol
i. parathyroid hormone
j. thyroxine
k. calcitonin (lowers)
l. antidiuretic hormone

Functions:

11. ___ raises blood calcium level
12. ___ reduces stress
13. ___ maintains secondary female sex characteristics
14. ___ involved in circadian rhythms
15. ___ stimulates water reabsorption by kidneys

Fourth goal: Match the glands to the hormones (*a–l*). Some glands require two answers.

Glands:

16. ___ testes
17. ___ adrenal cortex
18. ___ pancreas
19. ___ thyroid
20. ___ adrenal medulla

Fifth goal: Select five hormones secreted by the anterior pituitary by answering yes or no to each of these.

21. ___ thyroid-stimulating hormone
22. ___ androgens
23. ___ gonadotropic hormones
24. ___ glucagon
25. ___ oxytocin
26. ___ growth hormone
27. ___ prolactin
28. ___ antidiuretic hormone
29. ___ estrogens
30. ___ adrenocorticotropic hormone

How many goals did you make? __________

## DEFINITIONS WORDSEARCH

Review key terms by using the following alphabetized list of terms to fill in the blanks. Then complete the wordsearch.

```
C R E T I N I S M N U J L I P
O A R H G T Z B E S A I L H R
R A L P O X Y T O C I N H U O
T M L C F R S H U K N H S R S
I E R G I U M N V Y S A F T T
S D G T R T D O F W U J H U A
O E P O L I O U N Y L R E R G
L X H Y T F G N B E I S E F L
P Y J T D H S E I D N S G B A
I M P H E R O M O N E B G F N
N O R E P I N E P H R I N E D
S G W Y T J T U D H R T K F I
P T Y E R W Q A S E D F R G N
```

calcitonin
cortisol
cretinism
hormone
insulin
myxedema
norepinephrine
oxytocin
prostaglandin

a. ______________ Thyroid hormone that regulates blood calcium.

b. ______________ Chemical signal.

c. ______________ Condition due to improper development of thyroid in infants.

d. ______________ Posterior pituitary hormone causing uterine contractions and milk letdown.

e. ______________ Pancreatic hormone that lowers blood glucose.

f. ______________ Adrenal gland hormone that increases blood glucose.

g. ______________ Condition caused by lack of thyroid hormone in adult.

h. ______________ Stress hormone from adrenal medulla.

i. ______________ Local tissue hormone.

# Chapter Test

## Objective Questions

Do not refer to the text when taking this test.

____ 1. All hormones are believed to
- a. have plasma membrane receptors.
- b. affect cellular metabolism.
- c. increase the amount of cAMP.
- d. increase the amount of protein synthesis.

____ 2. The adrenal glands are
- a. at the base of the brain.
- b. on the trachea.
- c. on the kidney.
- d. beneath the stomach.

____ 3. Which statement is NOT true about hormones?
- a. Hormones search throughout the bloodstream for their receptors.
- b. They act as chemical signals.
- c. They are released by endocrine glands.
- d. They can affect our appearance, our metabolism, or our behavior.

____ 4. Hormonal secretions are most often controlled by
- a. negative feedback mechanisms.
- b. positive feedback mechanisms.
- c. the hormone insulin.
- d. the cerebrum of the brain.

____ 5. Steroid hormones
- a. combine with hormone receptors in the plasma membrane.
- b. pass through the membrane.
- c. activate genes leading to protein synthesis.
- d. Both *b* and *c* are correct.

____ 6. Which gland produces the greatest number of hormones?
- a. posterior pituitary
- b. anterior pituitary
- c. thymus
- d. pineal gland

____ 7. The hypothalamus controls the anterior pituitary via
- a. nervous stimulation.
- b. the midbrain.
- c. vasopressin.
- d. releasing hormones.

____ 8. ADH and oxytocin are secreted by the
- a. hypothalamus.
- b. posterior pituitary.
- c. thyroid gland.
- d. parathyroids.

____ 9. Which hormone is involved with milk production and nursing?
- a. prolactin
- b. androgens
- c. antidiuretic hormone
- d. growth hormone

____ 10. The anterior pituitary stimulates the
- a. thyroid.
- b. adrenal cortex.
- c. adrenal medulla.
- d. pancreas.
- e. Both *a* and *b* are correct.

____ 11. Too much urine matches too
- a. little ADH.
- b. much ADH.
- c. little ACTH.
- d. much ACTH.

____ 12. Thyroxine
- a. increases metabolism.
- b. stimulates the thyroid gland.
- c. lowers oxygen uptake.
- d. All of these are correct.

____ 13. The adrenal cortex produces hormones affecting
- a. glucose metabolism.
- b. amino acid metabolism.
- c. sodium balance.
- d. All of these are correct.

____ 14. Which hormone regulates blood calcium levels?
- a. calcitonin
- b. parathyroid hormone
- c. cortisol
- d. Both *a* and *b* are correct.

____ 15. Which gland produces sex hormones?
- a. anterior pituitary
- b. posterior pituitary
- c. adrenal cortex
- d. Both *a* and *b* are correct.

____ 16. Tetany occurs when there is too
- a. little calcium in the blood.
- b. much calcium in the blood.
- c. little sodium in the blood.
- d. much sodium in the blood.

____ 17. Cushing syndrome is due to a malfunctioning
- a. thyroid.
- b. adrenal cortex.
- c. adrenal medulla.
- d. pancreas.

____ 18. A simple goiter is caused by
- a. too much salt in the diet.
- b. too little iodine in the diet.
- c. too many sweets in the diet.
- d. a bland diet.

____ 19. Acromegaly might be due to a tumor of the
- a. pancreas.
- b. anterior pituitary.
- c. thyroid.
- d. adrenal cortex.

____20. If a person is suffering from insulin shock, he or she should
a. be given some sugar.
b. sit with the head down.
c. be given insulin.
d. not eat fatty foods.

____21. Diabetes insipidus is a disease of the
a. pancreas.
b. adrenal cortex.
c. posterior pituitary.
d. Both *a* and *b* are correct.

____22. In which case is insulin not produced?
a. type 1 diabetes
b. type 2 diabetes
c. type 3 diabetes
d. diabetes insipidus

____23. Which of these is not a similarity between the nervous and endocrine systems?
a. Both use chemical signals.
b. Both use hormones.
c. Both use nerve impulses.
d. Both act from a distance or act locally.

____24. The system most directly affected by the secretion of epinephrine for blood pressure adjustments is the ______________ system.
a. respiratory
b. cardiovascular
c. urinary
d. reproductive

____25. The gonads that produce sex hormones also belong to which system?
a. lymphatic system
b. nervous system
c. reproductive system
d. urinary system

## Critical Thinking Questions

Answer in complete sentences.

26. Explain the occurrence of a goiter when an individual does not receive enough iodine in the diet.

27. Why does the release of renin by the kidneys cause the blood pressure to rise?

**Test Results:** ______ number correct ÷ 27 = ______ × 100 = ______ %

## Exploring the Internet

The Online Learning Center at *www.mhhe.com/maderbiology8* has additional study material and practice quizzes that can help you master the content of this chapter. You can also find links to websites exploring additional topics in biology. Access to the Online Learning Center is free for those who have purchased a new textbook.

## Answer Key

### Study Exercises

**1. a.** steroid hormone **b.** hormone-receptor complex **c.** protein synthesis **d.** peptide hormone **e.** hormone-receptor complex **f.** cyclic AMP **g.** active **2. a.** ES, NS **b.** ES, NS **c.** NS **d.** NS **e.** ES **3. a.** hypothalamus, hypothalamic-releasing hormone **b.** pituitary gland, growth hormone, ACTH **c.** pineal gland, melatonin **d.** thyroid gland, thyroxine, calcitonin **e.** parathyroid, parathyroid hormone **f.** thymus, thymosin **g.** adrenal gland, cortisol, aldosterone, epinephrine, norepinephrine **h.** pancreas, insulin, glucagon **i.** ovary, estrogen, progesterone **j.** testis, testosterone **4. a.** decrease **b.** negative **c.** antagonistic **5. a.** (2) **b.** (4) **c.** (6) **d.** (1) **e.** (8) **f.** (3) **g.** (5) **h.** (7) **6. a.** TRH (thyroid-releasing hormone) **b.** TSH (thyroid-stimulating hormone) **c.** thyroxine **d.** ACRH (adrenocorticoid-releasing hormone) **e.** ACTH (adrenocorticotropic hormone) **f.** cortisol **7. a.** PP **b.** AP **c.** PP **d.** AP **8. a.** thyroid **b.** thyroxine **c.** adrenal cortex **d.** cortisol **e.** ovaries **f.** estrogen, progesterone **g.** testes **h.** testosterone **9. a.** prolactin **b.** Melanocyte-stimulating **c.** skeletal **d.** dwarf **e.** giant **f.** acromegaly **g.** enlarge **10. a.** cretinism, simple goiter, myxedema **b.** exophthalmic goiter **c.** cretinism **d.** myxedema **e.** simple goiter **11. a.** (1) and (3) **b.** (1) and (3) **c.** (2) **d.** (1) **e.** (2) and (4) **f.** (1) **g.** (2) **12. a.** AM **b.** AC **c.** AM **d.** AC **e.** AC **f.** AM **g.** AM **h.** AC **13. a.** yes, no **b.** yes, no **c.** no, yes **d.** no, yes **e.** yes, no **f.** no, yes **14. a.** renin **b.** angiotensin I **c.** angiotensin II **d.** aldosterone **e.** rises **f.** atrial natriuretic hormone (ANH) **15. a.** AD **b.** AD **c.** CS **d.** AD **e.** CS **f.** CS **g.** AD **h.** CS **16. a.** insulin **b.** glucagon **17. a.** increases **b.** decreases **c.** decreases **18. a.** M **b.** F **c.** F **d.** M **e.** M **f.** F **19. a.** 3, 5, 9 **b.** 6, 7, 9 **c.** 1, 8 **d.** 2, 4 **20. a.** F **b.** T **c.** T **d.** F

## Game: Hormone Hockey

First goal: **1.** f **2.** i **3.** h **4.** d **5.** b. Second goal: **6.** f **7.** b **8.** d **9.** d **10.** a. Third goal: **11.** i **12.** h **13.** b **14.** a **15.** l. Fourth goal: **16.** c **17.** g and h **18.** d and e **19.** j **20.** f. Fifth goal: **21.** yes **22.** no **23.** yes **24.** no **25.** no **26.** yes **27.** yes **28.** no **29.** no **30.** yes

## Definitions Wordsearch

```
C R E T I N I S M               P
O A   H                         R
R A L   O X Y T O C I N         O
T M   C   R         N           S
I E     I   M       S           T
S D       T   O     U           A
O E         O   N   L           G
L X           N   E I           L
  Y             I   N           A
  M                 N E         N
N O R E P I N E P H R I N E D
                                I
                                N
```

**a.** calcitonin **b.** hormone **c.** cretinism **d.** oxytocin **e.** insulin **f.** cortisol **g.** myxedema **h.** norepinephrine **i.** prostaglandin

## Chapter Test

**1.** b **2.** c **3.** a **4.** a **5.** d **6.** b **7.** d **8.** b **9.** a **10.** e **11.** a **12.** a **13.** d **14.** d **15.** c **16.** a **17.** b **18.** b **19.** b **20.** a **21.** c **22.** a **23.** b **24.** b **25.** c **26.** When an individual does not receive enough iodine in the diet, the thyroid is unable to produce thyroxine. The lack of thyroxine in the blood causes the anterior pituitary to produce more TSH, and this hormone promotes increase in the size of the thyroid. **27.** Renin leads to the formation of angiotensin II, which stimulates the adrenal cortex to secrete aldosterone. Aldosterone causes sodium to be reabsorbed by the kidneys, and this leads to an increase in blood volume and blood pressure.

# 43

# Reproduction

## Chapter Review

**Asexual reproduction** involves one parent producing a large number of offspring in a constant environment. **Sexual reproduction** produces genetic variations because two parents contribute genes to the offspring. Therefore, it has advantages in a changing environment. Fertilization tends to be external in aquatic species; it is internal for land species.

In the human male, spermatogenesis occurring in **seminiferous tubules** of the **testes** produces **sperm** that mature in the epididymides and may be stored in the vasa deferentia.

**Semen,** which contains mature sperm as well as secretions produced by seminal vesicles, the prostate gland, and bulbourethral glands, enters the urethra and is ejaculated during male orgasm when the **penis** is erect.

Hormonal regulation in males, involving secretions from the hypothalamus, the anterior pituitary, and the testes, maintains the male hormone testosterone. **Testosterone** brings about and maintains the secondary sex characteristics of males.

In females, an **oocyte** (i.e., ovum or egg) produced by an ovary enters an oviduct, which leads to the uterus. The uterus opens into the vagina. The external genital area includes the vaginal opening, the clitoris, the labia minora, and the labia majora.

In the nonpregnant female, the **ovarian** and **uterine cycles** are under hormonal control of the hypothalamus; the anterior pituitary; and the female sex hormones, **estrogens** and **progesterone.** Estrogen and progesterone maintain the secondary sex characteristics of females.

If fertilization occurs, the **corpus luteum** is maintained because of **human chorionic gonadotropin (HCG)** production. Progesterone production does not cease, and the embryo implants in the thick uterine lining.

Numerous birth-control methods and devices are available for those who wish to prevent pregnancy. Infertile couples are increasingly using assisted reproductive technologies to increase the chances of pregnancy.

Sexually transmitted diseases include **AIDS**, genital warts, genital herpes, hepatitis, chlamydia, gonorrhea, and syphilis. Many sexually transmitted diseases (STDs) are easily transferred from one person to the next through sexual contact. Some STDs can lead to discomfort, others to sterility or even death.

## Study Exercises

Study the text section by section as you answer the questions that follow.

### 43.1 How Animals Reproduce (pp. 774–75)

- Among animals, there are two patterns of reproduction: asexual reproduction and sexual reproduction.
- Sexually reproducing animals have gonads for the production of gametes, and many have accessory organs for the storage and passage of gametes into or from the body.
- Animals have various means of assuring fertilization of gametes and protecting immature stages.

1. Label each of the following as describing asexual reproduction (A) or sexual reproduction (S):
   a. _____ Budding is one type.
   b. _____ Gametes are produced by the same or different individuals.
   c. _____ Offspring have a different combination of genes than either parent.
   d. _____ Offspring tend to have the same genotype and phenotype as the parents.
   e. _____ Hermaphroditic earthworms practice it.
   f. _____ It produces offspring that may be better adapted to a new environment.
   g. _____ Regeneration is one type.
   h. _____ Usually a large number of offspring are produced.

2. Complete the following statements:

a. Yolk is ______________________.

b. Metamorphosis is ______________________.

c. Extraembryonic membranes ______________________.

d. Ovoviviparous means ______________________.

e. Viviparous means ______________________.

## 43.2 Male Reproductive System (pp. 776–79)

- The human male reproductive system continually produces a large number of sperm transported within a fluid medium.
- Hormones control the production of sperm and maintain the primary and secondary sex characteristics of males.

3. Using the alphabetized list of terms and the blanks provided, identify and state a function for the parts of the human male reproductive system shown in the following diagram.

bulbourethral gland  epididymis  penis  prostate gland  seminal vesicles
testis  urethra  urinary bladder  vas deferens

| | Structure | Function |
|---|---|---|
| a. | ______________ | ______________________ |
| b. | ______________ | ______________________ |
| c. | ______________ | ______________________ |
| d. | ______________ | ______________________ |
| e. | ______________ | ______________________ |
| f. | ______________ | ______________________ |
| g. | ______________ | ______________________ |
| h. | ______________ | ______________________ |
| i. | ______________ | ______________________ |

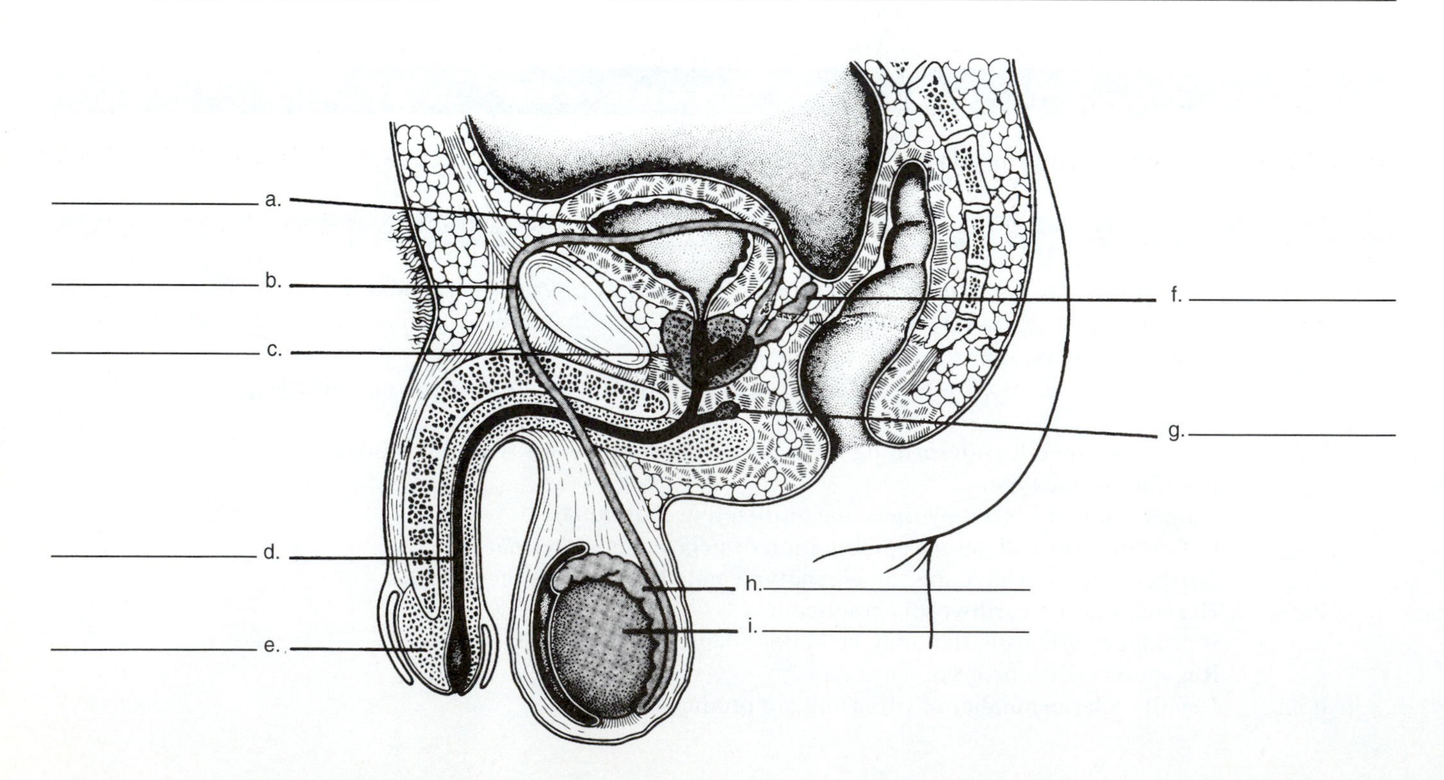

4. Trace the path of sperm through the male reproductive system:
testis, a.__________________, b.__________________, and c.__________________.

5. The three organs that add secretions to seminal fluid are a.__________________, b.__________________, and c.__________________. Do they also produce hormones? d.__________________

6. Indicate whether these statements are true (T) or false (F). Rewrite the false statements to make true statements.

a. _____ Testosterone exerts negative feedback control over the anterior pituitary secretion of LH.

Rewrite:__________________________________________________

b. _____ Inhibin exerts negative feedback control over the anterior pituitary secretion of FSH.

Rewrite: __________________________________________________

7. Place the appropriate letters next to each statement.

ST—seminiferous tubules IC—interstitial cells

a. _____ produce androgens
b. _____ produce sperm
c. _____ controlled by FSH
d. _____ controlled by LH

8. Describe the effect of testosterone on the following:

a. sexual organs __________________
b. facial hair __________________
c. larynx __________________
d. muscular strength __________________
e. oil and sweat glands __________________

In question 9, fill in the blanks.

9. As the level of testosterone rises in the blood, the secretion of LH a.__________________. As the level of testosterone falls in the blood, the hypothalamus b.__________________ the secretion of the gonadotropic-releasing hormone. Secretion of inhibin c.__________________ FSH secretion. Greatly decreased testosterone secretion d.__________________ beard and pubic hair growth. The presence of testosterone e.__________________ secretion of oil and sweat glands.

## 43.3 Female Reproductive System (pp. 780–84)

- The female reproductive system produces one egg monthly and prepares the uterus to house the developing fetus.
- Hormones control the monthly reproductive cycle in females and play a significant role in maintaining pregnancy, should it occur.

10. Using the alphabetized list of terms and the blanks provided, identify and state a function for the human female reproductive structures and urinary structures shown in the following diagram.

cervix ovary oviduct urethra urinary bladder uterus vagina

| | Structure | Function |
|---|---|---|
| a. | ______________ | ______________________________ |
| b. | ______________ | ______________________________ |
| c. | ______________ | ______________________________ |
| d. | ______________ | ______________________________ |
| e. | ______________ | ______________________________ |
| f. | ______________ | ______________________________ |
| g. | ______________ | ______________________________ |

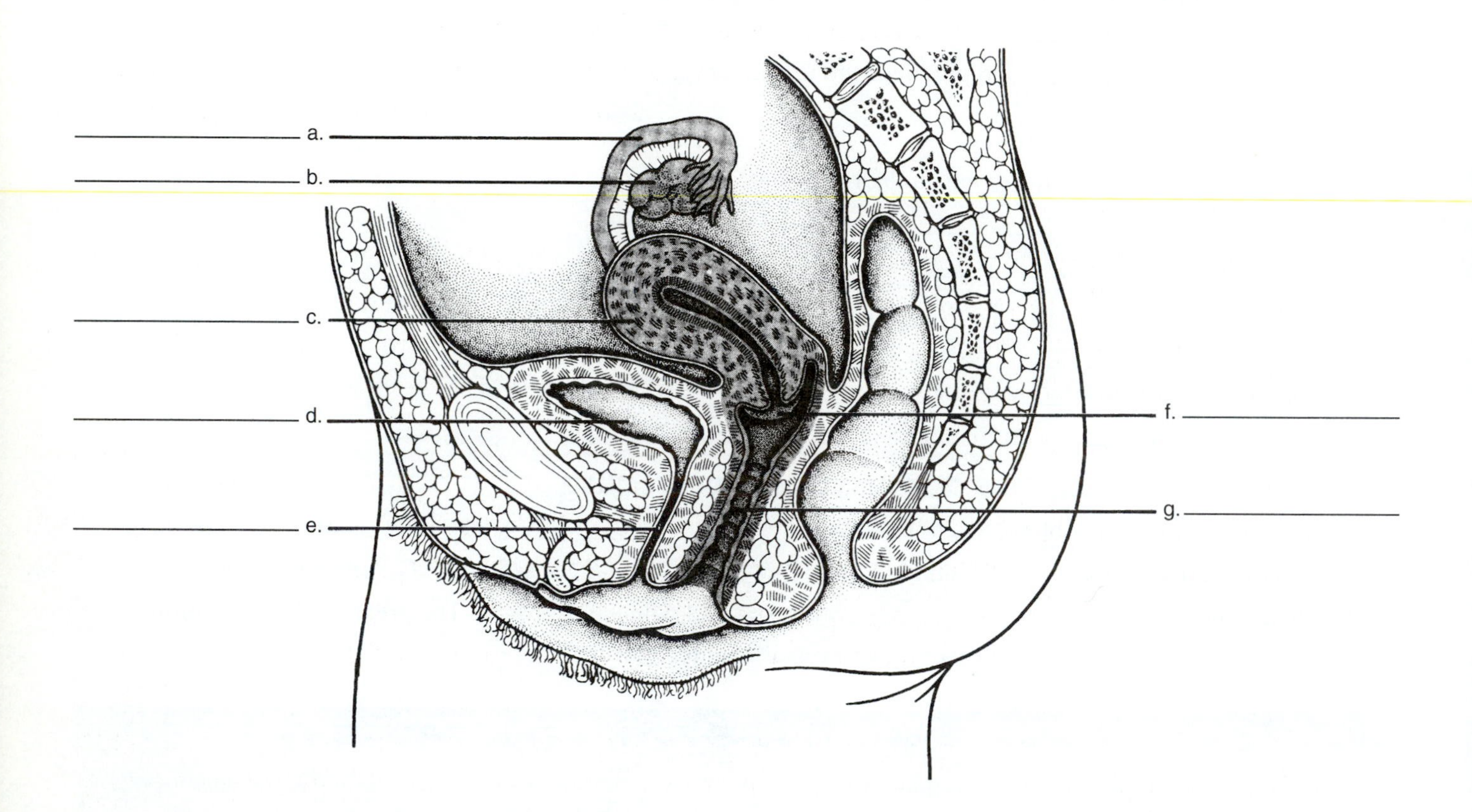


In question 11, fill in the blanks.

11. When sperm enter the female reproductive tract, they are deposited into the a. ______________. From there, they pass through the b. ______________ of the uterus. They swim up through the c. ______________ until they reach the egg cell.

12. Label this diagram of the vulva using the following alphabetized list of terms.
    anus
    glans clitoris
    labia majora
    labia minora
    mons pubis
    urethra
    vagina

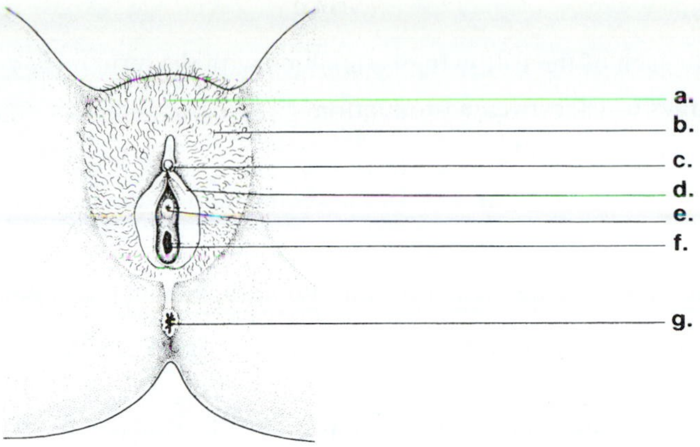

a. ____
b. ____
c. ____
d. ____
e. ____
f. ____
g. ____

13. Each [a] ____ in the ovary contains an oocyte. The secondary follicle contains the [b] ____ and produces the female sex hormones. A secondary follicle develops into a(n) [c] ____ follicle. [d] ____ is the release of the secondary oocyte (egg) from the ovary. A follicle that has lost its oocyte becomes a(n) [e] ____.

14. Rearrange the letters to indicate the sequence of organs along the female reproductive tract.

    ____

    a. oviduct
    b. ovary
    c. vagina
    d. uterus

15. Match the definitions to these terms:
    estrogen FSH LH progesterone

    a. ____ gonadotropic hormones
    b. ____ female sex hormones
    c. ____ primarily secreted by follicle
    d. ____ primarily secreted by corpus luteum

16. Fill in the following table to indicate the events in the ovarian and uterine cycles (simplified, and assuming a 28-day cycle).

| Ovarian Cycle | Events | Uterine Cycle | Events |
|---|---|---|---|
| Follicular phase—Days 1–13 | a. ____<br>Follicle maturation<br>Estrogen secretion | b. ____—Days 1–5<br>d. ____—Days 6–13 | c. ____<br>e. ____ |
| Ovulation—Day 14 | LH spike | | |
| Luteal phase—Days 15–28 | Corpus luteum forms<br>h. ____ | f. ____—Days 15–28 | g. ____<br>____ |

17. What hormonal changes occur if pregnancy takes place?

______________________________________________

18. Complete each of the following statements with the term *increase(s)* or *decrease(s):*

During days 6–13, estrogen production a.__________________. A buildup in blood estrogen during the first half of the ovarian cycle b.__________________ FSH secretion. Ovulation is keyed by a(n) c.__________________ in LH. In the second half of the ovarian cycle, the developing corpus luteum d.__________________ its secretion of progesterone. Increased progesterone production e.______________ LH production. During menstruation, blood flow from the vagina f.________________.

19. Indicate whether the following statements are true (T) or false (F). Rewrite the false statements to make true statements.

a. ______ Implantation occurs as soon as fertilization occurs. Rewrite: ______________________________

b. ______ HCG prevents degeneration of the corpus luteum. Rewrite: ______________________________

c. ______ During pregnancy, ovulation continues because estrogen and progesterone are still present. Rewrite: ______________________________

d. ______ Menopause is the time in a woman's life when the ovarian and uterine cycles cease. Rewrite: ______________________________

20. Describe the effect of the female sex hormones on the following:

a. body fat ______________________________

b. sex organs______________________________

c. breast development ______________________________

## 43.4 Control of Reproduction (pp. 784–86)

- Birth control measures vary in effectiveness from those that are very effective to those that are minimally effective.
- Assisted reproductive technologies consist of techniques used to increase the chances of pregnancy.

21. Following are two groups of birth control measures. Rank the members of each group from the most effective (1) to the least effective (4).

| A | | B | |
|---|---|---|---|
| a. _____ | coitus interruptus | e. _____ | vasectomy |
| b. _____ | spermicidal jelly/cream | f. _____ | natural family planning |
| c. _____ | condom + spermicide | g. _____ | diaphragm + spermicide |
| d. _____ | natural family planning | h. _____ | IUD |

In questions 22–24, fill in the blanks.

22. The two common causes of infertility in females are a.______________________________ and b.______________________.

23. The most common cause of infertility in males is a.________________ caused by b.__________________.

24. In which assisted reproductive technologies is the egg fertilized in laboratory glassware?

______________________________________________

## 43.5 Sexually Transmitted Diseases (pp. 788–91)

- Sexually transmitted diseases include AIDS, genital warts, genital herpes, hepatitis, chlamydia, gonorrhea, and syphilis.

25. AIDS, genital warts, and genital herpes are caused by a. ____________. Among the STDs caused by viruses, treatments are only available for b. ____________ and c. ______________________. Chlamydia, gonorrhea, and syphilis are caused by d. ______________________.

26. AIDS (acquired immunodeficiency syndrome) is caused by a virus known as a. ____________ (human immunodeficiency viruses). HIV attacks b. ___________ cells. When these cells c. ___________ in number, the person becomes susceptible to other types of infections. During the asymptomatic category A stage, the person infected is a d. ___________, and therefore is highly infectious. The category e. ________________ stage may last six to eight years. The category C stage of an HIV infection is called f. ____________________ and is characterized in part by the development of opportunistic diseases.

27. Genital warts are caused by the human ______________________ (HPVs), sexually transmitted and associated with cancer of the cervix and other tumors.

28. Genital herpes is caused by the ______________________ viruses. After the ulcers heal, the disease is dormant, and blisters can reoccur repeatedly.

29. Hepatitis is caused by a a. ______________________. Hepatitis A is usually acquired by b. ______________________________________________________ and can be sexually transmitted. Hepatitis infects the c. __________________.

30. *Chlamydia* is named for the bacterium a. ___________________________. Chlamydial infections may result in pelvic b. ______________________.

31. Gonorrhea is caused by the bacterium a. ___________________________. In the male, a typical symptom of gonorrhea is a thick, greenish-yellow urethral discharge 3–5 days after contact. In females, it may spread to the oviducts, causing b. ___________________________ disease.

32. Syphilis is caused by a bacterium called a. ______________________ and can be treated with penicillin. During the b. ____________ stage of syphilis, a hard chancre (ulcerated sore with hard edges) indicates the site of infection. During the secondary stage, the victim breaks out in a c. ________ that does not itch. During the d. ____________ stage, syphilis may affect the cardiovascular and nervous systems.

# Chapter Test

## Objective Questions

Do not refer to the text when taking this test.

____ 1. The interstitial cells
   a. are located in the testes.
   b. secrete male sex hormones.
   c. store sperm.
   d. Both *a* and *b* are correct.

____ 2. The vas deferens
   a. becomes erect.
   b. carries sperm.
   c. is surrounded by the prostate gland.
   d. All of these are correct.
   e. None of these are correct.

____ 3. The prostate gland
   a. is removed when a vasectomy is performed.
   b. is not needed to maintain the secondary sexual characteristics.
   c. receives urine from the bladder.
   d. All of these are correct.
   e. None of these are correct.

____ 4. Which gland is an endocrine gland?
   a. seminal vesicles
   b. prostate gland
   c. bulbourethral gland
   d. testes
   e. All of these are correct.

____ 5. FSH
   a. stimulates sperm production.
   b. is found in the female.
   c. is produced by the anterior pituitary gland.
   d. All of these are correct.

____ 6. Gonadotropic hormones are produced by the
   a. testes.
   b. ovaries.
   c. anterior pituitary.
   d. Both *a* and *b* are correct.

____ 7. Which of these is a gonadotropic hormone?
   a. testosterone
   b. FSH
   c. estrogen
   d. All of these are correct.

____ 8. In the human male, hormones from the ____________________ stimulate production of testosterone by secreting ____________________.
   a. testis; seminal fluid
   b. hypothalamus; tropic hormones
   c. pituitary gland; luteinizing hormone
   d. seminal vesicles; follicle-stimulating hormone
   e. prostate gland; releasing factors

____ 9. Testosterone is necessary for
   a. the development of the male sex organs.
   b. sperm maturation.
   c. the development of the secondary sex characteristics.
   d. All of these are correct.

____ 10. The urethra is part of the reproductive tract in the
   a. female.
   b. male.
   c. Both *a* and *b* are correct.
   d. None of these are correct.

____ 11. The endometrium
   a. lines the vagina.
   b. breaks down during menstruation.
   c. produces estrogen.
   d. Both *a* and *b* are correct.

____ 12. The uterus
   a. is connected to both the oviducts and vagina.
   b. is not an endocrine gland.
   c. contributes to the development of the placenta.
   d. All of these are correct.

____ 13. Which structure is present after ovulation?
   a. primary follicle
   b. secondary follicle
   c. vesicular follicle
   d. corpus luteum

____ 14. Ovulation occurs
   a. due to hormonal changes.
   b. always on day 14.
   c. in postmenopausal women.
   d. as a result of sexual intercourse.
   e. Both *a* and *b* are correct.

____ 15. Which of these secretes hormones involved in the ovarian cycle?
   a. hypothalamus
   b. anterior pituitary gland
   c. ovary
   d. All of these are correct.

____ 16. Follicle-stimulating hormone stimulates the
   a. release of the oocyte from the follicle.
   b. development of a follicle.
   c. development of the endometrium.
   d. beginning of the menstrual flow.
   e. Both *a* and *c* are correct.

____ 17. Secretions from which of the following structures are required before implantation can occur?
   a. the ovarian follicle
   b. the pituitary gland
   c. the corpus luteum
   d. Both *a* and *b* are correct.
   e. All of these are correct.

____18. Human chorionic gonadotropin (HCG) is different from other gonadotropic hormones because it
a. is produced by the maternal part of the placenta.
b. is not produced by a female endocrine gland.
c. does not act upon the ovaries or the corpus luteum.
d. does not stimulate any tissue in the body.
e. does not enter the bloodstream.

____19. What do FSH, LH, testosterone, progesterone, and estrogen have in common?
a. They occur only in the female.
b. They occur only in the male.
c. All of them directly affect the uterine lining.
d. All of them are necessary for sexual reproduction.
e. Both *a* and *c* are correct.

____20. Menstruation begins in response to
a. an increase in circulating estrogen levels.
b. a decrease in circulating progesterone levels.
c. rupture of the ovarian follicle.
d. changes in the blood $CO_2$ level.
e. secretion of FSH by the pituitary.

____21. In vitro fertilization takes place in
a. the vagina.
b. a surrogate mother.
c. laboratory glassware.
d. the uterus.

____22. What do all methods of birth control have in common?
a. They all use some device.
b. They all interrupt sexual intercourse.
c. They are all fairly expensive.
d. None of these are correct.

____23. A vasectomy
a. prevents the egg from reaching the oviduct.
b. prevents sperm from reaching seminal fluid.
c. prevents the release of seminal fluid.
d. inhibits sperm production.

____24. Which of these means of birth control prevents implantation?
a. diaphragm
b. IUD
c. cervical cap
d. female condom

____25. Which method of birth control also gives 100% assurance of protection from sexually transmitted diseases?
a. diaphragm and spermicide
b. birth control pill
c. condom and spermicide
d. None of these offer 100% protection from sexually transmitted diseases.

## Critical Thinking Questions

Answer in complete sentences.

26. Speculate on the benefits and drawbacks of hermaphroditism.

27. Contrast aspects of external fertilization with internal fertilization.

**Test Results:**______ number correct ÷ 27 = ______ × 100 = ______ %

## Exploring the Internet

The Online Learning Center at *www.mhhe.com/maderbiology8* has additional study material and practice quizzes that can help you master the content of this chapter. You can also find links to websites exploring additional topics in biology. Access to the Online Learning Center is free for those who have purchased a new textbook.

# ANSWER KEY

## STUDY QUESTIONS

**1. a.** A **b.** S **c.** S **d.** A **e.** S **f.** S **g.** A **h.** A **2. a.** stored food for embryo **b.** a dramatic change in shape **c.** serve the needs of the embryo **d.** eggs are retained in the body until they hatch **e.** offspring are born alive **3. a.** urinary bladder; stores urine **b.** vas deferens; conducts and stores sperm **c.** prostate gland; contributes to semen **d.** urethra; conducts both urine and sperm **e.** penis; organ of sexual intercourse **f.** seminal vesicles; contribute to semen **g.** bulbourethral gland; contributes nutrients and fluid to semen **h.** epididymis; stores sperm as they mature **i.** testis; production of sperm and male sex hormones **4. a.** epididymis **b.** vas deferens **c.** urethra **5. a.** seminal vesicles **b.** prostate gland **c.** bulbourethral glands **d.** no **6. a.** T **b.** T **7. a.** IC **b.** ST **c.** ST **d.** IC **8. a.** promotes development **b.** causes growth **c.** larger larynx, longer vocal cords result in deep voice **d.** promotes development **e.** increases secretions **9. a.** decreases **b.** increases **c.** decreases **d.** decreases **e.** increases **10. a.** oviduct; conduction of egg **b.** ovary; production of eggs and sex hormones **c.** uterus; houses developing fetus **d.** urinary bladder; storage of urine **e.** urethra; conduction of urine **f.** cervix; opening of uterus **g.** vagina; receives penis during sexual intercourse and serves as birth canal **11. a.** vagina **b.** cervix **c.** oviduct **12. a.** mons pubis **b.** labia majora **c.** glans clitoris **d.** labia minora **e.** urethra **f.** vagina **g.** anus **13. a.** follicle **b.** secondary oocyte **c.** Graafian **d.** Ovulation **e.** corpus luteum **14.** b, a, d, c **15. a.** FSH and LH **b.** progesterone and estrogen **c.** estrogen **d.** progesterone **16. a.** FSH secretion begins **b.** menstruation **c.** endometrium breaks down **d.** proliferative phase **e.** endometrium rebuilds **f.** secretory phase **g.** endometrium thickens and glands are secretory **h.** progesterone secretion is prominent **17.** The placenta adds to the production of estrogen and progesterone. It also produces HCG. **18. a.** increases **b.** decreases **c.** increase **d.** increases **e.** decreases **f.** increases **19. a.** F, Implantation occurs several days after fertilization. **b.** T **c.** F, During pregnancy, ovulation discontinues because estrogen and progesterone are secreted by the corpus luteum and the placenta exert feedback control over the hypothalamus and the anterior pituitary **d.** T **20. a.** increases **b.** maturation and maintenance **c.** promotes **21. a.** 3 **b.** 2 **c.** 1 **d.** 4 **e.** 1 **f.** 4 **g.** 3 **h.** 2 **22. a.** blocked oviducts **b.** endometriosis **23. a.** low sperm count or large proportion of abnormal sperm **b.** environmental factors **24.** IVF and GIFT **25. a.** viruses **b.** AIDS **c.** genital herpes **d.** bacteria **26 a.** HIV **b.** helper T **c.** decline **d.** carrier **e.** B **f.** full-blown AIDS **27.** papillomaviruses **28.** herpes simplex **29. a.** virus **b.** drinking sewage-contaminated water **c.** liver **30. a.** *Chlamydia trachomatis* **b.** inflammatory disease (PID) **31. a.** *Neisseria gonorrhoeae* **b.** pelvic inflammatory **32. a.** *Treponema pallidum* **b.** primary **c.** rash **d.** tertiary

## CHAPTER TEST

**1.** d **2.** b **3.** b **4.** d **5.** d **6.** c **7.** b **8.** c **9.** d **10.** b **11.** b **12.** d **13.** d **14.** a **15.** d **16.** b **17.** e **18.** b **19.** d **20.** b **21.** c **22.** d **23.** b **24.** b **25.** d **26.** When animals are hermaphroditic, it's easier to find a partner because any two animals that meet can reproduce sexually. In some instances hermaphroditic animals self-fertilize (e.g., tapeworm) and then variability would most likely be reduced in the next generation. **27.** External fertilization, which usually occurs in aquatic or moist habitats, requires that males and females release their gametes at the same time. Many offspring are produced at the same time. Internal fertilization requires compatible copulatory organs and fewer offspring are produced. Typically the embryo is protected and the offspring is well cared for.

# 44

# Development

## Chapter Review

During **fertilization,** the egg and sperm interact. The acrosome of a sperm releases enzymes that digest a pathway through the zona pellucida. The sperm binds to the plasma membrane of the egg, and then enters the egg. The sperm nucleus fuses with the egg nucleus.

During the early developmental stages, **cleavage** leads to a **morula,** which becomes the **blastula** when an internal cavity (the **blastocoel**) appears. Then, at the **gastrula** stage, invagination of cells into the blastocoel results in formation of the germ layers: **ectoderm, mesoderm,** and **endoderm.** The establishment of **germ layers** leads to the formation of organs.

Development of the lancelet, frog, and chick over the first three stages of embryo formation differs due to the amount of **yolk** in the egg.

During neurulation, the nervous system develops from midline ectoderm, just above the **notochord.** At this point, it is possible to draw a typical cross section of a chordate embryo.

**Cellular differentiation** begins with cleavage when maternal determinants are parceled out. **Morphogenesis** produces the shape and form of the body. **Induction** is the ability of one embryonic tissue to influence the development of another tissue by giving off chemical signals.

Work with the roundworm, *C. elegans,* and the fruit fly, *Drosophila,* has shown that development is orderly because genes are turned on in sequence. **Apoptosis** is also necessary to development. Master genes code for morphogen gradients, which contain transcription factors that turn on other master genes. **Homeotic genes** are master genes that control pattern formation.

Human development can be divided into **embryonic development** (months 1 and 2) and **fetal development** (months 3–9). The **extraembryonic membranes,** including the **chorion, amnion, allantois,** and **yolk sac,** appear early in human development. In mammals, the placenta functions in gas, nutrient, and waste exchange between fetal blood and maternal blood.

Organ development begins with neural tube and heart formation. There follows a steady progression of organ formation during the embryonic development. During fetal development, refinement of features occurs and the fetus adds weight. The process of birth includes three stages.

## Study Exercises

Study the text section by section as you answer the questions that follow.

### 44.1 Early Development Stages (pp. 796–99)

- Development begins when a sperm fertilizes an egg.
- The first stages of embryonic development in animals lead to the establishment of the embryonic germ layers.
- The presence of yolk affects the manner in which animal embryos go through the early developmental stages.

1. What is the proper sequence for the following events that describe fertilization? Indicate by letters. ____________________
   a. When released, these enzymes digest a pathway for the sperm through the zona pellucida. The sperm binds to the plasma membrane of the egg.
   b. The head of a sperm has a membrane-bounded acrosome filled with digestive enzymes.
   c. The sperm enters the egg, the sperm nucleus fuses with the egg nucleus, and fertilization is complete.
   d. Several sperm penetrate the corona radiata.

2. Label this diagram of early development using the following alphabetized list of terms.

archenteron
blastocoel (used twice)
blastopore
blastula
ectoderm
endoderm
gastrula
morula

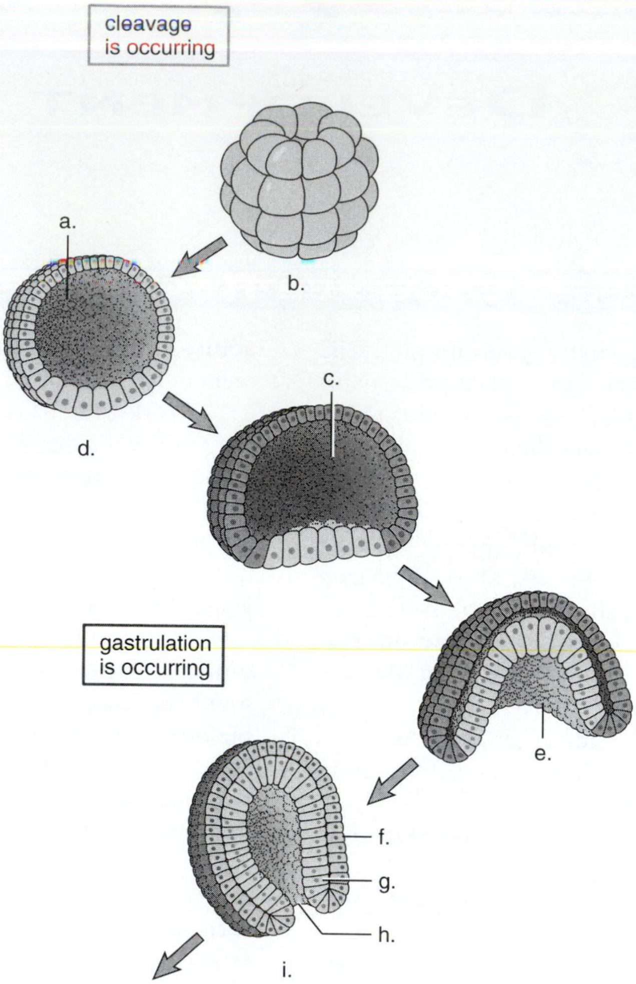

cleavage is occurring

a.

b.

c.

d.

gastrulation is occurring

e.

f.

g.

h.

i.

3. Indicate whether these statements are true (T) or false (F).
   a. _____ Cell division during cleavage does not produce growth.
   b. _____ The blastula is a solid ball of cells.
   c. _____ The ectoderm and endoderm form after the mesoderm in the gastrula.
   d. _____ The development of organs can be related to germ layers.
4. Indicate the germ layer (ectoderm, endoderm, mesoderm) of the vertebrate gastrula stage that is the source of the following:
   a. ____________________epidermis of the skin
   b. ____________________nervous tissue
   c. ____________________lining of the stomach
   d. ____________________muscles of the upper limb
   e. ____________________blood
5. Complete these sentences by using the terms *lancelet, frog,* or *chick.* In the a.____________________ embryo, the cells have little yolk, and cleavage is equal. In the b.____________________, the cells at the animal pole are smaller than those at the vegetal pole because those at the vegetal pole contain yolk. In the c.____________________, the cells with yolk cleave more slowly than those without yolk. Still, in both the d.____________________ and the e.____________________, the blastula is a hollow ball of cells. In the f.____________________, there is so much yolk that the embryo forms on top of the yolk and the blastocoel is created when the cells lift up from the yolk.

6. Indicate which of these describes formation of mesoderm and the coelom in the *lancelet, frog,* or *chick.*
   a. Invagination of cells along the edges of primitive streak is followed by a splitting of the mesoderm.
   _______________
   b. Migration of cells from the dorsal lip of the blastopore is followed by a splitting of the mesoderm.
   _______________
   c. Outpocketings of the primitive gut form two layers of mesoderm and the coelom. _______________

## Neurulation and the Nervous System (p. 799)

- In vertebrates, the nervous system develops above the notochord after formation of a neural tube.

7. Label this diagram of a vertebrate embryo using the alphabetized list of terms.
   coelom
   ectoderm
   endoderm
   gut
   mesoderm
   neural tube
   notochord
   somite

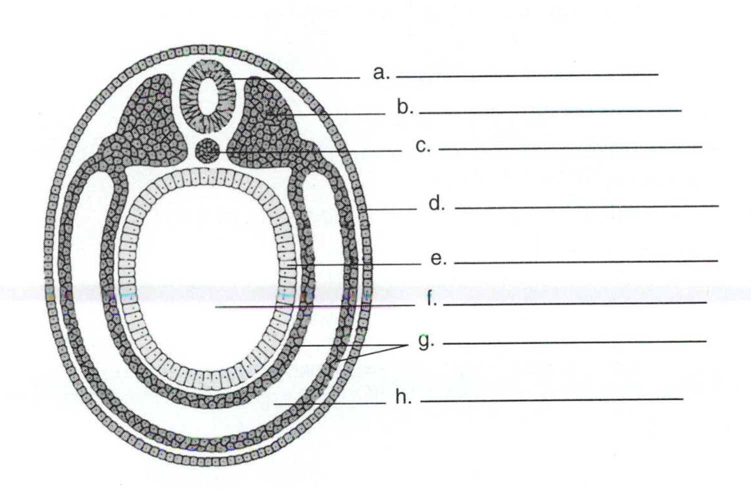


8. The diagram in question 7 shows the neural tube above the notochord. Explain the significance of this tissue relationship. _______________________________________________
_______________________________________________

## 44.2 Developmental Processes (pp. 800–4)

- Cytoplasmic segregation and induction help bring out cellular differentiation and morphogenesis.
- Developmental genetics has benefited from research into the development of *Caenorhabditis elegans*, a roundworm, and *Drosophila melanogaster*, a fruit fly.
- Apoptosis and homeotic genes are involved in shaping the outward appearance of animals.
- A hierarchy of gene activity causes development to be orderly.

9. Place the appropriate letter(s) next to each statement.

C—cytoplasmic segregation    I—induction    C, I—both

a. _____ the parceling out of maternal determinants during cleavage and thereafter
b. _____ embryos that receive a portion of the gray crescent develop normally
c. _____ ability of one embryonic tissue to influence the development of another tissue
d. _____ the nervous system develops above the notochord
e. _____ the reciprocal development of the lens and optic vesicle
f. _____ its importance is dependent on chemical signals that influence developing

10. Match the definitions to the terms. Answers can be used more than once, and there can be more than one answer per phrase.

fate map    apoptosis    homeotic genes    morphogen gradients

a. _____ *C. elegans*
b. _____ induction is an ongoing process
c. _____ during development, some cells die
d. _____ proteins coded for by master genes
e. _____ genes important in pattern formation
f. _____ *Drosophila*
g. _____ transcription factors; proteins that bind to DNA

11. Indicate whether these statements are true (T) or false (F).
a. _____ Work with model organisms has shown that morphogen gradients coded for by master genes turn on the next set of master genes and so forth.
b. _____ During induction, chemical signals pass from one tissue to the next.
c. _____ Homeotic genes are restricted to *Drosophila*.
d. _____ If development proceeds normally, each new cell is expected to have a particular fate—be a part of a particular organ and perform a particular function.

## 44.3 Human Embryonic and Fetal Development (pp. 805–12)

- Humans, like chicks, are dependent upon extraembryonic membranes that perform various services and contribute to development.
- During the embryonic period of human development, all systems appear.
- The placenta is a unique mammalian structure where exchange between fetal blood and maternal blood takes place.

12. Label this diagram of the extraembryonic membranes of the human embryo using the following alphabetized list of terms.
allantois
amnion
chorion
embryo
fetal portion of placenta
maternal portion of placenta
yolk sac

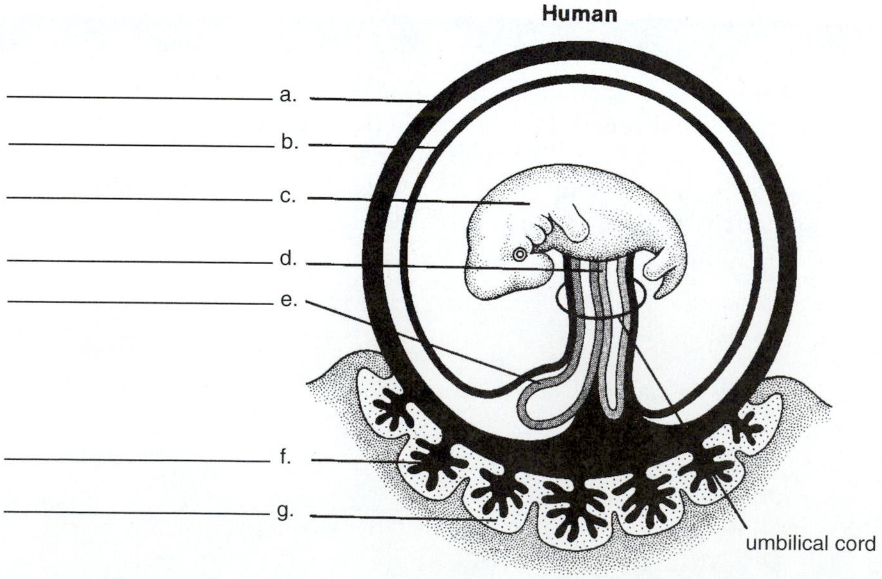

Human

a. _______________
b. _______________
c. _______________
d. _______________
e. _______________
f. _______________
g. _______________

umbilical cord

13. Complete the following table:

| Membrane | Chick Function | Human Function |
|---|---|---|
| Chorion | | |
| Amnion | | |
| Allantois | | |
| Yolk sac | | |

## Embryonic Development (pp. 806–8)

14. To describe human embryonic development, complete the following table with the number of the event that occurs at each time indicated.
    1. all internal organs formed; limbs and digits well formed; recognizable as human although still quite small
    2. fertilization; cell division begins
    3. limb buds begin; heart is beating; embryo has a tail
    4. implantation; embryo has tissues; first two extraembryonic membranes
    5. fingers and toes are present; cartilaginous skeleton
    6. nervous system begins; heart development begins
    7. head enlarges; sense organs prominent

| Time | Events |
|---|---|
| a. First week | |
| b. Second week | |
| c. Third week | |
| d. Fourth week | |
| e. Fifth week | |
| f. Sixth week | |
| g. Two months | |

15. Indicate whether these statements about fetal development (months 3–9) are true (T) or false (F).
    a. _____ It is possible to distinguish sex.
    b. _____ The notochord is replaced by the spinal column.
    c. _____ Limb buds are still present.
    d. _____ Fingernails and eyelashes appear.

16. What events are associated with the three stages of birth (parturition)?
    a. first stage: ____________________
    b. second stage: ____________________
    c. third stage: ____________________

# CHAPTER TEST

## OBJECTIVE QUESTIONS

Do not refer to the text when taking this test.

____ 1. The ______ develops first.
a. morula
b. blastula
c. blastocoel
d. gastrula

____ 2. The ______ is a hollow ball.
a. morula
b. blastula
c. gastrula
d. Both *a* and *b* are correct.

____ 3. The ______ contains germ layers.
a. morula
b. blastula
c. gastrula
d. All of these are correct.

____ 4. The ______ undergoes cleavage but lacks a morula.
a. lancelet
b. frog
c. chick
d. Both *a* and *b* are correct.

____ 5. The ____________ has a notochord during development.
a. lancelet
b. frog
c. human
d. All of these are correct.

____ 6. The nervous system develops from the
a. ectoderm.
b. mesoderm.
c. endoderm.
d. notochord.

____ 7. Cellular differentiation is due to
a. parceling out of cytoplasm.
b. activation of particular genes.
c. parceling out of genes.
d. Both *a* and *b* are correct.
e. Both *a* and *c* are correct.

____ 8. What induces the development of the nervous system?
a. endoderm
b. presumptive notochord
c. ectoderm
d. presumptive neural tube

____ 9. Homeodomain proteins
a. occur in the nucleus.
b. regulate transcription.
c. regulate translation.
d. Both *a* and *b* are correct.

For questions 10–14, match the descriptions to the extra-embryonic membranes.
a. chorionic villi
b. chorion
c. amnion
d. allantois
e. yolk sac

____ 10. Placenta
____ 11. Umbilical blood vessels
____ 12. Watery sac
____ 13. First site of red blood cell formation
____ 14. Treelike extensions that penetrate the uterine lining

____ 15. The zygote begins to undergo cleavage in the
a. cervix.
b. ovary.
c. oviduct.
d. uterus.

____ 16. Select an incorrect association.
a. cleavage—cell division
b. morphogenesis—fertilization
c. differentiation—specialization of cells
d. growth—increase in size

____ 17. The placenta
a. brings blood to the developing fetus.
b. allows exchanges of substances between maternal blood and fetal blood.
c. forms the umbilical cord.
d. Both *a* and *b* are correct.
e. All of these are correct.

____ 18. When an embryo is clearly recognizable as a human being, it is called a
a. developed embryo.
b. fetus.
c. newborn.
d. blastocyst.

____ 19. Which system is the first to be visually evident?
a. nervous
b. respiratory
c. digestive
d. skeletal

____ 20. During which stage of parturition is the baby born?
a. first
b. second
c. third

## Critical Thinking Questions

Answer in complete sentences.

21. Should the chemicals functioning during induction be considered hormones?

22. What is the significance of the homeobox?

**Test Results:** ______ number correct ÷ 22 = ______ × 100 = ______ %

## Exploring the Internet

The Online Learning Center at *www.mhhe.com/maderbiology8* has additional study material and practice quizzes that can help you master the content of this chapter. You can also find links to websites exploring additional topics in biology. Access to the Online Learning Center is free for those who have purchased a new textbook.

## Answer Key

### Study Exercises

**1.** d, b, a, c **2. a.** blastocoel **b.** morula **c.** blastula **d.** blastocoel **e.** archenteron **f.** ectoderm **g.** endoderm **h.** blastopore **i.** gastrula **3. a.** T **b.** F **c.** F **d.** T **4. a.** ectoderm **b.** ectoderm **c.** endoderm **d.** mesoderm **e.** mesoderm **5. a.** lancelet **b.** frog **c.** frog **d.** lancelet **e.** frog **f.** chick **6. a.** chick **b.** frog **c.** lancelet **7. a.** neural tube **b.** somite **c.** notochord **d.** ectoderm **e.** endoderm **f.** gut **g.** mesoderm **h.** coelom **8.** The notochord induces the formation of the neural tube. See Figure 44.4, p. 799, in text. **9. a.** C **b.** C **c.** I **d.** I **e.** I **f.** C, I **10. a.** fate map and apoptosis **b.** fate map **c.** apoptosis **d.** morphogen gradients **e.** homeotic genes **f.** homeotic genes and morphogen gradients **g.** morphogen gradients **11. a.** T **b.** T **c.** F **d.** T **12.** See Figure 44.11, p. 805, in text.

**13.**

| Chick Function | Human Function |
|---|---|
| gas exchange | exchange with mother's blood |
| protection; prevention of desiccation and temperature changes | protection; prevention of temperature changes |
| collection of nitrogenous wastes | blood vessels become umbilical blood vessels |
| provision of nourishment | first site of blood cell formation |

**14. a.** 2 **b.** 4 **c.** 6 **d.** 3 **e.** 7 **f.** 5 **g.** 1 **15. a.** T **b.** F **c.** F **d.** T **16. a.** dilation of cervix **b.** The mother pushes as the baby moves down the birth canal. **c.** Afterbirth is expelled.

### Chapter Test

**1.** a **2.** b **3.** c **4.** c **5.** d **6.** a **7.** c **8.** b **9.** d **10.** b **11.** d **12.** c **13.** e **14.** a **15.** c **16.** b **17.** b **18.** b **19.** a **20.** b **21.** Traditionally, a hormone is considered to be a secretion of an endocrine gland carried in the bloodstream to a target organ. According to this definition, the chemicals that function during induction are not hormones. In recent years, some scientists have broadened the definition of a hormone to include all types of chemical signals. Therefore, in the broadest sense these chemicals are hormones. **22.** A particular sequence of DNA nucleotides, called the homeobox, occurs in homeotic genes in almost all eukaryotic organisms. This suggests that this sequence is important to development because it has been conserved for quite some time.

# PART VIII BEHAVIOR AND ECOLOGY

# 45

# ANIMAL BEHAVIOR

## CHAPTER REVIEW

Biologists ask *mechanistic questions* and *survival value questions* about **behavior.** Mechanistic questions pertain to mechanisms of behavior, and survival value questions pertain to the adaptive nature of behavior. Hybrid studies with warblers show that behavior has a genetic basis. The nervous and endocrine systems control behavior, as shown by garter snake experiments and sea slug (*Aplysia*) DNA studies.

A behavior sometimes undergoes development after birth, as exemplified by improvement in laughing gull chick begging behavior. **Learning** occurs when a behavior changes with practice. Experiments teaching male birds to sing in their species dialect show that various factors—such as social experience—influence whether learning takes place.

Genes influence behavior, so it is reasonable to assume that adaptive behavioral traits will evolve. Both sexes are expected to behave in a manner that will raise their reproductive success. Females who produce only one egg a month are expected to choose the best mate, and males who produce many sperm are expected to inseminate as many females as possible. **Sexual selection** is natural selection due to mate choice by females and competition among males. Do females choose mates who have the best traits for survival or the ones to whom they are attracted? Or are these hypotheses one and the same? Experiments with satin bowerbirds and birds of paradise have been inconclusive. Showy plumage displays in birds may attract females because health and vigor are indicated or possibly because sons of these males will have a greater chance of being selected by females.

A cost-benefit analysis is particularly applicable to male competition. A **dominance hierarchy,** as seen in baboons, and establishment of a **territory,** as seen in red deer, are two ways in which strong males get to reproduce more than weaker males. Behaviors will only continue if the benefits, such as greater access to females, outweigh the costs, such as shorter life expectancy, and therefore fewer offspring.

**Communication** between animals consists of chemical, auditory, visual, and tactile signals. Social living can help an animal avoid predators, rear offspring, and find food. Disadvantages include fighting among members, spread of a contagious disease, and the possibility of subordination to others.

**Altruistism** seems self-sacrificing until the concept of **inclusive fitness,** which includes personal reproductive success and the reproductive success of relatives, is examined. Among social insects, sisters share 75% of their genes rather than 50%. This makes it more likely that they will help raise siblings.

## STUDY EXERCISES

Study the text section by section as you answer the questions that follow.

### 45.1 BEHAVIOR HAS A GENETIC BASIS (PP. 818–19)

- Behaviors have a genetic basis but can also be influenced by environmental factors.
- The nervous and endocrine systems have immediate control over behaviors.

1. Label each of the following as describing a mechanistic question (M) or a survival value question (S):
   a. _____ How do insects spread their wings before they begin to fly?
   b. _____ Do male robins attack any other male that attempts to mate with their partner?
   c. _____ How do garter snakes sense slugs?
   d. _____ Why do some garter snakes eat slugs and others eat frogs?

2. Describe the experiment and results concerning Cape Verde blackcap warblers, which do not migrate, and German blackcap warblers, which migrate to Africa.

   a. experiment: ___

   b. results: ___

   c. conclusion: ___

3. Inland garter snakes eat frogs and fish, and coastal garter snakes eat slugs. Investigators discovered that inland snakes do not respond to the smell of slugs, and hybrids generally have only an intermediate ability to respond to the smell of slugs. What conclusion was reached? ___

4. Egg-laying behavior of *Aplysia* involves a set sequence of movements. Investigators found that the gene that controls behavior codes for hormones. What conclusion was reached? ___

## 45.2 Behavior Undergoes Development (pp. 820–21)

- Behaviors sometimes undergo development after birth, as when learning affects behavior.

5. Indicate whether these statements about the experiment with laughing gull chicks are true (T) or false (F).
   a. ___ Laughing gull chicks seek their own food.
   b. ___ Motor development helps explain why the pecking behavior of chicks improves.
   c. ___ Operant conditioning—a form of learning—helps explain why older chicks choose a model that looks more like the parent.

6. Explain statement 5*c* here: ___

7. Due to imprinting, chicks follow the first moving object they see. In relation to this observation, explain the following:

   a. sensitive period ___

   b. need for social interaction ___

8. The following diagram illustrates experiments studying how birds learn to sing:

   Explain each of the frames.

   a. first ___

   b. second ___

   c. third ___

   d. What conclusion is appropriate? ___

## 45.3 Behavior Is Adaptive (pp. 823–27)

- Natural selection influences such behaviors as methods of feeding, avoiding predators, and reproducing.
- A cost-benefit analysis is a tool for testing adaptive hypotheses about behavior.
- Female selectivity of mates and male competition to secure a mate influence behavior and evolution.

9. Indicate whether these statements about the adaptive nature of behavior are true (T) or false (F).
   a. _____ Behavior has a genetic basis.
   b. _____ Certain behaviors can improve reproductive success.
   c. _____ The nervous and endocrine systems control behavior.
10. With reference to the reproductive behavior of satin bowerbirds, females chose males with well-kept bowers that contained blue objects. Why might this support the good genes hypothesis? ______________________
    ______________________________________________
11. With reference to the reproductive behavior of birds of paradise, perhaps females choose the males with spectacular plumes because it signifies a. ____________ or because their sons will be b. ____________ to females also. It was also found that raggiana offspring are fed a more c. ____________ food than those of the related species, the trumpet manucode. This seems to correlate with the fact that the male raggiana birds of paradise are d. ____________, while the trumpet manucode birds are e. ____________. Birds are monogamous when it takes two parents to f. ____________ the offspring.
12. a. In terms of a cost-benefit analysis, what is the benefit to dominant males in a baboon troop? ____________
    ______________________________________________
    b. What are the costs? ______________________________________________
    ______________________________________________
    c. What is the benefit to subordinate males in a baboon troop? ______________________
    d. What are the costs? ______________________________________________
    e. Why do you predict that the benefits for each must outweigh the costs? ______________________

## 45.4 Animal Societies (pp. 828–29)

- Animals living in societies have various means of communicating with one another.

13. Match the descriptions to these terms:
    1. chemical communication
    2. auditory communication
    3. visual communication
    4. tactile communication

    a. _____ Honeybees do a waggle dance in a dark hive.
    b. _____ Male raggiana birds of paradise do spectacular courtship dances.
    c. _____ Birds sing songs.
    d. _____ Cheetahs spray a pheromone onto a tree.

## 45.5 Sociobiology and Animal Behavior (pp. 830–31)

- Group living is adaptive in some situations and not in others.
- Behaviors that appear to reduce fitness may be found to increase fitness on closer examination.

14. Indicate whether these statements are true (T) or false (F). Rewrite any false statements to make them true.

    a. _____ Subordinate males have less chance to mate, but group living may help them survive. Rewrite: _____

    ______________________________________________

    b. _____ Animals that live alone may have to spend less time grooming. Rewrite: ______________

    ______________________________________________

    c. _____ Animals that capture large prey tend to live alone. Rewrite: ______________

    ______________________________________________

    d. _____ The cost of social living outweighs the benefits, but animals like being with others. Rewrite: _____

    ______________________________________________

15. Match the statements to these terms (multiple answers are possible):

    1. altruism
    2. inclusive fitness
    3. helpers at the nest
    4. reciprocity

    a. _____ Males do not prevent receptive female chimpanzees from copulating with several members of a group.
    b. _____ A behavior seems to be self-sacrificing.
    c. _____ Older siblings take care of younger siblings.
    d. _____ Worker bees do not reproduce and instead help raise siblings.
    e. _____ A younger bird helps an older bird raise its young but takes over the territory when the older bird dies.

## Definitions Wordmatch

Review key terms by completing this matching exercise, selecting from the following alphabetized list of terms:

altruism
behavior
communication
imprinting
inclusive fitness
operant conditioning
pheromone
sociobiology
territoriality

a. ______________ Behavior related to the act of marking or defending a particular area against invasion by another species member; area often used for the purpose of feeding, mating, and caring for young.

b. ______________ Social interaction that has the potential to decrease the lifetime reproductive success of the member exhibiting the behavior.

c. ______________ Signal by a sender that influences the behavior of a receiver.

d. ______________ Chemical substance secreted by one organism that influences the behavior of another.

e. ______________ Increase in reproduction that results from direct selection and indirect selection.

f. ______________ Observable, coordinated responses to environmental stimuli.

g. ______________ Application of evolutionary principles to the study of social behavior of animals, including humans.

# Chapter Test

## Objective Questions

Do not refer to the text when taking this test.

____ 1. Which of these pertain(s) to behavior?
   a. The heart pumps blood into the arteries.
   b. Ants lay a pheromone trail to guide other ants.
   c. Birds have warning calls.
   d. Both *b* and *c* are correct.

____ 2. Which of these statements are supported by data?
   a. Males compete for mates.
   b. Operant conditioning brings about learning.
   c. Social interactions control behavior.
   d. Females carry out sexual selection.
   e. All of these statements are supported by data.

For questions 3–7, match the statements to these descriptions of behavior:
   a. Behavior has a genetic basis.
   b. The nervous and endocrine systems control behavior.
   c. Behavior undergoes development.

____ 3. Hybrid studies with warblers reveal this.
____ 4. Garter snakes differ in their ability to smell slugs.
____ 5. Laughing gull chicks improve in their ability to recognize their parent.
____ 6. Egg-laying behavior in *Aplysia* reveals this.
____ 7. Caged birds can learn to sing their species' song if they hear a recording of it during a sensitive period.

____ 8. The pecking improvement of laughing gull chicks
   a. can be explained by operant conditioning.
   b. correlates with improved motor skills.
   c. is a form of learning.
   d. All of these are correct.

____ 9. A sensitive period for learning was observed when
   a. hybrid garter snakes were intermediate in their ability to smell slugs.
   b. captive birds learned to sing a more developed song by hearing a recording.
   c. imprinting occurred.
   d. Both *b* and *c* are correct.

____ 10. The adaptiveness of behavior may be associated with which statement(s)?
   a. Behavior has a genetic basis.
   b. The nervous and endocrine systems control behavior.
   c. Altruism is involved in inclusive fitness.
   d. Both *a* and *c* are correct.

____ 11. Which of these is NOT consistent with reproduction in females?
   a. selecting the best mate possible
   b. having a higher potential to produce many offspring
   c. nurturing offspring until they can care for themselves
   d. producing few eggs over a lifetime

____ 12. Male competition leads to
   a. dominance hierarchies and reduction in fighting.
   b. defense of a territory.
   c. neglect of the young.
   d. Both *a* and *b* are correct.

____ 13. A cost-benefit analysis can explain why
   a. subordinate males remain in a group.
   b. red deer males are large despite the chances of it shortening their life span.
   c. older siblings take care of younger siblings.
   d. All of these are correct.

____ 14. Inclusive fitness explains
   a. seemingly altruistic behavior.
   b. why older siblings help raise younger siblings.
   c. the benefit of being a worker bee.
   d. All of these are correct.

## Critical Thinking Questions

Answer in complete sentences.

15. What evidence shows that behavior is inherited?

16. According to the tenets of sociobiology, is the behavior of animals altruistic?

**Test Results:** _____ number correct ÷ 16 = _____ × 100 = _____ %

## Exploring the Internet

The Online Learning Center at *www.mhhe.com/maderbiology8* has additional study material and practice quizzes that can help you master the content of this chapter. You can also find links to websites exploring additional topics in biology. Access to the Online Learning Center is free for those who have purchased a new textbook.

## Answer Key

### Study Exercises

**1. a.** M **b.** S **c.** M **d.** S **2. a.** Mate the two types of warblers. **b.** Hybrids show migratory restlessness. **c.** Hybrids inherit genes from both parents, and therefore show behavior intermediate between the two. **3.** The nervous system controls the eating behavior of garter snakes. **4.** Hormones also control behavior. **5. a.** F **b.** T **c.** T **6.** Due to operant conditioning, chicks learn to peck correctly (i.e., only at models that closely resemble the parent) because in that way they are rewarded with food. **7. a.** Behavior is best learned during a sensitive period immediately after birth. **b.** Clucking by a hen that has recently had chicks can bring about the behavior even outside the sensitive period. **8. a.** Isolated bird sings but does not learn to sing the species' song. **b.** Bird learns to sing the song if a recording is played during a sensitive period. **c.** Bird learns to sing the song of a social tutor of another species outside a sensitive period. **d.** Social interactions help learning take place. **9. a.** T **b.** T **c.** T **10.** Aggressive males are able to have well-kept bowers, and this behavior, which may be inherited, may lead to reproductive success. **11. a.** health **b.** attractive **c.** nutritious **d.** polygynous **e.** monogamous **f.** feed **12. a.** first chance to mate **b.** might be injured protecting the troop **c.** protection **d.** less frequent chance to mate **e.** because the behavior evolved through natural selection **13. a.** 4 **b.** 3 **c.** 2 **d.** 1 **14. a.** T **b.** T **c.** F, . . . tend to live in a group **d.** F, The benefits of social living outweigh the costs or else animals would not live in a group. **15. a.** 2 **b.** 1, 3 **c.** 1, 2, 3 **d.** 1, 2, 3 **e.** 4

### Definitions Wordmatch

**a.** territoriality **b.** altruism **c.** communication **d.** pheromone **e.** inclusive fitness **f.** behavior **g.** sociobiology

### Chapter Test

**1.** d **2.** e **3.** a **4.** b **5.** c **6.** b **7.** c **8.** d **9.** d **10.** d **11.** b **12.** d **13.** d **14.** d **15.** Experimentation has shown that behavior has a genetic basis. Hybrid warblers show migratory restlessness, a trait intermediate to both parents, indicating that behavior is inherited. Hybrid garter snakes generally have an intermediate ability to smell slugs. Because behavior has a genetic basis, it has to be inherited. **16.** It may appear to be altruistic but may be explainable by inclusive fitness—which depends not only on the number of direct descendants due to personal reproduction but also on the number of offspring produced by relatives that the individual has helped nurture.

# 46

# Ecology of Populations

## Chapter Review

**Ecology** is the study of the interactions of organisms with other organisms and with the physical environment. Ecology encompasses several levels of study: **organism; population; community; ecosystem;** and finally, the **biosphere.** The interactions of organisms with the abiotic and biotic environment affect the **population density** and **distribution.**

Populations have a certain size that depends, in part, on their **intrinsic rate of natural increase** (*r*). There are two patterns of population growth: **exponential growth** results in a J-shaped growth curve and **logistic growth** results in an S-shaped growth curve. Exponential growth can only occur when resources are abundant; otherwise, logistic growth occurs. When population size reaches the **carrying capacity** of the environment, **environmental resistance** opposes **biotic potential.**

A **survivorship** curve describes the mortality (deaths per capita) of a population. There are three idealized survivorship curves: with type I, most individuals survive well past the midpoint of the life span; with type II, survivorship decreases at a constant rate throughout the life span; and with type III, most individuals die young. **Age structure diagrams** tell what proportion of the population is prereproductive, reproductive, and postreproductive.

Density-dependent and density-independent factors regulate population size. **Density-independent factors,** such as weather and fire, and density-dependent factors, such as predation and competition, are both extrinsic factors. Intrinsic factors such as territoriality may also be involved.

Life history patterns have been related to the logistic growth curve. So-called ***r*-selection** occurs in unpredictable environments (density-independent factors regulate population size) and favors small adults that reproduce early and do not invest in parental care. So-called ***K*-selection** occurs in stable environments (density-dependent factors regulate population size) and favors large adults that reproduce repeatedly during a long life span and invest much energy in parental care.

The human population is currently in the exponential part of its growth curve. **More-developed countries (MDCs)** experienced **demographic transition** some time ago, and the populations of most are either not growing or are decreasing. **Less-developed countries (LDCs)** are only now undergoing demographic transition, but will still experience much growth because of their pyramid-shaped age structure diagram. Control of population size in LDCs and reduction of resource consumption in MDCs are necessary to protect the environment.

## Study Exercises

Study the text section by section as you answer the questions that follow.

### 46.1 Scope of Ecology (pp. 836–37)

- Ecology is the study of the interactions of organisms with other organisms and with the physical environment.
- The interactions of organisms with the abiotic and biotic environment affect their distribution and abundance.

1. Match the statement to these levels of ecological study:

   community ecosystem
   population biosphere

   a. ______________ a group of populations interacting in an area
   b. ______________ a community interacting with its physical environment
   c. ______________ all the individuals of the same species in an area
   d. ______________ portion of earth's surface where living things exist
   e. ______________ focus of ecological study is growth and regulation of size
   f. ______________ focus of ecological study is interactions such as predation and competition between populations

2. Populations vary in density (number of individuals per area) and distribution. Use these labels to describe distribution:

clumped
random
uniform

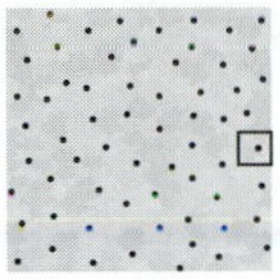

a. ____________________

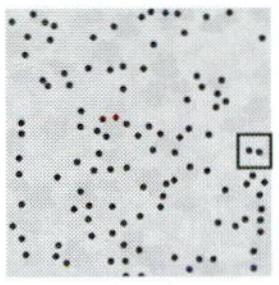

b. ____________________

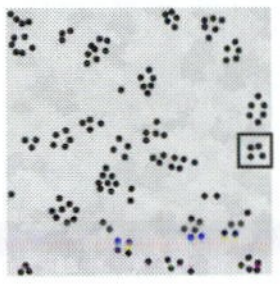

c. ____________________

3. Which pattern of distribution is most common? ______________

4. [a.] ______________ factors determine where an individual organism can live. Temperature is an example of a(n) [b.] ______________ factor, and availability of prey is an example of a(n) [c.] ______________ factor that can limit where types of organisms are found.

## 46.2 Characteristics of Populations (pp. 838–43)

- Population size is dependent upon natality, mortality, immigration, and emigration.
- Population growth models predict changes in population size over time under particular conditions.

5. Considering natality (birthrate), mortality (death rate), immigration, and emigration, which two lead to an increase in population size and which two lead to a decrease in population size?

| **Increase** | **Decrease** |
|---|---|
| a. ____________________ | c. ____________________ |
| b. ____________________ | d. ____________________ |

6. Calculate the intrinsic rate of natural increase ($r$) when the birthrate is .06 per capita per unit time and the death rate is .04 per capita per unit time.

$r$ = [a.] ______ – [b.] ______ = [c.] ______ per capita unit time

d. Given a population of panthers in the Florida Everglades, and per unit time = 1 year and population = 50 panthers, how many panthers would there be after a year? ______________

Questions 7–12 pertain to the following two population growth curves:

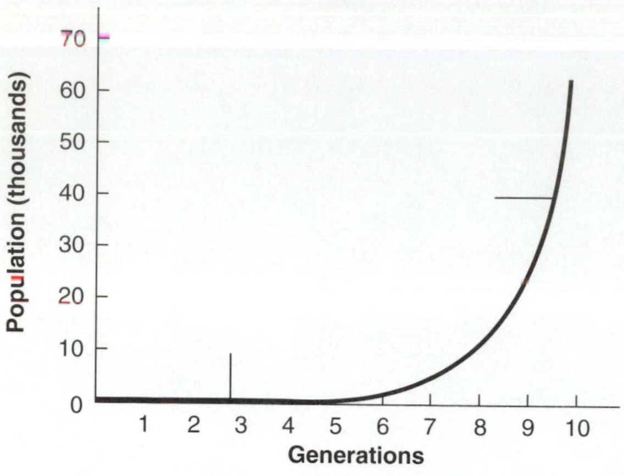

a. 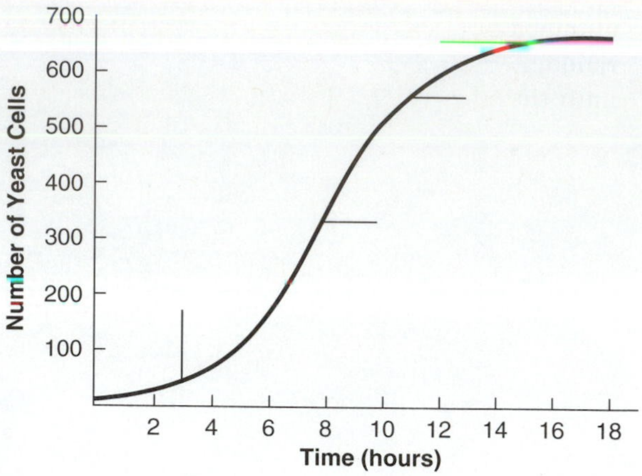 ______________________ b. ______________________

7. Place the label *ep* for exponential growth pattern or *lg* for logistic growth pattern beneath the appropriate figure.

8. Which of these two growth curves has a *lag phase*? ____________ Place this label where appropriate on the curve(s).

9. Which of these two growth curves has an *exponential growth phase*? ____________ Place this label where appropriate on the curve(s).

10. During the lag phase, growth is a. ____________; during the exponential growth phase, growth is b. ____________.

11. Which of these two growth curves has a *deceleration phase*? ____________ Place this label where appropriate on the curve(s).

12. Which of these two growth curves has a *stable equilibrium phase*? ____________ Place this label where appropriate on the curve(s).

13. Which of the phases of a growth curve (lag, exponential, deceleration, or stable equilibrium) best represents the biotic potential of a population? ______________________

14. a. Which of these phases (lag, exponential, deceleration, or stable equilibrium) best represents environmental resistance? ______________________

    b. What is environmental resistance? ______________________

    ______________________

    c. At what point will the stable equilibrium phase occur? ______________________

    ______________________

Questions 15–17 pertain to the following equation: $\frac{dN}{dt} = rN\left(\frac{K - N}{K}\right)$

15. When $N$ is much smaller than $K$, the term $K - N/K$ is approximately 1. Is the opportunity for population growth maximal or minimal? ____________

16. When $N$ is about equal to $K$, the term $K - N/K$ is zero. Is the opportunity for population growth maximal or minimal? ____________

17. Which of the following can be associated with biotic potential, and which can be associated with carrying capacity?

    a. ____________ $K$

    b. ____________ $r$

## Mortality Patterns (p. 842)

- Mortality within a population is recorded in a life table and illustrated by a survivorship curve.

18. Study the following diagram of survivorship curves and then answer the questions:

a. Which curve shows that the members of a cohort die at a constant rate? _____________

b. Which curve shows that the members of a cohort tend to die early in life? _____________

c. Which curve shows that the members of a cohort usually live through their entire allotted life span? _____________

19. What is a cohort? _______________________________________________

## Age Distribution (p. 843)

- Populations have an age distribution consisting of prereproductive, reproductive, and postreproductive age groups.

20. a. If the age structure diagram has a pyramid shape and the prereproductive population is largest, what do you predict about population growth?

_______________________________________________

b. Why? _______________________________________________

21. a. If the age structure diagram has an urn shape and the postreproductive group is largest, what do you predict about population growth?

_______________________________________________

b. Why? _______________________________________________

## 46.3 Regulation of Population Size (pp. 844–45)

- Factors that affect population size are classified as density-independent factors and density-dependent factors.

22. In general, density-independent factors are [a.] _____________ (*abiotic/biotic*), such as [b.] _____________, and density-dependent factors are [c.] _____________ (*abiotic/biotic*), such as [d.] _____________. Intrinsic factors such as [e.] _____________ may also be involved in regulating population size.

## 46.4 Life History Patterns (pp. 846–47)

- Life histories range from discrete reproductive events resulting in rapid population growth to repeated reproduction resulting in a stable population size.

Match questions 23–26 to these descriptions:

a. unpredictable environment and density-independent regulation of population size
b. stable environment and density-dependent regulation of population size

23. _____ large adults, long life span, slow to mature, repeated reproduction, and much care of offspring
24. _____ opportunistic pattern
25. _____ equilibrium pattern
26. _____ small adults, short life span, fast to mature, many offspring during a burst of reproduction, and little or no care of offspring

## 46.5 Human Population Growth (pp. 849–51)

- The human population is still growing exponentially, and how long this can continue is not known.

27. If the human population outstrips its carrying capacity, what will happen to the size of the population?
_____________________________________________

28. If the human population is curtailed by environmental resistance, what will happen to the size of the population? _____________________________________________

For the short answers in questions 29–32, use the following:

1. More-developed countries (MDCs)
2. Less-developed countries (LDCs)

29. a. Which type of country has a pyramid-shaped age structure diagram? _______________
b. With such an age distribution, growth is expected for some time. Why? _______________
_____________________________________________

30. Which type of country has a high per capita consumption rate? _______________
_____________________________________________

31. a. Which type of country has undergone demographic transition? _______________
b. What is demographic transition? _______________

32. Which type of country has a high net reproductive rate? _______________

# Chapter Test

## Objective Questions

Do not refer to the text when taking this test.

_____ 1. Select the incorrect association.
a. population—all the members of a species in same area
b. community—populations interacting with the physical environment
c. biosphere—surface of the earth where organisms live
d. ecosystem—energy flow and chemical cycling occur

_____ 2. Distribution of organisms tends to be
a. clumped.
b. determined by limiting factors.
c. the same as the population density.
d. Both *a* and *b* are correct.

____ 3. If the birthrate is 10 per 1,000 and the death rate is 10 per 1,000, then the net reproductive rate is
a. 0.
b. 10.
c. 20.
d. 100.

____ 4. If the intrinsic rate of natural increase is positive, then
a. population growth will occur.
b. the size of the population will increase.
c. environmental resistance is likely to come into play.
d. All of these are correct.

____ 5. In the equation $N_{t+1} = RN_t$, $R$ equals
a. environmental resistance.
b. biotic potential.
c. net reproductive rate.
d. Both *b* and *c* are possible.

____ 6. To calculate $R$,
a. assume immigration = emigration.
b. subtract death rate per capita per unit time from birthrate per capita per unit time.
c. Both *a* and *b* are correct.
d. Neither *a* nor *b* is correct.

____ 7. During exponential growth, growth
a. remains steady.
b. is accelerating.
c. is declining.
d. depends on the environment.

____ 8. During the stable equilibrium phase of logistic growth,
a. the term $(K–N)/K = 0$.
b. the term $(K–N)/K = 1$.
c. the population has outstripped the carrying capacity.
d. biotic potential is in full force.

____ 9. Which sentence is most appropriate?
a. Environmental resistance encourages biotic potential.
b. If population size is at carrying capacity, growth is unlikely.
c. Environmental resistance consists only of density-independent factors.
d. Exponential growth can usually occur indefinitely.

____10. Which is true of an S-shaped growth curve?
a. represents the exponential growth pattern
b. represents the logistic growth pattern
c. does not usually occur in nature
d. Both *b* and *c* are correct.

____11. Survivorship in a population is related to
a. age of death.
b. biotic potential.
c. age structure diagram.
d. All of these are correct.

____12. If the survivorship curve is a straight diagonal line, then
a. the rate of death is constant, regardless of age.
b. most individuals live out the expected life span.
c. most individuals die early.
d. environmental resistance has occurred.

____13. Select the density-dependent effect.
a. climate
b. predation
c. natural disaster
d. weather

____14. Density-dependent effects
a. increase as density increases.
b. tend to be biotic factors.
c. tend to be abiotic factors.
d. Both *a* and *b* are correct.

____15. *K*-strategists tend to have a(n) ____________ growth curve.
a. J-shaped
b. S-shaped

____16. Select the characteristic that is NOT consistent with *r*-selection.
a. large body size
b. many offspring
c. short life span
d. fast to mature

____17. The countries in Asia and Africa are
a. MDCs experiencing rapid growth.
b. LDCs experiencing rapid growth.
c. MDCs experiencing slow growth.
d. LDCs experiencing slow growth.

____18. The doubling time for the world's population will most likely
a. always remain the same.
b. become longer because of demographic transition.
c. become shorter and shorter regardless.
d. fluctuate because of depressions.

____19. The world population increases by the number of people found in a medium-size city (200,000) every
a. year.
b. six months.
c. month.
d. day.

## Critical Thinking Questions

Answer in complete sentences.

20. Under what conditions could the growth of a population be infinite?

21. How is the exponential growth of a population similar to the effect of compound interest on money saved in a bank?

**Test Results:** ______ number correct ÷ 21 = ______ × 100 = ______ %

## Exploring the Internet

The Online Learning Center at *www.mhhe.com/maderbiology8* has additional study material and practice quizzes that can help you master the content of this chapter. You can also find links to websites exploring additional topics in biology. Access to the Online Learning Center is free for those who have purchased a new textbook.

## Answer Key

### Study Exercises

**1. a.** community **b.** ecosystem **c.** population **d.** biosphere **e.** population **f.** community **2. a.** uniform **b.** random **c.** clumped **3.** clumped **4. a.** Limiting **b.** abiotic **c.** biotic **5. a.** natality **b.** immigration **c.** mortality **d.** emigration **6. a.** .06 **b.** .04 **c.** .02 **d.** 51 **7. a.** ep **b.** lg **8.** both; see Figures 46.4 *b* and 46.5*b*, pages 839–40, in text **9.** both; see Figures 46.4 *b* and 46.5*b*, pages 839–40, in text **10. a.** slow **b.** accelerating **11.** b; see Figure 46.5*b*, p. 840, in text **12.** b; see Figure 46.5*b*, p. 840, in text **13.** exponential **14. a.** deceleration **b.** encompasses all environmental factors that oppose biotic potential **c.** when biotic potential and environmental resistance are equal **15.** maximal **16.** minimal **17. a.** carrying capacity **b.** biotic potential **18. a.** II **b.** III **c.** I **19.** an original group of individuals born at the same time **20. a.** It will continue for some time. **b.** A large number of women will be entering their reproductive years. **21. a.** It will decline. **b.** Most of the population is postreproductive. **22. a.** abiotic **b.** weather, fire **c.** biotic **d.** predation, competition **e.** territoriality **23.** b **24.** a **25.** b **26.** a **27.** A crash will occur. **28.** It will become steady. **29. a.** 2 **b.** A large number of women are entering their reproductive years. **30.** 1 **31. a.** 1 **b.** decreased death rate followed by decreased birthrate **32.** 2

### Chapter Test

**1.** b **2.** d **3.** a **4.** d **5.** d **6.** c **7.** b **8.** a **9.** b **10.** b **11.** a **12.** a **13.** b **14.** d **15.** b **16.** a **17.** b **18.** b **19.** d **20.** If environmental resistance were absent, the population size could continue to increase without end. This is unlikely, because the increase in population size without an increase in resources would force environmental resistance to be present. **21.** Interest in the bank is paid on both the principal (initial amount of money) and also on the interest generated from and added to that base amount. During exponential growth, new individuals are generated from the original base population, from their descendants, and so on.

# 47

# Community Ecology

## Chapter Review

A **community** is an assemblage of populations interacting with one another within the same environment. Communities are characterized by their composition (types of species) and by their diversity, which includes richness (number of species) and evenness (relative abundance). Both abiotic and biotic factors control community composition.

In a community, each population occupies a **habitat** and also has an **ecological niche,** the role an organism plays in its community, including its habitat and its interactions with other organisms. Relationships between populations in a community are defined by such interactions as **competition, predation, parasitism, commensalism,** and **mutualism.** The **competitive exclusion principle** states that no two species can occupy the same niche at the same time. **Resource partitioning** is observed when **character displacement** occurs, as in Galápagos finches, but is also believed to be present whenever similar species feed on slightly different foods or occupy slightly different habitats. Predation reduces the size of the **prey** population but can have a feedback effect that limits the **predator** population. Prey species have evolved various means to escape predators—for example, chemical defenses in plants and **mimicry** in animals. Coevolution is observed between predators and prey and between parasite and host. A **parasite** derives nourishment from the host.

Species in a community may exhibit several types of **symbiosis.** In parasitism, the fitness of the parasite increases, and that of the host decreases. **Commensalism** has a neutral effect on both species, in that neither is benefited nor harmed. In mutualism, the fitness of both species increases.

Communities are dynamic and undergo **ecological succession,** a change in a community following a disturbance. The process of succession is complex and may not always reach particular ends of community composition and diversity. Habitat patches at different stages of succession produce greater species diversity within the community.

To increase biodiversity, intermediate levels of disturbance may be desirable, predation and competition may be beneficial (except when exotic species are introduced), and a size large enough for greater degree of diversity is helpful.

## Study Exercises

Study the text section by section as you answer the questions that follow.

### 47.1 Concept of the Community (pp. 858–60)

- Communities are assemblages of interacting populations, which differ in composition and diversity.
- Environmental factors influence community composition and diversity.

1. Draw a line between the terms in the second column that pertain to those in the first column, and a line between those in the third column that pertain to those in the second column.

| | | |
|---|---|---|
| a. composition | c. richness | e. relative abundance |
| b. diversity | d. evenness | f. number of species |

2. Label each of these findings with (1) for the individualistic model of community structure and (2) for the interactive model of community structure.
   a. _____ Five different coral reefs all contain the same species in the same relative numbers.
   b. _____ There was so much overlapping between species between the forest and field that determining where one ended and the other began was impossible.

3. Species richness and islands. Study this diagram.

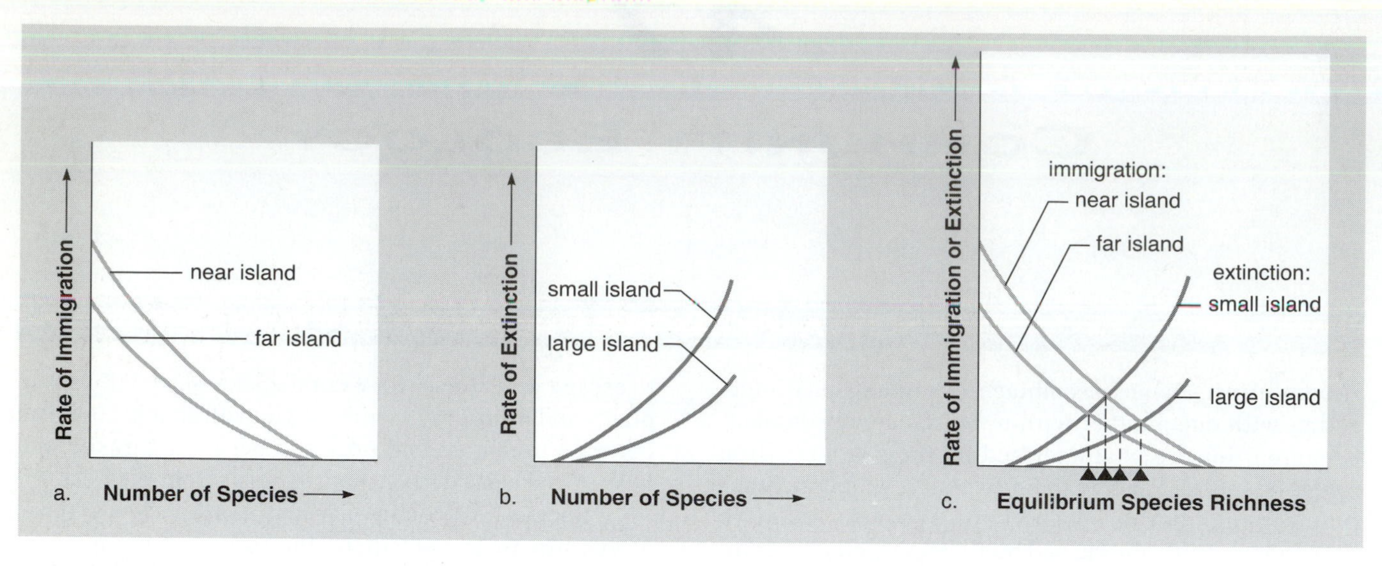


According to the diagram, distance from the mainland affects immigration.

a. Which type of island will have the largest amount of immigration? ______________________

b. According to the diagram, size of the island affects extinction. Which type of island will have the least amount of extinction? ______________________

c. Combining the two concepts, list the four types of islands in terms of species richness, from the greatest to the smallest, and explain:

(1) ______________________

(2) ______________________

(3) ______________________

(4) ______________________

## 47.2 Structure of the Community (pp. 861–71)

- Community organization involves the interactions among species, such as competition, predation, parasitism, commensalism, and mutualism.
- The ecological niche is the role an organism plays in its community, including its habitat and its interactions with other organisms.
- Competition leads to resource partitioning, which reduces competition between species.
- Predation reduces prey population density but also can lead to a reduction in predator population density.
- There are a number of different types of prey defenses, including mimicry.
- Symbiotic relationships include parasitism, commensalism, and mutualism.

4. a. The habitat of an organism is ______________________, but the ecological niche of an organism includes its habitat and its interactions with other organisms.

b. What type of interactions? ______________________

______________________

5. a. Which would you expect to be larger—the fundamental niche of an organism or its realized niche?

______________________

b. Why? ______________________.

6. You would expect competition to be a (– –) interaction because both species are a. ______________________. Gause's laboratory experiments supported the b. ______________________, which states that no two species can occupy the same niche. c. ______________________ is a way for two species to ensure different niches. For example, five species of warblers can coexist because each species d. ______________________.

7. Study this graph, which describes experimental results obtained by G. F. Gause, and then answer the questions that follow.

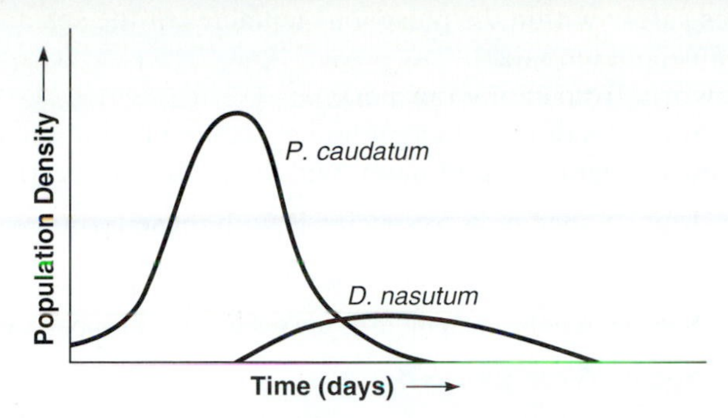


a. What happened to the prey population of *Paramecium?* ____________________

b. What happened to the predator population of *Didinium?* ____________________

c. In nature, both predator and prey populations continue to exist but in reduced densities. Explain.

____________________

8. The Canadian lynx and the snowshoe hare population sizes cycle.

a. Why would they cycle if the predator-prey relationship is causing the cycling? ____________________

____________________

b. Why would they cycle if a hare-food relationship is involved? ____________________

____________________

____________________

9. Identify the antipredator device in the following:

a. Inchworms resemble twigs. ______________

b. Dart-poison frogs are brightly colored. ______________

c. Frilled lizards open up folds of skin around the neck. ______________

10. If appropriate, label the following as describing Batesian (B) or Müllerian (M) mimicry:

a. _____ A predator mimics another species that has a successful predatory style.
b. _____ A prey mimics another species that has a successful defense.
c. _____ A predator captures food.
d. _____ Several different species with the same defense mimic one another.

11. a. ______________ occurs when two species evolve in response to one another. Give an example from the text.

b. ____________________

____________________

12. Commensalism, mutualism, and parasitism are all different types of ________________________ relationships.

13. Complete this table to compare types of symbiosis by marking a plus (+) if the species *benefits,* a minus (–) if the species is *harmed,* and a zero (0) if the symbiotic relationship has *no effect* on the species.

| | **First Species** | **Second Species** |
|---|---|---|
| **Parasitism** | | |
| **Commensalism** | | |
| **Mutualism** | | |

14. Place the appropriate letter next to each statement.

C—commensalism M–mutualism P—parasitism

a. _____ The clownfish lives safely within the poisonous tentacles of the sea anemone, which other fish avoid.
b. _____ Certain bacteria cause pneumonia.
c. _____ Humans get a tapeworm from eating raw pork.
d. _____ Epiphytes grow in branches of trees but get no nourishment from the trees.
e. _____ Flowers provide nourishment to a pollinator, and the pollinator carries pollen to another flower.

15. Label the symbiotic relationships of species A to species B with these terms:

commensalism competition mutualism parasitism predation

a. ______________________ Species A consumed more of the resource than species B.
b. ______________________ Species A eats species B.
c. ______________________ Species A is cultivated by species B as a source of food.
d. ______________________ Species A infects species B.
e. ______________________ Species A rides along with species B to get food while species B hunts.

## 47.3 Community Development (pp. 872–73)

- Ecological succession is a change involving a series of species replacements in a community following a disturbance.

16. Draw a series of stages that illustrate the changes in plant species composition during secondary succession, from early succession to climax community.

17. Label each of the following as being characteristic of an early successional stage (E) or a late successional stage (L):

a. _____ equilibrium species
b. _____ opportunistic species
c. _____ *K*-strategists
d. _____ *r*-strategists
e. _____ few species present
f. _____ many species present
g. _____ climax community

## 47.4 Community Biodiversity (pp. 874–75)

- The intermediate disturbance hypothesis suggests that a moderate amount of disturbance leads to a diverse environment, and therefore more biodiversity.
- Predation and competition can help maintain biodiversity, and island biogeography suggests how to maintain species richness.

18. Place a check by each factor that increases community diversity.

a. _____ high or severe levels of disturbance
b. _____ low levels of disturbance
c. _____ intermediate levels of disturbance
d. _____ uniformity of habitat
e. _____ patchiness of habitat
f. _____ competition between species
g. _____ human interference

# Chapter Test

## Objective Questions

Do not refer to the text when taking this test.

____ 1. A community is made up of all the
   a. members of a given population.
   b. plant populations of a given area.
   c. populations of a given area.
   d. populations of a given area plus the abiotic habitat in which they live.

____ 2. Which is a facet of species diversity?
   a. species richness
   b. species evenness
   c. composition only
   d. Both *a* and *b* are correct.

____ 3. Which of these is likely to have the greatest number of species?
   a. a small island with many patches
   b. a small, homogenous island
   c. a large island with many patches
   d. a large, homogenous island

____ 4. Gause showed that two paramecia species
   a. cannot occupy the same test tube.
   b. can exist in the same test tube.
   c. can exist in two different areas of the same test tube.
   d. None of these are correct.

____ 5. Competition
   a. always eliminates one or the other species.
   b. widens niche breadth.
   c. narrows niche breadth and increases species diversity.
   d. None of these are correct.

____ 6. The niche that an organism occupies while interacting with all others in its community is its
   a. realized niche.
   b. fundamental niche.
   c. habitat.
   d. patch.

____ 7. Predator population size is limited in part by available ______________, while prey population size is limited by ______________.
   a. living space; food
   b. predators; prey
   c. food; predators
   d. food; food

____ 8. Plants produce hormone analogues that
   a. interfere with metabolism of the adult insect.
   b. inhibit egg production in insects.
   c. interfere with the development of insect larvae.
   d. None of these are correct.

____ 9. Antipredator defenses may include
   a. camouflage.
   b. fright.
   c. warning.
   d. All of these are correct.

____ 10. Mimicry can help
   a. a predator capture food.
   b. prey avoid capture.
   c. Both *a* and *b* are correct.
   d. None of these are correct.

____ 11. Bees, wasps, and hornets are examples of
   a. Batesian mimicry.
   b. Müllerian mimicry.
   c. mimicry for predation.
   d. Both *a* and *c* are correct.

For questions 12–14, match the organisms with these terms:
   a. parasitism
   b. commensalism
   c. mutualism

____ 12. virus

____ 13. termites and protozoans

____ 14. barnacle

____ 15. In which relationship do both species benefit?
   a. mutualism
   b. commensalism
   c. symbiosis
   d. Both *b* and *c* are correct.

____ 16. Which can be a parasite?
   a. bacteria
   b. plants
   c. animals
   d. All of these are correct.

____ 17. What parasite is a vector for Lyme disease?
   a. virus
   b. bacterium
   c. tick
   d. fungus

____ 18. Choose the scenario that best represents secondary succession.
   a. trees—shrubs—grasses—perennial grasses
   b. annual weeds and grasses—perennial grasses—shrubs—trees
   c. shrubs—annual weeds—perennial grasses—trees
   d. All of these are correct.

____ 19. Which characteristic is typical of a late successional community?
   a. comprised of *K*-strategists
   b. many species present
   c. climax community
   d. All of these are correct.

____ 20. Which of the following factors increases the diversity of a community?
   a. patchiness
   b. high levels of disturbance
   c. little or no competition
   d. All of these are correct.

## Critical Thinking Questions

Answer in complete sentences.

21. How does niche diversity affect species diversity in a community?

22. Explain how intermediate levels of disturbance maintain diversity.

**Test Results:** _____ number correct ÷ 22 = _____ × 100 = _____ %

## Exploring the Internet

The Online Learning Center at *www.mhhe.com/maderbiology8* has additional study material and practice quizzes that can help you master the content of this chapter. You can also find links to websites exploring additional topics in biology. Access to the Online Learning Center is free for those who have purchased a new textbook.

## Answer Key

### Study Exercises

**1.** Draw lines between: *b* and *c, b* and *d, e* and *d, f* and *c* **2. a.** 2 **b.** 1 **3. a.** near the mainland **b.** large island **c.** (1) large, near island because it has good immigration and minimal extinction (2) large, far island does not have as much immigration but still has minimal extinction (3) small, near island does have some immigration but also has extinction (4) small, far island has limited immigration and much extinction **4. a.** where it lives **b.** predation, competition, symbiosis (parasitism, commensalism, mutualism) **5. a.** fundamental **b.** because it includes the full range of species potential **6. a.** competing for the same resource **b.** competitive exclusion principle **c.** Resource partitioning **d.** occupies a different spruce tree zone **7. a.** died out **b.** died out **c.** prey defenses reduce predation **8. a.** The predator overkills the prey, and as the prey population declines, so does the predator population. **b.** As the hares die off due to lack of food, the predator population would decline, and as the hare population recovers, so would the predator population recover. **9. a.** camouflage **b.** warning coloration **c.** fright **10. b.** B **d.** M (both *a* and *c* should be left blank) **11. a.** Coevolution **b.** The cuckoo lays its eggs in the nests of other birds. To do so, it must lay an egg similar in appearance to that of the host and do so rapidly, and it must leave most of the host bird's eggs in the nest so the cuckoo egg goes unnoticed. The cuckoo hatches first and removes the host's eggs from the nest. **12.** symbiotic

**13.**

| First Species | Second Species |
|---|---|
| + | – |
| + | 0 |
| + | + |

**14. a.** C **b.** P **c.** P **d.** C **e.** M **15. a.** competition **b.** predation **c.** mutualism **d.** parasitism **e.** commensalism **16.** See Figure 47.19, page 873, in text **17. a.** L **b.** E **c.** L **d.** E **e.** E **f.** L **g.** L **18.** c, e, f

### Chapter Test

**1.** c **2.** d **3.** c **4.** c **5.** c **6.** a **7.** c **8.** c **9.** d **10.** b **11.** b **12.** a **13.** c **14.** b **15.** a **16.** d **17.** c **18.** b **19.** d **20.** a **21.** The greater the number of niches, the greater the number of possible species that can fill them, and therefore the greater the diversity. **22.** At low levels of disturbance, the organisms that dominate the community (*K*-strategists) become more abundant, and other species have fewer opportunities to become established. At high levels of disturbance, only the *r*-strategists, which reproduce quickly, survive. At intermediate levels of disturbance, both *r*- and *K*-strategists maintain their populations at a smaller size, but overall diversity increases.

# 48

# Ecosystems and Human Interferences

## Chapter Review

The **biosphere** consists of the atmosphere, the hydrosphere, and the lithosphere. All the ecosystems of the world comprise the biosphere. An **ecosystem** is a community of organisms plus the physical and chemical environment. Some populations are **producers** and some are **consumers.** Producers are **autotrophs** that produce their own organic food. Consumers are **heterotrophs** that take in organic food. Consumers may be **herbivores, carnivores, omnivores,** or **decomposers.**

Energy flows through an ecosystem. Producers transform solar energy into food for themselves and all consumers. As herbivores feed on plants and carnivores feed on herbivores, energy is converted to heat. Feces, urine, and dead bodies become food for decomposers. Eventually, all the solar energy that enters an ecosystem is converted to heat, which dissipates. Therefore, ecosystems require a continual supply of solar energy.

Inorganic nutrients are not lost from the biosphere as is energy. They recycle within and between ecosystems. Decomposers return some portion of inorganic nutrients to autotrophs, and other portions are imported or exported between ecosystems in global cycles.

The **food webs** of ecosystems contain **grazing food chains** (begin with a producer) and **detrital food chains** (begin with detritus). A **trophic level** includes all the organisms that feed at a particular link in food chains. In general, biomass and energy content decrease from one trophic level to the next, as is depicted in an **ecological pyramid.**

The global cycling of inorganic elements involves the biotic and abiotic parts of an ecosystem. Cycles usually contain (1) a reservoir (a source normally unavailable to organisms), (2) an exchange pool (a source available to organisms), and (3) the biotic community.

In the **water (hydrologic) cycle,** evaporation of ocean waters and transpiration from plants contribute to atmospheric moisture. Rainfall over land results in bodies of fresh water and groundwater. Eventually, all water returns to the oceans.

In the **carbon cycle,** respiration by organisms adds as much carbon dioxide to the atmosphere as photosynthesis removes. Human activities, such as the burning of fossil fuels and trees, add carbon dioxide to the atmosphere. Carbon dioxide and other gases trap heat, leading to **global warming.** The effects of global warming could be a rise in sea level and a change in climate patterns, with disastrous effects.

In the **nitrogen cycle,** the biotic community keeps nitrogen recycling back to producers. Human activities convert atmospheric nitrogen to fertilizer, which when broken down by soil bacteria adds nitrogen oxides to the atmosphere. Nitrogen oxides and sulfur dioxide react with water vapor to form acids that contribute to **acid deposition.** Acid deposition is killing lakes and forests, and it also corrodes marble, metal, and stonework. Nitrogen oxides and hydrocarbons (HC) react to form **photochemical smog,** which contains ozone and **PAN.** These oxidants are harmful to animal and plant life.

In the **phosphorus cycle,** the biotic community recycles phosphorus back to the producers, and only limited quantities are made available by the weathering of rocks. Phosphates are mined for fertilizer production, and fertilizer runoff can result in **eutrophication** of lakes and ponds.

Global warming, acid deposition, and water pollution reduce biodiversity. **Ozone shield** depletion, associated with CFCs, is expected to result in decreased productivity of the oceans. Without an adequate ozone shield to absorb UV radiation, health, food sources, and climate are seriously affected.

## Study Exercises

Study the text section by section as you answer the questions that follow.

### 48.1 The Nature of Ecosystems (pp. 880–84)

- The biosphere, where all living things exist, consists of ecosystems.
- In an ecosystem, organisms interact with one another and with the physical and chemical environments.
- Trophic (feeding) relationships are essential to the workings of an ecosystem.
- Autotrophs are producers; photoautotrophs capture solar energy and produce organic nutrients. Heterotrophs are consumers, and take in preformed organic nutrients.

1. Match the descriptions to these biotic components of an ecosystem:

   carnivores consumers decomposers herbivores omnivores autotrophs

   a. ____________________ organisms of decay
   b. ____________________ feed only on other animals
   c. ____________________ producers in an ecosystem
   d. ____________________ heterotrophs
   e. ____________________ feed directly on green plants
   f. ____________________ feed on both plants and animals

2. Label this diagram using these terms:

   decomposers consumers producers inorganic nutrient pool

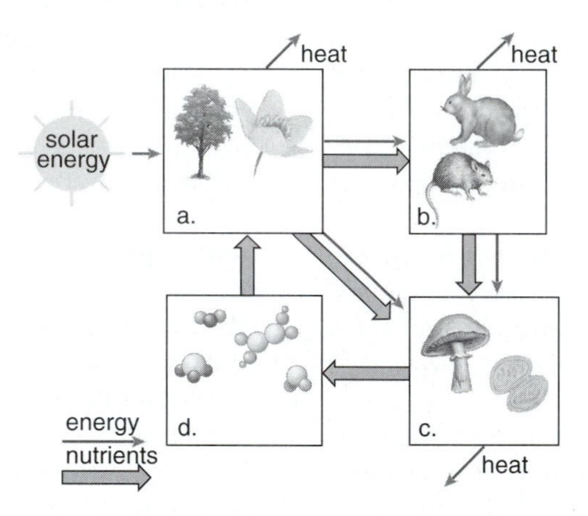


3. Energy doesn't cycle in ecosystems.
   a. Which populations contain the most energy? ____________________
   b. Which populations contain the least energy? ____________________
   c. What happened to the energy? Explain on the basis of the second law of thermodynamics. ____________________
   ____________________

4. Chemicals do cycle in an ecosystem.
   a. What two molecules do plants use to make glucose? ____________________
   b. Glucose could conceivably pass from a producer population to the ____________________ populations to the ____________________ populations.
   c. Eventually, the glucose is broken down, and what is returned to plants? ____________________
   ____________________

## 48.2 Energy Flow (pp. 885–86)

- Various paths of energy flow are represented by food webs, which describe trophic relationships.
- Solar energy enters biotic communities via photosynthesis, and as organic nutrients pass from one trophic level to another, heat is returned to the atmosphere.
- Energy is lost between trophic levels.

These are food web diagrams.

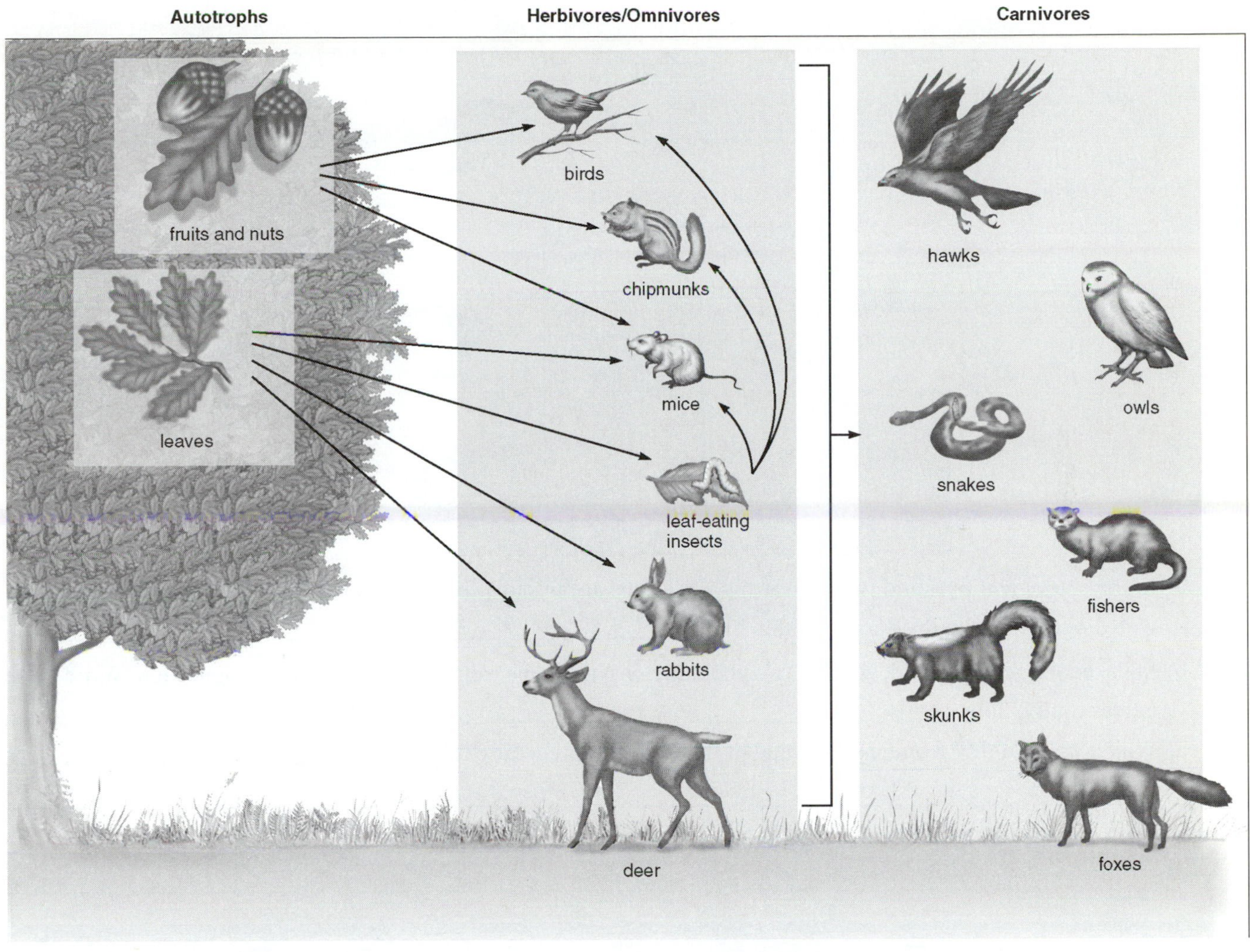


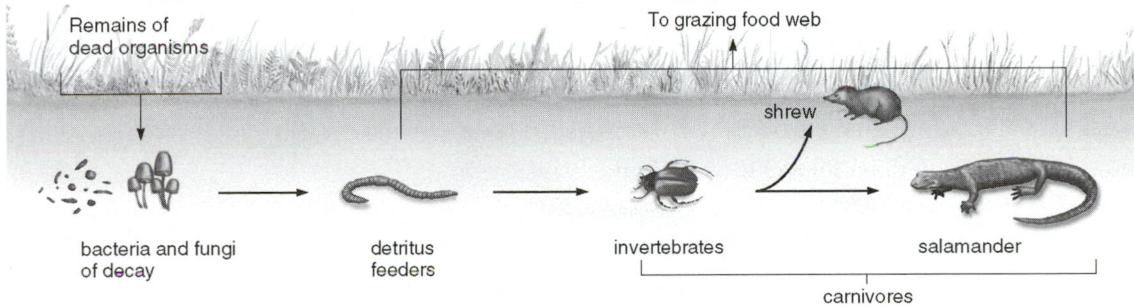


5. a. How many trophic levels do you see in the grazing portion of this food web diagram? ______________

   b. Name a population at the first trophic level. ______________

   c. Name two populations at the second trophic level. ______________

   d. Name two populations at the third trophic level. ______________

   e. From these trophic levels, construct a grazing food chain. ______________

6. Construct a detrital food chain from the food web diagram.

7. Explain one way in which the detrital food web and the grazing food web are always connected.

This is an ecological pyramid diagram.

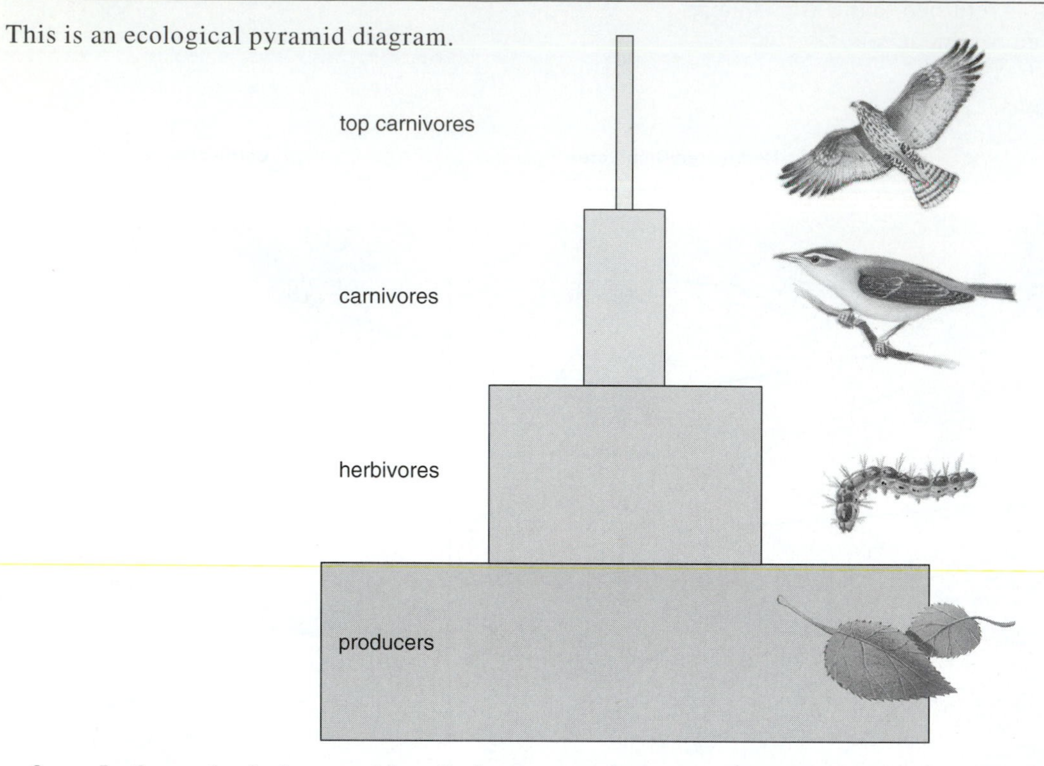


8. a. In the ecological pyramid, write in two populations at the same trophic level as the caterpillar from the food web diagram. Write the names on this line. ______

   b. In the ecological pyramid, write in two populations at the same trophic level as the bird population. Write these names on this line. ______

9. a. Why is each higher trophic level smaller than the one preceding it? ______

   b. What is the so-called 10% rule of thumb? ______

## 48.3 Global Biogeochemical Cycles (pp. 886–93)

- Nutrients cycle within and between ecosystems in global biogeochemical cycles.
- The biogeochemical cycles are: the water cycle, the carbon cycle, the nitrogen cycle, and the phosphorus cycle.

10. Examine the following diagram and then fill in the blanks.

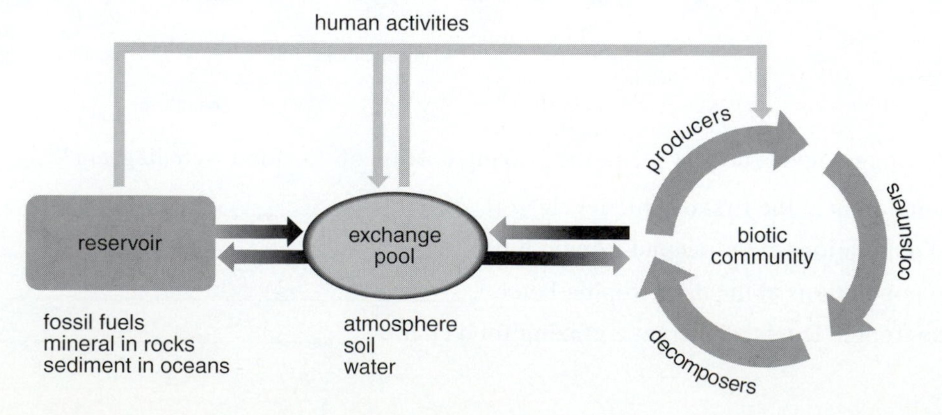

a. What is a reservoir? ______________________________

b. What is an exchange pool? ______________________________

c. What is a biotic community? ______________________________

d. Explain the arrows labeled *human activities.* ______________________________

______________________________

11. Complete this diagram of the water cycle by filling in the boxes, using these terms:

ice    $H_2O$ in the atmosphere    ocean    groundwaters

Label the arrows using these terms:

precipitation (twice)    transpiration from plants and evaporation from soil    evaporation
transport of water vapor by wind

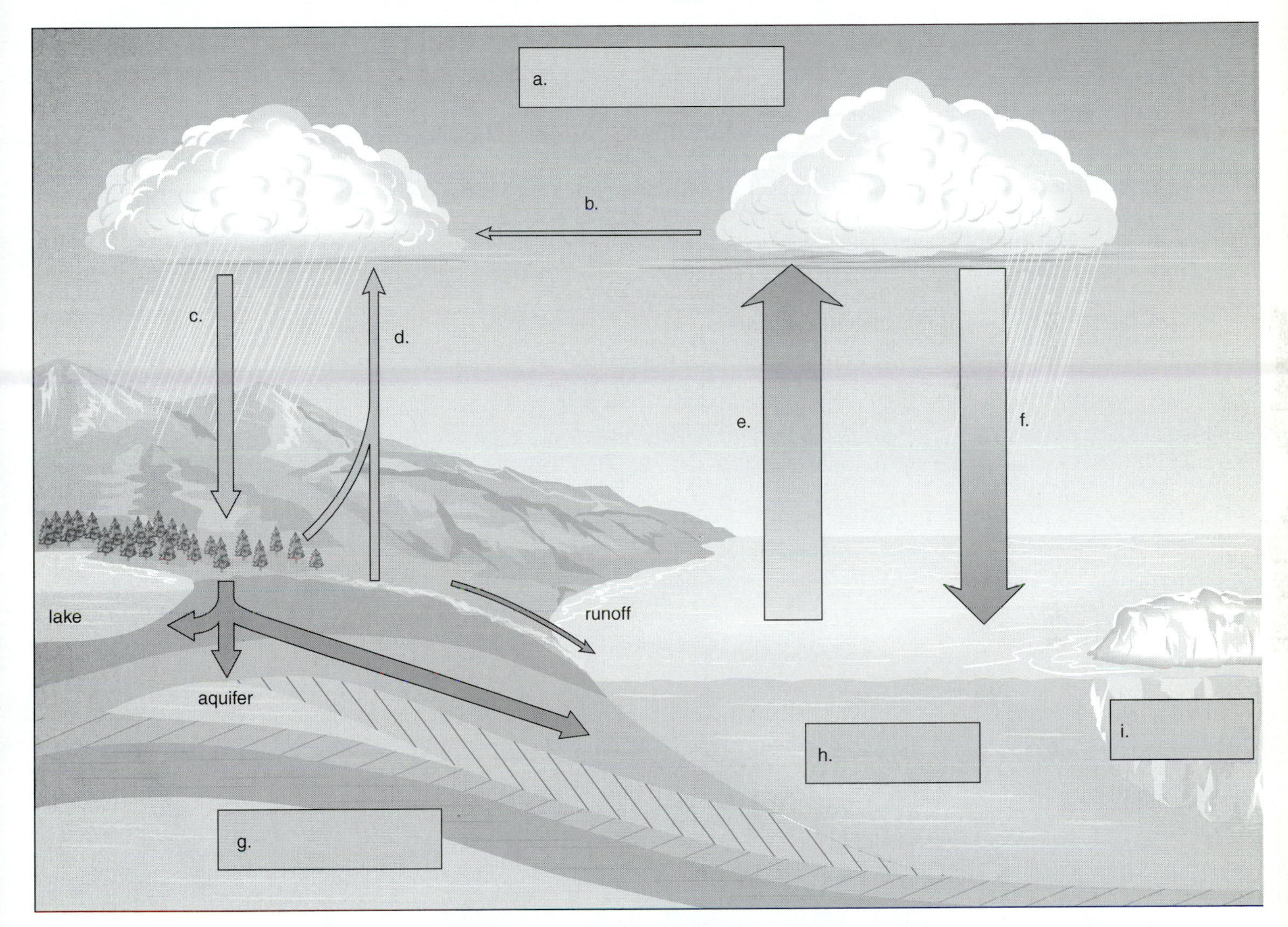


12. Place a check beside the statements that correctly describe the water cycle.
    a. _____ Water cycles between the land, the atmosphere, and the ocean, and vice versa.
    b. _____ We could run out of fresh water.
    c. _____ The ocean receives more precipitation than the land.
    d. _____ Water that is in the aquifers never reaches the oceans.

Fill in the blanks.

13. In the carbon cycle, carbon dioxide is removed from the atmosphere by the process of a. ____________ but is returned to the atmosphere by the process of b. ____________. Living things and dead matter in soil are carbon c. ____________ and so are the d. ____________ because of shell accumulation. In aquatic ecosystems, carbon dioxide from the air combines with water to produce e. ____________ that algae can use for photosynthesis. In what way do humans alter the transfer rates in the carbon cycle? f. ____________

14. Fill in the table to indicate the source of gases that cause the greenhouse effect.

| Gas | From |
|---|---|
| Carbon dioxide ($CO_2$) | a. ____________ |
| Nitrous oxide ($N_2O$) | b. ____________ |
| Methane ($CH_4$) | c. ____________ |

d. Why are these gases called the greenhouse gases? ____________

____________

15. Place a check beside all those statements that may be expected because of global warming.
   a. _____ a global temperature increase by as much as 4.5°C
   b. _____ melting of glaciers and a rise in sea level
   c. _____ massive fish kills and plant destruction
   d. _____ drier conditions inland where droughts may occur
   e. _____ expansion of forests into arctic areas

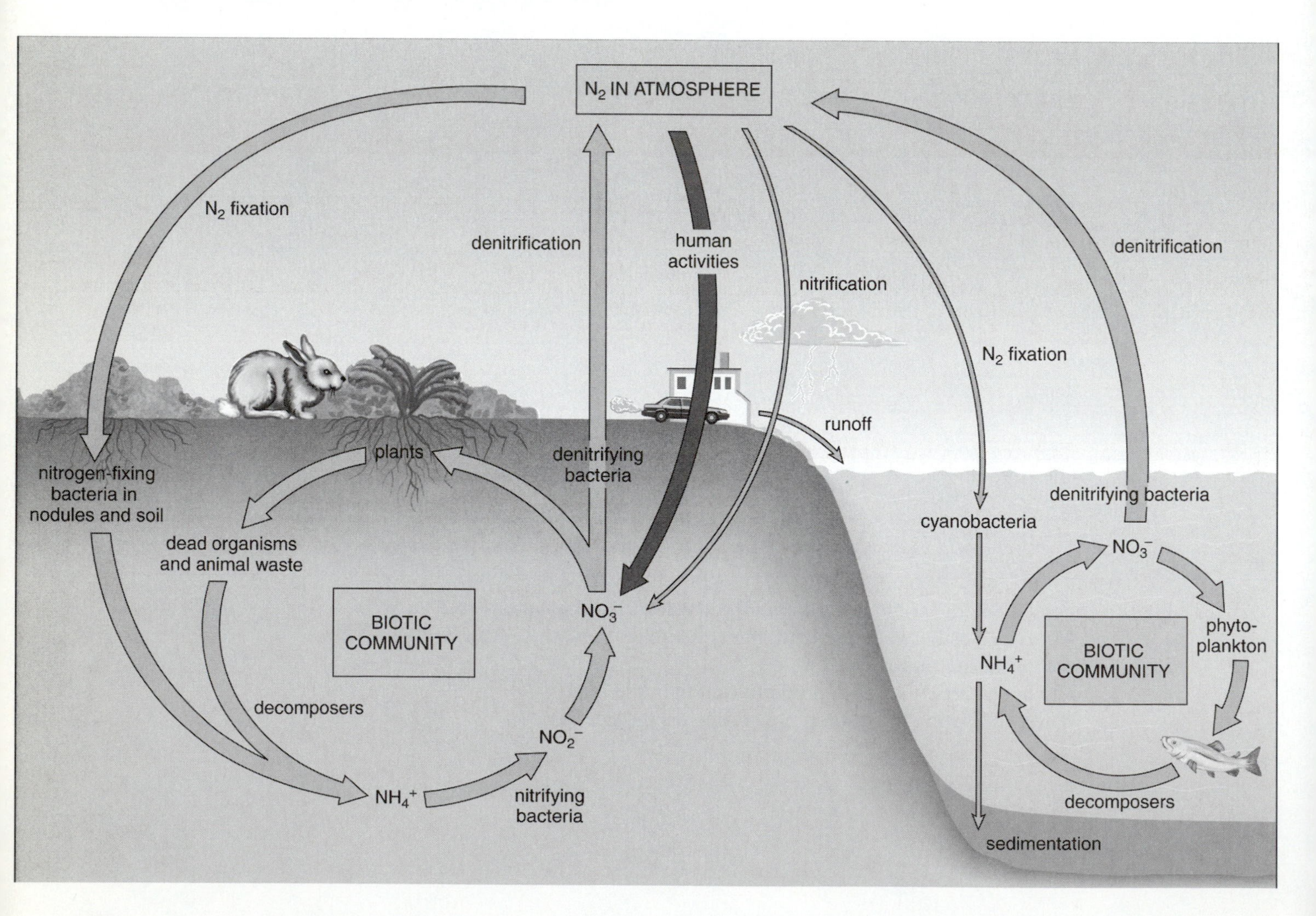

Questions 16 and 17 are based on the preceding diagram of the nitrogen cycle.

16. Match the descriptions to these types of bacteria:
    denitrifying
    nitrifying
    nitrogen-fixing

    a. ____________________ bacteria that convert nitrate to nitrogen gas
    b. ____________________ bacteria that convert ammonium to nitrate
    c. ____________________ bacteria in legume nodules that convert nitrogen gas to ammonium

17. Plants cannot use nitrogen gas. What are two ways in which plants receive a supply of nitrogen for incorporation into proteins and nucleic acids? ____________________

18. When humans produce fertilizers, the gas a.____________ is removed from the atmosphere and changed to b.____________, which enters the atmosphere. Acid deposition occurs when nitrogen oxides and c.____________ in the atmosphere are converted to acids that return to Earth.

19. Place a check beside those statements that may be expected because of acid deposition.
    a. _____ dying forests
    b. _____ lower agricultural yields
    c. _____ sterile lakes
    d. _____ corroded marble, metal, and stonework

20. Photochemical smog arises when a.____________ and b.____________ react with one another in the presence of sunlight. Smog contains the pollutants c.____________ and d.____________.

21. Place a check beside all those effects that may be expected from the occurrence of smog.
    a. _____ breathing difficulties
    b. _____ damage to plants
    c. _____ thermal inversions
    d. _____ cleaner air than usual

22. Place a check beside the statements that correctly describe the results when producers take up phosphate.
    a. _____ becomes a part of phospholipids
    b. _____ becomes a part of ATP
    c. _____ becomes a part of nucleotides
    d. _____ becomes a part of the atmosphere

23. Indicate whether these statements are true (T) or false (F). Rewrite all false statements to be true statements.
    a. _____ Excess phosphate in bodies of water may cause radiation poisoning. Rewrite: ____________________

    ____________________

    b. _____ Most ecosystems have plenty of phosphate. Rewrite: ____________________

    ____________________

    c. _____ The phosphorus cycle is a sedimentary cycle. Rewrite: ____________________

    ____________________

    d. _____ Phosphate enters ecosystems by being taken up by animals. Rewrite: ____________________

    ____________________

24. Place a check beside those items that correctly describe biological magnification.
    a. _____ pertains to organic poisons like pesticides and herbicides
    b. _____ affects other organisms but not humans
    c. _____ refers to an accumulation of foreign molecules in the body because they are not excreted
    d. _____ refers to an enlargement of certain parts of the body due to evolution

25. What is the ozone shield, and why is it important? ____________________

____________________

26. Explain the significance of the following:

$$Cl + O_3 \rightarrow ClO + O_2$$

_______________________________________________

27. What are some of the possible effects of increased ultraviolet radiation on humans and other organisms? _____

_______________________________________________

## Definitions Wordsearch

Review key terms by using the following alphabetized list of terms to fill in the blanks below. Then complete the wordsearch.

R G L P T S Q Z I F A G A C N V
B E W D O O F R B A M N C G F D
G U P P Y L R O X A Q U I F E R
F T O D P G M U E F C T D H C L
L R E M U S N O C O R A D R O T
R O S A T D D E F U N N E L S I
P P A N V T N I C H E R P N Y M
N H T E N B T S A X L A O T S J
N I A H C D O O F Q S U S O T E
S C Q B E O R B A U F A I F E R
C A R N I V O R E E O C T I M U
I T F O R D O Q Y P O A I T A M
L I G L U N J U L P H N O L H Z
N O I T A C I F I R T I N E T F
I N N X C B L R Y T S F J R S R

acid deposition
aquifer
carnivore
CFC
consumer
ecosystem
eutrophication
food chain
food web
nitrification
PAN

a. ______________ Organic compounds containing carbon, chlorine, and fluorine atoms.

b. ______________ Rock layers that contain water and release it in appreciable quantities to wells or springs.

c. ______________ The order in which one population feeds on another in an ecosystem.

d. ______________ Organism that feeds on another organism in a food chain.

e. ______________ In a food chain, a consumer that eats other animals.

f. ______________ Biological community together with the associated abiotic environment; characterized by a flow of energy and a cycling of inorganic nutrients.

g. ______________ Type of chemical found in photochemical smog.

h. ______________ Process by which nitrogen in ammonia and organic compounds is oxidized to nitrites and nitrates by soil bacteria.

i. ______________ In ecosystems, complex pattern of interlocking and crisscrossing food chains.

j. ______________ Enrichment of water by inorganic nutrients used by phytoplankton; over-enrichment often caused by human activities leading to excessive bacterial growth and oxygen depletion.

k. ______________ The sulfate or nitrate salts of acids produced by commercial and industrial activities return to Earth as rain or snow.

# Chapter Test

## Objective Questions

Do not refer to the text when taking this test.

____ 1. A complex of interconnected food chains in an ecosystem is called a(an)
  a. ecosystem.
  b. ecological pyramid.
  c. trophic level.
  d. food web.

____ 2. Wolves and lions are at the same trophic level because they both
  a. are large mammals.
  b. eat only when they are hungry.
  c. eat primary consumers.
  d. live on land.
  e. Both *b* and *c* are correct.

____ 3. The biomass of herbivores is smaller than that of producers because
  a. herbivores are not as efficient in converting energy to biomass.
  b. some energy is lost during each energy transformation.
  c. the number of plants is always more than the number of herbivores.
  d. woody plants live longer than omnivores and herbivores.
  e. All of these are correct.

In questions 4–7, indicate whether the statements are true (T) or false (F).

____ 4. Energy flows through a food chain because it is constantly lost from organic food as heat.

____ 5. A food web contains many food chains.

____ 6. The weathering of rocks is one way that phosphate ions are made available to plants.

____ 7. Respiration returns carbon to the atmosphere.

____ 8. About ________ of the energy available at a particular trophic level is incorporated into the tissues at the next trophic level.
  a. 1%
  b. 10%
  c. 25%
  d. 50%
  e. 75%

Questions 9–11, refer to the following food chain: grass → rabbits → snakes → hawks

____ 9. Each population
  a. is always larger than the one before it.
  b. supports the next level.
  c. is an herbivore.
  d. is a carnivore.

____ 10. Rabbits are
  a. consumers.
  b. herbivores.
  c. more plentiful than snakes.
  d. All of these are correct.

____ 11. Hawks
  a. are top carnivores.
  b. give off $O_2$ that will be taken up by rabbits.
  c. contain phosphate that will eventually be taken up by grass.
  d. All of these are correct.

____ 12. Which of the following contribute(s) to the carbon cycle?
  a. respiration
  b. photosynthesis
  c. fossil fuel combustion
  d. All of these are correct.

____ 13. The largest reserve of unincorporated carbon is in
  a. the soil.
  b. the atmosphere.
  c. the ocean.
  d. deep sediments.

____ 14. The greenhouse effect
  a. is caused by particles in the air.
  b. is caused in part by carbon dioxide.
  c. will cause temperatures to increase.
  d. will cause temperatures to decrease.
  e. Both *b* and *c* are correct.

____ 15. The form of nitrogen most plants make use of is
  a. atmospheric nitrogen.
  b. nitrogen gas.
  c. organic nitrogen.
  d. nitrates.

In questions 16–18, match the air pollutants to these conditions:
  a. ozone shield destruction
  b. global warming
  c. acid deposition
  d. photochemical smog

____ 16. CFCs

____ 17. $SO_2$

____ 18. $CO_2$

____ 19. UV radiation
  a. causes mutations.
  b. impairs crop growth.
  c. kills plankton.
  d. All of these are correct.

____ 20. Nitrogen oxides and hydrocarbons react in the presence of sunlight to produce
a. acid particles.
b. ground level ozone.
c. greenhouse gases.
d. All of these are correct.

____ 21. What may occur as a result of the greenhouse effect?
a. coastal flooding
b. loss of food
c. excess plant growth
d. Both *a* and *b* are correct.

____ 22. What contributes to the greenhouse effect?
a. nuclear power
b. burning of fossil fuels
c. geothermal energy
d. Both *a* and *c* are correct.

____ 23. Which is the cause of stratospheric ozone depletion?
a. chlorine
b. PANs
c. nitrates
d. Both *b* and *c* are correct.

____ 24. Acid deposition is associated with
a. dying lakes.
b. dying forests.
c. dissolving of copper from pipes.
d. All of these are correct.

## Critical Thinking Questions

Answer in complete sentences.

25. Why is a food chain normally limited to four or five links?

26. How would the shortage of an element in the exchange pool affect an ecosystem? Explain.

**Test Results**: ______ number correct ÷ 26 = ________ × 100 = ______%

## Exploring the Internet

The Online Learning Center at *www.mhhe.com/maderbiology8* has additional study material and practice quizzes that can help you master the content of this chapter. You can also find links to websites exploring additional topics in biology. Access to the Online Learning Center is free for those who have purchased a new textbook.

## Answer Key

### Study Exercises

**1. a.** decomposers **b.** carnivores **c.** autotrophs **d.** consumers **e.** herbivores **f.** omnivores **2. a.** producers **b.** consumers **c.** decomposers **d.** inorganic nutrient pool **3. a.** producers **b.** consumers **c.** It dissipates. With every transformation, as when the energy in food is converted to ATP, there is always a loss of usable energy. Eventually, all solar energy taken in by plants becomes heat. **4. a.** carbon dioxide and water. **b.** consumer, decomposer **c.** carbon dioxide and water **5. a.** three **b.** tree **c.** birds, chipmunks **d.** foxes, fishers **e.** tree → rabbits → snakes **6.** old leaves and dead twigs → bacteria and fungi of decay → mice → hawks **7.** Members of the grazing food web die and are decomposed by bacteria and fungi. **8. a.** rabbits and deer **b.** foxes and snakes **9. a.** Less energy is available to be passed on. **b.** In general, only about 10% of the energy of one trophic level is available to the next trophic level. **10. a.** a source usually available to the biotic community **b.** a source from which organisms do generally take chemicals, such as the atmosphere or soil. **c.** producers, consumers, and decomposers that interact through nutrient cycling and energy flow **d.** Humans remove elements from reservoirs and exchange pools and make them available to producers. For example, humans convert $N_2$ in the air to make fertilizer, and they mine phosphate to make fertilizer. **11.** See Figure 48.9, page 887, in text. **12.** a, b, c **13. a.** photosynthesis **b.** respiration **c.** reservoirs **d.** oceans **e.** bicarbonate ion **f.** by burning fossil fuels that add carbon to the atmosphere **14. a.** fossil fuel and wood burning **b.** fertilizer use and animal wastes **c.** bacterial decomposition (in guts of animals), in sediments, and in flooded rice paddies **d.** These gases are called greenhouse gases because, like the panes of a greenhouse, they allow solar radiation to pass through but hinder the escape of heat. **15.** a, b, d, e **16. a.** denitrifying **b.** nitrifying **c.** nitrogen fixing **17.** nitrogen-fixing bacteria in nodules and nitrate in soil **18. a.** $N_2$ **b.** $NO_3$ **c.** sulfur dioxide **19.** a, b, c, d **20. a.** $NO_x$ **b.** HC **c.** PAN **d.** ozone **21.** a, b, c **22.** a, b, c **23. a.** F, . . . may cause algal bloom **b.** F, . . . have a limited supply of phosphate **c.** T **d.** F, . . . taken up by plants **24.** a, c **25.** Ozone is a layer within the stratosphere that protects the Earth's surface from ultraviolet radiation. **26.** The chlorine breaks down the ozone releasing ClO and oxygen. ClO also combines with ozone, releasing more oxygen. Oxygen does not absorb UV radiation. **27.** It will increase the incidence of skin cancer and decrease the productivity of living systems. Loss of oceanic plankton will disrupt marine ecosystems.

## Definitions Wordsearch

```
                        A
B E W D O O F           C
  U               A Q U I F E R
  T                     D   C
  R E M U S N O C       D   O
  O             F       E   S
  P A N         C       P   Y
  H                     O   S
N I A H C D O O F       S   T
  C                     I   E
C A R N I V O R E       T   M
  T                     I
  I                     O
N O I T A C I F I R T I N
  N
```

**a.** CFC **b.** aquifer **c.** food chain **d.** consumer **e.** carnivore **f.** ecosystem **g.** PAN **h.** nitrification **i.** food web **j.** eutrophication **k.** acid deposition

## Chapter Test

**1.** d **2.** c **3.** b **4.** F **5.** T **6.** T **7.** T **8.** b **9.** b **10.** d **11.** b **12.** d **13.** c **14.** e **15.** d **16.** a **17.** c **18.** b **19.** d **20.** b **21.** d **22.** b **23.** a **24.** d **25.** By the laws of thermodynamics, energy conversion at each link of a food chain results in nonusable heat. Too little useful energy remains for more links. **26.** A shortage of an element such as nitrogen or phosphorus would reduce the biomass of the producer population. Therefore, the biomass of each succeeding population in the ecosystem would most likely be smaller than it otherwise would be.

# 49

# The Biosphere

## Chapter Review

Because the Earth is a sphere, the sun's rays are more direct at the equator, and so temperature is warmest at the equator, cooling toward the poles. The tilt of the Earth on its axis, along with the Earth's rotation around the sun, creates the seasons. Because the oceans are warmer at the equator than the poles, air rises at the equator and moves toward the poles; these air currents in turn cause ocean currents that affect **climate** around the world.

Warm air rising at the equator loses its moisture and then descends at about 30° north and south latitude and so forth to the poles. This movement of air in general accounts for different amounts of rainfall at different latitudes. Topography also plays a role in the distribution of rainfall.

Among terrestrial biomes, the **tundra** just south of the North Pole has cold winters and short summers; the vegetation consists largely of short grasses and sedges, and dwarf woody shrubs. Proceeding southward, the **taiga** is a coniferous forest, the **temperate deciduous forest** has well-defined seasons, and the **tropical rain forest** is a broad-leafed evergreen forest.

Among grasslands, which have less rainfall than forests, the savanna is a grassland that supports the greatest variety and number of large herbivores. The prairie found in the United States has a limited variety of vegetation and animal life. In **deserts,** some plants such as cacti are succulents, and others are shrubs with thick leaves they often lose during dry periods.

Among aquatic biomes, freshwater communities include streams, rivers, lakes, and ponds. Lakes experience spring and fall overturns. Lakes and ponds have rooted plants in the littoral zone, plankton and fishes in the sunlit limnetic zone, and bottom-dwelling organisms in the profundal zone. **Estuaries** near the mouth of rivers are the nurseries of the sea. Marine communities include coastal communities and the oceans. An ocean has a **pelagic division** (open waters) and **benthic division** (ocean floor). **Coral reefs** are productive communities found in shallow tropical waters.

## Study Exercises

Study the text section by section as you answer the questions that follow.

### 49.1 Climate and the Biosphere (pp. 900–2)

- Solar radiation provides the energy that drives climate differences in the biosphere.
- Global air circulation patterns and physical features produce the various patterns of temperature and rainfall about the globe.

1. Indicate whether these statements are true (T) or false (F).
   a. _____ Because the Earth is a sphere, solar energy hitting Earth is uniformly distributed.
   b. _____ The distribution of rainfall is partially due to topography.
   c. _____ Heat always passes from warm areas to colder areas.
   d. _____ The great deserts of the world lie at the equator.
   e. _____ Warm air moves from the equator to the poles.
   f. _____ Arctic winds across the Great Lakes produce lake-effect snows.
   g. _____ Rain shadows always form on the leeward side of the mountains.

- The Earth's major terrestrial biomes are tundra, forests, shrublands, grasslands, and deserts.

2. Label this diagram that compares the effects of altitude and latitude on vegetation using the following alphabetized list of terms (some terms are used more than once).
   coniferous forest
   deciduous forest
   tropical forest
   tundra

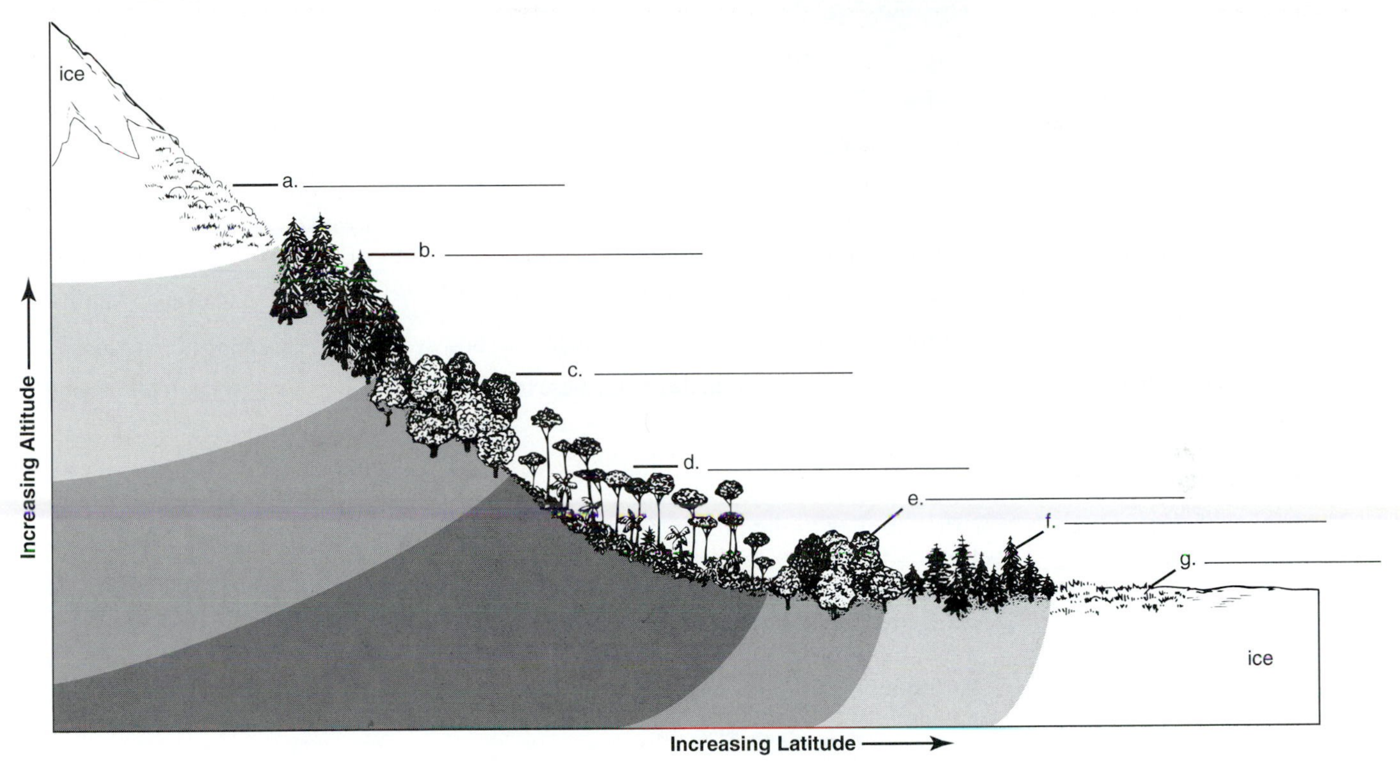


3. The diagram in question 2 emphasizes that vegetation is determined in part by a.__________________. b.__________________ also plays a major role, and therefore, tropical rain forests are found at the equator where both c.__________________ and d.__________________ are maximal.
4. For each biome listed, write a one- or two-word description for the temperature and rainfall.

| Biome | Temperature | Rainfall |
|---|---|---|
| Tundra | | |
| Desert | | |
| Grassland | | |
| Taiga | | |
| Temperate deciduous forest | | |
| Tropical rain forest | | |

5. Label the soil diagram using the following alphabetized list of terms. Then indicate the layer in which *leaching* occurs.

parent material
topsoil
subsoil

A ____________________

B ____________________

C ____________________

Soil Horizons

6. Because of limited leaching (due to limited rainfall), the A horizon is deep in [a.]____________________, and this made the prairies of the United States good agricultural lands. Generally, in [b.]____________________, both the A and B horizons supply inorganic nutrients for tree root growth. In tropical rain forests, however, because leaching is extensive, there is only a shallow [c.]____________________ horizon; therefore, these forests [d.]____________________ (can, cannot) support crops for many years.

## Tundra (p. 904)

7. From each pair, select the one that applies to the tundra, and write it on the answer line.
   a. ____________________ light—dark
   b. ____________________ cold—hot
   c. ____________________ short grasses—trees
   d. ____________________ musk-ox—horses
   e. ____________________ epiphytes—permafrost

## Coniferous Forests (p. 905)

8. From each pair, select the one that applies to the taiga.
   a. ____________________ broad-leafed trees—narrow-leafed trees
   b. ____________________ cold—hot
   c. ____________________ cool lakes—pools and mires
   d. ____________________ zebras—moose

9. From each pair, select the one that applies to the temperate rain forest on the west coast of Canada and the United States.
   a. ____________________ short trees—tall trees
   b. ____________________ old trees—young trees
   c. ____________________ ferns and mosses—sedges and grasses

## Temperate Deciduous Forests (p. 906)

10. From each pair, select the one that applies to temperate deciduous forests.
   a. ____________________ conifers only—oak and maple trees
   b. ____________________ flowering shrubs—short grasses
   c. ____________________ caribou—white-tailed deer
   d. ____________________ rabbits and skunks—lemmings and prairie chickens

## Tropical Forests (pp. 908–9)

11. From each pair, select the one that applies to tropical rain forests.
    a. ______________________ deciduous broad-leafed trees—evergreen broad-leafed trees
    b. ______________________ lianas and epiphytes—pine needles
    c. ______________________ few insects—many insects
    d. ______________________ colorful birds—drab birds
    e. ______________________ horses and zebras—monkeys and large cats

## Shrublands (p. 910)

12. From each pair, select the one that applies to shrublands.
    a. ______________________ rainfall in summer—rainfall in winter
    b. ______________________ shrubs with thick roots—trees with shallow roots

## Grasslands (p. 910)

13. From each pair, select the one that applies to United States prairies.
    a. ______________________ rabbits—prairie dogs
    b. ______________________ hawks—parakeets
    c. ______________________ trees—grasses

14. From each pair, select the one that applies to the African savanna.
    a. ______________________ elephants—moose
    b. ______________________ even rainfall—severe dry season
    c. ______________________ herds of herbivores—large primates

## Deserts (p. 912)

15. From each pair, select the one that applies to North American deserts.
    a. ______________________ cool days—cool nights
    b. ______________________ cacti—broad-leafed evergreen trees
    c. ______________________ lizards and snakes—elephants and zebras

## *Hoop Dreams*

Each biome is a hoop. The plants and animals are balls. Try to get the balls in the right hoops.

| Hoops (biomes) | Plants (balls) | Animals (balls) |
|---|---|---|
| desert | lichens | moose |
| taiga | spruce trees | beaver and muskrat |
| U.S. prairie | epiphytes | lemming |
| temperate deciduous forest | oak trees | monkey |
| tundra | grasses | lizard |
| African savanna | acacia trees | buffalo |
| tropical rain forest | cacti | wildebeest |

Possible number of baskets is 14. How many baskets did you make? ______________________

## 49.3 Aquatic Communities (pp. 913–21)

- The Earth's major aquatic biomes are of two types: freshwater and saltwater.
- Ocean currents also affect the climate and the weather over the continents.

16. Aquatic communities can be divided into two major types: the a. ______________________ communities that consist of lakes, ponds, rivers, and streams, and the b. ______________________ communities along the coast and in the ocean.

## Lakes (pp. 914–15)

17. Lakes occur as nutrient-poor or [a] __________ lakes and nutrient rich or [b] __________ lakes. In the temperate zone, deep lakes are stratified. In the fall, as the top layer called [c] __________ cools, and in the spring as it warms, a (an) [d] __________ occurs. During this time, a mixing of [e] __________ and [f] __________ takes place.

18. List and describe the three life zones of a lake, noting the types of organisms found in each.

    a. ______________________________

    ______________________________

    b. ______________________________

    ______________________________

    c. ______________________________

    ______________________________

## Coastal Communities (pp. 916–17)

19. Indicate whether these statements are true (T) or false (F).

    a. _____ Salt marshes in the tropics and mangrove swamps in the temperate zone occur at the mouth of a river.

    b. _____ Estuaries offer protection and nutrients to immature forms of marine life.

    c. _____ Rocky coasts are protected, but sandy shores are bombarded by the seas as the tides roll in and out.

    d. _____ There are different types of shelled and algal organisms at the upper, middle, and lower portions of the littoral zone of a rocky coast.

## Oceans (pp. 918–21)

20. Place the appropriate letter next to each statement.

    P—pelagic division B—benthic division

    a. _____ has the sublittoral and abyssal zones

    b. _____ has the greater overall diversity of organisms

    c. _____ includes neritic and oceanic provinces

    d. _____ includes organisms living on the continental shelf and slope

    e. _____ has organisms that depend on floating debris from above for food

    f. _____ is penetrated by sunlight

21. Complete the following table by noting the amount of light present (bright/semidark/dark) and the types of organisms found (phytoplankton/strange-looking fish/filter feeders/carnivores/sea urchins) in each ocean zone.

| Ocean Zone | Amount of Light | Organisms |
|---|---|---|
| Epipelagic | | |
| Mesopelagic | | |
| Bathypelagic | | |
| Abyssal | | |

22. a. What supports life in the epipelagic zone? __________

    b. What supports life in the abyssal zone? __________

# Chapter Test

## Objective Questions

Do not refer to the text when taking this test.

For questions 1–10, indicate whether the statements are true (T) or false (F).

____ 1. Climate determines the geographic location of a biome.

____ 2. Grasslands usually receive a greater annual rainfall than deserts.

____ 3. The taiga is the northernmost forested biome.

____ 4. Temperate deciduous forests show the greatest species diversity of all forested biomes.

____ 5. The leaves of tropical rain forest evergreen trees are needlelike.

____ 6. The profundal zone of a lake is the zone closest to the shore.

____ 7. An estuary acts as a nutrient trap, existing where a large river flows into an ocean.

____ 8. The solid part of a coral reef consists of the skeletons of dead coral.

____ 9. A food chain in the pelagic division could be: phytoplankton, zooplankton, small fish, herring.

____10. The benthic division receives less light penetration than the pelagic division.

____11. Which of the following phrases is NOT true of the tundra?
- a. low-lying vegetation
- b. northernmost biome
- c. few large mammals
- d. short growing season
- e. many different types of species

____12. A temperate deciduous forest will
- a. be warm and moist.
- b. be hot and dry.
- c. be cold and have limited rain.
- d. have moderate temperatures and moderate rain.
- e. have moderate temperatures and little rain.

____13. A tropical rain forest will typically
- a. be warm and moist.
- b. be hot and dry.
- c. be cold and have limited rain.
- d. have moderate temperatures and moderate rain.
- e. have moderate temperatures and little rain.

____14. A desert will typically
- a. be warm and moist.
- b. be hot and dry.
- c. be cold and have limited rain.
- d. have moderate temperatures and moderate rain.
- e. have moderate temperatures and little rain.

____15. The biome that best supports grazing animals is a
- a. tropical rain forest.
- b. coniferous forest.
- c. grassland.
- d. desert.

____16. Which biome has most of the animals living in trees?
- a. taiga
- b. temperate deciduous forest
- c. tropical rain forest
- d. savanna
- e. grassland

____17. Which type of biome has succulent, leafless plants with stems that store water and roots that can absorb great quantities of water in a brief period?
- a. tropical rain forest
- b. tundra
- c. temperate deciduous forest
- d. desert
- e. savanna

____18. Large grazing animals are most numerous in which biome?
- a. tundra
- b. grassland
- c. coniferous forest
- d. deciduous forest
- e. tropical rain forest

____19. Which zone of the ocean is the deepest?
- a. epipelagic
- b. mesopelagic
- c. bathypelagic
- d. abyssal
- e. estuarial

____20. Which zone in the ocean receives the most sunlight?
- a. epipelagic
- b. mesopelagic
- c. bathypelagic
- d. abyssal
- e. estuarial

## Critical Thinking Questions

Answer in complete sentences.

21. Both the temperate rain forest and the chaparral occur in California. Explain various differences in climate and vegetation.

22. Explain why estuaries are the nurseries of the sea.

**Test Results:** _______ number correct ÷ 22 = _______ × 100 = _______ %

## Exploring the Internet

The Online Learning Center at *www.mhhe.com/maderbiology8* has additional study material and practice quizzes that can help you master the content of this chapter. You can also find links to websites exploring additional topics in biology. Access to the Online Learning Center is free for those who have purchased a new textbook.

## Answer Key

### Study Exercises

**1. a.** F **b.** T **c.** T **d.** F **e.** T **f.** T **g.** T **2.** See Fig. 49.5, page 903, in text. **3. a.** temperature **b.** Rainfall **c.** temperature **d.** rainfall

**4.**

| Temperature | Rainfall |
|---|---|
| cold | little |
| hot | little |
| moderate | limited |
| cool | moderate |
| moderate | rather high |
| hot | high |

**5. a.** topsoil, leaching **b.** subsoil **c.** parent material **6. a.** temperate grasslands **b.** forests **c.** A **d.** cannot **7. a.** dark **b.** cold **c.** short grasses **d.** musk-ox **e.** permafrost **8. a.** narrow-leafed trees **b.** cold **c.** cool lakes **d.** moose **9. a.** tall trees **b.** old trees **c.** ferns and mosses **10. a.** oak and maple trees **b.** flowering shrubs **c.** white-tailed deer **d.** rabbits and skunks **11. a.** evergreen broad-leafed trees **b.** lianas and epiphytes **c.** many insects **d.** colorful birds **e.** monkeys and large cats **12. a.** rainfall in winter **b.** shrubs with thick roots **13. a.** prairie dogs **b.** hawks **c.** grasses **14. a.** elephants **b.** severe dry season **c.** herds of herbivores **15. a.** cool nights **b.** cacti **c.** lizards and snakes

**Hoop Dreams** desert: cacti, lizard; taiga: spruce trees, moose; U.S. prairie: grasses, buffalo; temperate deciduous forest: oak trees, beaver and muskrat; tundra: lichens, lemming; African savanna: acacia trees, wildebeest; tropical rain forest: epiphytes, monkey

**16. a.** freshwater **b.** saltwater **17. a.** oligotrophic **b.** eutrophic **c.** epilimnion **d.** overturn **e.** oxygen **f.** nutrients **18. a.** littoral zone—aquatic plants, microscopic organisms **b.** limnetic zone, sunlit main body—some surface organisms and plankton **c.** profundal zone, depths where sunlight does not reach—molluscs, crustaceans, worms **19. a.** F **b.** T **c.** F **d.** T **20. a.** B **b.** P **c.** P **d.** B **e.** B **f.** P

**21.**

| Amount of Light | Organisms |
|---|---|
| bright | phytoplankton |
| semidark | carnivores |
| dark | strange-looking carnivores |
| dark | filter feeders and sea urchins |

**22. a.** photosynthesis by algae **b.** debris floating down from above

## CHAPTER TEST

**1.** T **2.** T **3.** T **4.** F **5.** F **6.** F **7.** T **8.** T **9.** T **10.** T **11.** e **12.** d **13.** a **14.** b **15.** c **16.** c **17.** d **18.** a **19.** d **20.** a **21.** The temperate rain forest lies along the coast and has much rainfall; the old trees are covered by ferns and mosses and grow very tall. The chaparral occurs among hills and has limited rainfall; the shrubs that occur there are adapted to arid conditions and regrowth after fire. **22.** An estuary, a partially enclosed body of water where fresh water and seawater meet, is a nutrient trap. An estuary offers a protective environment where larval marine forms can mature before moving out to other coastal areas and the open sea.

# 50

# Conservation Biology

## Chapter Review

**Conservation biology** is the scientific study of biodiversity and its management for sustainable human welfare. **Biodiversity** is the variety of life on Earth; the exact number of species is unknown, but there are many more insects than other types of organisms. Biodiversity must be preserved at the genetic, community or ecosystem, and landscape levels of organization. Conservationists have discovered that biodiversity is not evenly distributed in the biosphere, and therefore saving particular areas may protect more species than saving other areas.

Biodiversity has both direct and indirect value. Direct value is the observable services of individual wild species. Wild species are the best source of new medicines. They also meet other medical needs; for example, the bacterium that causes leprosy grows naturally in armadillos. Wild species also have agricultural value. Domesticated plants and animals are derived from wild species, a source of genes for the improvement of their phenotypes. They can also be used as biological pest controls, and most flowering plants make use of animal pollinators. Additionally, wild species are still used for food (i.e., fish and shellfish). Hardwood trees from natural forests supply lumber for various purposes.

The indirect services provided by ecosystems are largely unseen but absolutely necessary for our well-being. These services include the workings of biogeochemical cycles, waste disposal, provision of fresh water, prevention of soil erosion, and regulation of climate. Additionally, people enjoy vacationing in natural settings.

**Habitat loss** is the most frequent cause of extinction, followed by the introduction of **alien species**, **pollution**, **overexploitation**, and **disease**. Habitat loss has occurred in all parts of the biosphere, but concern has now centered upon tropical rain forests and coral reefs, where biodiversity is especially high. Alien species have been introduced into foreign ecosystems because of colonization, horticulture and agriculture, and accidental transport. Global warming is the form of pollution most responsible for extinction. Overexploitation can often be explained by the positive feedback cycle. For example, commercial fishing has become so efficient that fisheries of the world are collapsing.

Some researchers emphasize the need to preserve **biodiversity hotspots** because of their richness. When preserving ecosystems, it is wise to first determine which are the **keystone species**. Often it is necessary to save **metapopulations** because of habitat fragmentation. This requires saving **source populations** over **sink populations**.

Conservation biology is assisted by two types of computer analysis. A **gap analysis** tries for a fit between biodiversity concentrations and land still available to be preserved. A **population viability analysis** indicates the minimum size of a population needed to prevent extinction from happening. Because many ecosystems have been degraded, **habitat restoration** may be necessary before **sustainable development** is possible. Three principles of restoration are: (1) start before sources of wildlife and seeds are lost; (2) use simple biological techniques that mimic natural processes; and (3) aim for sustainable development so that the ecosystem fulfills the needs of humans.

## Study Exercises

Study the text section by section as you answer the questions that follow.

### 50.1 Conservation Biology and Biodiversity (pp. 926–27)

- Conservation biology addresses a crisis–the loss of biodiversity.
- Conservation biology is an applied, goal-orientated, multidisciplinary field.
- Extinction rates have risen to many times their natural levels, and many types of ecosystems are disappearing.
- Biodiversity includes species diversity, genetic diversity, community diversity, and landscape diversity in marine, freshwater, and terrestrial habitats.

1. Place an X beside those fields involved in conservation biology:

   a. _____ genetics
   b. _____ physiology
   c. _____ behavior
   d. _____ veterinary science
   e. _____ range management

2. You can deduce from your answer to question 1 that conservation biology involves the development of a. ________________ concepts and the b. _______________ of these concepts to the everyday world.

3. Place an X beside the value principles that guide the field of conservation biology:

   a. _____ Extinctions due to human activities are undesirable.
   b. _____ Biodiversity is desirable to the health of the biosphere.
   c. _____ Biodiversity has no value in and of itself.
   d. _____ Extinctions due to human activities are of little concern.

4. Biodiversity involves the number of different organisms on Earth. Name two other types of diversity:

   a. ________________________________________

   b. ________________________________________

5. Biodiversity is not evenly distributed, as exemplified by regions of the biosphere where a wide variety of species are found. These regions are called ________________________________.

## 50.2 Value of Biodiversity (pp. 928–31)

- Biodiversity has both direct and indirect value.

6. Give an example of the value of biodiversity to:

   a. medicine ________________________________________

   b. agriculture ________________________________________

   c. consumptive use ________________________________________

7. Place an X beside all those areas that are NOT at all dependent on biodiversity:

   a. _____ ecotourism
   b. _____ prevention of soil erosion
   c. _____ biogeochemical cycles
   d. _____ provision of fresh water
   e. _____ waste disposal
   f. _____ regulation of climate

8. Based on your answer to question 7, the conclusion is that biodiversity has ____________ (much or little) indirect value.

## 50.3 Causes of Extinction (pp. 932–36)

- Habitat loss, introduction of alien species, pollution, overexploitation, and disease are now largely responsible for biodiversity loss.
- Global warming will shift the optimal range of many species northward and disrupt many coastal ecosystems.

9. Rank in order these threats to biodiversity from the greatest threat to the least threat.

a. _____ pollution
b. _____ habitat loss
c. _____ disease
d. _____ alien species
e. _____ overexploitation

10. Indicate whether these statements are true (T) or false (F).

a. _____ A species is often affected by more than one of the threats mentioned in question 9.
b. _____ An alien species is sometimes brought into a new area by horticulturists.
c. _____ Tropical rain forest destruction should be associated with the introduction of disease into an area.
d. _____ Although much of the interior of continents have suffered loss of habitat, the coastline has not suffered as much loss.

11. Match the concerns to the type of pollution:

acid deposition    eutrophication    ozone depletion    organic chemicals    global warming

a. ____________________ forests and lakes are dying
b. ____________________ increased amount of UV radiation
c. ____________________ glaciers are melting and the sea level is rising
d. ____________________ pesticides and detergents can have hormonal effects
e. ____________________ massive fish kills

## 50.4 Conservation Techniques (pp. 937–40)

- Identifying and conserving biodiversity hotspots and/or keystone species can save many other species.
- Because of habitat fragmentation, it is often necessary to conserve metapopulations today.
- Computer analyses can be done to select areas for preservation and to determine the minimal population size needed for survival.
- Landscape preservation often involves restoration of habitats today.

12. Sometimes it is necessary to decide the location and extent of an area to preserve. With that in mind what is a

a. keystone species? ____________________

b. metapopulation? ____________________

c. sink population? ____________________

d. edge effect? ____________________

13. a. A computer analysis called a gap analysis helps scientists find "______________," areas where biodiversity is high outside already preserved areas.

b. A population viability analysis helps researchers determine first how large a population has to be to maintain itself, and then how much ______________ a species needs to remain at this size.

14. Sometimes it is necessary to restore an area so that it is suitable for the maintenance of species. What three principles guide restoration efforts?

a. It's best to begin ____________________.

b. It's best to use biological techniques that ______________.

c. The goal is ____________ development, so that an ecosystem can maintain itself.

# Definitions Wordmatch

Review key terms by completing this matching exercise, selecting from the following alphabetized list of terms:

biodiversity hotspot
conservation biology
gap analysis
global warming
landscape
keystone species
metapopulation
population
restoration ecology
sink population
source population
sustainable development

a. ______________ Use of computers to discover places where biodiversity is high outside of preserved areas.

b. ______________ A number of interacting ecosystem fragments.

c. ______________ Population subdivided into several small and isolated populations due to habitat fragmentation.

d. ______________ Subdiscipline of conservation biology that seeks ways to return ecosystems to their former state.

e. ______________ Species whose activities have a significant role in determining community structure.

f. ______________ Region of the world that contains unusually large concentrations of species.

# Chapter Test

## Objective Questions

Do not refer to the text when taking this test.

____ 1. What is defined as the scientific study of biodiversity and its management for human welfare?
a. biology
b. environmental biology
c. conservation biology
d. None of the above are correct.

____ 2. A population viability analysis includes calculating the
a. minimum population size needed to prevent extinction.
b. amount of habitat needed to maintain this population size.
c. how much pollution this species can withstand.
d. Both *a* and *b* are correct.

____ 3. What kind of change is aggravating the biodiversity crisis?
a. genetic drift
b. global environmental change
c. excessive reproduction
d. None of these are correct.

____ 4. One objective of ____________________ is to maintain the capacity of species to adapt to changing conditions through evolution.
a. conservation biology
b. global environmental change
c. habitat fragmentation
d. ecosystems

____ 5. To maintain biodiversity, a gap analysis tells us
a. which regions should be conserved.
b. how large each preserved population is.
c. if we are preserving the correct regions.
d. Both *a* and *c* are correct.

____ 6. The greatest majority of known species are:
a. onychophorans.
b. enchinoderms.
c. insects.
d. chordates.

____ 7. Humans have
a. not influenced extinction rates.
b. reduced extinction rates.
c. greatly increased extinction rates.
d. reduced all activities that cause extinction.
e. Both *b* and *d* are correct.

____ 8. Biodiversity, a rich and valuable resource, includes
a. genetic diversity.
b. species richness.
c. diverse communities and ecosystems.
d. All of these are correct.

____ 9. Which two ecosystems have the most species?
a. tropical forests and coral reefs
b. tropical forests and prairies
c. coral reefs and the seashores
d. rivers and lakes

____ 10. Periodic mass extinctions have occurred due to
a. lunar eclipses.
b. geological events.
c. astrophysical events.
d. Both *b* and *c* are correct.

____ 11. Humans threaten to cause a mass extinction when they participate in which habitat-destroying activities?
a. deforestation of tropical rain forests
b. river damming and diversion
c. lake pollution and eutrophication
d. All of these are correct.

____12. Ecosystems are critically endangered in the United States largely due to
a. pollution.
b. habitat loss.
c. alien species.
d. overexploitation.

____13. Habitat fragmentation leads to
a. demographic impacts.
b. genetic consequences.
c. extinctions.
d. All of these are correct.

____14. Prehistoric humans caused many extinctions by
a. hunting.
b. habitat transformation.
c. introduction of alien predators.
d. All of these are correct.

____15. Climate shifts in terrestrial environments may
a. eliminate habitats of some species.
b. favor alien species.
c. favor marine species.
d. Both *a* and *b* are correct.

____16. Which environment has an uneven distribution of species?
a. terrestrial
b. freshwater
c. marine
d. All of these are correct.

____17. At the current species level, recent estimates suggest that the number of described species is about
a. 1.4 – 1.6 million.
b. 2.3 – 3.1 million.
c. 4.1 – 5.2 million.
d. 50 – 60 million.

____18. Which of these is an indirect value of biodiversity?
a. medicinal value
b. consumptive use value
c. regulation of climate
d. Both *a* and *b* are correct.

____19. Enforcement of biodiversity management must ultimately be defined by
a. biological experts.
b. the government.
c. society.
d. None of these are correct.

____20. The field of conservation biology became a separate field during the
a. late 1890s.
b. 1950s.
c. 1990s.
d. It has always been around.

## Critical Thinking Questions

Answer in complete sentences.

21. Taking into consideration what you have learned about the value of biodiversity, what are the key reasons for saving tropical rain forests?

22. What steps need to be taken if an ecosystem is to be restored?

**Test Results:** ______ number correct ÷ 22 = ______ × 100 = ______ %

## Exploring the Internet

The Online Learning Center at *www.mhhe.com/maderbiology8* has additional study material and practice quizzes that can help you master the content of this chapter. You can also find links to websites exploring additional topics in biology. Access to the Online Learning Center is free for those who have purchased a new textbook.

# ANSWER KEY

## STUDY EXERCISES

**1.** a,b,c,d,e **2. a.** scientific **b.** application **3.** a, b, c **4. a.** genetic **b.** community **5.** hot spots **6. a.** Rosy periwinkle is a source of a cancer medicine. **b.** Crops are derived from wild species. **c.** Fish are caught in the wild. **7.** None should be checked because all are dependent on biodiversity. **8.** much **9.** b, d, a, e, c **10. a.** T **b.** T **c.** F **d.** F **11. a.** acid deposition **b.** ozone depletion **c.** global warming **d.** organic chemicals **e.** eutrophication **12. a.** influence the viability of a community more than you would suspect from their numbers **b.** a population subdivided into several small and isolated populations **c.** a population that loses members because the environment is not favorable **d.** reduces the amount of habitat typical of an ecosystem. **13. a.** gap **b.** habitat **14. a.** right away **b.** mimic nature **c.** sustainable

## DEFINITIONS WORDMATCH

**a.** gap analysis **b.** landscape **c.** metapopulation **d.** restoration ecology **e.** keystone species **f.** biodiversity hotspot

## CHAPTER TEST

**1.** c **2.** d **3.** b **4.** a **5.** d **6.** c **7.** c **8.** d **9.** a **10.** d **11.** d **12.** b **13.** d **14.** d **15.** d **16.** d **17.** a **18.** c **19.** c **20.** c **21.** Answers may vary, but may include medicines yet to be discovered, ecotourism, formation of rain clouds, and so on. **22.** You must save sources of wildlife and seeds before they are lost. You should use simple biological techniques that mimic natural processes. You should aim for sustainable development so that the ecosystem still fulfills the needs of humans.